World Sustainability Series

Series Editor

Walter Leal Filho, European School of Sustainability Science and Research, Research and Transfer Centre "Sustainable Development and Climate Change Management", Hamburg University of Applied Sciences, Hamburg, Germany

Due to its scope and nature, sustainable development is a matter which is very interdisciplinary, and draws from knowledge and inputs from the social sciences and environmental sciences on the one hand, but also from physical sciences and arts on the other. As such, there is a perceived need to foster integrative approaches, whereby the combination of inputs from various fields may contribute to a better understanding of what sustainability is, and means to people. But despite the need for and the relevance of integrative approaches towards sustainable development, there is a paucity of literature which address matters related to sustainability in an integrated way.

Notes on the quality assurance and peer review of this publication

Prior to publication, the works published in this book are initially assessed and reviewed by an in-house editor. If suitable for publication, manuscripts are sent for further review, which includes a combined effort by the editorial board and appointed subject experts, who provide independent peer-review. The feedback obtained in this way was communicated to authors, and with manuscripts checked upon return before finally accepted. The peer-reviewed nature of the books in the "World Sustainability Series" means that contributions to them have, over many years, been officially accepted for tenure and promotion purposes.

Walter Leal Filho · Sandeep Kautish ·
Vijay Prakash Gupta
Editors

Metaverse and Sustainability: Business Resilience Towards Sustainable Development Goals

Springer

Editors
Walter Leal Filho
Hamburg University of Applied Sciences
Hamburg, Germany

Vijay Prakash Gupta
Institute of Business Management
GLA University
Mathura, Uttar Pradesh, India

Sandeep Kautish
Institute of Innovation
PhysicsWallah Private Limited
Noida, Uttar Pradesh, India

ISSN 2199-7373 ISSN 2199-7381 (electronic)
World Sustainability Series
ISBN 978-3-031-89547-0 ISBN 978-3-031-89545-6 (eBook)
https://doi.org/10.1007/978-3-031-89545-6

This Springer imprint is published by the registered company Springer Nature Switzerland AG
The registered company address is: Gewerbestrasse 11, 6330 Cham, Switzerland

Preface

The Metaverse is a collective virtual space created by the convergence of virtually enhanced physical reality and physically persistent virtual reality. It encompasses a broad range of digital environments, social interactions, and experiences facilitated by augmented reality (AR), virtual reality (VR), and the internet. In the Metaverse, users can interact with each other and virtual objects in real time, often through the use of avatars. Some of the key characteristics of the Metaverse include:

Interactivity: Users can engage with the environment and interact with other users in various ways, such as socialising, playing games, or collaborating on projects.

Persistence: The Metaverse continues to exist and evolve even when users are not logged in, similar to the real world where events and changes occur continuously.

Immersive Experiences: By utilising VR and AR technologies, the Metaverse offers immersive experiences that can simulate real-world scenarios or create entirely new virtual environments.

User-Generated Content: A significant part of the Metaverse is composed of content and experiences created by users, allowing for creativity and personalisation.

Economy: The Metaverse may support the economy, including virtual goods, services, and currencies, enabling transactions and commerce within these digital spaces.

In summary, the Metaverse represents a digital universe where individuals can connect, create, and engage in a variety of activities through a blend of physical and digital elements.

In an era defined by rapid technological advancement and growing concerns over environmental sustainability, the intersection of the Metaverse and sustainable development presents both remarkable opportunities and formidable challenges. This book, *Metaverse and Sustainability: Business Resilience Towards Sustainable Development Goals*, aims to explore this dynamic relationship and provide readers with insights into how businesses can leverage the Metaverse to not only thrive but also contribute positively towards the Sustainable Development Goals (SDGs).

As the concept of the Metaverse evolves, it offers a digital realm where immersive experiences can reshape our interactions, economies, and even our perceptions of sustainability. It opens the door for innovative solutions that can transform traditional business models, making them more efficient, inclusive, and environmentally friendly. By harnessing the potential of virtual environments, organisations can engage stakeholders, reduce their carbon footprint, and promote sustainable practices in unprecedented ways.

This book is structured to guide readers through the fundamental concepts of the Metaverse, its implications for various industries, and strategies that businesses can adopt to build resilience in face of sustainability challenges. Each chapter combines theoretical foundations with practical applications, case studies, and insights from industry leaders who are already paving the way towards a more sustainable future.

As we embark on this journey through the synergy of the Metaverse and sustainability, we invite Readers to envision how emerging technologies can serve as catalysts for positive change. Together, we can unlock the potential of a digital future that prioritizes inclusivity, innovation, and environmental stewardship, ensuring that our pursuit of progress aligns with the broader goals of humanity.

We thank all authors for sharing their knowledge and their experience by means of their chapters, and those colleagues who have contributed to it by assisting with the reviews. Thanks to its scope and interdisciplinarity, this books makes an excellent reading to everyone interested on the Metaverse and sustainability.

Hamburg, Germany — Walter Leal Filho
Noida, India — Sandeep Kautish
Mathura, India — Vijay Prakash Gupta

Contents

Metaverse Enabled Sustainable HRM Practices: Managing ESG for Business Resilience and Corporate Performance

Abhishek Kumar Pandey, Pankaj Tripathi, Parag S. Shukla, and Pramod Kumar Upadhyay

Abstract Human resource strategies are prominent in ensuring employee performance and operational effectiveness in corporate companies. In the realm of business, the Environmental, Social, and Governance (ESG) framework serves as a catalyst for the promotion of sustainability. Business strategies, day to day practices, its model, and core policy can only be fruitful once employees are ready to understand and imbibe its essence in them. In this chapter, the researchers aim to explore and highlight the critical facets linked to ESG-driven HRM practices and corporate performance. The findings will have greater significance for HR policymakers and regulatory bodies, which can formulate specific policies in the HRM domain linked to long-term corporate performance. Present Chapter explore various prominent discussion on green practices in HRM area which directly and indirectly adhere to a sound corporate performance. It also exclusively offers imperatives for corporate and business level managers to draw attention on SDG, ESG and Green HRM practices which has manifold positive outcomes.

A. K. Pandey (✉)
Assistant Professor, Department of Management and Commerce, School of Management Sciences, Lucknow, India
e-mail: abhishekmba0811@gmail.com; abhishekpandey@smslucknow.ac.in

P. Tripathi
Assistant Professor, Department of Accounting and Financial Management, Faculty of Commerce, The Maharaja Sayajirao University of Baroda, Vadodara, India
e-mail: pankaj.tripathi-afm@msubaroda.ac.in

P. S. Shukla
Assistant Professor, Department of Commerce and Business Management, The Maharaja Sayajirao University of Baroda, Vadodara, India
e-mail: parag.shukla.msu@gmail.com

P. K. Upadhyay
Assistant Professor (Commerce), Maharaja Bijli Pasi Government P.G. College, University of Lucknow, Lucknow, India
e-mail: pramodkrupadhyay76@rediffmail.com

W. Leal Filho et al. (eds.), *Metaverse and Sustainability: Business Resilience Towards Sustainable Development Goals*, World Sustainability Series,
https://doi.org/10.1007/978-3-031-89545-6_1

Keywords Metaverse · Sustainable HRM practices · Business resilience · Corporate performance · SDGs

1 Introduction

Human resource management (HRM) is one of the key functions in any business organization whether it works for profit and not for profit. Human resource management practices cater a variety of activities, procedures, and regulations that pertain to the management of personnel in a professional environment, including recruitment and selection, training and development, performance management, employee relations, strategic talent management, transfer and promotion management, and grievance handling.

HRM as a discipline experiences many new domains like talent management, HR analytics, green HRM etc. In green HRM, HR action takers focus to formulate and implement such practices and policies which promote as well as include requirements of sustainability at workplace. Sustainable development goals (SDG) Dirven HR practices emphasize to formulate a workplace culture which is oriented towards sustainable development goals accomplishments in directly or indirectly manner. Several prominent firms have fully incorporated environmental, social, and governance (ESG) principles into their business strategies and organizational values, recognizing it as a crucial driver of innovation and a means to gain a competitive edge. These organizations not only prioritized sustainability in their technology and product innovation, but also demonstrated superior standards and accomplishments in corporate governance, employee relations, and social responsibility (Li et al. 2024).

1.1 Understanding Meaning and Origin of ESG Framework

It is important to highlight that ESG (Environment, Society and Governance) is a term first time highlighted in the report of "The Coining of ESG: The Who Cares Wins Report" by the United Nations' Chief Kofi Annan made significant efforts by aligning corporate sector companies to come ahead to incorporate such initiatives which ensures solid growth and development with emphasis on three important words i.e. environment, society and governance (Pollman 2024).

1.2 Scope and Connections of ESG and SDG

The term ESG, which stands for environmental, social, and governance, is a framework that is used to measure the social effect and sustainability of corporations. According to Friede et al. (2015), these criteria consist of environmental elements, social ties, and governance systems. The Sustainable Development Goals (SDGs) are a set of seventeen objectives that were presented by the United Nations in the year 2015 with the intention of handling the global issues such as poverty, inequality, climate change, and peace (United Nations 2015). According to Sachs et al. (2019), businesses that integrate their environmental, social, and governance (ESG) strategies with the Sustainable creation Goals (SDGs) contribute to the creation of a more sustainable global economy. Both the framework has common agenda to achieve sustainable development goal through different process and operating at different levels.

1.2.1 Environment Aspects in Sustainable Development Goals (SDGs) Report

Goals- 6 and 7 of the Sustainable Development Goals (SDGs) caters issues related to water and sanitation, while Goal -13 addresses climate change (United Nations 2015). ESG's environmental perspective is in line with these SDGs. Sustainable practices and a smaller company environmental impact can be achieved if businesses work toward these environmental objectives (Friede et al. 2015).

1.2.2 The Social Pillar in SDGs

As per exploration of Sachs et al. (2019), SDG 1 (No Poverty in world), SDG 3 (Good Health and Well-being of citizens), and SDG 8 (Decent Work and Economic Growth) are all SDGs that are relevant to social aspects and realted with the social components of ESG. The working conditions of employees, the quality of ties with local communities, and society as a whole are all enhanced when businesses adopt robust social policies (KPMG 2020). It is also the key aspects of Human resource management philosophy of any organisation.

1.2.3 Components of Governance in SDGs

Goals -16 and 17 (Peace, Justice, and Strong Institutions) and Partnerships for the Goals (United Nations 2015) are related to ESG's governance perspectives. Corporate governance of an organisation that promotes ethical leadership, openness, and accountability is an important factor in achieving these SDGs (KPMG 2020).

1.3 SDG Driven HR Practices and Its Consequences to Corporate Performance

The concept of sustainability has experienced a resurgence in popularity and has become a prominent subject in the business world when it was first introduced by the Brundtland Commission Report of the United Nations in the 1980s. It highlights the influence of enterprises on the social and environmental aspects and evaluates the performance of organizations based on sustainable development goals (Kramar 2022; Chang and Ke 2024). Given that the Sustainable Development Goals (SDGs) necessitate altered conduct from individuals in various types of businesses, such as public sector business operations, private work, and third sector enterprises, the SDGs have a direct correlation with human resource management (HRM) (Chams 2019; Kramar 2022; Brewster 2024). Consequently, numerous recent publications have established a strong linkage between the SDGs and HRM. In the field of human resource management, for instance, Sustainable Development Goal 8 (SDG 8), which is titled "Decent Work and Economic Growth," profoundly demonstrates an undeclared debate between those who believe that economic growth is essential to progress, regardless of the short-term trouble inherited, and those who think that no economic progress is acceptable if it requires people to be working in conditions that are deplorable or with contracts that are questionable from an ethical standpoint (Brewster 2024).

Around ninety-six percent of the research, according to Muhmad and Muhmad (2020), revealed a positive association between sustainable initiatives of the business and its financial, notably following the adoption of the Sustainable Development Goals (Ramos et al. 2022). The ability of a firm to efficiently manage its resources and generate value to the organization is what is meant by the term "corporate performance." It incorporates a number of different characteristics, such as financial performance, operational performance, economic performance, as well as environmental and social performance. A multifaceted perspective is required in order to determine the success of a corporation in modern time highly competitive and risk-prone business scenario. Historically, the performance of a corporation has been evaluated based on profitability and return on investment (What Is Corporate Performance? 2021).

2 Review of Literature

2.1 A Brief Review on Metaverse and HRM

The Metaverse, a virtual environment enabling user interaction within immersive digital settings, has lately gained significance in the business sector as a resource for human resource management. The Metaverse offers enterprises the chance to transform conventional HRM activities, including recruitment, training, cooperation, and employee engagement (Lee et al. 2022). Studies indicate that virtual environments enhance employee learning, enable remote collaboration, and promote engagement via immersive simulations and interactions (Aguinis et al. 2021). The Metaverse is a novel method for addressing geographical and resource constraints in sustainability, enabling firms to adopt sustainable HR practices with less environmental effect. This digital change diminishes the necessity for physical infrastructure, transport, and resources, hence promoting environmental sustainability (Kang and Lee 2023). Furthermore, the Metaverse can facilitate awareness and comprehension of environmental, social, and governance (ESG) issues via virtual training sessions and workshops, rendering it an effective instrument for incorporating sustainability into human resource management methods.

2.2 The Metaverse as a Platform for ESG Integration

The Metaverse provides a unique platform for organizations to incorporate ESG principles into their HRM practices, especially concerning employee engagement, training, and development. Virtual environments enable firms to conduct interactive and engaging training programs centered on ESG subjects, like climate change, social justice, and ethical governance (Xu et al. 2023). This immersive learning experience can foster a deeper understanding of ESG issues among employees and help build a culture of sustainability within the organization.

Moreover, the Metaverse enables companies to simulate real-world scenarios related to sustainability challenges, allowing employees to practice decision-making in a risk-free environment (Hodgson 2023). This virtual training can improve problem-solving abilities and equip employees to address ESG-related difficulties in practical scenarios. The Metaverse enables virtual cooperation among international teams, thereby diminishing the carbon footprint linked to corporate travel while preserving productivity and engagement (Aust 2024).

2.3 Theoretical Background of the Study

2.3.1 A Brief Review on Diversity and Inclusion

It is necessary to promote diversity and inclusion within the workforce in order to achieve many of the Sustainable Development Goals (SDGs), including Goal 10, which aims to reduce inequalities and Goal 5, which focuses on gender equality. It is possible for human resource practices to contribute to social development and organizational success by fostering a diverse and inclusive workplace. Some examples of such practices include the implementation of bias-free recruitment processes, the provision of diversity training, and the creation of support networks for underrepresented groups (Fig. 1).

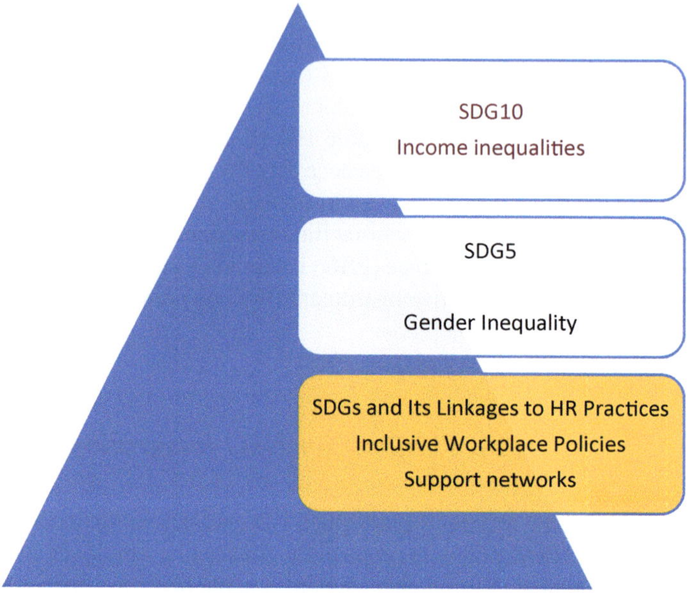

Fig. 1 SDGs, HR practices and inclusive workplace policies. *Source* Model created by authors based on Review of Literature

2.3.2 SDGs and Corporate Performance

There is a favourable impact on the financial performance of corporations when SDGs are reported. Regulatory agencies, various stakeholders, policymakers, board members, and managers can all benefit from the findings because they have significant practical implications. Present research work offers a significant and unique contribution to the existing body of literature on corporate disclosure (Al Lawati and Hussainey 2022). A research study offers a practical way for companies to find out how the SDGs relate to their ESG performance and how they may monitor their progress towards the SDGs in a tangible way (Khaled et al. 2021). In realistic and different sense, SDGs cannot be achieved until industry actors will not contribute it. SDGs related achievable targets set forth by Industry people will improve the overall performance of the firm. Hence, parameters associated to corporate performance can be measured and monitored in terms of planned interventions on SDGs. Companies have the option to allocate resources and provide incentives for human resource management (HRM) to support sustainability efforts, or they may deliberately impede or limit significant actions in this area (Aust 2024). Such prior studies highlight the linkages of SDGs and firm performance in the corporate world. For various reasons, the SDGs hold great significance for organizations. Organizations can use this all-encompassing framework to coordinate their sustainability initiatives and address the world's most critical problems. This is leading to a shift in the focus of studies toward holistic approaches to sustainability (Ranjbari et al. 2021; Hofer et al. 2024).

2.3.3 ESG Driven HRM Practices at a Glance

A research report has found that having a diverse gender representation in management teams is positively associated with performance in terms of labor and human rights. This performance acts as a mediator by promoting more disclosure of information regarding these concerns. Further, it has been stated that such a performance facilitates a higher level of information disclosure. The results of this study consequently appear to suggest that the inclusion of female candidates in management teams serves as an accelerating force for increased social responsibility (Monteiro et al. 2022). In a study, it has been indicated that labor factors have the greatest impact on job engagement within a sustainable human resources management system (Navajas-Romero et al. 2022). In an article, it has been provided that an understanding of the ways in which businesses can improve their social reputation by implementing various Green Human resource management (GHRM) programs. These initiatives can help businesses promote their employees by educating them on green human resource management initiatives and by encouraging greener work culture within the organization and society. In order for businesses, governments, and industries to improve their social performance through the use of GHRM principles, the study is extremely beneficial (Jiang et al. 2023).

Workplace justice, sustainability, and business ethics are commonly linked to human resource management (HRM) practices within the framework of corporate social responsibility (CSR) initiatives. These practices are frequently referred to as sustainable HRM. It is common knowledge that human resources (HR) may play a significant part in strengthening the sustainability of a company by implementing policies and practices that are beneficial to the company's employees and society as a whole. Furthermore, it is well acknowledged that all of these activities need to be recorded and reported in order to achieve exposure and credibility. In recent times, there has been a significant focus on the Sustainable Development Goals (SDGs) as well as the elements of Environmental, Social, and Governance, which are the parameters for documenting and reporting relevant practices that are adopted by commercial entities. Human resource management has the potential to play a significant part in the process of establishing and executing such practices, particularly those that are associated with society (Papalexandris 2022). In this way, it is widely accepted that SDG driven HR practices emphasize to formulate and implement sustainable way to manage HR functions in workplace setting. There is a further scope to investigate the SDG driven HR practices prevalent in global organizations across the international markets.

2.3.4 The Stakeholder Theory of Freeman (1984)

According to Edward Freeman (1984), "The twentieth-first century is one of "Managing for Stakeholders." The task of executives is to create as much value as possible for stakeholders without resorting to trade-offs. Great companies endure because they manage to get stakeholder interests aligned in the same direction (Freeman 2018)." In this statement, there is a no. of great insights are present which can be utilized by organizations to improve their employees and corporate performance. This theory can be considered to understand the role of HRM in Shaping Corporate performance through imbibing ESG practices.

2.3.5 Resource Based Theory (RBT)

This theory is first time put forth by Penrose in the year 2009. A significant method in strategic management subject is the resource-based theory (RBT). A lot of managers and professionals use it as a guide to identify what their company requires to stay ahead of the competition. To understand and foretell the bedrock of a business's performance and competitive advantage, the theory offers a crucial foundation and understandings. This is extension of the work of Barney (1991) (Barney 1991; Penrose 2009; Utami and Alamanos 2023). This theory can be utilized to work on strategic HRM practices which are also contributing to become a firm more ESG driven and improving its performance. Even, these are connected with key agendas of SDG-17 goals.

2.3.6 Carroll's CSR Pyramid

"Corporate social responsibility encompasses the economic, legal, ethical, and discretionary (philanthropic) expectations that society has of organizations at a given point in time" (Carroll 1979, 1991). This is, the original statement of Carroll's four-part definition of CSR. It is one of the broad framework and theoretical understanding of key components of CSR process for any organisation (Fig. 2).

2.3.7 Psychological Contract Theory (PCT) and Its Unnoticed Outcomes on ESG and SDG Framework

The psychological contract refers to an implicit expectation between employees and management (Argyris 1960; Levinson et al. 1980; Rajalaxmi and Naresh 2018). The relevance of this contract and its theoretical prominence can be extended to ESG and SDG framework. In the context of Environmental, Social, and Governance (ESG) frameworks and the Sustainable Development Goals (SDGs), the idea of psychological contracts is becoming an increasingly important notion. Not only can businesses that actively incorporate environmental, social, and governance (ESG)

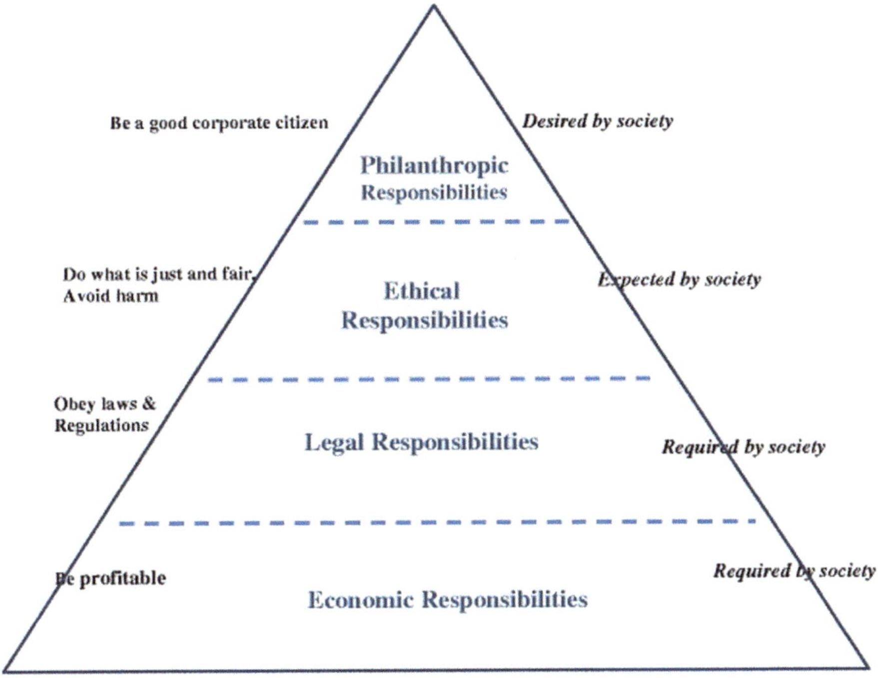

Fig. 2 Pyramid model of CSR. *Source* Springer Open Access

standards into their operations contribute to sustainability, but they also contribute to holding the expectations and commitments of their employees. The employees of a company may have the expectation that the company would display excellent environmental stewardship, respect social duties, and maintain ethical governance procedures when the company makes a commitment to ESG principles. The need is to utilise metaverse technologies to create work space which accelerates ESG and SDGs driven business resilience strategies and corporate performance.

3 Discussion

Metaverse based Sustainable HRM practices demonstrate the importance of HR strategies in advancing ESG principles in business models. ESG in HRM practises promotes daily sustainability, aligning firms with the global sustainable development agenda (Li et al. 2024). ESG goals and a more sustainable workplace are promoted by the Metaverse's virtual platform for employee participation, training, and cooperation. As companies realise the importance of business resilience for social and environmental sustainability as well as financial performance (Brewster 2024), this is especially important. ESG-driven HRM practises improve business performance by integrating sustainability into corporate culture (Chams 2019). This helps organizations acquire a competitive edge, recruit and retain top personnel, and improve stakeholder relations. The environmental aspects of sustainability, such as Green HRM policies that reduce the workforce's environmental imprint, improve long-term corporate performance (Ranjbari et al. 2021). Diversity, equity, and inclusion support ESG social goals and create a more engaged and innovative workforce (Monteiro et al. 2022). Businesses can improve operational efficiency and global sustainability by integrating HR practices to the SDGs (Sachs et al. 2019). The research shows that SDG-driven HRM solutions including responsible labor and ethical governance increase stakeholder interactions and corporate performance. These initiatives promote openness, accountability, and ethical decision-making, which aligns with ESG governance (KPMG 2020). ESG-driven HRM on the Metaverse is a novel way to overcome traditional hurdles to sustainable business operations. Interactive training and collaborative workspaces can help employees accept sustainability goals in the virtual world (Aust 2024). This promotes ongoing learning and ESG engagement among employees.

There has been considerable discussion on incorporating Sustainable Development Goals into corporate strategy, since it has been shown to have a positive effect on the financial outcomes of companies when they report on SDGs. This can provide benefits to a range of individuals and groups involved in the process, such as regulatory agencies, policymakers, board members, and managers. It is important to emphasize the practical consequences and the impact it has on the existing body of knowledge related to corporate disclosure (Al Lawati and Hussainey 2022). This analysis explores the impact of HRM practices driven by the parameters of SDGs on company performance, basically by ESG related variables and sustainable HRM practices. Corporations are becoming more aware of the significance of aligning their operations with the SDGs practices in order to channelize their overall performance. The SDGs function as a worldwide model for sustainable development, with a focus on promoting economic growth, social inclusivity, and environmental preservation. Companies that regularly disclose their progress towards these objectives exhibit transparency and accountability, which can bolster their reputation and financial performance (Khaled et al. 2021).

Studies suggest that when businesses link their operations with Sustainable Development Goals (SDGs), especially by implementing Environmental, Social, and Governance (ESG) policies, it leads to enhanced corporate performance. The measurement and monitoring of company performance, including financial returns, operational efficiency, and stakeholder satisfaction, can be effectively achieved by specific SDG interventions (Navajas-Romero et al. 2022). This connection promotes a business strategy that is both sustainable and socially responsible, which in turn attracts investment and supports long-term successes criteria are crucial for assessing the long-term viability and ethical implications of a firm. ESG-focused HRM strategies are essential for attaining the SDGs by incorporating sustainability principles into human resource management. Gender diversity in management teams has been found to have a favourable correlation with improved performance in labor and human rights. This, in turn, encourages increased transparency in disclosing information related to these issues (Monteiro et al. 2022). By incorporating a range of viewpoints in decision-making processes, a firm can bolster its commitment to social responsibility and sustainability.

In addition, ESG-focused HRM approaches include Green Human Resource Management (GHRM) efforts, which seek to integrate sustainability into HR responsibilities. GHRM practices encompass the dissemination of knowledge among employees on environmental concerns, fostering a corporate culture that prioritizes sustainability, and enacting rules that facilitate the adoption of sustainable practices (Jiang et al. 2023). These measures not only support environmental objectives but also improve the company's social performance by promoting a responsible and ethical work environment. Sustainable human resource management practices are intricately linked to CSR efforts. CSR entails the implementation of policies that specifically target social, environmental, and ethical issues, thereby offering a significant contribution to the overall sustainability of the firm. Sustainable human resource management (HRM) practices, including the promotion of workplace justice, business ethics, and social responsibility, are essential components of corporate social responsibility (CSR) programs (Papalexandris 2022). These methods guarantee that organizations function in a way that is advantageous to both employees and society, thereby improving their social standing and overall effectiveness. An essential element of sustainable HRM is the requirement for meticulous documentation and comprehensive reporting of operations to guarantee transparency and credibility. Precise reporting of HR practices that contribute to the goals of SDGs and ESG is necessary due to the emphasis placed on these areas. By enhancing the firm's social and environmental impact, this not only benefits the company but also offers stakeholders vital insights about the company's dedication to sustainability.

Implementing HR practices that are aligned with the SDGs is crucial for cultivating a workplace culture that is focused on attaining sustainable development objectives. These practices entail the establishment of policies and processes that directly or indirectly endorse SDGs, such as advocating for fair employment (SDG 8) and mitigating disparities (SDG 10). By incorporating these objectives into human resources operations, organizations can improve their social impact and make a meaningful contribution to worldwide sustainability initiatives (Brewster 2024). Studies demonstrate a robust correlation between SDGs and Human HRM practices, indicating that organizations must embrace a comprehensive strategy to evaluate success in the current cutthroat economic landscape. Implementing sustainable human resource management (HRM) strategies, such as efforts focused on diversity and inclusion, is essential for attaining the Sustainable Development Goals (SDGs) pertaining to societal development and organizational prosperity (Muhmad and Muhmad 2020) (Table 1).

It can be discussed further that Freeman's Stakeholder Theory (1984) asserts that executives should prioritize the creation of maximum value/wealth for all their stakeholders instead than concentrating on trade-offs among conflicting interests. This idea posits that corporations achieve long-term success by harmonizing the objectives of

Table 1 Key components in ESG—driven HRM practices affecting the corporate performance

SDG reporting	ESG and corporate performance	ESG-Driven HRM practices	Sustainable HRM and CSR
• Improves financial performance • Enhances transparency and accountability • It Benefits regulatory agencies, stakeholders, and managers	• Alignment with SDGs and ESG practices leads to better corporate performance • Factors related to financial returns and operational efficiency can be monitored through SDG interventions	• Gender diversity in management teams improves labor and human rights performance • Green HRM initiatives promote environmental awareness and green culture	• Involves policies addressing social, environmental, and ethical concerns • Enhances social reputation and overall performance

Source Author's compilation on ESG Driven HRM Practices and Corporate Performance based on Literature Review

diverse stakeholders, hence improving corporate performance. By adopting Environmental, Social, and Governance (ESG) principles, firms can align stakeholder interests, consequently influencing human resource management (HRM) and enhancing corporate performance. It is also the need of modern age that people should be sensitize to take strong and impactful decision which can stimulate sustained growth and development for upcoming generations. It should be the moto of executive or core committees of each multinational companies. This will be promoted ESG led HR interventions in highly perform centric organisations.

Further, the Resource-Based Theory (RBT), initiated by Penrose in 2009 and further developed by Barney, offers a framework for comprehending competitive advantage via the efficient utilization of organizational resources. It pertains to strategic HRM, which can propel firms towards ESG targets and aid in achieving Sustainable Development Goals (SDGs). This approach assists firms in identifying the resources required for maintaining competitiveness, while simultaneously promoting ESG objectives. Generating such Human resources pool /talented individuals who are bound by ESG framework and philosophy will be the game changer approach for upcoming generations.

Policy makers and corporate core committees can utilise the Carroll's CSR Pyramid (1979, 1991) model to delineates the four principal obligations of enterprises: economic, legal, ethical, and discretionary (philanthropic). This model offers a comprehensive framework for comprehending corporate social responsibility (CSR) and its influence by public expectations. So, corporate performance metrics must consider CSR related contributions and related factors which will be encouraging for such organizations and their related stakeholders.

The Psychological Contract Theory (PCT) pertains to the implicit expectations between employees and management. PCT is becoming increasingly significant within ESG and SDG frameworks, as employees frequently anticipate that companies would adhere to ESG standards. This implicit contract can affect employee engagement and organizational performance when it aligns with environmental, social, and governance criteria. HR professionals need to devise such HRD interventions which fulfils the aspirations of ESG and SDG at one instance and also incorporate hidden issues and challenges explored by psychological contract between corporation and its stakeholders.

4 Imperatives for Corporate Managers

Corporate managers have a leading role in incorporating Sustainable Development Goals into corporate policy interventions and key business processes, especially in the area of Human Resource Management (HRM). Managers must ensure that their HR policies are in line with the SDGs, specifically by supporting fair and productive employment (SDG 8) and minimizing disparities (SDG 10). This caters establishing precise and quantifiable objectives that connect human resources initiatives to overarching sustainability objectives, fostering a culture of responsibility and intention. It is essential to highlight diversity and inclusion as fundamental elements of HRM initiatives. Studies indicate that the presence of both male and female individuals in management committees has a positive impact on labor rights and promotes a sound work environment and well-structured social responsibility. Managers ought to adopt impartial recruitment procedures, offer diversity training, and create support networks for marginalized groups, thereby fostering a fairer work environment. Implementing Green Human Resource Management (GHRM) techniques can profoundly improve a company's environmental performance. Managers should prioritize the education of staff regarding sustainability, fostering a culture of environmental consciousness inside the firm, and incorporating environmental factors into HR policy, such as implementing sustainable travel practices and energy-saving efforts. Precise recording and reporting of HR practices pertaining to SDGs are crucial for ensuring transparency and establishing confidence. Managers ought to establish strong and reliable systems for monitoring the success of Sustainable Development Goal (SDG) efforts, ensuring adherence to Environmental, Social, and Governance (ESG) criteria, and effectively communicating accomplishments to stakeholders. The act of being transparent fosters trust and improves the company's standing.

Business and corporate managers ought to develop human resources strategies that tackle social, environmental, and ethical issues, including equitable labor practices, community involvement, and ethical procurement. Integrating HRM with CSR improves the company's social impact and promotes long-term viability. Managers need to consistently oversee the influence of HRM practices guided by SDGs on the overall performance of the company. This includes the examination of financial performance, effectiveness in operations, and the contentment of stakeholders. Regular evaluations allow managers to adjust methods as necessary, guaranteeing that HRM processes stay efficient and in line with developing SDG goals. By incorporating these essential principles and HRM practices, business managers may actively promote significant advancements in attaining Sustainable Development Goals (SDGs), improve corporate efficiency, and make valuable contributions to global sustainable development. Managers working in different capacity in business organizations can imitate small steps which promote sustainable and green practices in terms of managing people. Even, a feeling of green and sustainable work processes must be incorporated to each step in business operations.

Some are the imperative and directives for business professionals derived from the present research work:

(1) Corporate managers need to work upon creating a sound metaverse based work infrastructure which promotes business resilience and sustainability.
(2) Redirect emphasis from immediate profits to sustainable value generation for all stakeholders (workers, customers, investors, communities). Guarantee that decisions advantage not only shareholders but also enhance the wider stakeholder ecosystem.
(3) Integration of Environmental, Social, and Governance (ESG) considerations into fundamental business strategies.
(4) Setting the quantifiable ESG targets that align with stakeholder interests and global sustainability aims.
(5) Formulate such HR policies and practices that emphasize ESG objectives.
(6) Develop a talent pool of employees dedicated to sustainability, ethical governance, and social responsibility.
(7) Strategically utilize internal resources (human, intellectual, and physical) to achieve a competitive advantage while promoting ESG and SDG objectives.
(8) Identify and allocate resources essential for maintaining enduring competitiveness. Integrate Corporate Social Responsibility into Performance Metrics.
(9) Incorporate company Social Responsibility (CSR) into company performance assessment frameworks.
(10) Utilize Carroll's CSR Pyramid to equilibrate economic, legal, ethical, and charitable obligations, guaranteeing that business actions align with public expectations.
(11) Identify and Regulate Psychological Contracts like comprehend the unspoken anticipations personnel possess regarding ESG obligations. Formulate HR solutions that address psychological contracts, ensuring corporate actions correspond with employee expectations around ethical governance and social responsibility.
(12) Urge leadership to make decisions that reconcile immediate benefits with enduring sustainability.
(13) Participate in significant decision-making that fosters enduring progress for future generations.
(14) Ensure that corporate strategies and actions are congruent with the global Sustainable Development Goals agenda.

(15) Develop a strategy that aligns firm performance with essential Sustainable Development Goals, particularly in domains such as environmental stewardship and social equality.
(16) These directives will assist managers in constructing responsible, high-performing, and resilient organizations.
(17) ESG and SDG cell should be created which try to metaverse space for achieving major corporate actions and strategies.

5 Conclusions

In this chapter, it has been concluded that ESG principles in HRM enabled by the Metaverse, can help firms achieve sustainability, business resilience, and improved performance in long run (Khaled et al. 2021). As companies realize the importance of sustainability, HR practices must align with ESG and SDG frameworks. These approaches benefit business, society, and the environment. This chapter provides a basis for HR policymakers to create sustainable HR strategies that boost business performance and resilience. In a competitive and sustainability-driven business climate, organizations must implement Green HRM, diversity and inclusion, and ethical governance (Ramos et al. 2022). This chapter appropriately contains the inter linkages of SDG driven HRM practices, ESG interventions and overall influence on corporate functionality and performance. On summarizing, it can be explored that implementing SDGs into organization plans and policies with ESG-driven Human Resource Management (HRM) practices greatly improves corporate performance. Companies that synchronize their operations with SDGs and embrace sustainable Human Resource Management practices exhibit enhanced financial performance, operational efficiency, and stakeholder satisfaction. Human resource management strategies that are driven by environmental, social, and governance (ESG) factors, such as initiatives focused on global human resource management (GHRM) and programs promoting diversity, play a significant role in enhancing organizations focus towards to social responsibility and sustainability management. Thorough documentation and reporting of these practices provide transparency and reliability, promoting a responsible and ethical work environment. As business organizations prefer sustainability then the significance of HR practices driven by SDGs in attaining corporate success and supporting global sustainability objectives will grow further. It is again important to highlight that corporate managers have a crucial responsibility in incorporating SDGs into HR practices in order to improve organizational performance and create a sound environment. Managers can achieve considerable increases in financial returns, operational efficiency, and stakeholder trust by aligning HRM strategies with SDGs, supporting diversity, implementing Green HRM, and fostering openness. Integrating Corporate Social Responsibility (CSR) into HR policies and consistently overseeing these efforts guarantees that organizations not only achieve their sustainability goals but also keep a competitive edge and long-term prosperity. By adopting these strategies, firms can effectively contribute to global sustainability objectives while also attaining strong corporate performance. Corporate managers are crucial in relating the Human Resource Management (HRM) practices with Sustainable Development Goals (SDGs). Through the implementation of ESG-oriented HR policies, the promotion of diversity, the cultivation of environmental consciousness, and the establishment of open reporting mechanisms, they guarantee long-term value generation, social accountability, and sustainable business practices, thereby advancing both corporate and global development.

References

Aguinis H et al (2021) Harnessing the metaverse for human resource development. J Organ Behav 42(2):156–171
Al Lawati H, Hussainey K (2022) Does sustainable development goals disclosure affect corporate financial performance? Sustainability 14(13):7815. https://doi.org/10.3390/su14137815
Aust I (2024) Human resource management for sustainability: Strategic perspectives and corporate governance. J Sustain Bus 15(3):225–240
Aust I, Cooke FL, Muller-Camen M, Wood G (2024) Achieving sustainable development goals through common-good HRM: context, approach and practice. Ger J Human Resour Manage 38(2):93–110
Barney J (1991) Firm resources and sustained competitive advantage. J Manag 17(1):99–120. https://doi.org/10.1177/014920639101700108 (Original work published 1991)
Brewster C (2024) Economic growth and sustainable HRM: Balancing the scale. Sustain Manage Rev 11(2):145–167
Brewster C, Brookes M (2024) Sustainable development goals and new approaches to HRM: why HRM specialists will not reach the sustainable development goals and why it matters. Ger J Human Resour Manage 23970022241229037
Carroll AB (1979) A three-dimensional conceptual model of corporate social performance. Acad Manag Rev 4:497–505
Carroll AB (1991) The pyramid of corporate social responsibility: toward the moral management of organizational stakeholders. Bus Horizs 34(4):39–48
Carroll AB (2016) Carroll's pyramid of CSR: taking another look. Int J Corporate Soc Responsib 1:3. https://doi.org/10.1186/s40991-016-0004-6
Chams N (2019) Sustainable human resource management and SDGs: an integrative review. J Clean Prod 236:117–125. https://doi.org/10.1016/j.jclepro.2019.117-125
Chang YL, Ke J (2024) Socially responsible artificial intelligence empowered people analytics: a novel framework towards sustainability. Hum Resour Dev Rev 23(1):88–120
Dumont J, Shen J, Deng X (2017) Green HRM practices: How to create a sustainable workplace. J Bus Ethics 156(4):755–774
Ehnert I, Harry W, Zink KJ (2016) Sustainability and HRM: challenges and future directions. Int J Human Resour Manage 27(2):203–219
Freeman RE (1984) Strategic management: a stakeholder approach. Pitman, Boston
Freeman RE (2018) About |Stakeholder theory. Retrieved 12 Oct 2024, from http://stakeholdertheory.org/about/
Friede G, Busch T, Bassen A (2015) ESG and financial performance: aggregated evidence from more than 2000 empirical studies. J Sustain Finance Investment 5(4):210–233. https://doi.org/10.1080/20430795.2015.1118917
Friede G, Busch T, Bassen A (2015) ESG and financial performance: Aggregated evidence from more than 2000 empirical studies. J Sustain Finan Investment 5(4):210–233

Hodgson C (2023) Virtual workspaces and the future of human resource management. Hum Resour Dev Q 34(1):123–135

Hofer A, Kortsch T, Kauffeld S (2024) Shaping organizations sustainably: insights and perspectives. Gruppe Interaktion Organisation Zeitschrift für Angewandte Organisationspsychologie (GIO) 55(2):115–125

Jiang Y, Zaman SI, Jamil S et al (2023) A triple theory approach to link corporate social performance and green human resource management. Environ Dev Sustain. https://doi.org/10.1007/s10668-023-03272-3

Kang J, Lee H (2023) Metaverse as a platform for sustainable business practices: opportunities and challenges. Bus Horiz 66(3):345–354

Khaled M, Rashed T, Amr M (2021) Monitoring ESG and SDG performance: practical tools for businesses. J Bus Ethics 162(1):65–80. https://doi.org/10.1007/s10551-020-04638-6

Khaled R, Ali H, Mohamed EKA (2021) The sustainable development goals and corporate sustainability performance: mapping, extent and determinants. J Clean Prod 311:127599. https://doi.org/10.1016/j.jclepro.2021.127599

KPMG (2020) The ESG imperative: corporate governance and sustainable development goals. KPMG Global Rep 10(4):1–12

KPMG (2020) The time has come: the KPMG survey of sustainability reporting 2020. KPMG International. https://home.kpmg/xx/en/home/insights/2020/11/the-time-has-come-survey-of-sustainability-reporting.html

Kramar R (2022) Green human resource management: an integrative model of corporate sustainability. Int J Human Resour Manage 33(2):220–240

Lee Y, Lee J, Lee S (2022) Exploring the future of HRM in the metaverse. J Bus Manag 8(4):240–257

Li J, Pan Z, Sun Y, Zheng W, Zhang Z, Wang Z (2024) From compliance to strategy: paradigm shift in corporate ESG practices. Acad J Humanit Soc Sci 7(2):185–193

Li X, Wang Z, Chen Y (2024) ESG and human resource management: a strategic alignment for resilience and performance. Corp Sustain Stud 8(2):55–77. https://doi.org/10.1007/s40570-023-00124-5

Monteiro AP, García-Sánchez IM, Aibar-Guzmán B (2022) Labour practice, decent work and human rights performance and reporting: the impact of women managers. J Bus Ethics 180(2):523–542

Monteiro P, Sousa C, Marques J (2022) Gender diversity in management and social responsibility disclosure: a review. J Corp Responsib 18(1):89–101

Muhmad S, Muhmad N (2020) Corporate performance and sustainability: the role of ESG. J Bus Res 123:485–496

Navajas-Romero V, Ceular-Villamandos N, Pérez-Priego MA, Caridad-López del Río L (2022) Sustainable human resource management the mediating role between work engagement and teamwork performance. PLoS ONE 17(8):e0271134

Papalexandris N (2022) Sustainable development and the critical role of HRM. Studia Universitatis Babes Bolyai-Oeconomica 67(3):27–36

Penrose E (2009) The theory of the growth of the firm, 4th edn. Oxford University Press, Oxford

Pollman E (2024) The making and meaning of ESG (October 31, 2022). U of Penn, Inst for Law & Econ Research Paper No. 22–23, European Corporate Governance Institute—Law Working Paper No. 659/2022, Harvard Business Law Review, vol 14, p 403. Available at SSRN: https://ssrn.com/abstract=4219857

Rajalakshmi M, Naresh B (2018) Influence of psychological contract on workplace bullying. Aggress Violent Beh 41:90–97. https://doi.org/10.1016/j.avb.2018.05.001

Ramos DL, Chen S, Rabeeu A, Abdul Rahim AB (2022) Does sdg coverage influence firm performance? Sustainability 14(9):4870. https://doi.org/10.3390/su14094870

Ramos J, Teixeira M, Silva A (2022) ESG initiatives and corporate performance: the role of SDGs. Sustain Account Manage J 12(3):303–319

Ranjbari M, Kuijk R, Lopez S (2021) The holistic approach to sustainable development: integrating SDGs and ESG in HR practices. Sustainability 13(4):10034. https://doi.org/10.3390/su130410034

Renwick DWS, Redman T, Maguire S (2013) Green human resource management: a review and research agenda. Int J Manag Rev 15(1):1–14
Sachs J, Schmidt-Traub G, Kroll C (2019) The sustainable development goals and pathways to sustainability: global perspectives and corporate responsibilities. Nature Sustain 2(9):789–795. https://doi.org/10.1038/s41893-019-0294-7
Sachs J, Schmidt-Traub G, Kroll C, Lafortune G, Fuller G, Woelm F (2019) Sustainable development report 2019. Sustain Dev Solutions Netw. https://doi.org/10.1017/9781108649241
United Nations (2015) Transforming our world: the 2030 agenda for sustainable development. United Nations. https://doi.org/10.18356/6e556a0a-en
Utami H, Alamanos E (2023) Resource-based theory: a review. In: Papagiannidis S (Ed) TheoryHub Book. Available at https://open.ncl.ac.uk. ISBN: 9781739604400
What is corporate performance ? | 4 Answers from Research papers (2021) SciSpace—Question. https://typeset.io/questions/what-is-corporate-performance-5ercnuwv7l
Xu L, Chen S, Zhang Y (2023) The role of the Metaverse in promoting corporate sustainability and resilience. Sustainability 15(6):2408

Leveraging Emerging Digital Ecosystem to Attain Long-Term Sustainable Growth

Manali Agrawal and Prateek Kumar Bansal

Abstract *Introduction*: India's booming tech industry and entrepreneurial culture make it well-positioned to contribute to the development of the metaverse. India's government has been working to develop a $1 trillion digital economy that is fueled by the country's sizable and rapidly expanding youth population. India is an important player for offering digital labor to the metaverse due to its population competitive advantage, growing stature in product development, and well-established reputation as a global center for IT outsourcing. India is an excellent destination to offer digital services because of its strong IT infrastructure and thriving IT industry. By establishing itself as an important factor in the creation of the global metaverse, India can draw in foreign investment and collaboration. *Material and Methods*: In order to determine how current trends in technology adoption, virtual worlds, digital economies, and user behavior might affect integrated sustainable growth through metaverse in the future, a number of studies and publications are reviewed. *Results*: By utilizing virtual technology in novel ways, nations can not only promote sustainable development but also establish a model for a more egalitarian and sustainable future. *Conclusion and Recommendation*: A country might leverage the metaverse to achieve sustainable growth by carefully utilizing its resources to advance social, economic, and environmental sustainability. To reduce emissions associated with travel, promote remote work and virtual gatherings. Encourage the use of digital resources to reduce waste production and the use of natural resources. Encourage the development of new e-commerce, virtual service, and digital content creation jobs and business models in the metaverse economy.

Keywords Metaverse · Sustainable growth · Economy · Digital ecosystem

M. Agrawal (✉)
NSB Academy, Bangalore, Karnataka, India
e-mail: manaliagrawal14@gmail.com

P. K. Bansal
Institute of Business Management, GLA University, Uttar Pradesh, Mathura, India
e-mail: prateek.bansal@gla.ac.in

W. Leal Filho et al. (eds.), *Metaverse and Sustainability: Business Resilience Towards Sustainable Development Goals*, World Sustainability Series,
https://doi.org/10.1007/978-3-031-89545-6_2

1 Introduction

A "metaverse" is a collective virtual shared environment that is formed when virtually optimized physical reality and physically sustained virtual space—which includes the internet, augmented reality, and all virtual worlds combined—converge. Through avatars and other digital representations, users can engage in interactive and immersive interactions within a vast and virtual environment. Virtual reality (VR) and augmented reality (AR), as well as social media, online gaming, and digital economy, are all a part of the metaverse (Mystakidis 2022).

India is well-positioned to play a major role in the creation of the metaverse due to its expanding tech industry and entrepreneurial zeal. This potential is highlighted by the government's ambitious objective of building a $1 trillion digital economy, which is driven by the large and youthful population of the nation. India is a great place for digital services because of its robust IT infrastructure, competitive talent pool, and reputation as a global center for IT outsourcing. India can attract large foreign investment and develop collaboration with global IT giants by positioning itself as a prominent player in the metaverse. The metaverse presents enormous promise for social innovation, economic growth, and sustainable development. It is a virtual reality environment where users can connect and participate in a variety of activities. Science fiction has contributed to promoting the idea of the metaverse, which is now being developed by a number of tech businesses to produce new kinds of digital experiences and interactions. Even when users are not signed in, the metaverse is still dynamic and constantly morphing. Interactions between various virtual economies, platforms, and surroundings are possible. Digital assets and material are created, owned, and exchanged by users. With virtual avatars, the metaverse enables social interactions and teamwork among users. Users can experience a sense of presence and immersion with the help of cutting-edge technology like VR and AR (Peukert et al. 2024).

The metaverse is a virtual world where users can interact with each other and digital objects. It has applications in various areas, such as gaming, education, healthcare, and social interactions. The metaverse can contribute to "tech for good" initiatives by promoting sustainable development, equality, and inclusivity. While gaming and social interactions are prominent applications of the metaverse, its potential extends far beyond these areas. It offers immersive and interactive learning experiences, allowing students to explore virtual classrooms, historical events, or scientific concepts in a more engaging way.Healthcare professionals can use the metaverse for virtual consultations, surgeries, and training. Patients can also benefit from virtual therapy sessions and rehabilitation programs. Metaverse platforms can host virtual shopping experiences, allowing customers to try on clothes, explore virtual stores, and interact with products in a 3D environment (Iqbal and Campbell 2023).

2 Impact of Metaverse on Economic System

The economic situation of a country can be greatly impacted by the metaverse's use in a number of ways, posing both opportunities and concerns. Avatar's attire, virtual real estate, and other digital assets become increasingly popular as a result of the metaverse, opening up new commercial opportunities (Fig. 1).

In order to promote entrepreneurial activity and economic diversity, the metaverse offers startups and small and medium-sized enterprises (SMEs) a platform to develop innovative products and services. Investing more in technology related to the metaverse can promote expansion in the tech industry, incorporating hardware, software, and content production. By enabling flexible hours and lowering the requirement for actual office space, the metaverse can improve remote work capabilities, which can boost output and save operating expenses. Better educational opportunities and training programs can improve the skills and knowledge of the workforce, which will increase output and stimulate the economy. Businesses can access new revenue streams through e-commerce in the metaverse, which enables them to reach a worldwide audience without being limited by physical geography. Virtual visits to historical landmarks, museums, and natural wonders can increase tourism-related income without having the same negative environmental effects as traditional travel.

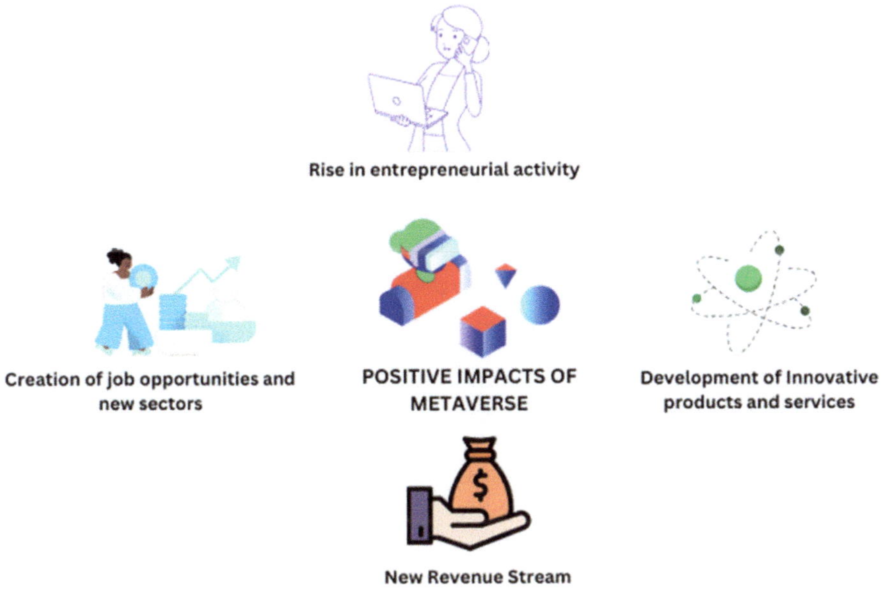

Fig. 1 Positive impacts of metaverse. *Source* Compiled by Author

Because the technology is still in its early stages, it is unclear how much the metaverse will influence economies in Asia and around the world. According to preliminary estimations gathered from existing literature, the metaverse has the potential to significantly impact the world economy and increase in size.

- According to the estimates of Grand View Research the worldwide market size for the metaverse (i.e., revenue) might reach US$678.8 billion to US$13 trillion annually by 2030 (Metaverse Market Report, Grand View Research)
- The estimated global GDP impact of the metaverse by 2030 is estimated to be between US$1.5 trillion and US$3 trillion annually by 2031 (The Potential Global Economic Impact of the Metaverse n.d.-a).
- The Analysis Group's analysis projects that by 2031, Asia Pacific's net GDP contribution might reach US$1.28 trillion annually (The Potential Global Economic Impact of the Metaverse n.d.-b).

3 Literature Review

Significant advancements in ubiquitous access and identity management, scalability, interoperability, and immersive realism are necessary to create a functional and integrated Metaverse. Although there are many positive aspects, such increasing interest and improvements in hardware, overcoming computational constraints and encouraging stakeholder engagement still pose significant obstacles. By addressing these issues, a Metaverse that improves our interactions, education, and engagement with the digital world will become possible (Dionisio et al. 2013).

The four main categories of the metaverse are mirror world, virtual reality (VR), augmented reality (AR), and lifelogging. The metaverseis a complex digital ecosystem. The metaverse exhibits great potential as a cutting-edge pedagogical instrument that can revolutionize the way students engage and learn. Teachers can successfully incorporate the metaverse into their lesson plans and support long-term educational growth by being aware of its possibilities and limitations. The secret is striking a balance between the metaverse's immersive and interactive features with security, privacy, and social adaption measures (Kye et al. 2021).

With its ability to create creative and captivating learning experiences through a variety of world kinds, including racing/jump, survival, maze, and multi-choice, the Metaverse presents a promising medium for sustainable education. The Metaverse can promote equitable educational opportunities and help achieve SDG 4 by improving learning motivation through engaging experiences that are game-like (Park and Kim 2022).

Future research is anticipated to go on diverse paths as the metaverse continues to develop, investigating novel avenues, tackling obstacles, and enhancing the amalgamation of technology and applications. It's possible that this dynamic environment will inspire creative breakthroughs and fresh uses of the metaverse (Cheng et al. 2022).

A prudent approach to digitization is required for the shift to Industry 5.0, especially with the introduction of metaverse technologies. Businesses can deploy and manage the metaverse responsibly by using frameworks like the Triple Bottom Line and ESG views, and connecting them to the Sustainable Development Goals. With the use of these technologies, organizations may recognize and mitigate any potential negative effects, ensuring that the metaverse's advantages are distributed fairly and sustainably (De Giovanni 2023).

The study emphasizes how the metaverse may drive sustainable development and resilience in green supply chains in a transformational way. Supply chain companies may efficiently traverse hurdles and promote environmentally friendly practices by utilizing virtual environments to foster mutual trust, ease knowledge sharing, and enhance collaboration. To provide real advantages for businesses and the environment, future research should examine workable implementations and strategies for incorporating the metaverse into green supply chain management (Chen and Huang 2024). With the use of avatars and holograms, the metaverse—a concept that combines augmented and virtual reality technologies—has the potential to greatly expand the physical world and enable seamless interactions in both real and virtual worlds. Distinguishing hype from current reality is still difficult, even if virtual worlds and immersive games like Roblox, Second Life, Fortnite, and VRChat offer insights into the potential socio-economic implications of the metaverse (Dwivedi et al. 2022).With the use of immersive experiences and the latest developments in artificial reality, connectivity, and interaction technology, the Metaverse is set to transform our society. The Metaverse notion has emerged from fiction thanks to the combination of 5G technology, high-speed connectivity, low latency, sophisticated hardware and software capabilities, graphics processing power, and blockchain technology. The study highlights important issues and concerns for the ethical application of the Metaverse while offering insights into its potential as a revolutionary technology for good (Iqbal and Campbell 2023).

"Big tech" promotes the metaverse's transformative qualities and emphasizes how it improves social connections, work, and leisure. This change has the potential to blur the boundaries between the physical and digital worlds, changing shared experiences, brand engagements, and business operations. The metaverse is a promising but complicated technical development with drawbacks as well as advantages. Even though it can cut down on physical travel and provide immersive experiences, there are environmental considerations that need to be taken into consideration, especially with regard to energy use and carbon emissions. To lessen the metaverse's environmental impact and guarantee a more sustainable digital future, future research and development should concentrate on integrating sustainable practices and technology in the metaverse (Spiegel and Wang 2024).

The applications and usage patterns of the metaverse have a considerable impact on how it affects the environment. Industrial metaverse applications have promise for favorable environmental outcomes, especially in minimizing physical travel and boosting resource efficiency. Consumer-oriented activities, on the other hand, may contribute to negative energy consumption and environmental repercussions. In order to reduce environmental difficulties and optimize the metaverse's potential for good impact, stakeholders must prioritize sustainability in the creation of the metaverse by incorporating energy-efficient technologies, renewable energy sources, and responsible usage behaviors (Kshetri and Dwivedi 2023).

The paper provides a thorough examination of the metaverse's environmental effects, recognizing both its possible advantages in terms of less energy use and travel as well as its difficulties in controlling energy-intensive devices. It emphasizes how crucial it is to use sustainable methods and tools while creating and deploying the metaverse in order to lessen its negative environmental effects and make a constructive contribution to efforts to combat climate change (Palak et al. 2023).

The study emphasizes how the metaverse can help with decarbonization, energy efficiency, better air quality, and mitigating the effects of climate change. It highlights how crucial it is to implement ecologically friendly procedures and guidelines in order to maximize the metaverse industry's potential and reduce its environmental impact (Zhao and You 2023).

A new sociotechnical phenomenon called the Metaverse is being shaped by a range of individuals with different goals. We can comprehend how communal sensemaking processes impact its evolution by applying SCOT theory. It is important to view the Metaverse as a dynamic, ever-evolving phenomenon, with studies concentrating on how it keeps changing. The Metaverse can become a useful place for all parties involved by reorienting towards humanist values and resolving conflicts between competing agendas (What Is the Metaverse and Who Seeks to Define It? Mapping the Site of Social Construction—Mateusz Dolata, Gerhard Schwabe 2023, n.d.).

The Metaverse has enormous potential to transform traditional methods of healthcare, education, environmental preservation, and economic empowerment in order to further the Sustainable Development Goals. The Metaverse has the potential to significantly contribute to tackling global issues and advancing sustainable development by supporting economic inclusion, improving healthcare delivery, democratizing access to high-quality education, and advancing environmental sustainability (Ritterbusch and Teichmann 2023).

The case study's empirical evidence demonstrates how digital transformation may greatly assist businesses in reaching carbon neutrality and green, sustainable development, particularly when utilizing the Metaverse's possibilities. The fact that environmental benefits have the highest growth rate (19.6% monthly) indicates how much opportunity there is for digital solutions to support sustainability initiatives. Subsequent investigations ought to concentrate on delving deeper into the ways in which the Metaverse is employed across diverse industries and formulating all-encompassing approaches to reconcile digital transformation with objectives of sustainable development. Businesses that use this approach can significantly contribute to solving environmental issues and advancing a sustainable future (Sustainability | Free Full-Text | Digital Transformation to Help Carbon Neutrality and Green Sustainable Development Based on the Metaverse n.d.).

As a new technology, the metaverse has immense potential to promote social, economic, and environmental sustainability. A large and varied body of literature has been produced as a result of the explosion in scholarly interest in the use of the metaverse for sustainable development. Three conceptual clusters, three fundamental themes, one motor theme, and two emergent themes were found when the literature was synthesized and analyzed (Johri et al. 2024).

Ireland leads in the metaverse research impact, demonstrating the caliber and importance of its contributions to this area. Despite having the most papers in the metaverse field, China's research is not thought to have had enough of an impact, indicating a discrepancy between quantity and quality. The majority of the widely cited work in the topic of metaverse is devoted to historical viewpoints, technological reviews, and educational innovations (Li and Li 2024).

4 Potential Impact of Metaverse on GDP in Asia

The impact of the metaverse in Asia will probably fluctuates throughout nations owing to differences in their technological infrastructure, legislative frameworks, and levels of economic growth. Nations like South Korea, Japan, and Singapore, which have highly developed internet infrastructure, might reap more immediate and noticeable economic rewards. On the other hand, in order to fully exploit the promise of the metaverse, developing countries will need to concentrate on constructing the required infrastructure and regulatory frameworks. With the potential to significantly increase GDP by 2035, the metaverse offers Asia a great chance for innovation and economic prosperity. But in order to fully realize this promise, significant issues with regulation, infrastructure, and inclusion must be resolved. By 2035, investment alternatives might result in an annual economic impact for Asia of between US$0.8 trillion and US$1.4 trillion.29 This is roughly approximately 1.3–2.4% of Asia's estimated GDP in 2035 (The Metaverse in Asia n.d. Deloitte Report).

Between 2020 and 2030, the Asia Pacific metaverse market is anticipated to expand at a substantial rate, with a compound annual growth rate (CAGR) of 45.4% (Fig. 2). By the end of the decade, the metaverse will have become a significant economic force in the area primarily to exponential growth that will be fueled by rising infrastructure investments, advances in hardware and software, and an expansion of service offerings. The period of fastest development occurs from 2027 to 2030, suggesting a phase of strong adoption and expansion in the second half of the decade. The market is expanding slowly, which suggests that metaverse technologies are being adopted and invested in initially. The market has grown significantly due to widespread acceptance in a number of industries, strong infrastructure development, and rising hardware, software, and service demand.

The global metaverse market is segmented by end-use (Fig. 3), which includes retail, media and entertainment, education, automotive, aerospace and military, banking, financial services, and insurance (BFSI), and others. Furthermore, the digital twin technology is shaking up the entire product lifecycle management (PLM) in the industrial and automotive sectors, from designing to engineering to operations

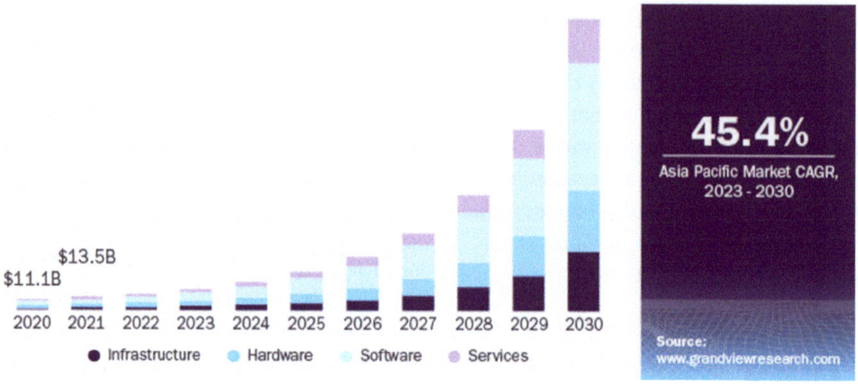

Fig. 2 Asia Pacific metaverse market size, by product, 2020–2030(USD Billion). *Source* Metaverse Market Report/ grandviewresearch.com

and servicing. Rather of assessing the manufacturing process's historical performance, digital twins offer a substantial business opportunity through future prediction. Furthermore, autonomy enables the production system to intelligently and effectively react to unforeseen circumstances. When a product can recognize a problem before it breaks, customer satisfaction and loyalty may rise as a result. According to estimates, the BFSI sector will generate 24.4% of total income in 2022. However, because of the growing global gaming market, the media and entertainment sector is anticipated to grow at a CAGR of 43.5% over the course of the projection year. Companies are putting more and more emphasis on hosting virtual music concerts, and as more companies embrace this idea, the revenue growth of this market sector is expected to increase.

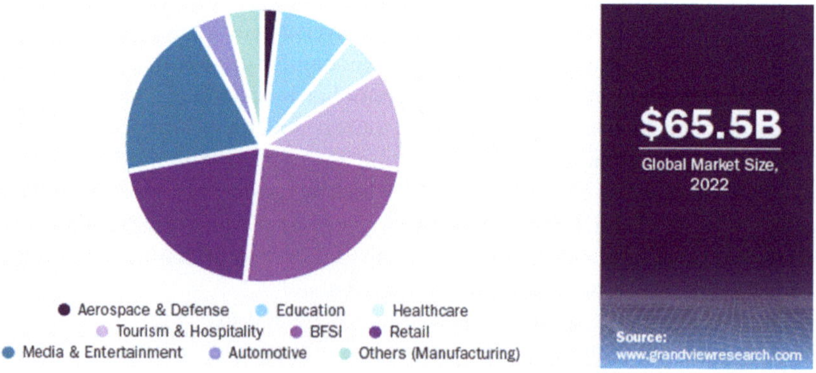

Fig. 3 Global metaverse market share, by end-use, 2022(%). *Source* Metaverse Market Report/ grandviewresearch.com

5 Is India Ready to Adapt Metaverse?

India is in a unique position to reap advantages from the metaverse considering its young, tech-savvy population and its trend towards becoming the third-largest economy by 2030. India has the largest number of STEM graduates worldwide and a population under thirty, signifying it has a large pool of talent eager to contribute to the metaverse. Given its well-established reputation as a hub for IT outsourcing and its recent emergence as a center for product development, India is well-positioned to meet the technological demands of the metaverse. Though there is a lot of talent, brain drain is still an issue since many highly educated Indians are looking for work abroad. In order to ensure that all social and economic groups have access to and may profit from metaverse innovations, it is imperative that concerns of technology equity be addressed in order to fully fulfill the potential of the metaverse. In order to guarantee that everyone may benefit from the metaverse, it is imperative to close the digital divide while making efficient use of India's enormous human capital. It will be essential to create laws and regulations that protect data privacy, security, and fair access. India is well-positioned to lead in this developing industry thanks to its economic growth, youthful, educated population, good cultural views toward the metaverse, and strong corporate engagement.

6 Macroeconomic Determinants of Metaverse in India

The metaverse, an emerging digital frontier, is poised to significantly impact India's economy, driven by various macroeconomic factors. One of the primary determinants is India's expanding digital economy, fueled by rapid internet penetration, increasing smartphone usage, and a robust technology sector. As the world's largest producer of STEM graduates, India possesses a skilled talent pool essential for the development of metaverse-related industries. The government's push for digitalization, through initiatives like Digital India, further supports the creation of a conducive environment for virtual ecosystems. Moreover, rising disposable income and the burgeoning middle class are driving demand for immersive digital experiences in entertainment, education, and retail. However, inflationary pressures, currency fluctuations, and the regulatory framework around data privacy and digital transactions will play critical roles in shaping the metaverse's growth in India. Addressing these challenges while fostering innovation can position India as a global leader in the metaverse space.

A vigorous economy is indicated by a greater GDP growth rate, which also increases consumer spending and disposable income. These factors can encourage the adoption of innovative technologies, such as the metaverse. A conducive investment environment attracts in both international and domestic capital for the infrastructure and technologies essential to the metaverse. For the metaverse to function, high-speed internet connectivity is required. This necessity is supported by India's growing internet user base, which is powered by reasonably priced data plans. For the metaverse to function safely and securely, clear laws pertaining to data privacy, cybersecurity, and digital transactions are essential. Venture capital and other forms of funding are available to help startups and companies working on technology relevant to the metaverse.

- India has a services-led economy (55.0% of GVA), with the three sectors that contribute the most to the country's GDP are manufacturing (14.7%), wholesale and retail (11.3%), and agriculture (18.4%).
- GDP per capita was US$6,187, indicating that a sizeable portion of the population may not be able to access the metaverse due to the average cost of the necessary immersion gear. In 2021, India recorded a record $38.5 billion in venture capital investments. The thriving Indian start-up ecosystem, valued at US$400 billion and home to 50,000 active start-ups, including 107 unicorns, has been bolstered by a record amount of investments. Notably, consumer technology, fintech, and software as a service (SaaS) constituted 75% of the total value of venture capital investments.

In addition to digitalization, foreign direct investment (FDI) and public–private partnerships are likely to play crucial roles in accelerating the development of India's metaverse infrastructure. Investment in technologies such as 5G, AI, and blockchain will be pivotal in creating seamless and secure virtual experiences. Furthermore, India's demographic advantage—with a young, tech-savvy population—provides a ready market for metaverse adoption across various sectors, from virtual real estate to e-commerce. The government's proactive stance in encouraging innovation through schemes like Startup India is fostering an ecosystem that can fuel metaverse-related ventures. However, to sustain this growth, policymakers will need to address the digital divide, ensuring equitable access to high-speed internet and advanced technologies across rural and urban areas. Additionally, macroeconomic stability, coupled with a clear regulatory framework for virtual assets and digital ownership, will be essential for long-term sustainable growth of the metaverse in India (Vig 2022).

7 India's Digital Infrastructure

In order to trade digital assets in the metaverse, digital payments shall be critical (Fig. 4). India may play a significant role in this growth, given that it leads the globe in real-time digital payment rates. The government-led India Stack initiative, which aimed towards bringing in the country's populations into the digital era, is responsible for developing the digital infrastructure that fosters this. Projects like Aadhar, the largest biometric ID system in the world, and Unified Payments Interface, the largest real-time digital payment system worldwide, are part of the India Stack. India is seeing remarkable progress in financial inclusion with the help of its digital infrastructure. The number of adults with bank accounts has increased significantly in just 7 years because to advancements in digital infrastructure (The Metaverse in Asia n.d.).

Fig. 4 India's digital infrastructure. *Source* Compiled by Author

With its current infrastructure, India should be able to quickly achieve scale considering its sizable population if it can expand the India Stack into the digital components of the metaverse like blockchain and NFTs. First, in December 2021, the government unveiled a national blockchain plan. The Reserve Bank of India will release a Digital Rupee between 2022 and 2023, according to official plans. India's developed digital payment systems can be used to plan the metaverse's economic infrastructure if there is greater regulatory clarity and a defined legal framework for virtual digital assets. In terms of app downloads, India is the largest mobile gaming market globally. Since games are a gateway to the metaverse, it is projected that the Indian gaming market would more than treble to US$7 billion by 2026. Indian game developers have already started to make inroads into the metaverse, including Altergame and The Spartan Group. India is making progress in the film business in addition to gaming; Pooja Entertainment purchased its first virtual plot in the metaverse and intends to provide viewers with realistic and engaging experiences (*Exploring the Possibilities to Implement Metaverse in Higher Education Institutions of India | Education and Information Technologies*, n.d.). Zee Entertainment, an Indian media conglomerate, has also welcomed its new hires through the metaverse and wants to launch NFTs from original web series, TV episodes, movies, and music (Figs. 5 and 6).

The metaverse involves the exchange of personal and financial data, making it a target for cybercriminals. Data breaches can lead to identity theft, financial loss, and reputational damage.

Phishing scams can trick users into revealing sensitive information or downloading malicious software. Themetaverse can be a breeding ground for counterfeit goods and fraudulent transactions. Metaverse platforms may track user data to personalize experiences and target advertisements. This can raise concerns about privacy and surveillance. Users may not have sufficient control over their data or understand how it is being used. India has a relatively robust regulatory framework for digital payments, which includes:

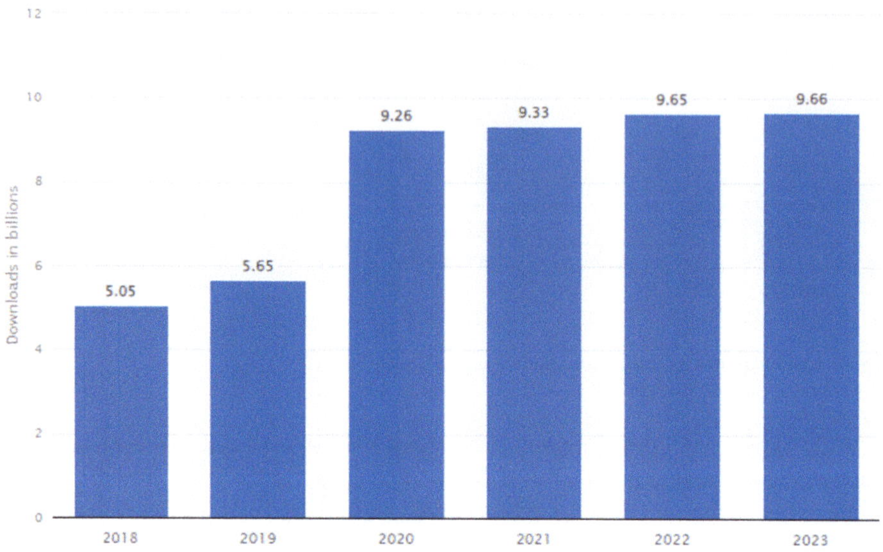

Fig. 5 Annual number of mobile game downloads made in India from 2018 to 2023(in billions). *Source* https://www.statista.com/statistics/1327738/india-number-of-mobile-game-downloads/India (n.d.)

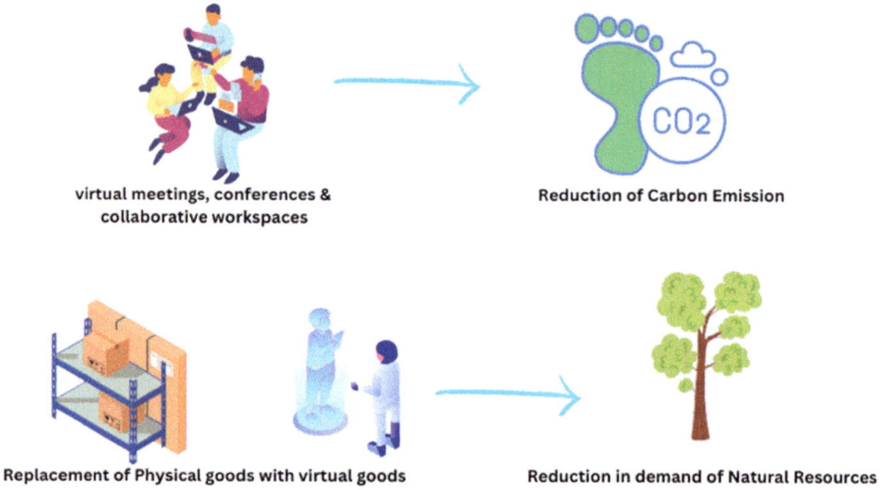

Fig. 6 Attainment of sustainable growth through use of virtual reality. *Source* Compiled by Author

The RBI is the central bank of India and oversees the digital payments ecosystem. It has issued guidelines and regulations for various digital payment methods, such as UPI, IMPS, and NEFT.

Payment and Settlement Systems Act (PSS Act): This act provides the legal framework for payment systems in India. National Payments Corporation of India (NPCI): NPCI is an umbrella organization that operates several retail payment systems, including UPI.

8 Metaverse and Sustainable Growth

The metaverse, a collective virtual environment, holds immense potential for India to achieve sustainable growth and development. With its robust digital infrastructure, skilled workforce, and strong entrepreneurial culture, India is well-positioned to emerge as a global leader in this evolving domain. A country can supplement sustainable growth through the metaverse in a number of ways that take into consideration its unique attributes and possibilities.

- By facilitating virtual conferences, meetings, and collaborative workspaces, the metaverse lowers carbon emissions and the need for physical travel. Virtual events, exhibitions, and entertainment contribute to reducing the environmental damage that comes with big-scale in-person meetings.
- Physical goods can be replaced by digital assets and virtual goods in the metaverse, which lowers the demand for natural resources. Sustainability can be promoted by using virtual simulations to optimize resource allocation and energy use in real-world infrastructure.
- Through immersive and interactive virtual classrooms, the metaverse can improve education and encourage sustainability literacy and lifelong learning. Sustainable behaviors and environmental awareness can be promoted through virtual platforms.
- The employment and economic growth of digital producers, developers, and entrepreneurs are facilitated by the metaverse economy. Within the metaverse ecosystem, green technology innovation has the potential to propel sustainable growth and provide answers to pressing environmental issues.
- Creating sustainable communities in the metaverse encourages cooperation, information exchange, and group efforts to achieve sustainability objectives. Social impact campaigns, sustainable project financing, and community-driven sustainability initiatives can all be hosted on virtual platforms.

- Sustainability-focused decision-making, resource management, and environmental monitoring can all be enhanced by using data analytics and AI in the metaverse. Virtual environment user behavior analysis offers valuable information about consumer preferences and sustainable practices.

The creation of legal frameworks in the metaverse guarantees the implementation of sustainable and moral activities, including data privacy, carbon offsetting, and ethical virtual economies. Incorporating stakeholders from many sectors into virtual governance models fosters sustainability, inclusion, and transparency. By incorporating sustainability principles into the metaverse's design, development, and use, nations can take advantage of this new digital environment to solve global issues, spur innovation, and accomplish long-term sustainable growth. Harnessing the metaverse could open new economic avenues, drive innovation, and help address social challenges. The metaverse offers the opportunity to transform sectors such as education, healthcare, and social interactions by providing immersive and interactive experiences that can positively impact individuals and communities alike.

However, to fully unlock the potential of the metaverse, India must focus on key aspects such as investing in advanced digital infrastructure, building a capable workforce, and creating a supportive regulatory environment. Additionally, ethical concerns like privacy, inclusivity, and responsible AI usage need to be addressed to ensure that the metaverse develops in a sustainable and socially responsible manner.

By tackling these challenges and capitalizing on opportunities, India can establish itself as a global leader in the metaverse space, contributing to economic growth, social progress, and environmental sustainability.

9 Policy Implications

By implementing following policy measures, India can position itself as a global leader in the metaverse and leverage its potential for sustainable growth, economic development, and social impact.

- Country can focus on investing in robust broadband infrastructure to ensure widespread access to the metaverse.
- Development of data centers and cloud computing facilities should be promoted.
- Encourage the adoption of 5G technology to enable seamless connectivity and immersive experiences.
- Invest in education and training programs to develop a skilled workforce in metaverse technologies, such as virtual reality, augmented reality, and blockchain.
- Collaborate with industry and academia to foster innovation and entrepreneurship in the metaverse sector.
- Develop a clear and supportive regulatory framework for the metaverse, addressing issues such as data privacy, consumer protection, and intellectual property.
- Promote responsible AI development and usage within the metaverse.
- Establish guidelines for ethical metaverse development, focusing on inclusivity, accessibility, and avoiding discrimination.
- Promote the use of the metaverse for social good, such as education, healthcare, and disaster relief.

- Collaborate with other countries and international organizations to develop global standards and best practices for the metaverse.
- Participate in international initiatives to promote the ethical and sustainable development of the metaverse.

10 Conclusion

The metaverse, a collective virtual shared space, presents a significant opportunity for India to achieve sustainable growth and development. By leveraging its strong digital infrastructure, skilled talent pool, and entrepreneurial spirit, India can position itself as a global leader in the metaverse. India's digital economy has witnessed remarkable growth, driven by factors such as increased internet penetration, government initiatives, and a young, tech-savvy population. The adoption of metaverse technologies is rapidly increasing, with a growing number of users and applications. The metaverse offers significant economic opportunities, including job creation, innovation, and new business models. The metaverse can contribute to social development by providing access to education, healthcare, and virtual communities. The metaverse can be developed in a sustainable manner, reducing the need for physical resources and minimizing environmental impact. India's economy is predicted to grow at a rate that will make it the third largest in the world, offering an excellent foundation for innovations and technical advancements. In order to trade digital assets in the metaverse, digital payments will be essential. India may play a significant role in this growth, given that it leads the globe in real-time digital payment rates. The government-led India Stack initiative, which aimed to usher the country's populace into the digital era, is responsible for developing the digital infrastructure that is driving this. A country might leverage the metaverse to achieve sustainable growth by carefully utilizing its resources to advance social, economic, and environmental sustainability. To reduce emissions associated with travel, promote remote work and virtual gatherings. Encourage the use of digital resources to reduce waste production and the use of natural resources. Encourage the development of new e-commerce, virtual service, and digital content creation jobs and business models in the metaverse economy. Create online marketplaces that promote environmentally friendly goods and services so that customers may make decisions that will not harm the environment. Create and implement regulations that guarantee data privacy, environmental responsibility, and moral behavior in the metaverse. Involve stakeholders from many sectors in virtual governance models to improve accountability, transparency, and engagement. Countries may establish a dynamic digital ecosystem that promotes social well-being, economic resilience, and environmental conservation by incorporating the metaverse into their national goals for sustainable development. By utilizing virtual technology in novel ways, nations can not only promote sustainable development but also establish a model for a more egalitarian and sustainable future. To fully realize the potential of the metaverse, India should invest in Digital Infrastructure, continue to invest in broadband connectivity, data centers, and 5G networks to support the growth of the metaverse. Promote research and development in metaversetechnologies and applications. Establish clear guidelines and regulations for the metaverse to ensure its ethical and responsible development. Educate the population about the metaverse and its benefits, empowering them to participate and benefit from this emerging technology. Collaborate with other countries and international organizations to develop global standards and best practices for the metaverse. By implementing these policies and leveraging its existing strengths, India can position itself as a global leader in the metaverse. The metaverse offers a promising avenue for sustainable growth, economic development, and social innovation, and India is well-positioned to capitalize on this opportunity.

The metaverse can revolutionize education by providing immersive and interactive learning experiences, create opportunities for social connection and inclusion, especially for marginalized communities. Advances in hardware, such as virtual reality headsets and haptic devices, will enhance the user experience and make the metaverse more accessible.

References

Chen P-K, Huang X (2024) Enhancing supply chain resilience and realizing green sustainable development through the virtual environment of the metaverse. Sustain Dev 32(1):438–454

Cheng X, Zhang S, Fu S, Liu W, Guan C, Mou J, Ye Q, Huang C (2022) Exploring the metaverse in the digital economy: an overview and research framework. J Electron Bus Digit Econ 1(1/2):206–224. https://doi.org/10.1108/JEBDE-09-2022-0036

De Giovanni P (2023) Sustainability of the metaverse: a transition to industry 5.0. Sustainability 15(7), Article 7. https://doi.org/10.3390/su15076079

Dionisio JDN, Iii WGB, Gilbert R (2013) 3D virtual worlds and the metaverse: current status and future possibilities. ACM Comput Surv 45(3):34:1–34:38. https://doi.org/10.1145/2480741.2480751

Dwivedi Y, Hughes L, Baabdullah A, Ribeiro-Navarrete S, Giannakis M, Al-Debei M, Dennehy D, Metri B, Buhalis D, Felix R (2022) Metaverse beyond the hype: multidisciplinary perspectives on emerging challenges, opportunities, and agenda for research, practice and policy. Int J Inf Manage. https://doi.org/10.1016/j.ijinfomgt.2022.102542

Exploring the possibilities to implement metaverse in higher education institutions of India | Education and Information Technologies (n.d.) Retrieved 10 Oct 2024, from https://link.springer.com/article/https://doi.org/10.1007/s10639-024-12691-2

India: Number of mobile game downloads 2023 (n.d.) Statista. Retrieved 22 June 2024, from https://www.statista.com/statistics/1327738/india-number-of-mobile-game-downloads/

Iqbal MZ, Campbell AG (2023) Metaverse as Tech for good: current progress and emerging opportunities. Virtual Worlds 2(4), Article 4. https://doi.org/10.3390/virtualworlds2040019

Johri A, Joshi P, Kumar S, Joshi G (2024) Metaverse for sustainable development in a bibliometric analysis and systematic literature review. J Clean Prod 435:140610. https://doi.org/10.1016/j.jclepro.2024.140610

Kshetri N, Dwivedi YK (2023) Pollution-reducing and pollution-generating effects of the metaverse. Int J Inf Manage 69:102620. https://doi.org/10.1016/j.ijinfomgt.2023.102620

Kye B, Han N, Kim E, Park Y, Jo S (2021) Educational applications of metaverse: possibilities and limitations. J Educ Eval Health Prof 18:32. https://doi.org/10.3352/jeehp.2021.18.32

Li H, Li B (2024) The state of metaverse research: a bibliometric visual analysis based on citespace. J Big Data 11(1):14. https://doi.org/10.1186/s40537-024-00877-x

Mystakidis S (2022) Metaverse. Encyclopedia 2:486–497. https://doi.org/10.3390/encyclopedia2010031

Palak S, Gulia P, Gill NS, Chatterjee JM (2023). Metaverse and its impact on climate change. In: Hassanien AE, Darwish A, Torky M (eds) The future of metaverse in the virtual era and physical world. Springer International Publishing, vol 123, pp 211–222. https://doi.org/10.1007/978-3-031-29132-6_12

Park S, Kim S (2022) Identifying world types to deliver gameful experiences for sustainable learning in the metaverse. Sustainability 14(3), Article 3. https://doi.org/10.3390/su14031361

Peukert C, Qahri-Saremi H, Schultze U, Thatcher JB, Cheung CMK, Frenzel-Piasentin A, Greve M, Matt C, Trenz M, Turel O (2024) Metaverse: a real change or just another research area? Electron Mark 34(1):32. https://doi.org/10.1007/s12525-024-00711-5

Ritterbusch GD, Teichmann MR (2023) Defining the metaverse: a systematic literature review. IEEE Access 11: 12368–12377. https://doi.org/10.1109/ACCESS.2023.3241809

Spiegel S, Wang H (2024) Exploring climate science in the metaverse: interactive storytelling in immersive environments for deep learning and public engagement. In Coren E, Wang H (eds) Storytelling to accelerate climate solutions. Springer International Publishing, pp 365–378. https://doi.org/10.1007/978-3-031-54790-4_17

Sustainability | Free Full-Text | Digital Transformation to Help Carbon Neutrality and Green Sustainable Development Based on the Metaverse (n.d.) Retrieved June 22, 2024, from https://www.mdpi.com/2071-1050/15/9/7132

The Metaverse in Asia (n.d.) Deloitte Indonesia. Retrieved June 22, 2024, from https://www2.deloitte.com/id/en/pages/about-deloitte/articles/the-metaverse-in-asia.html

The Potential Global Economic Impact of the Metaverse (n.d.-a) Retrieved June 22, 2024, from https://www.analysisgroup.com/Insights/publishing/the-potential-global-economic-impact-of-the-metaverse/

The Potential Global Economic Impact of the Metaverse (n.d.-b) Retrieved June 22, 2024, from https://www.analysisgroup.com/Insights/publishing/the-potential-global-economic-impact-of-the-metaverse/

Vig S (2022) Intellectual property rights and the metaverse: an Indian perspective. J World Intellect Property 25(3):753–766. https://doi.org/10.1111/jwip.12249

What is the Metaverse and who seeks to define it? Mapping the site of social construction—Mateusz Dolata, Gerhard Schwabe, 2023 (n.d.) Retrieved June 22, 2024, from https://journals.sagepub.com/doi/full/10.1177/02683962231159927

Zhao N, You F (2023) The growing metaverse sector can reduce greenhouse gas emissions by 10 Gt CO_2e in the united states by 2050. Energy Environ Sci 16(6):2382–2397. https://doi.org/10.1039/D3EE00081H

Balancing Technological Innovation with Environmental Impact in Metaverse Infrastructure: A Critical Analysis

T. S. Ragi and Shruti Singh

Abstract This chapter critically examines the intricate relationship between technological innovation and environmental sustainability in Metaverse infrastructure development. As the Metaverse emerges, balancing rapid technological advancement with environmental responsibility becomes imperative. By scrutinizing the environmental impact of Metaverse infrastructure, including energy consumption and resource utilization, the chapter highlights the urgency of addressing sustainability challenges. It evaluates strategies like energy-efficient hardware and decentralized architectures, critically analyzing their efficacy in mitigating environmental impact. Challenges, limitations, and socioeconomic implications are explored, along with stakeholder perspectives and case studies showcasing successful approaches. Policy and governance frameworks are assessed, offering insights into regulatory approaches and policy incentives. Looking ahead, the chapter provides anticipatory insights and recommendations for stakeholders to advance sustainability while fostering innovation. In conclusion, it emphasizes the importance of steering the Metaverse towards ecological harmony for a sustainable digital future.

Keywords Technological Innovation · Environment · Sustainability · Metaverse

1 Introduction

The rapid progression of digital technologies has brought forth a new era in which virtual worlds are becoming essential parts of our everyday lives. Among these, the Metaverse stands out as a revolutionary concept, offering a fully immersive and interconnected digital realm that goes beyond traditional online experiences. The Metaverse, built on the progress in augmented reality (AR), virtual reality (VR),

T. S. Ragi · S. Singh (✉)
School of Management Sciences, Varanasi, India
e-mail: shrutisingh@smsvaranasi.in

T. S. Ragi
e-mail: ragits@smsvaranasi.in

W. Leal Filho et al. (eds.), *Metaverse and Sustainability: Business Resilience Towards Sustainable Development Goals*, World Sustainability Series,
https://doi.org/10.1007/978-3-031-89545-6_3

blockchain, and artificial intelligence (AI), has the potential to completely revolutionize our digital interactions, work, and recreational activities, opening up a world of exciting possibilities and a promising future (Johnson and Brown 2023). However, as the Metaverse expands, it raises significant questions about its ecological impact.

The infrastructure underpinning this extensive digital environment necessitates substantial processing capacity, data retention capabilities, and network bandwidth, all of which consume significant energy. Data centers play a crucial role in the functioning of the Metaverse, but they also require substantial quantities of electricity to operate servers and cooling systems, which leads to a considerable carbon footprint (Smith 2022; Davis and Green 2021). The environmental impact of such infrastructure, including the production and disposal of hardware and the energy needs of data centers, poses severe challenges to its sustainability.

This chapter delves into the delicate balance between technological advancement and environmental preservation within the context of Metaverse development. While the Metaverse presents significant opportunities for growth and progress, it is imperative to give precedence to mitigating the environmental impact of its infrastructure. The implementation of sustainable methods in hardware creation, such as the use of recycled materials and the prioritization of energy efficiency in design, is not just crucial, but a responsibility we all share for minimizing the negative effects of these practices. Furthermore, technologies like green computing, virtualization, and decentralized architectures offer promising avenues to reduce the energy consumption and resource utilization associated with Metaverse infrastructure (Anderson and Park 2022; Lee and Kim 2023).

In light of the global challenges posed by climate change and resource depletion, it is crucial to include sustainability in developing technologies' core (Anderson and Park 2022). Blockchain technology allows for decentralized infrastructures, which can disperse the storage and processing of data over numerous nodes. This helps alleviate the workload on individual data centers. Nevertheless, it is crucial to effectively control the environmental consequences of blockchain technology, especially for energy-intensive activities such as mining. Furthermore, adopting lifecycle assessment (LCA) approaches can assist developers in comprehending the ecological consequences of Metaverse technologies, including the entire lifecycle from production to disposal. This facilitates the ongoing enhancement of sustainable practices, giving us hope for a more environmentally conscious future (White and Black 2022).

The chapter begins by outlining the environmental implications associated with Metaverse infrastructure, laying the groundwork for understanding the scope of the issue. The text explores various sustainable technological solutions and assesses their effectiveness in mitigating the environmental impact (Lee and Kim 2023). The discussion also delves into the challenges and limitations of implementing these solutions, acknowledging the complexities involved in striking a balance between rapid technical progress and ecological responsibility (Thompson 2023).

This chapter offers a thorough analysis of the current state of affairs by studying the socioeconomic dynamics and stakeholder viewpoints related to the sustainability of the Metaverse. By examining specific instances and analyzing policies, this study provides valuable perspectives on practical methods and possible directions for the future. The consideration of the role of government and business laws in ensuring that technical improvements in the Metaverse align with environmental sustainability is also taken into account. Policy frameworks that provide financial rewards for using renewable energy require specific energy efficiency requirements and encourage using sustainable practices in creating technology can steer the industry in a more environmentally conscious direction (Green 2023).

The main objective of this chapter is to add to the continuing discussion on how to guide the development of the Metaverse in an inventive and environmentally sustainable way. The goal is to ensure that this digital frontier can live alongside the ecological limits of our planet.

2 Environmental Impact Assessment

Significant environmental challenges have arisen with the advent of the Metaverse as a groundbreaking digital ecosystem, calling for adopting technological solutions and effective management strategies. Environmental risks are associated with the massive amounts of energy, resources, and electronic trash produced by the ever-expanding Metaverse infrastructure. This section provides an in-depth analysis of the most pressing environmental concerns and helpful information regarding the management actions that will be required to lessen their impact.

2.1 A Review of Energy Use and Greenhouse Gas Emissions

One primary ecological concern associated with Metaverse infrastructure is energy use. Server operation, data administration, and cooling system maintenance in the data centers underpin this digital environment and consume much electricity. One major contributor to global carbon emissions and, by extension, to the acceleration of climate change is the energy demand, often met by non-renewable resources (Davis and Green 2021).

Deploying energy-efficient technologies, investigating renewable energy possibilities, and optimizing data center operations are all managerially required steps toward tackling this challenge. Renewable energy contracts, energy-efficient hardware, and cutting-edge cooling technology are all vital. Additionally, businesses can reduce their overall carbon emissions by using energy monitoring systems to track and enhance their energy consumption (Smith 2022).

2.2 *Maximizing Resource Utilization*

Rare earth metals and other hardware-related elements are among the many physical resources needed for Metaverse infrastructure construction and maintenance. Deforestation, pollution, and increased carbon emissions are only a few significant environmental impacts of these commodities' extraction and processing (Anderson and Park 2022).

Achieving environmental sustainability while keeping costs low requires careful management of resource utilization. Managers should prioritize sustainable sourcing approaches. These techniques entail working with suppliers who adhere to environmentally responsible practices and investigating alternatives to scarce resources. The total demand for resources might be reduced if funding were to go into R&D projects that sought to create sustainable and long-lasting hardware, such as energy-efficient servers or recyclable computer components. Recycling and reusing materials is at the heart of the circular economy, which aims to lessen human environmental impact by reducing resource consumption (Lee and Kim 2023).

2.3 *Electronic Waste*

Due to the rapid pace of technological advancement in the Metaverse, hardware becomes obsolete quickly, leading to massive amounts of electronic waste (e-waste). Because it contains toxic chemicals that can leach into groundwater and rivers, causing contamination and health problems, e-waste poses a severe environmental threat (Thompson 2023).

Strong IT asset lifecycle management rules are essential to any management strategy for dealing with e-waste. This includes hardware procurement, use, and disposal as part of its whole lifecycle, which is managed strategically. Verifying that the disposal procedures are eco-friendly is also part of this. Working with authorized e-waste recyclers, keeping gadgets up-to-date, and instituting take-back programs with manufacturers, where the manufacturer takes back the product at the end of its life for recycling or disposal, are all critical to consider. In addition, Green (2023) argues that the environmental impacts of electronic waste can be significantly reduced by implementing responsible procurement policies that prioritize items designed to be durable and recyclable.

2.4 *Water Use and Cooling Systems*

Not only do data centers use much power, but they also have complex cooling systems to keep them from getting too hot. Much water gets wasted and, in some cases, dangerous greenhouse gases get released since these cooling systems employ refrigerants or water. Particularly in areas where water is already scarce, data centers' water consumption significantly influences the environment (White and Black 2022).

In order to reduce water and energy consumption, it is essential from a management perspective to maximize cooling efficiency. This necessitates looking at state-of-the-art cooling techniques, such as using renewable energy sources for cooling or liquid cooling. Managers should also consider where data centers are located since the severity of water scarcity in a particular area could affect the consequences of data center water usage differently. Reduce environmental impact by switching to water-efficient cooling systems or waste heat recovery methods. The cooling infrastructure requires regular maintenance and updates to achieve peak performance while minimizing environmental impact. This includes regular cleaning of cooling systems, updating to more energy-efficient models, and implementing smart cooling strategies based on data center usage patterns (Johnson and Brown 2023).

Among the many ways in which the Metaverse affects the natural world are through its cooling systems, energy use, carbon emissions, resource consumption, and electronic waste. It is crucial to strike a balance between the demands of technological innovation and sustainability goals in order to manage these environmental challenges. Organizations can use energy-efficient technologies, sustainable resource management techniques, effective e-waste reduction programs, and cooling system optimization to lessen their ecological footprint and boost creativity. If we want the Metaverse to grow into a digital ecology that is both sustainable and environmentally conscious, we must adopt this all-encompassing approach.

3 Assessing Environmentally Friendly Technological Solutions

As the Metaverse grows into a major digital ecosystem, it causes major problems for the environment that need thorough assessments of their impacts and innovative solutions. The data centers, cooling systems, and hardware that keep the Metaverse running all have major impacts on the environment, particularly when it comes to the use of energy, the extraction of resources, electronic waste, and water.

Data centers, which use massive amounts of power to run server operations and cooling systems, make energy consumption a matter of paramount importance. The use of non-renewable energy sources to meet these demands is a major contributor to greenhouse gas emissions (Davis and Green 2021). To address this, it is essential to use energy-efficient hardware and software, such as low-power CPUs and GPUs, and to implement state-of-the-art cooling technologies, such as liquid cooling and waste heat recovery (Smith 2022; Johnson and Brown 2023). Data centers can significantly reduce their carbon emissions by switching to renewable energy sources and installing energy monitoring systems (Lee and Kim 2023).

Because of the abundance of commodities needed to build Metaverse infrastructure—including rare earth metals—resource allocation is another critical factor to think about. Damage to habitats and contamination are two outcomes of these resources' extraction and processing that worsen the environment (Anderson and Park 2022). In order to lessen our impact on the environment, it is essential to employ sustainable resource management practices, such as adopting circular economy principles, making use of recycled materials, and purchasing durable hardware (Green 2023). One way to reduce resource use is to find alternatives to rare materials and find better ways to recycle them.

The rapid degradation of technology and the presence of toxic compounds in electronic trash (e-waste) pose serious threats to the environment (Thompson 2023). Implementing stringent lifecycle management procedures, working with certified e-waste recyclers, and extending device lifecycles through maintenance and upgrades are all part of effective management strategies. One way to lessen the toll that electronic trash takes on the environment is to implement measures like taking part in take-back programs and making sure that hardware is long-lasting and recyclable (White and Black 2022) (Fig. 1).

Significant amounts of water and energy are used by data centers' cooling systems. Modern cooling systems and careful site selection of data centers in relation to water supply can mitigate these effects (Johnson and Brown 2023; White and Black

Fig. 1 Word Cloud representing the impact of sustainable technologies on environment

2022). Reduce your environmental effect even further by using water-saving cooling technology and installing waste heat recovery systems.

Integrating impact evaluation with sustainable technological treatments is essential for an effective plan to address the environmental effects of Metaverse infrastructure. Effective e-waste management, energy-efficient technologies, sustainable resource management, and appropriate cooling systems are essential to resolving Metaverse environmental challenges. For the sake of a greener digital future and the quick growth of virtual worlds, it is essential to implement this all-encompassing plan.

4 Socioeconomic Dynamics: A Critical Analysis and Potential Paths Forward

The Metaverse's socioeconomic dynamics show how technological progress and its broader effects on society and the economy interact in complicated ways. In order to understand how the Metaverse will influence social behavior, digital fairness, economic opportunity, and workforce change, it is essential to track its development closely. By understanding these processes, we can better plan for the future and come up with ideas for creating a more equitable and inclusive digital space.

4.1 Economic Growth and New Job Prospects

According to Kumar and Sharma (2023), the Metaverse has the potential to greatly impact the economy through facilitating new forms of commerce, such as digital commodities, virtual real estate, and immersive advertising. These innovations may promote innovation and economic growth, which in turn could generate new job opportunities in a wide range of sectors, including but not limited to virtual customer service, content production, and software development. The benefits, however, are not always distributed equitably because entry is often limited to those who already have the requisite knowledge and means.

Policies that support the development of skills and allocate resources to underprivileged groups are crucial if we are to see a more equitable sharing of the Metaverse's economic benefits. Narrowing the digital divide and promoting inclusive economic development can be achieved through initiatives that educate the public about emerging technologies and work to improve digital literacy.

Government officials should team up with trailblazers in the field to launch vocational and academic programs tailored to the needs of the Metaverse economy. Baker and Smith (2024) found that helping small businesses and entrepreneurs in the digital space can inspire new ideas and open up new opportunities for everyone involved.

4.2 *Making Sure that Internet Resources and Opportunities Are Available to Everyone*

A significant obstacle to digital justice is ensuring that all people have equal access to the resources needed to participate in the Metaverse, such as reliable, high-speed internet and robust computer devices. Vulnerable populations may be less able to take use of Metaverse technology since the digital divide exacerbates existing inequalities (Martin and Lopez 2023).

Improving access to technology and the internet is crucial to addressing the problem of digital fairness. Internet access, technological subsidies for underserved communities, and digital literacy education should be top priorities. It is recommended that public access points and community activities be funded by organizations and governments in order to increase digital accessibility. Working together with IT companies can help reduce costs and provide affordable solutions for people in need (Johnson and Brown 2023).

4.3 *Cultural Influence and Social Behavior*

By opening up new avenues for communication and individual expression, the Metaverse is revolutionizing established norms of social conduct and cultural practices. However, worries about privacy, security, and becoming too reliant on technology are heightened by these changes (Gonzalez and Patel 2024).

We must prioritize the creation of standards and recommendations for the promotion of positive online behavior. As part of this effort, frameworks for protecting privacy, strengthening cybersecurity, and encouraging responsible behavior online have been developed. Encouraging positive social ties and minimizing negative consequences must be prioritized. All parties involved should work together to establish standards for the safety and confidentiality of online interactions. Helping people learn how to be safe online and in virtual spaces is an important goal of many public health and education programs (Chen and Williams 2024).

4.4 *Labor Market Shifts and Economic Inequality*

The rise of the Metaverse could cause shifts in the economy and the job market, with more demand for highly skilled workers and less room for more conventional occupations (Davis and Green 2021). It is critical to prepare the workforce for these changes in order to ensure equal growth.

To meet the needs of the digital economy, workforce development programs should focus on reskilling and upskilling initiatives. Working together, the government, businesses, and universities can ease the transition. Some suggestions include launching vocational training programs and certifying digital skills to help people get ready for new jobs. Job displacement can have less of an impact if they are helped with workforce transition and career counseling (Martin and Lopez 2023).

5 Conclusion

Significant socioeconomic implications, including promising opportunities and difficult challenges, are brought about by the advent of the Metaverse. Addressing economic prospects, digital equality, social conduct, and workforce adaption is vital for cultivating a more inclusive and fair digital future as the Metaverse evolves. Stakeholders may make sure the Metaverse's benefits are shared and that its growth benefits society by putting focused policies and initiatives into action.

Finding a middle ground between economic growth and social responsibility should be a top priority for any future undertakings. This includes making technology more accessible, making it easier to learn new skills, promoting responsible behavior online, and preparing people for new jobs. The Metaverse may be a force for good by working together and using strategic foresight to overcome the social and economic challenges it presents while also encouraging innovation.

References

Anderson J, Park H (2022) Sustainable resource management in emerging technologies: challenges and strategies. J Environ Manage 150(2):198–210. https://doi.org/10.1016/j.jenvman.2022.150202

Baker T, Smith A (2024) Equitable access in the digital age: Bridging the gap in emerging technologies. J Digit Incl 18(1):45–58. https://doi.org/10.1002/digi.12456

Chen X, Williams R (2024) Navigating the digital world: strategies for positive social interactions and mental well-being in virtual environments. J Cyberpsychology 29(2):121–134. https://doi.org/10.1016/j.cyberpsych.2024.121234

Davis L, Green R (2021) The energy demands of virtual environments: assessing the carbon footprint of data centers. Energy Policy J 45(3):345–356. https://doi.org/10.1016/j.enpol.2021.230202

Gonzalez M, Patel S (2024) Digital addiction and privacy concerns: the societal impact of virtual worlds. J Soc Technol Stud 12(3):199–212. https://doi.org/10.1016/j.sotech.2024.199123

Green M (2023) E-waste management and sustainable IT practices: a global perspective. J Environ Sci Technol 60(1):44–58. https://doi.org/10.1021/acs.est.3c00001

Johnson K, Brown T (2023) Cooling technologies in data centers: Innovations for reducing environmental impact. Int J Energy Res 47(4):778–790. https://doi.org/10.1002/er.7009

Kumar R, Sharma P (2023) Economic impacts of virtual economies and digital goods. J Digital Econ 22(2):155–168. https://doi.org/10.1016/j.digecon.2023.155020

Lee S, Kim Y (2023) Circular economy in tech: Extending the lifecycle of hardware in the digital age. J Cleaner Prod 280(1):123–136. https://doi.org/10.1016/j.jclepro.2023.124002

Martin J, Lopez A (2023) Digital equity: Bridging the divide in emerging virtual environments. J Digit Access Incl 9(2):75–89. https://doi.org/10.1016/j.digiinclus.2023.075020

Smith A (2022) Transitioning to renewable energy in data centers: opportunities and challenges. Renew Energy Rep 32(2):101–113. https://doi.org/10.1016/j.rer.2022.101203

Thompson R (2023) Addressing the e-waste crisis: Strategies for sustainable digital development. Environ Policy Rev 52(2):201–215. https://doi.org/10.1007/s10584-023-10365-z

White C, Black A (2022) Water usage in data centers: environmental impacts and sustainable practices. Water Resour Manage 65(3):233–246. https://doi.org/10.1007/s11269-022-03199-2

Exploring a Decade of Educational Transformation with Metaverse for a Sustainable Future: A Review

Mrinal Kanti Das, Soumya Mukherjee, Indrajit Ghosal, Dipak Saha, Dipankar Dutta, and Anisha Biswas

Abstract This study thoroughly examines academic literature to investigate the educational transformation with Metaverse for a Sustainable Future. In the present analysis, bibliometric methods are used with 145 articles spanning from 2014 to 2023 and published in high impact journals to analyze the pattern, intellectual and conceptual structure among the articles. The study identifies the top authors, leading journals, and major themes. This is followed by an in-depth bibliographic-network-based content analysis. In addition, four major clusters which represent the central research dimensions of the discipline in question were discussed extensively. The experiences of Generation Z, the development of virtual human resources, the overall effects on education, and the psychological elements of virtual learning environments are all covered by thematic clusters. The study pointed out some pertinent issues which are under-researched at the moment that can be attended to in the days to come. The study also identifies key research gaps, providing direction for future research. The literatures were restricted to the fields of commerce, management, tourism, services, marketing, economics, applied economics, econometrics, human resources, and industrial relations. Future studies could expand to encompass

M. K. Das
Department of Commerce, Kanchrapara College, Kanchrapara, West Bengal, India
e-mail: mrinal@kpcoll.ac.in

S. Mukherjee (✉)
School of Management Studies, Swami Vivekananda University, Kolkata, West Bengal, India
e-mail: mukherjeedrsoumya@gmail.com

I. Ghosal
Brainware University, Kolkata, India

D. Saha
Department of Business Administration, Institute of Engineering and Management, Kolkata, India
e-mail: dipak.saha@iem.edu.in

D. Dutta
Department of Commerce, Balurghat, West Bengal, India

A. Biswas
Department of Hospital Management, Brainware University, Kolkata, India

W. Leal Filho et al. (eds.), *Metaverse and Sustainability: Business Resilience Towards Sustainable Development Goals*, World Sustainability Series,
https://doi.org/10.1007/978-3-031-89545-6_4

a broader perspective. This research also sheds light on the evolving story of the metaverse in the context of education.

Keywords Metaverse · Virtual reality · Augmented reality · Artificial intelligence · Virtual world · Education · Learning · Teaching · Paradigm shift · Bibliometric analysis · Thematic clusters · Educational technology · Emerging trends

1 Introduction

The term 'metaverse' was first popularized in the 1992 novel of the same name, however, there have been many definitions and understanding of the term ever since. The word can be dissected into the word meta which means beyond or virtuality and the word verse which means a world. The resultant word is one which in an educational setting indicates the existence of a virtual reality interface that is accessible via the internet. This is largely attributed to the fact that this region provides them with a high degree of interaction and immersion, hence enabling them to learn and contact their fellow learners at the same time with much more ease.

The Metaverse is a new dimension of learning that has a dramatic impact on education and the society at large—it envisions a better tomorrow. Unlike in the traditional classrooms where boring lectures are delivered metaverse facilitates an interesting way of learning that makes it possible to deal with complex issues with ease. Metaverse can also incorporate learning processes by integrating innovative technologies like; Virtual Reality (VR), Augmented Reality (AR) and Artificial Intelligence (AI) where learning experience will be personalized. The word enhances the quality of education but also encourages education for all ages and concern for the environment. The educational dimension of the Metaverse will become an important asset in overcoming the challenges of the present century by equipping future generations of people with the knowledge and tools necessary for creating a better world.

The metaverse—the virtual equivalent of the interactions we carry out offline—has taken the tech world by storm and has become a trending topic in a very short time. As noted by Hollensen et al. (2023: 2), a 3D digital universe is an environment with the physical features of the world. It enables its members to relate with one another and their surroundings through their visuals and actions via avatars. The Metaverse enhances the experience of 'going' into another world as one does not use a normal 2D device such as a laptop, smartphone and so on; rather, one uses 3D animated figures called avatars. Dhelim et al. (2015) explain that this radical change is not merely about usability for the purposes of enjoyment or engagement within a 3D environment—work, play, and learning takes place in such 3D environments Recent research conducted by Çelikkol in 2022 indicates that the use of augmented and virtual reality applications in the Metaverse enhances user experience by making the interaction significantly more fun and interesting. So although the Metaverse is often just referred to as a concept, it can be seemingly seen as an interactive arena that brings new means of exploring and experiencing the cyberspace, hence an era of active exploration and interaction.

Despite its promise, the word "metaverse" is vague because it covers so many things including concepts, idealogy, patterns of learning, and technology, particularly that of virtual reality and e-learning. The complexity of the metaverse within learning environments underscores the growth of the term to mean that the metaverse has moved from a purely entertainment based system that consisted only of games to a field that is more useful and practical in education. Both educators and learners will have to accept the 'wrongs' and the 'rights' that come with this change.

With the rise of virtual reality technology adoption, which is incorporated into new styles of learning, there has been a growing moat in the utilization of the metaverse for purposes of education. This modern application can revolutionize learning, however, the effective employability of this technology hinges upon clear ethical guidelines, inter-disciplinarity and training of both tutors and students themselves. Through extensive bibliographic inquiry, this research aims at addressing the existing gap in the literature by examining the interdisciplinary fields of the metaverse and education and their relation to the components of learning attitudes, interactions, and outcomes. Appraisal studies have focused on the evolution, geographic distribution, socio-technical factors, development constraints, and educational transformation about the metaverse, however there is still no full understanding of how it changes the learning environments.

In consideration of these aspects, this research focuses on exploring this predominantly educational and contextual aspect of the metaverse and its use in education with greater detail. This study also highlights the necessity of possessing sufficient knowledge about the effects of metaverse on learning attitudes and interactions, and educational achievements. Ultimately, it is provided suggestions for the academicians, practitioners, and decision makers who aspire to improve the quality of education, utilizing the possibilities offered by the metaverse.

2 Literature Review

The rising ubiquity of information technology within the dynamic context of education has attracted researchers' attention to the metaverse. Understanding the importance of this virtual world in revolutionizing education systems, the current section embarks on an extensive literature review in a bid to define pertinent research gaps. Scholars seek to clarify the relations between technology and learning, as well as the metaverse impacts. These questions address, inter alia, the issues of teaching strategies, the extent and quality of students' involvement, the measurable results of education, the principles of educating in a group, morality, teachers' skills advancement, students' expositions, and the appropriate technological set-up.

2.1 Governing Education's Confluence of Virtual and Real Environments

Owing to its innovative characteristics, the metaverse is heralding a new era in the designed educational spaces in that it has provided various spaces that improve the learning experience remarkably. From a cultural standpoint, the metaverse is an imagined spatial realm in which one can traverse in three dimensions superimposed on the earth with additional technologies like clouds, extensive data, virtual environments, and machine intelligence. The publication by Hines and Netland (2023) categorically states and appreciates the role of the metaverse in education as especially critical to blended learning and across multiple subjects. It is an expansive model of technological application that can evolve the traditional methods of education into superior structures.

In the study of Suh and Ahn in the year 2022, it is noted that the central reason for the influence exuded by the metaverse is its peculiar feature which is the existence of immersive interactive environments. This dynamism, which allows for instant fulfillment to both teachers and learners within physical and virtual spaces alike, transcends the limits of conventional systems of education. The metaverse as a teaching tool supports the engagement of learners in active involvement with digital replicas, thus, supporting a learner-based education system, which in turn, augments the level of education. In addition, the study of Guo and Gao (2022) revealed the significance of the metaverse in English language teaching and learning network in the emotional development and cognitive growth of the learners by keep monitoring how emotions of the students changed. In fact, the metaverse posits as a device that incorporates both the real and imaginary spheres of learning to expand the concept of education to a completely different level.

2.2 Interactions of the Metaverse Realm in the Dynamics of Education

The metaverse is perceived to serve multiple purposes whereby its aesthetics enhances the interactions within the learning spaces. In supporting this assertion, the authors Díaz et al. (2020) have observed the world of metaverse and virtual interactions as great tools in academic endeavors. The authors Díaz et al. (2020) have observed this transformative capability of the metaverse as being the state of the art, in which there is a blend of real-time and time-shifted modes of providing and consuming information. However, when learning activities enabled by the metaverse are employed with modern devices like collaborative, or flipped methods, the communications relating to learning activities are further enriched due to the flexible learning engagement offered by the metaverse transforming learning.

For example, Getchell's et al. in 2010 supported findings that second life and similar metaverse portals lend themselves well to the integration of gaming strategies. This intentional design aides in interactions centered on learning but also enhances the learning delivery systems. Metaverse interactions also arouse students' curiosity and provide an interactive platform for sharing information and views through experiences. As a result, this cooperative element foster encourages an elaborate learning environment which in the end promotes the academic performance. Thus, the construction of the metaverse is viewed as a fluid encapsulation where the two facets of technology and teaching coexist, leading to an improved creation that transcends the common limits experienced in traditional education.

2.3 Unveiling Transformative Educational Outcomes: Managing the Metaverse Experience

The metaverse is powerful because of its entrenchment of technology into both learning and implementation, enhancing outcomes in both physical spaces and in the digital realm. A case in point, as noted by Jaramillo-Mujica et al. (2017) at the Universidad Militar Nueva Granada is that spatial virtual technologies, including the immersive metaverse spaces for Physics Mechanics, were very readily adopted. The appealing characteristics of these advances are realized by the students who can afford a virtual environment capable of supporting rich learning experiences, thus helping a lot with the negative trends caused by the COVID-19 pandemic.

By addressing the limitations of existing methods, the Metaverse demonstrates its change abilities in the area of online learning. While conventional training methods may involve video-based teaching techniques that are less effective, the use of virtual reality within the metaverse has been shown to yield better educational results, Lee et al. (2022a, b). Similar to that, Jovanović and Milosavljević (2022) pointed out that the metaverse platform called VoRtex, in which students are mimicking the spatial occupancy of their learning activities, is a collocated view on how students can work within such environments. This advances the scope of online education while also reinforcing the quality of engagement that the metaverse affords to learners in terms of real immersive flexible work based learning environments.

2.4 Examining Positive Attitudes Around Learning in Metaverse Education

According to Lee et al. (2022a, b), the users have maintained a pronounced attitude of learning towards the use of metaverse technology for educational purposes, particularly mobile ones. Li et al. (2022) exemplifies the effect of AI Metaverse dualism in physical education using a 360° football coaching session thereby improving the teaching methods and the effectiveness of the session. Similarly, in a 2013 paper, Schlemmer and Marson present how computer games, along with the metaverse, are based on technological immersion into both virtual and physical worlds which increases one's interest and promotes learning. Lee and Hwang's 2022 research focuses on the fact that there are virtual reality and metaverse systems nowadays and these systems are conducive to sustaining positive learning attitudes, enhancing flexibility and elevating learning outcomes. While the metaverse applications are more and more accepted and used, there are still not enough bibliometric studies and evolvement histories on how they are implemented in education. In response, researchers offer important research topics to fill in these gaps while recognizing possible difficulties with the implementation of metaverse in the educational sector.

RQ1: How the chronology of the use of the metaverse in education has evolved over time?
RQ2: Which authors, organisations, and countries have made the biggest contributions to this research domain?
RQ3: Which journals, articles, and references are most frequently cited in this area of research?
RQ4: Which themes are most prevalent in this field of study?
RQ5: What role does the metaverse play in fostering positive attitudes toward learning in educational settings?
RQ6: What is the future scope in this area of study at present era?
RQ7: What are the main difficulties in incorporating the metaverse into teaching methods?

3 Methods

3.1 Defining the Appropriate Search Terms

A thorough investigation of the cross-disciplinary elements—education and the metaverse—was conducted in this study, which made the incorporation of particular keywords related to each domain necessary. In order to guarantee a comprehensive examination of every aspect related to the metaverse and education domains, a deliberate fusion of terminology was developed. The two different sets of strings and related keyword clusters used for document selection from the Dimensions database for Scopus indexed content are shown in Table 1.

The creation of these terms was the result of a thorough preliminary search that aimed to find important and pertinent papers. "Virtual Reality," "Augmented Reality," "Mixed Reality," "Extended Reality," "Avatar," "Second Life," "Artificial Intelligence," and "Virtual World" were among the phrases pertaining to the metaverse. There were four terms related to education at the same time: "Education," "Learn," "Teach," and "Paradigm Shift."

Table 1 Article inclusion and exclusion criteria

Selection Criteria	Exclude	Include
Database: "Dimension" Date of Search: "1st January 2024" Period of Publications: 2014–2023		
Search term: "Metaverse" OR "Virtual Reality" OR "Augmented Reality" OR "Mixed Reality" OR "Extended Reality" OR "Avatar" OR "Second Life" OR "Artificial Intelligence" OR "Virtual World" AND "Education" OR "Learn" OR "Teach" AND "Paradigm Shift"	–	12,348
Subject area: "Commerce, Management, Tourism and Services, Marketing, Economics, Applied Economics, Econometrics, Human Resources and Industrial Relations"	11,171	1171
Publication type: "Article, Chapter, Edited Book"	525	646
Journal List: "UGC Journal List Group II"	501	145
Language screening: "Include documents published in English only"	0	145

This two-pronged strategy not only guaranteed that various metaverse and educational elements would be inclusive, but it also enabled a thorough and sophisticated investigation of the points of intersection between these interdisciplinary elements. The terminology chosen was chosen according to the necessity of addressing most of the subjects within the disciplines of education and metaverse. Additionally, Table 1 services the inclusion and exclusion criteria in relation to the emerging field.

3.2 Data Collection

The study employed as its primary data source Dimensions' wide-ranging collection of multiple publications that have undergone double-blind peer-review and predominantly published by high-impact factor, and mostly Scopus indexed journals, as pointed out by Thelwall (2018). The study made a systematic effort to identify and select a total of 145 pertinent articles, which is presented in Table 1. All of them focus on English publications and a host of predetermined keywords such as "Metaverse," "Virtual Reality," "Augmented Reality" among others were used in the selection process across various fields, diverse in nature.

The study, however, proceeded with caution because of the limitations of online databases and the risk of integrating wrong information due to the referencing of recent publications after the study, as articulated by Donthu et al. (2021). In order to minimize risk of including masking data in the database, several procedures were undertaken to purge and enhance the dataset, following recommendations in Donthu et al. (2021) and Zupic and Cater (2015). It has been concluded that the primary purpose of such a comprehensive cleaning of data, was to improve the quality and trustworthiness of the analysis which follows.

In order to enhance the bibliographic and bibliometric data, as well as facilitate more advanced analysis, the research team made use of the "natural language processing" feature of the VOSviewer software. It was now possible to clean and standardise some terms which were found in article's title, abstract and keywords. This ensured that there was no altered data during the process of conducting datasets, thus improving the consistency of the study and the accuracy of its qualitative assessment. The processes of collection, cleansing and portrayal of data were arduous but helped to entrench the findings of this study and also provide a springboard for future researches.

3.3 Selecting the Techniques for Analysis

Mishra et al. (2018) and Goyal and Kumar (2021) assert that bibliometric analysis is a set of measurement techniques aimed at evaluating and measuring the text. According to Suominen et al. (2016) and Groff et al. (2020), this approach is particularly useful in the literature review process, enriching the existing research efforts rather than replacing them. The bibliometric analysis extends to the preparing and assembling of deep bibliographies on specific topics, active tracing of a particular research area, and evaluation of the research output as pointed out by Gao et al. (2021) and Hossain et al. (2022). Thus, it is a guide in understanding the ever-dynamic world of research.

As per Donthu et al. (2021), a wide range of bibliometric analysis techniques is used by researcher scholars including authorship, citation analysis, bibliographic coupling, co-citation, and co-word analysis. This is how biographical data present in scholars' work can be explored to detect aberrations, patterns, and relationships in published knowledge. In brief, the methods of bibliometric analysis make it possible to structure the information and provide it in numerical form making it easier to grasp the history as well as the current importance of research in relation to any theme.

4 Results and Discussions

4.1 The Transformation of the Term Metaverse in Education in Terms of Chronology

For the purpose of this investigation, the following keywords "Metaverse" OR "Virtual Reality" OR "Augmented Reality" OR "Mixed Reality" OR "Extended Reality" OR "Avatar" OR "Second Life" OR "Artificial Intelligence" OR "Virtual World" AND "Education" OR "Learn" OR "Teach" AND "Paradigm Shift" are highlighted to recover 145 results from the Core Collection of UGC Journal List Group II from Dimensions on January 1, 2024. The literature that was acquired includes Review Articles (n = 63), Highly Cited Chapters (n = 133), and Enriched Edited Books (n

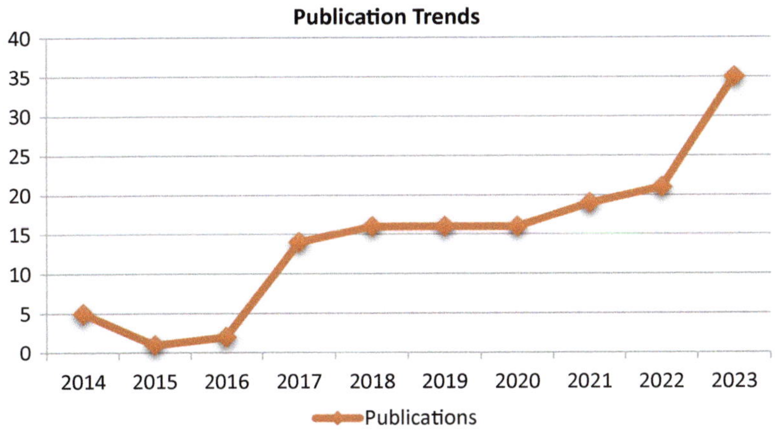

Fig. 1 Visualization of the yearly publication on metaverse in education

= 450) that were released between 2014 and 2023. Leading authors with notable connections to reputable universities have started their journey on these timely subjects. Commerce, Management, Tourism and Services, Marketing, Economics, Applied Economics, Econometrics, Human Resources, and Industrial Relations are among the primary subjects covered in citations.

However, research on using the metaverse has entered educational contexts a bit earlier, although according to the results shown in Fig. 1, this aspect has been observed since the onset of the pandemic and a particular growth has been witnessed from the year 2023.

4.2 Identifying Authors, Organizations, and Countries Having the Biggest Contributions to Research on the Use of the Metaverse in Education

The citation count for the top authors, organizations, and countries in the metaverse education field is important to understand the preferences of researchers involved in this developing area. Based on bibliometric and citation analyses, VOSviewer was utilized by the researchers to map and present the top ten authors, organizations, and countries (Table 2). It is noted that a small number of authors and accompanying institutions are heavily cited; this highlights their importance in the field and directs attention to certain sides of the research. Such findings help those researchers and practitioners who wish to study the field of education in metaverse.

Table 2 Most influential authors, organizations, and country

TC	Author(s)	TP	TC	Organisation	TP	TC	Country	TP
258	Hertogh, Marcel	1	258	Delft University Of Technology	1	947	United Kingdom	12
258	Liu, Yan	1	245	Newcastle University	2	884	United States	20
258	Van Nederveen, Sander	1	143	University of Bradford	1	800	Netherlands	3
143	Gupta, Suraksha	1	131	Tu Dortmund University	1	593	Germany	2
143	Irani, Zahir	1	112	University of Colorado Denver	1	554	Italy	5
143	Mahroof, Kamran	1	112	University of Tampa	1	547	China	6
143	Sivarajah, Uthayasankar	1	89	Hec Montréal	1	542	Denmark	2
131	Wilkesmann, Maximiliane	1	89	Université Laval	1	510	India	5
131	Wilkesmann, Uwe	1	88	Saint Joseph's University	2	509	Australia	6
112	Dincelli, Ersin	1	86	Northeastern University	2	501	France	3
112	Yayla, Alper	1	80	Capita (United Kingdom)	1	482	Poland	2
102	Alamanos, Eleftherios	1	80	Centre For Water Systems, College Of Engineering, Mathematics And Physical Sciences, University of Exeter	1	477	Norway	2
102	Papagiannidis, Savvas	1	80	Danish Hydraulic Institute	1	475	Spain	2
102	Tueanrat, Yanika	1	80	Eurecat, Lleida Agri-Food Science And Technology Park	1	470	South Africa	3
92	Bennett, Elisabeth E	3	80	Euro-Mediterranean Center On Climate Changes	1	467	Saudi Arabia	3

4.3 *Top Journals, Articles, and Referenceshaving an Impact in Metaverse in Imparting Education*

The current study determined the top 10 journals, articles, and references based on occurrences and citations to give researchers and practitioner's additional references

(Table 3). Researchers and practitioners can learn about the most recent research trends and important problems in the field by looking at these. Furthermore, a definitive perspective on the subject matter can be provided by the best papers and resources.

A pair of distinct viewpoints is clarified by the initial two journals that stem from the analysis. The "International Journal of Information Management" promotes international cooperation between academics and industry professionals by publishing scholarly research on the organization, retrieval, and utilization of information across a range of domains. It provides an environment for the advancement of information management ideas and methods.

An academic journal called "International Journal of Project Management" examines and publishes studies on innovative approaches, tactics, and strategies for efficient project management. For scholars, researchers, and professionals looking for new developments in the ever-evolving field of project management, it is an invaluable resource.

Table 3 Most influential journals for metaverse in education

Journals	Publisher	TP	TC	h-index	IF
International Journal of Information Management	Elsevier	3	536	152	18.958
International Journal of Project Management	Elsevier	1	258	167	8.000
Journal of Business Research	Elsevier	3	149	236	11.3000
Industrial Marketing Management	Elsevier	1	143	161	8.890
Vine Journal of Information and Knowledge Management Systems	Emerald	1	131	37	3.640
Advances in Developing Human Resources	SAGE	4	118	53	3.100
The Journal of Strategic Information Systems	Elsevier	1	112	104	7.000
European Journal Of Information Systems	*Taylor and Francis*	1	89	119	9.011
Journal of Marketing Management	*Taylor and Francis*	2	88	82	4.400
Advances in Intelligent Systems and Computing	Springer	18	87	58	0.215

Regarding how metaverse has investigated the field of education, the first two pertinent articles make a significant contribution. In Dwivedi et al. (2023), diverse viewpoints on prospects, difficulties, and consequences are provided as the influence of generative conversational AI on research, practice, and policy is examined. Liu et al. (2017) reports an empirical study conducted in China that looks at how collaborative design and construction processes are affected by BIM (Building Information Modelling).

The initial pair of the referenced articles examine several aspects related to the growth of HR. Toward this end, Bennett (2009, 2010) explore various aspects of human resource development in their separate contributions to the journal "Advances in Developing Human Resources." The publication outlines the evolution of the metaverse into different spheres of life (Tables 4 and 5).

4.4 The Contribution of Metaversein Learning Environments

Table 6 shows the theme clusters on the metaverse in education as produced through bibliographic coupling. This type of analysis organizes and displays the significant concepts and the relevant issues that form the academic debate around the integration of metaverse into education by using references of different scholars. Researchers and educators can use the table as a useful tool to comprehend recurring themes and patterns in the changing relationship between metaverse technology and educational approaches.

Cluster 1 (Gen Z: Navigating AI, Metaverse): This cluster investigates the particular possibilities, challenges, and learning processes that this demographic experiences in the context of education system-in the age of artificial intelligence and the metaverse. By applying a targeted lens, the analysis seeks to reveal the complex relationships forming this group's educational path. It looks at how new technologies affect their educational environment and mould their own learning experiences in the rapidly developing fields of artificial intelligence and the metaverse.

Cluster 2 (Virtual HRD): This cluster examines human resource development (HRD) from the angle of a virtual) approach. That is, how the metaverse can improve the employee's learning experience, skill enhancement, and training. This close look at these aspects of research is aimed at providing useful insights on how human resources practices are being adapted in a virtual setting. On the other hand, this approach to virtual HRD shows the paradigm shift that the metaverse is causing in the understanding of aspects of technology, talent management, and talent development in virtual settings.

Cluster 3 (Metaverse Effects in Other Areas): This cluster is expected to assess the metaverse in relation to many aspects of education and its models. It can be suggested that the study looks at the concerns regarding curriculum development and implementation, involvement of students, approaches to instruction, and the

Table 4 Most influential articles on metaverse in education

Authors	Title	TC
Dwivedi et al. (2023)	So what if ChatGPT wrote it? Multidisciplinary perspectives on opportunities, challenges and implications of generative conversational AI for research, practice and policy	462
Liu et al. (2017)	Understanding effects of BIM on collaborative design and construction: An empirical study in China	258
Sivarajah (2020)	Role of big data and social media analytics for business to business sustainability: A participatory web context	143
Wilkesmann (2018)	Industry 4.0—organizing routines or innovations?	131
Sivarajah et al. (2020)	Immersive virtual reality in the age of the Metaverse: A hybrid-narrative review based on the technology affordance perspective	112
Tueanrat et al. (2021)	Going on a journey: A review of the customer journey literature	102
Templier and Paré (2017)	Transparency in literature reviews: an assessment of reporting practices across review types and genres in top IS journals	89
Sušnik et al. (2018)	Multi-Stakeholder Development of a Serious Game to Explore the Water-Energy-Food-Land-Climate Nexus: The SIM4NEXUS Approach	80
Tynan et al. (2014)	Interpreting value in the customer service experience using customer-dominant logic	80
Huang et al. (2017)	Antecedents of student MOOC revisit intention: Moderation effect of course difficulty	74
Hew et al. (2016)	Predicting instructional effectiveness of cloud-based virtual learning environment	61
Cosenza et al. (2014)	Credibility in the blogosphere: A study of measurement and influence of wine blogs as an information source	59
Bennett and McWhorter(2021)	Virtual HRD's Role in Crisis and the Post Covid-19 Professional Lifeworld: Accelerating Skills for Digital Transformation	50
Lyytinen et al. (2021)	Metahuman systems = humans + machines that learn	44
Lee (2022)	Application of Metaverse Service to Healthcare Industry: A Strategic Perspective	40
Bennett (2014)	Introducing New Perspectives on Virtual Human Resource Development	36
Foroudi (2020)	Co-creating brand image and reputation through stakeholder's social network	36

Table 5 Most influential articles and cited references

Authors	Title	Year	TC
Bennett, EE (2009). Advances In Developing Human Resources, 11(3), 362–374	Virtual HRD: The Intersection of Knowledge Management, Culture, and Intranets	2009	8
Bennett, EE (2014). Advances In Developing Human Resources, 16(3), 263–280	Introducing New Perspectives on Virtual Human Resource Development	2014	8
McWhorter, RR (2010). Advances In Developing Human Resources, 12(6), 623–631	Exploring the Emergence of Virtual Human Resource Development	2010	5
Anderson, JC, et al. (1988). Psychological Bulletin, 103(3), 411–423	Structural Equation Modeling in Practice: A Review and Recommended Two-Step Approach	1988	4
Ausburn, LJ, et al. (2014). Advances In Developing Human Resources, 16(3), 371–390	Virtual Human Resource Development: Definitions, Challenges, and Opportunities	2014	4
Bennett, EE, et al. (2014). Handbook of Human Resource Development, 567–589	Virtual HRD	2014	4
Davis, FD (1989). MIS Quarterly, 13(3), 319–340	Perceived Usefulness, Perceived Ease of Use, and User Acceptance of Information Technology	1989	4

system of education in general. By examining these components in detail, the cluster analyzes the effects of the metaverse on the educational sector and clarifies how this technological innovation is reshaping the processes of teaching and learning.

Cluster 4 (Virtual Introspection: Exploring Environments): This group, which centres on virtual introspection, is prepared to examine the uses of virtual environments for education in a psychological and experiential manner. The purpose of this research is to delve into the specifics of education in the metaverse by analyzing the learning psychological processes and users' emotional experiences. Such research on virtual introspection can help in understanding the engagement demographics have towards the learning process in 3D environments and their in-depth personal views on other facets of education delivered in such settings (Fig. 2).

4.5 Implementation of Metaverse in Improving Education

The metaverse is capable of pedagogy enhancement (Liu and Zhang 2011) by providing social networking services which help the foreign language learning process and kindle imagination of the learners. Engaged learning through the metaverse interactions raises reasons for desire and achievement within Teams (Tarouco et al. 2013). Joining up interactions gives rise to close-knit learning communities

Table 6 Thematic clusters of metaverse in SMEs through bibliographic coupling

Clusters	Authors	Total citations
Cluster 1: Gen Z: Nevigating AI, Metaverse	Allal-Chérif (2022)	28
	Cheung and Wu (2023)	3
	Cho et al. (2023)	6
	Dabić et al. (2023)	11
	Dincelli et al. (2022)	112
	Dobrowolski et al. (2022)	20
	Dwivedi et al. (2023)	462
	Ghouri et al. (2022)	11
	Kaur and Kumar (2021)	19
	Kulkov et al. (2021)	15
	Liu et al. (2017)	258
	Lyytinen et al. (2021)	44
	Ng et al. (2023)	6
	Rejeb et al. (2021)	21
	Sivarajah et al. (2020)	143
	Templier and Paré (2017)	89
	Tueanrat et al. (2021)	102
	Zachmeier and Cho (2014)	15
Cluster 2: Virtual HRD	Bennett (2014)	36
	Bennett and McWhorter(2021)	50
	Bennett and McWhorter (2022)	6
	Khandelwal and Upadhyay (2019)	28
	Mcwhorter (2014)	26
	Mysirlaki and Paraskeva (2019)	12
	Oh and Huang (2018)	15
	Park (2018)	6
	Rahimi et al. (2022)	3
Cluster 3: Metaverse impacts across spheres	Alturas (2021)	9
	Ghorbani et al. (2019)	29
	Grijalvo et al. (2022)	13
	Huang et al. (2017)	74
	Jo and Park (2023)	2
	Peng and Abdullah (2017)	12
	Sušnik et al. (2018)	80
	White et al. (2021)	1
	Xu et al. (2017)	1

(continued)

Table 6 (continued)

Clusters	Authors	Total citations
Cluster 4: Virtual Introspection: Exploring Environments	Brown and Patterson (2021)	8
	Carnegie et al. (2023)	3
	Cosenza et al. (2014)	59
	Foroudi et al. (2020)	36
	Hew et al. (2016)	61
	Lee (2022)	40
	Porath (2022)	1
	Tynan et al. (2014)	80
	Uzir et al. (2023)	1

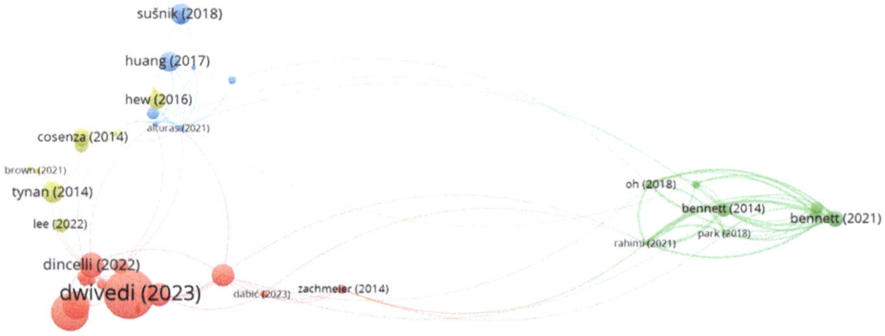

Fig. 2 Visualization of thematic clusters

which are normally not restricted by borders Geographical (Araya et al. 2018). Multilingual engagement is effectively amplified by utilities in the form of SecondLife and Solipsis which plays a key role in the learning of languages (Cruz-Lara et al. 2011). The progression of one commonplace makes such two-dimensional 'metaverse' harder even to imagine creating instead of dozens of fusions of supplemental and virtual reality, video games and augmented game designs all integrating real space and existing in different aspects. All things considered, the marauding education provides stupendous benefits for scholars to interact, engage and ultimately improve their academic performances.

4.6 Metaverse in Fostering Positive Attitudes Toward Learning in Educational Settings

The metaverse employs virtual reality technology with the aim of enhancing modern educational approaches and improving students' achievement levels (Anacona et al. 2019). Also, the use of social networks in a metaverse is expected to play a vital role in enhancing educational seek outcomes (Ribeiro et al. 2013). Such platforms that combine VR, AR, deep learning and metaverse contents can measure mobile learner satisfaction, experiences as well as presence (Cho et al. 2022). The improvement of students' performance in mathematics comes with the usage of augmented reality and application of handheld devices in the virtual space (Reyes 2020). Emotion sensitive Avatars in PBL metaverse classes teach students better through effective personalized teaching (Barry et al. 2015). Consequently, it may be stated that technological breakthroughs in the metaverse can revolutionize the conventional methods of imparting knowledge to students and have significant improvements on the performance of the students.

4.7 Hardships of Metaverse in Imparting Education

Positive learning attitudes towards the metaverse in education have been documented in the literature; where student satisfaction has been found to positively predict students' intentions to pursue medical education (Almarzouqi et al. 2022). Based on student experiences, metaverse-based learning is evaluated more favorably than traditional synchronous learning (Nishino et al. 2012). Chinese culture is successfully transmitted through metaverse-based virtual reality, offering an animated cultural experience (Shen 2022). By enhancing learners' self-assurance in their cultural identity, the incorporation of culture into learning experiences reinforces favorable views regarding the metaverse in education.

Despite the possible benefits, educators are faced with metaverse challenges as well, including concerns regarding negative implications of improper engagement and device use (Cheong and Lee 2022). It is all the more important to resolve such issues because there is a possibility of evolution of cultural differences, teachers will not be prepared in advance, lack of inter-discipline collaboration, and there are ethical issues surrounding it. Nevertheless, appropriate social, political, technological, and ethical issues should also be addressed by the researchers. They should also focus on how important it is to have codes of ethics enforced, avatar systems to be checked, and users to be cautioned at all times (Hendaoui et al. 2008; Park and Kim 2022a, b).

This is important because the high probability that use of metaverse will be expensive, the hassle of constantly having to switch from virtual to augmented reality, the insecurity of personal data, and the risk of compromise of data privacy. Asking users to adhere to certain ethical principles can be enforced through methods such as creating avatars in the metaverse as a way of monitoring user activities, while a double-check avatar system can be used in order to prevent the occurrence of certain suspicious activities (Park and Kim 2022a, b). It is obvious that there are issues which must be resolved in an effort to improve the safety and efficiency of education given in the metaverse.

5 Conclusion

The results of the current study show the dynamic shift in the degree of engagement with the metaverse in educational institutions over the years, with a steep growth from the year 2020 and a climax in usage in the year 2023. The implementation of the Metaverse into educational curricula has the potential to radically transform the existing conditions of any teaching/learning process to the positive side. In fact, through the use of VR, AR and AI, a teacher is able to create active personalized learning environments that are not only effective in retention of information but also enhances the creativity of the learners. In addition, the enhanced learning practices focus in part on the promotion of skills that are necessary for the tackling of sustainability issues. With the development of these frontiers still underway, the world is indeed promisingly becoming a global village, an enlightened society and an eco-friendly one at that.

Final remarks are offered in this paper in which the evolution of the educational sector through necessitated by the metaverse is placed in its greater context. Data indicates there is an increasing amount of literature concerning the use of analysis which suggests visual and immersive transformation of learning in a more about engagement and inclusiveness of the practice. In doing so, we are able to address those trends, major works, and cutting-edge issues which reveal the subsectoral influence of the Metaverse on education. This reassessment contains some useful information about the present trends but also shows that research development and collaboration is still needed in the area. It has been shown from this and other studies that there is a scope to readily politically transform the education system. In future, we have to work towards understanding these tools to build a better and fairer education system.

6 Implication of Future Research

The inferences derived from the study provide several avenues for further study on the emerging topic of metaverse in education. First and foremost, it is recommended to researchers to seek a more comprehensive understanding of the existing literature by looking for research articles outside of the purview of Dimensions' Core Collection of UGC Journal List Group II. Also, for the purposes of research into the metaverse in education, future studies may also consider a more innovative and flexible approach to the application of keywords in the research process in order to introduce new concepts and their corresponding variations.

That some degree of qualitative assessment may be possible is implied by the focus on quantitative figures regarding the input of authors, institutions and countries. Scholars are free to look at the qualitative dimensions of the contributions of which the impact on teaching methods, creative teaching methods, and application of metaverse technologies in different types of teaching settings is an example.

These identified thematic clusters are quite useful but more studies may aim to expand the dimensions of the boundaries of the trans-disciplinary perspectives explored and examine the metaverse in relation to other emerging technologies such as block-chain, AI or even extended reality. An investigation into this domain, however, would also benefit from examining the social consequences and ethical concerns that arise from the incorporation of the metaverse in education.

There is a need to persist with research in order to cope with the present pace of change in technology and emerging trends concerning metaverse applications in education. Such studies tend to be longitudinal which can provide insights into the effects of the metaverse on paradigms of education in the long term.

Lastly, future studies could examine possible biases in the literature, taking into account elements like institutional, cultural, or geographic biases that could affect how different viewpoints on the metaverse are portrayed in the classroom. The resolution of these issues will enable us to have a more in-depth and holistic view of the possible impact that the metaverse may have on education in the future.

References

Allal-Chérif Q (2022) Intelligent cathedrals: Using augmented reality, virtual reality, and artificial intelligence to provide an intense cultural, historical, and religious visitor experience. Technol Forecast Soc Chang 178:121604. https://doi.org/10.1016/j.techfore.2022.121604

Almarzouqi A, Aburayya A, Salloum SA (2022) Prediction of user's intention to use metaverse system in medical education: a hybrid SEM-ML learning approach. IEEE Access: Pract Innovations Open Solutions 10:43421–43434. https://doi.org/10.1109/ACCESS.2022.3169285

Alturas B (2021) Models of acceptance and use of technology research trends: literature review and exploratory bibliometric study. In: Al-Emran M, Shaalan K (eds) Recent advances in technology acceptance models and theories. Studies in systems, decision and control, vol 335. Springer, Cham. https://doi.org/10.1007/978-3-030-64987-6_2

Anacona JD, Millán EE, Gómez CA (2019) Application of Metaverses and the virtual reality in teaching. Electr J Educ Technol 13(25):59–67. https://doi.org/10.31908/19098367.4015

Anderson JC, Gerbing DW (1988) Structural equation modeling in practice: a review and recommended two-step approach. Psychol Bull 103(3):411–423. https://doi.org/10.1037/0033-2909.103.3.411

Ausburn LJ, Ausburn FB (2014) Technical perspectives on theory in screen-based virtual reality environments: leading from the future in VHRD. Adv Dev Hum Resour 16(3):371–390. https://doi.org/10.1177/1523422314532125

Barry DM, Ogawa N, Dharmawansa A, Kanematsu H, Fukumura Y, Shirai T, Yajima K, Kobayashi T (2015) Evaluation for students' learning manner using eye blinking system in metaverse. Procedia Comput Sci 60:1195–1204. https://doi.org/10.1016/j.procs.2015.08.181

Bennett EE, McWhorter RR (2014) Virtual human resource development. In: Chalofsky NE, Rocco TF, Morris ML (eds) Handbook of human resource development. Wiley, pp 567–589

Bennett EE (2014) Introducing new perspectives on virtual human resource development. Adv Dev Hum Resour 16(3):263–280. https://doi.org/10.1177/1523422314532091

Bennett EE, McWhorter RR (2021) Virtual HRD's role in crisis and the post covid-19 professional lifeworld: accelerating skills for digital transformation. Adv Dev Hum Resour 23(1):5–25. https://doi.org/10.1177/1523422320973288

Bennett EE, McWhorter RR (2022) Dancing in the paradox: virtual human resource development, online teaching, and learning. Adv Dev Hum Resour 24(2):99–116. https://doi.org/10.1177/15234223221079440

Bhugaonkar K, Bhugaonkar R, Masne N (2022) The trend of metaverse and augmented & virtual reality extending to the healthcare system. Cureus 12, 14(9):e29071. https://doi.org/10.7759/cureus.29071.

Brown S, Patterson A (2021) Me-search? Search me! A new twist in the tale of introspection. J Mark Manag 37(13–14):1343–1373. https://doi.org/10.1080/0267257x.2021.1928268

Cai W, Liu Y (2022) The value of immersive media in expanding Chinese public cultural participation and its realization path from the perspective of cultural education. Front Psychol 13:915913. https://doi.org/10.3389/fpsyg.2022.915913

Carnegie GD, Gomes D, McBride K (2023) COVID-19 and accounting as multidimensional technical, social and moral practice: a framework for future research. Meditari Accountancy Res 31(1):1–26. https://doi.org/10.1108/MEDAR-10-2022-1826

Çelikkol Ş (2022) Metaverse Dünyasi'nin, Tüketici Satin Alma Davranişlari Açisindan Değerlendirilmesi. İstanbul Kent Üniversitesi İnsan Ve Toplum Bilimleri Dergisi 3(1):64–75

Cheong Y, Lee YH (2022) A case study on elementary convergence education using metaverse platform. J Learner-Centered Curriculum Instr 22(16):561–580

Cheung KL, Wu H (2023) The use of blended synchronous learning for property education in and beyond the COVID-19 pandemic. Prop Manag 41(2):228–243. https://doi.org/10.1108/PM-11-2021-0086

Cho Y, Hong S, Kim M, Kim J (2022) DAVE: deep learning-based asymmetric virtual environment for immersive experiential metaverse content. Electronics 11(16):2604. https://doi.org/10.3390/electronics11162604

Cho J, tom Dieck MC, Jung T (2023) What is the Metaverse? Challenges, opportunities, definition, and future research directions. In: Jung T, tom Dieck MC, Correia Loureiro SM (eds) Extended reality and metaverse. XR 2022. Springer proceedings in business and economics. Springer, Cham. https://doi.org/10.1007/978-3-031-25390-4_1

Clark KR (2018) Learning theories: constructivism. Radiol Technol 90(2):180–182

Codish D, Ravid G (2014) Academic course gamification: the art of perceived playfulness. Interdisc J e-Skills Lifelong Learn 10:131–51. https://doi.org/10.28945/2066

Cosenza TR, Solomon MR, Kwon W-S (2014) Credibility in the blogosphere: a study of measurement and influence of wine blogs as an information source. J Consum Behav 14(2). https://doi.org/10.1002/cb.1496

Cruz-Lara S, Osswald T, Guinaud J, Bellalem N, Bellalem L, Camal JP (2011) A chat interface using standards for communication and e-learning in virtual worlds. In: Filipe J, Cordeiro J (eds) Enterprise information systems. ICEIS 2010. Lecture notes in business information processing, vol 73. Springer, Berlin, Heidelberg. https://doi.org/10.1007/978-3-642-19802-1_37

Dabić M, Maley JF, Švarc J, Poček J (2023) Future of digital work: challenges for sustainable human resources management. J Innov Knowl 8(2):100353. https://doi.org/10.1016/j.jik.2023.100353

Dahan NA, Al-Razgan M, Al-Laith A, Alsoufi MA, Al-Asaly MS, Alfakih T (2022) Metaverse framework: a case study on e-learning environment (ELEM). Electronics 11(10):1616. https://doi.org/10.3390/electronics11101616

Davis FD (1989) Perceived usefulness, perceived ease of use, and user acceptance of information technology. MIS Q 13(3):319–340. https://doi.org/10.2307/249008

Dhelim S, Kechadi T, Chen L, Aung N, Ning H, Atzori L (2015) Edge-enabled Metaverse: the convergence of metaverse and mobile edge computing. J Latex Class Files 14(8). https://doi.org/10.48550/arXiv.2205.02764.

Díaz JEM, Saldaña CAD, Ávila CAR (2020) Virtual world as a resource for hybrid education. Int J Emerging Technol Learning (iJET) 15(15):94–109. https://doi.org/10.3991/ijet.v15i15.13025

Dincelli E, Yayla A (2022) Immersive virtual reality in the age of the Metaverse: a hybrid-narrative review based on the technology affordance perspective. J Strateg Inf Syst 31(2):101717. https://doi.org/10.1016/j.jsis.2022.101717

Dobrowolski Z, Drozdowski G, Panait M (2022) Understanding the impact of generation Z on risk management—a preliminary views on values, competencies, and ethics of the generation Z in public administration. Int J Environ Res Public Health 19(7):3868. https://doi.org/10.3390/ijerph19073868

Donthu N, Kumar S, Mukherjee D, Pandey N, Lim WM (2021) How to conduct a bibliometric analysis: an overview and guidelines. J Bus Res 133:285–296. https://doi.org/10.1016/j.jbusres.2021.04.070

Dwivedi YK, Kshetri N, Hughes L, Slade EL, Jeyaraj A, Kar AK, Baabdullah AM, Koohang A, Raghavan V et al (2023) "So what if ChatGPT wrote it?" Multidisciplinary perspectives on opportunities, challenges and implications of generative conversational AI for research, practice and policy. Int J Inf Manage 71:102642. https://doi.org/10.1016/j.ijinfomgt.2023.102642

Foroudi P, Nazarian A, Ziyadin S, Kitchen P, Hafeez K, Priporas C, Pantano E (2020) Co-creating brand image and reputation through stakeholder's social network. J Bus Res 114:42–59. https://doi.org/10.1016/j.jbusres.2020.03.035

Gao P, Meng F, Mata MN, Martins JM, Iqbal S, Correia AB, Dantas RM, Waheed A, Xavier Rita J, Farrukh M (2021) Trends and future research in electronic marketing: a bibliometric analysis of twenty years. J Theor Appl Electron Commer Res 16(5):1667–1679. https://doi.org/10.3390/jtaer16050094

Getchell K, Oliver I, Miller A, Allison C (2010) Metaverses as a platform for game based learning. In: 24th IEEE international conference on advanced information networking and applications (AINA) 20–23 Apr 2010, Perth, Australia. https://doi.org/10.1109/AINA.2010.125

Ghorbani A, Danaei A, Zargar SM, Hematian H (2019) Designing of smart tourism organization (STO) for tourism management: a case study of tourism organizations of South Khorasan province, Iran. Heliyon 5(6):e01850. https://doi.org/10.1016/j.heliyon.2019.e01850

Ghouri AM, Mani V, Amin ul Haq M, Kamble SS (2022) The micro foundations of social media use: artificial intelligence integrated routine model. J Bus Res 144:80–92. https://doi.org/10.1016/j.jbusres.2022.01.084

Goyal K, Kumar S (2021) Financial literacy: a systematic review and bibliometric analysis. Int J Consum Stud 45(1):80–105. https://doi.org/10.1111/ijcs.12605

Grijalvo M, Segura A, Núñez Y (2022) Computer-based business games in higher education: a proposal of a gamified learning framework. Technol Forecast Soc Change 178:121597. https://doi.org/10.1016/j.techfore.2022.121597

Guo H, Gao W (2022) Metaverse-powered experiential situational English-teaching design: an emotion-based analysis method. Front Psychol 13:859159. https://doi.org/10.3389/fpsyg.2022.859159

Hendaoui A, Limayem M, Thompson CW (2008) 3D social virtual worlds-research issues and challenges. IEEE Internet Comput 12(1):88–92. https://doi.org/10.1109/MIC.2008.1

Hew T-S, Syed Abdul Kadir SL (2016) Predicting instructional effectiveness of cloud-based virtual learning environment. Ind Manage Data Syst 116(8):1557–1584. https://doi.org/10.1108/IMDS-11-2015-0475

Hines P, Netland TH (2023) Teaching a lean masterclass in the metaverse. Int J Lean Six Sigma 14(6):1121–1143. https://doi.org/10.1108/IJLSS-02-2022-0035

Hollensen S, Kotler P, Opresnik MO (2023) Metaverse—the new marketing universe. J Bus Strateg 44(3):119–125. https://doi.org/10.1108/JBS-01-2022-0014

Hossain MT, Nishu NA, Jahan N, Saadia H, Ela MZ, Moshrur-Ul-Alam Q, Hossain MZ (2022) Challenges of participating in online education/distance learning amidst the COVID-19 in Bangladesh: the response from students and teachers. In: Islam MR, Santosh Kumar B, Lamhot N (eds) Handbook of research on Asian perspectives of the educational impact of COVID-19. IGI Global, Hershey, PA, USA, pp 254–281

Huang L, Zhang J, Liu Y (2017) Antecedents of student MOOC revisit intention: moderation effect of course difficulty. Int J Inf Manage 37(2):84–91. https://doi.org/10.1016/j.ijinfomgt.2016.12.002

Iwanaga J, Muo EC, Tabira Y, Watanabe K, Tubbs SJ, D'Antoni AV, Rajaram-Gilkes M, Loukas M, Khalil MK, Tubbs RS (2023) Who really needs a Metaverse in anatomy education? A review with preliminary survey results. Clin Anat 36(1):77–82. https://doi.org/10.1002/ca.23949

Jaramillo-Mujica JA, Morales-Avella LF, Coy-Mondragon DM (2017) An experience using metaverses for teaching mechanical physics to engineering students. RevistaEducacion En Ingenieria 12(24):20–30. https://doi.org/10.26507/rei.v12n24.778

Jo H, Park DH (2023) Affordance, usefulness, enjoyment, and aesthetics in sustaining virtual reality engagement. Sci Rep 13:15097. https://doi.org/10.1038/s41598-023-42113-1

Jovanović A, Milosavljević A (2022) VoRtex Metaverse platform for gamified collaborative learning. Electronics 11(3):317. https://doi.org/10.3390/electronics11030317

Kanematsu H, Kobayashi T, Barry DM, Fukumura Y, Dharmawansa A, Ogawa N (2014) Virtual STEM class for nuclear safety education in metaverse. Procedia Comput Sci 35:1255–1261. https://doi.org/10.1016/j.procs.2014.08.224

Kaur K, Kumar P (2021) Social media usage in Indian beauty and wellness industry: a qualitative study. TQM J 33(1):17–32. https://doi.org/10.1108/TQM-09-2019-0216

Khandelwal K, Upadhyay AK (2019) Virtual reality interventions in developing and managing human resources. Hum Resour Dev Int 24(2):219–233. https://doi.org/10.1080/13678868.2019.1569920

Kulkov I, Berggren B, Hellström M, Wikström K (2021) Navigating uncharted waters: designing business models for virtual and augmented reality companies in the medical industry. J Eng Tech Manage 59:101614. https://doi.org/10.1016/j.jengtecman.2021.101614

Kye B, Han N, Kim E, Park Y, Jo S (2021) Educational applications of metaverse: possibilities and limitations. J Educ Eval Health Prof 18(32):1–13. https://doi.org/10.3352/jeehp.2021.18.32

Lee CW (2022) Application of metaverse service to healthcare industry: a strategic perspective. Int J Environ Res Public Health 19:13038. https://doi.org/10.3390/ijerph192013038

Lee H, Hwang YH (2022) Technology-enhanced education through VR-making and metaverse-linking to foster teacher readiness and sustainable learning. Sustainability 14(8):4786. https://doi.org/10.3390/su14084786

Lee H, Woo D, Yu S (2022) Virtual reality metaverse system supplementing remote education methods: based on aircraft maintenance simulation. Appl Sci 12(5):2667. https://doi.org/10.3390/app12052667

Lee SH, Lee H, Kim JH (2022) Enhancing the prediction of user satisfaction with metaverse service through machine learning. Comput Mater Continua 72(3):4983–4997. https://doi.org/10.32604/cmc.2022.027943

Li H, Cui C, Jiang S (2022) Strategy for improving the football teaching quality by AI and metaverse-empowered in mobile internet environment. Wireless Netw. https://doi.org/10.1007/s11276-022-03000-1

Liu X, Zhang J (2011) Foreign language learning through virtual communities. Energy Procedia 17:737–740. https://doi.org/10.1016/j.egypro.2012.02.165

Liu Y, Nederveen SV, Hertogh M (2017) Understanding effects of BIM on collaborative design and construction: an empirical study in China. Int J Project Manage 35(4):686–698. https://doi.org/10.1016/j.ijproman.2016.06.007

Lyytinen K, Nickerson JV, King JL (2021) Metahuman systems = humans + machines that learn. J Inf Technol 36(4):427–445. https://doi.org/10.1177/0268396220915917

Marmaridis I, Griffith S (2009) Metaverse services: extensible learning with mediated teleporting into 3D environments. In: Yang J, Ginige A, Mayr HC, Kutsche RD (eds) Information systems: modeling, development, and integration. UNISCON 2009. Lecture notes in business information processing, vol 20. Springer, Berlin, Heidelberg. https://doi.org/10.1007/978-3-642-01112-2_24

McWhorter RR (2010) Exploring the emergence of virtual human resource development. Adv Dev Hum Resour 12(6):623–631. https://doi.org/10.1177/1523422310395367

McWhorter RR (2014) A synthesis of new perspectives on virtual HRD. Adv Dev Hum Resour 16(3):391–401. https://doi.org/10.1177/1523422314532126

Mishra D, Gunasekaran A, Papadopoulos T (2018) Big data and supply chain management: a review and bibliometric analysis. ANN Oper Res 270:313–336

Mysirlaki S, Paraskeva F (2019) Virtual team effectiveness: insights from the virtual world teams of massively multiplayer online games. J Leadersh Stud 13(1):36–55. https://doi.org/10.1002/jls.21608

Melendez Araya NM, Hidalgo Avila RS (2018) Collaborative learning through integration of environments real and virtual-immersive. In: 37th International conference of the Chilean computer science society (SCCC), Santiago, Chile, pp 1–8. https://doi.org/10.1109/SCCC.2018.8705260

Ng DTK, Ching ACH, Law SW (2023) Online learning in management education amid the pandemic: a bibliometric and content analysis. Int J Manage Educ 21(2):100796. https://doi.org/10.1016/j.ijme.2023.100796

Nishino K, Mayumi T, Iribe Y, Mizuno S, Ogawa N, Yajima K, Aoki K, Fukumura Y (2012) Consideration on the relationship between changes in learners' learning preferences and the differences in e-learning modes of a course. Front Artif Intell Appl 243:1121–1130. https://doi.org/10.3233/978-1-61499-105-2-1121

Oh EG, Huang W-HD (2018) A review of technology research in HRD from a design-based research perspective. Hum Resour Dev Rev 17(3):258–276. https://doi.org/10.1177/1534484318784798

Ortega-Rodríguez PJ (2022) From extended reality to the metaverse: a critical reflection on contributions to education. Teoría de la Educación. Revista Interuniversitaria 34(2):189–208. https://doi.org/10.14201/teri.27864

Park J (2021) Exploring the possibility of using Metaverse in Korean language education. J Int Netw Korean Lang Cult 18(3):117–146. https://doi.org/10.15652/ink.2021.18.3.117

Park SM, Kim YG (2022) A Metaverse: taxonomy, components, applications, and open challenges. IEEE Access 10:4209–4251. https://doi.org/10.1109/ACCESS.2021.3140175

Park S, Kim S (2022) Identifying world types to deliver gameful experiences for sustainable learning in the Metaverse. Sustainability 14(3):1361. https://doi.org/10.3390/su14031361

Park S, Jeong S, Ju B (2018) Employee learning and development in virtual HRD: focusing on MOOCs in the workplace. Ind Commer Train 50(5):261–271. https://doi.org/10.1108/ICT-03-2018-0030

Peng J, Abdullah I (2017) Building a market simulation to teach business process analysis: effects of realism on engaged learning. Acc Educ 27(2):208–222. https://doi.org/10.1080/09639284.2017.1407248

Pigultong M (2022) Cognitive impacts of using a Metaverse embedded on learning management system for students with unequal access to learning resources. In: 2022 10th International conference on information and education technology (ICIET), pp 27–31. https://doi.org/10.1109/ICIET55102.2022.9779045

Pinchuk OP, Lytvynova SG, Burov OY (2017) Synthetic educational environment—a footpace to new education. Inf Technol Learn Tools 60(4):28–45. https://doi.org/10.33407/itlt.v60i4.1831

Porath CL, Gibson CB, Spreitzer GM (2022) To thrive or not to thrive: pathways for sustaining thriving at work. Res Organ Behav 42:100176. https://doi.org/10.1016/j.riob.2022.100176

Rahimi S, Khorasani A, Rezaeizadeh M, Waterworth J (2022) The virtual human resource development (VHRD) approach: an integrative literature review. Eur J Training Dev 46(5/6):484–503. https://doi.org/10.1108/EJTD-08-2020-0131

Rapanotti L, Hall JG (2010) Lessons learned in developing aSecond Life educational environment. In: Proceedings of the 2nd international conference on computer supported education, vol 1, pp 33–38. https://doi.org/10.5220/0002777200330038

Rejeb A, Rejeb K, Keogh JG (2021) Enablers of augmented reality in the food supply chain: a systematic literature review. J Foodserv Bus Res 24(4):415–444. https://doi.org/10.1080/15378020.2020.1859973

Reyes CEG (2020) Perception of high school students about using Metaverse in augmented reality learning experiences in mathematics. Pixel-Bit-Revista De Medios Y Educacion 58:143–159. https://doi.org/10.12795/pixelbit.74367

Ribeiro R, Mandaji M, Camas N, Tercariol A, Silva R, Mengalli N (2013) Paths to innovative pedagogical practices for learning and teaching: an analysis considering the I and II Web curriculum seminars. In: Proceedings of ICERI2013 conference, pp 2589–2597

Schaf FM, Paladini S, Pereira CE (2012) 3D AutoSysLab prototype—a social, immersive and mixed reality approach for collaborative learning environments. Int J Eng Pedagogy (iJEP) 2(2):15–22. https://doi.org/10.3991/ijep.v2i2.2083

Schlemmer E, Marson F (2013) Immersive learning: metaverses and digital games in education. In: Proceedings of the 2013 8th Iberian conference on information systems and technologies (CISTI 2013)

Shen S (2022) Metaverse-driven new energy of Chinese traditional culture education: edge computing method. Evol Intel. https://doi.org/10.1007/s12065-022-00757-4

Sivarajah U, Irani Z, Gupta S, Mahroof K (2020) Role of big data and social media analytics for business to business sustainability: a participatory web context. Ind Mark Manage 86:163–179. https://doi.org/10.1016/j.indmarman.2019.04.005

Siyaev A, Jo GS (2021) Neuro-symbolic speech understanding in aircraft maintenance metaverse. IEEE Access 9:154484–154499. https://doi.org/10.1109/ACCESS.2021.3128616

Siyaev A, Jo GS (2021) Towards aircraft maintenance Metaverse using speech interactions with virtual objects in mixed reality. Sensors (Basel) 21(6):2066. https://doi.org/10.3390/s21062066

Stephenson N (1992) Snow crash: a novel. Del Rey, Penguin Random House LLC

Suh W, Ahn S (2022) Utilizing the metaverse for learner-centered constructivist education in the post-pandemic era: an analysis of elementary school students. J Intelligence 10(1):17. https://doi.org/10.3390/jintelligence10010017

Sušnik J, Chew C, Domingo X, Mereu S, Trabucco A, Evans B, Vamvakeridou-Lyroudia L, Savić DA, Laspidou C, Brouwer F (2018) Multi-stakeholder development of a serious game to explore the water-energy-food-land-climate nexus: the SIM4NEXUS approach. Water 10(2):139. https://doi.org/10.3390/w10020139

Tarouco L, Gorziza B, Correa Y, Amaral EMH, Muller T (2013) Virtual Laboratory for teaching calculus: an immersive experience. In: 2013 IEEE global engineering education conference (EDUCON), IEEE global engineering education conference (EDUCON). Tech Univ Berlin, Berlin, Germany

Templier M, Paré G (2017) Transparency in literature reviews: an assessment of reporting practices across review types and genres in top IS journals. Eur J Inf Syst 27(5):503–550. https://doi.org/10.1080/0960085X.2017.1398880

Thelwall M (2018) Dimensions: a competitor to scopus and the web of science? J Informet 12(2):430–435. https://doi.org/10.1016/j.joi.2018.03.006

Tlili A, Huang R, Shehata B et al (2022) Is Metaverse in education a blessing or a curse: a combined content and bibliometric analysis. Smart Learn Environ 9:24. https://doi.org/10.1186/s40561-022-00205-x

Tueanrat Y, Papagiannidis S, Alamanos E (2021) Going on a journey: a review of the customer journey literature. J Bus Res 125:336–353. https://doi.org/10.1016/j.jbusres.2020.12.028

Tynan C, McKechnie S, Hartley S (2014) Interpreting value in the customer service experience using customer-dominant logic. J Mark Manag 30(9–10):1058–1081. https://doi.org/10.1080/0267257X.2014.934269

Uzir MdUH, Bukari Z, Halbusi HA, Lim R, Wahab SN, Rasul T, Thurasamy R et al (2023) Applied artificial intelligence: acceptance-intention-purchase and satisfaction on smartwatch usage in a Ghanaian context. Heliyon 9(8):e18666. https://doi.org/10.1016/j.heliyon.2023.e18666

van Eck NJ, Waltman L (2010) Software survey: VOSviewer, a computer program for bibliometric mapping. Scientometrics 84(2):523–538. https://doi.org/10.1007/s11192-009-0146-3. PMID:20585380

van Eck NJ, Waltman L (2014) CitNetExplorer: a new software tool for analyzing and visualizing citation networks. J Informet 8(4):802–823. https://doi.org/10.1016/j.joi.2014.07.006
van Eck NJ, Waltman L (2014b) Visualizing bibliometric networks. In Ding Y, Rousseau R, Wolfram D (eds) Measuring scholarly impact. Springer, Cham. https://doi.org/10.1007/978-3-319-10377-8_13
Wagner R, Passerino LM, Piovesan SD, de Lima JV (2013) Using 3D virtual learning environments in new perspective of education. In: 2013 12th international conference on information technology based higher education and training (ITHET). https://doi.org/10.1109/ITHET.2013.6671019
Wang Y, Lee LH, Braud T, Hui P (2022) Re-shaping post-COVID-19 teaching and learning: a blueprint of virtual-physical blended classrooms in the metaverse era. In: 2022 IEEE 42nd international conference on distributed computing systems workshops (ICDCSW), Italy. https://doi.org/10.1109/ICDCSW56584.2022.00053.
White T, Brody RG, Gupta G (2021) Who's got next? An analysis of the inhibitors to mobile game adoption in an introductory accounting class. In: Calderon TG (ed) Advances in accounting education: teaching and curriculum innovations (advances in accounting education, vol 25. Emerald Publishing Limited, Leeds, pp 23–47. https://doi.org/10.1108/S1085-462220210000025002
Wilkesmann M, Wilkesmann U (2018) Industry 4.0—organizing routines or innovations? VINE J Inf Knowl Manage Syst 48(2):238–254. https://doi.org/10.1108/VJIKMS-04-2017-0019
Xu Y, Johnson PM, Lee GE, Moore CA, Brewer RS (2017) Design and evaluation of the Makahiki open source serious game framework for sustainability education. In: Sustainability, green IT and education strategies in the twenty-first century. Springer, Cham, pp 503–523
Yang YQ, Shi QF, Zhang ZX, Shan XCA, Salam B, Lee C (2022) Robust triboelectric informationmat enhanced by multi-modality deep learning for smart home. Informat 5(1):e12360. https://doi.org/10.1002/inf2.12360
Zachmeier A, Cho Y (2014) Taking stock of the literature on HRD education. Eur J Training Dev 8(4):347–363. https://doi.org/10.1108/EJTD-07-2013-0080
Zupic I, Čater T (2015) Bibliometric methods in management and organization. Organ Res Methods 18(3):429–472. https://doi.org/10.1177/1094428114562629

Cognitive and Emotional Dimensions of Sustainability: Building Business Resilience in the Virtual World

Anu Dandona and Sandhya Bhatt

Abstract This chapter explores the cognitive and emotional dimensions of sustainability in the virtual world, emphasizing their crucial role in enhancing business resilience. In the metaverse, businesses can foster resilience by promoting cognitive engagement through sustainability awareness and informed decision-making, while emotional dimensions strengthen consumer loyalty and stakeholder collaboration. Technologies like virtual reality (VR) and augmented reality (AR) are discussed as effective tools for sustainability training and immersive consumer experiences. The chapter examines theoretical frameworks, including cognitive theories and emotional engagement perspectives, highlighting their support for sustainable practices. Additionally, narrative-driven experiences and virtual communities are identified as mechanisms for collective sustainability efforts, further contributing to organizational resilience. By integrating cognitive and emotional engagement into sustainability initiatives, businesses can achieve long-term adaptability and success while navigating the complexities of the digital landscape.

Keywords Sustainable behaviour · Digital environments · Virtual reality · Augmented reality · Environmental awareness · Behavioral change

1 Introduction

The metaverse offers unique opportunities for businesses to engage with sustainability in ways that were not possible before. However, to thrive in this evolving digital landscape, businesses must build resilience—not only through traditional financial and operational means but also by leveraging the cognitive and emotional dimensions of sustainability.

A. Dandona
D Y Patil International University, Akurdi, Pune, India

S. Bhatt (✉)
Amity University, Noida, India
e-mail: sbhatt@amity.edu

W. Leal Filho et al. (eds.), *Metaverse and Sustainability: Business Resilience Towards Sustainable Development Goals*, World Sustainability Series,
https://doi.org/10.1007/978-3-031-89545-6_5

Technological advancement has made it possible for establishment of virtual and augmented realities that are pivotal for enhancing interaction in various critical global concerns like the sustainability. In these virtual worlds, the characterization of psychology and emotions attached to sustainability is crucial for encouraging substantial engagement and improvements in customers' behaviours.

Therefore, this analysis is centred on examining psychological factors that influence people's attitudes and behaviours toward sustainable digital environment. It is the study of such cognitive factors as perception, attention, memory and emotional factors is possible to identify the set of interrelated factors that can drive a user to practice sustainability. In addition, when individuals develop a sense of presence and get involved in virtual environments they gain increased environmental sensitivity and become more likely to adopt pro-environmental behaviours. Thus, identifying guidelines on how to convey sustainability ideas through these media becomes important.

In this discourse, their purpose and effectiveness of virtual environment can be best understood when a multidisciplinary approach is employed which includes but not limited to psychology, technology and sustainability education. These platforms beyond offering innovative and entertaining solutions also support collaborative efforts towards the attainment of different Sustainable Development Goals (SDGs). Therefore, knowledge of factors that pertain to cognition and affect is critical in order to optimise the potential of virtual worlds as key drivers of sustainability communication and change.

2 Cognitive Aspects of Sustainability

Cognition plays a critical role in how businesses and consumers understand and engage with sustainability. Businesses that foster cognitive awareness around sustainability in virtual spaces help shape informed decision-making processes, promoting sustainable practices. In the metaverse, this awareness can directly contribute to business resilience by enabling businesses to adapt to environmental challenges, anticipate consumer trends, and innovate in ways that reduce waste and conserve resources.

Perception relates to the ability of the users to attend, conceive and comprehend the sustainability related information and stimuli in the form of the virtual worlds. Some aspects are unique signs, interactive components, and gamification, or interactive narrative that can influence the extent to which users actually receive messages concerning the environment and sustainability. It is thus possible to assert that the design of virtual experiences can significantly impact users' level of awareness and knowledge of sustainability objectives (Laukkanen et al. 2022).

Another cognitive process that influences users' aptitude to pay **attention** to sustainability information is attention. Where there are all sorts of stimuli, several of which could lead to sustainability issues. As Laukkanen et al. (2022) and Gesthuizen et al. (2024), for extended time durations, it is crucial to comprehend how to engage attention in these settings for maintaining behaviours.

Memory also came into play in the manner and the way that sustainability is stored and materialized. When learners apply various principles in sustainability ideas, then it is likely that they have strong memory recall of the relevant concepts as compared to a normal learning technique. It was understood that the degree of memory effectiveness is enhanced if the individual has the level of immersion and emotional experience in the virtual environment (Spangenberger et al. 2024; Wong 2024).

3 Emotional Dimensions of Sustainability

Emotional engagement with sustainability—especially through virtual platforms—can strengthen the connection between businesses and their stakeholders. When consumers and employees feel emotionally invested in a company's sustainability efforts, their loyalty and commitment increase, contributing to business resilience.

In the metaverse, businesses can use emotionally driven content and experiences to create deeper connections with their audiences, encouraging sustainable consumer behavior while building a loyal customer base that supports the company during challenging times.

Emotional facets of sustainability are critical for people's interactions with sustainability and decision-making with regard to sustainable goals and actions in contexts such as digital media. Empathy, guilty, and pride as examples of affect can create a positive or negative attitude towards sustainability concepts and influences the willingness of the users to embrace sustainable behaviour.

4 Empathy

Compassion is a strong human feeling which binds the person and the world together. When users feel empathy for the natural environment, they are more inclined to be involved in sustainable projects. This connection may be especially enhanced in virtual environments where people are able to embody the view of stakeholders of the environment or entities other than humans. For example, an actualization of the impacts of environmental pollution through the use of virtual glasses as a tool promotes users' responsibility to conserve the environment and supports greater adherence to sustainable practices.

5 Guilt

Another important role belongs to guilty, they also push people towards sustainable solutions and policies. Hypothesized perceptions of eco-guilt occur when people observe the inconsistency between the environmental beliefs and behaviour or when they think that their actions are detrimental to the environment. Such an emotional response may lead to behavioural change by the individuals with a view of eradicating guilt feelings. In digital settings, these feelings can be orchestrated by stories and situations that display the emergent consequences of unwise decisions. For instance, such visualization techniques as simulations depicting the implications of pollution create feelings of guilt, which leads the users to use environment friendly methods to reduce on guilt detected.

6 Pride

There is another emotion, pride that too can positively affect making a positive impact to the sustainability goals. As and when people get a feeling of pride in taking environment friendly actions or in achieving an 'environmental goal', they are more inclined to perpetuate this and even motivate others do so. Pride can also be nurtured through awards and incentives that are given within digital domains that are sustainable. For example, when learners use sustainability-oriented applications, points that would grant badges can be awarded in line with completion of sustainable activities since such configurations tend to elicit feelings of pride and thus longer-term engagement.

7 Virtual Reality (VR) and Augmented Reality (AR)

The immersive nature of VR and AR allows businesses to implement sustainability strategies in virtual environments that would be difficult to replicate in the real world. By using VR/AR technologies, companies can train employees on sustainable practices, simulate supply chain innovations, and even engage customers with interactive, sustainability-focused experiences. These tools not only enhance sustainability education but also equip businesses to build resilience by preparing for environmental disruptions, testing new strategies, and optimizing operations in a virtual setting.

Virtual Reality (VR) and Augmented Reality (AR) are the technologies that have the potential to deliver the engaging experience that makes a drastic difference on the level of environmental sensitivity and pro-ecological action. These technologies enable the visualisation of environmental problems within interfaces through which users can interact in ways that are difficult or impossible through the conventional means. Besides, encouraging people to learn about these problems, VR and AR allow the users to have an emotionally related experience, which is critical for behavioural change.

1. **Immersive Environments**

VR offers sealed and corrected digital spaces in which climate change and environment destruction or low impact simulation is possible to give users a feel of scenarios that may exist in the natural world. Such an engagement level assists the users to realize the severity and the need for change by demonstrating environmental challenges.

2. **Interactive Simulations**

AR makes real-life experiences better because it overlays real-world information on top of real-world scenes. Utilizing this capability enable users test potential effects of certain action towards the environment. For instance, the effects of pollution on wildlife or the need to recycle can be depicted by AR applications and it can be created by simulations that enable the recreation of a real life view (Augmented Island Studios 2023; Mosca et al. 2024). Environmental sustainability is among the most significant areas benefiting for the fact that AR provides a tangible representation for abstract ideas.

3. **Educational Impact**

Education becomes always effective through the use of VR as well as AR technology. With the help of VR technology it is possible to organize virtual field trips to distant ecosystems whereas AR can be helpful in making dull texts of books to come alive. These technologies elicit active participation from users thus enhancing the appreciation of sustainable practices in the corporate world (The Learning Guild 2024). The adoption of the AR technology tap into gamified application is preferably the right way, which engulfs the needed enhancement of commitment towards environmentally sustainable practices among users (INAUGMENT.com. 2024).

4. **Emotional Engagement**

Inculcating emotional bonds towards the problems of the environment is imperative facilitate behaviour change towards sustainability. There are also advantages of VR experiences such as the 'walking in a shoe' of endangered species or the consequences of natural disasters which go a long way in changing the user's perception and possible actions towards ecological conservation (Augmented Island Studios 2023; Scurati et al. 2021).

5. **Community Engagement**

Moreover, through VR and AR, the users are engaged to perform the action or make a decision affects the environmental topic collectively. For instance, VR platforms can mimic the town hall meetings, in which several users can discuss solutions to local environmental issues. Such approach does not only ensure awareness but also fosters the capacity of community members to own their environment's future (Virtual Reality 2024).

6. **Future Directions**

Hence, in the future, VR and AR appear to be even more decisive for the promotion of sustainable behaviours. Developments in these fields may result to more complex interactions and simulations that enhance the understanding of ecological matters and sustainable measures. More effort must be dedicated to constant improvement to harness the potential of immersive technologies in steering change for the environment.

8 Theoretical Frameworks

The knowledge of the levels of cognitive and emotional involvement in sustainability activities can be boosted through a various theoretical frameworks. These frameworks make it easier to dissect behaviours that is individual and collective towards sustainable practices. The synthesis of cognitive theories, emotionally related approaches, environmental education theories and models, behavioural change and participatory models offers a strong platform for promoting positive attitudes towards sustainable development.

1. **Cognitive Theories**

Cognitive theories largely centre on how psychological/dental processes affect sustainable behaviour or sustainable beliefs. One of the famous theories in this field is the Theory of Planned Behaviour, which notes that the persons would adopt the sustainable behaviour if they had the perceived behavioural control, Prototype and Consistent Social Norms. This specific model highlights the issues of knowledge, attitude and perceived control as the major factors that would prompt people into taking behaviours that are pro-environmental. Such insights encourage the stewardship approach by educating people on consequences of their behaviours on environmental systems in order to enhance change to sustainable behaviours.

Cognitive theories that emphasize knowledge acquisition and mental processing of sustainability information can inform businesses on how to design resilient strategies. Businesses can use these theories to understand how consumers and employees perceive and internalize sustainability efforts, which in turn influences long-term resilience.

2. **Emotional Engagement Perspectives**

There is enormous emphasis on people's emotional involvement for engaging in sustainable activities throughout their lives. Theories like the Affective Events Theory will show how people's reactions do affect the level of participation. When people are made to emotionally care through narratives, appeal to empathy, touch feelings and use of related incidences that embrace environmental concerns, changes are likely to occur. Praising feelings such as hope and pride regarding sustainability makes people shift their behaviours and become more supportive of sustainable practices. According to this approach, campaigns that seek to appeal to people's emotions may improve commitment and participation towards sustainable development (Papaevangelou et al. 2024).

Emotional engagement is critical for businesses that aim to build resilience in the metaverse. Companies can use insights from emotional engagement frameworks to craft experiences that not only educate but also motivate stakeholders to support sustainability initiatives, reinforcing the business's adaptive capacity.

3. **Environmental Education Frameworks**

The frameworks for EE involve both intellectual and affective components in order to raise awareness and to foster positive involvement in ecological concerns. These frameworks use on-line experiential understanding in which learner engages in projects that create understanding in the environment. Vital sustainability problems are usually complicated and personal so that this method enables obtaining a profound individual understanding of these issues. When people can view the occurrence of events in the light of big environmental themes, one will experience preparedness and willingness to be responsible for sustainable processes. Ensuring that the environmental education delivered at the formal and informal levels is harmonised, improves on the creation of the ecosystems supporting awareness of sustainable living (Chen and Law 2023).

Businesses that engage in virtual environmental education help shape the sustainability mindset of future consumers and employees. By fostering a culture of sustainability through education, businesses can ensure their resilience in the face of changing environmental standards and consumer expectations.

4. **Behavioural Change Frameworks**

Behavioural change frameworks are used to establish interventions towards overcoming the barriers to engaging in pro-environmental behaviours. For instance, there is the Transtheoretical Model that defines stages of change that assess a person's willingness and preparedness to adopt sustainable behaviours (pre-contemplation, contemplation, preparation, action, maintenance). This model is essential when coming up with a communication plan which can make a positive impact on people at different levels of change. These frameworks may also be helpful for educators and policymakers to use in order to improve motivation and to design interventions for promoting behavioural change for sustainability (Zhao 2024).

Encouraging sustainable behaviors within virtual environments can drive long-term resilience for businesses. Behavioral change frameworks allow businesses to create metaverse-based experiences that promote sustainable consumption, reduce environmental impact, and enhance adaptive capabilities.

5. **Participatory Frameworks**

Contribution-based frameworks are based on the principle of getting the communities involved in sustainable practices. Because they promote dialogue between the different stakeholders, these frameworks facilitate decision making and decision execution, which might result into sustainable practices. For example, participatory action research allows people of a given community to be empowered with instruments to estimate their needs and preferences towards sustainability. However, it empowers the owners of initiatives and boosts the emotional and cognitive involvement of the individuals in sustainability endeavours, which will ensure that they are interested in the results of the processes. In general, the work with participation adds to the discourse regarding sustainability and enables communities to act in unison (Engel et al. 2023).

Participatory frameworks, which emphasize collective action, are particularly important in building business resilience. By involving consumers, employees, and virtual communities in sustainability initiatives, businesses can leverage diverse perspectives to co-create solutions that ensure both environmental and organizational resilience.

9 Narrative-Driven Experiences

Torytelling within the metaverse is a powerful tool for businesses to engage stakeholders emotionally with sustainability. Businesses that craft narrative-driven experiences focused on sustainability not only educate but also inspire action, creating a resilient customer base that supports the company's sustainability journey. These stories help businesses align themselves with consumer values, thus enhancing brand loyalty and adaptability.

1. **Importance of Narrative-Driven Experiences**

One of the great tools of implementing concepts related to sustainability is storytelling because it emphasizes the narrative characteristics of the experience. It can bring about emotional appeal of the material to the target audience; this will make them understand and even remember more about sustainability. These may be in the form of games, media that embed the students into the story being told or even the conventional narratives of a story with purpose of placing the student.

2. **Storytelling Techniques for Engagement**

Using stories and particularly characters, it is possible to grab attention and explain such important and often cumbersome sustainability issues. Cox (2024) shows that the use of character-based storytelling can enhance people's perceived risk towards themselves, helping them to get more involved with sustainability issues and practices. It is therefore important for narrators to incorporate perceivable characters and issues which will enable the audience to change their behaviour towards environmental issues.

3. **Behavioural Change through Education**

Popular narrative approaches have been useful at learning institutions especially for the purpose of improving pro-environmental behaviours. Storytelling enables place-related value socialisation and empathy toward nature which are important in creating sustainable behaviour change (Liu 2023; Ouariachi et al. 2020). As the audiences hear and 'feel' with the tales, the subject opens an avenue in which the population is willing to embrace the behavioural shift required in environmental conservation.

4. **Case Studies and Applications**

There are many examples of successful use of storytelling approaches we can submit in sustainability projects that have yielded positive results. For example, it has been established that sustainable cantered approaches that encompasses storytelling can bring in the short term shift in behaviour, which is a demonstration of how theory can be put into practice (Parnell 2022). All these case-studies pointed to the possibility of such strategies which will create awareness and also ready-made models to sensitise the target-group to make better eco-friendly decisions.

5. **Challenges and Considerations**

As with any type of brand communication, narrative-driven experiences have two great stumbling blocks: impact and target reach. Effectiveness of such experiences can be moderate or high depending on context and per attendees' of the events' values or beliefs. In addressing the masses, it is important to frame stories in ways that appeal to different demographic populations as Klaebe suggested in 2012.

10 Virtual Communities and Collective Efforts

Virtual communities offer businesses opportunities to co-create sustainability solutions, which in turn build resilience. By working collaboratively with stakeholders in the metaverse, businesses can crowdsource ideas, test sustainable innovations, and form partnerships that strengthen their capacity to withstand disruptions. Collective efforts help businesses remain agile, adaptable, and resilient in a rapidly changing digital and environmental landscape.

1. **Importance of Virtual Communities**

Virtual communities can be considered essential since people with like-minded interests can join in the communities and work together. These communities let the participants to come together and get acquainted with one another since they form part of different communities and this is very important for group work (Law and Chang 2008). They not only help those concerned to enrich the γνώση they possess as discrete gregarious individuals but do even foster the internal interferential solidarity and effectiveness of the group.

2. **Trust in Virtual Communities**

Trust is assumed to be one of the major prerequisites of virtual communities. It would be a factor encouraging sharing of knowledge, resources and support within and between users, which is vital in collaborative work in realization of the SDGs (Hsu et al. 2011). This means that there can never be unity and coherence in the provision of services and delivery of tasks since trust is an essential component that should be encouraged among the people in the community by the leaders.

3. **Development of Social Identity**

Social identity is an important determinant of the willingness to contribute to virtual communities. According to Zhao et al (2012), presence of positive attitude towards the activity and positive self identification of the member with the community will lead to active participation in the activity in order to achieve set goals. This sense of belonging not only promotes commitment from a particular individual but also goals of the community and thereby unearths productivity in group settings as people strive to meet community goals.

4. **Facilitation of Collaborative Efforts**

Knowledge sharing is made easier with the help of virtual communities as these give a framework to work in collaboration. It is through the technology-based social capital builders such as identity profiling and community sub-group that members are able to build good working relationships, which in turn support the intended knowledge-sharing among members (Law and Chang 2008). These interactions can convert to tangible goals and policies within the framework of the SDGs hence increasing the collaboration impact.

5. **Achieving Sustainable Development Goals (SDGs)**

The fit of trust and social identity can together positively impact actions towards realizing the set SDGs in the virtual communities. With participants working and learning together, they can meet higher order world problems such as poverty, inequity and environmental degradation. Such a collaborative strategy not only tools up people's functions, and enhances an individual's sense of responsibility to contribute to the society; it also creates public participation and ensures that everyone plays a role to contribute towards sustainable development (Zhao et al. 2015).

11 Strategies for Businesses and Policymakers

As businesses navigate the complexities of the virtual world, it is crucial to adopt sustainability strategies that not only address environmental challenges but also enhance organizational resilience. By integrating cognitive and emotional dimensions into their approaches, businesses can foster deeper connections with stakeholders and create impactful sustainability initiatives. Policymakers also play a vital role in this process, as effective regulations and incentives can support businesses in adopting sustainable practices. The following strategies outline how both businesses and policymakers can strengthen resilience through cognitive and emotional engagement.

1. **Utilizing Behavioral Nudges**

Organizations can give directions that will result in the adoption of sustainable behaviour while at the same time offering choices to people. For instance, presets in virtual environments can be set to the ecological parameters, for example, choosing efficient energy avatars or encouraging the presence of green zones in virtual environments. A study shows that the minor changes of environment can have a strong impact on decision-making and encourage people to make sustainable decisions (Anshari et al. 2022). The laws that policymakers can foster for this kind of practice within the metaverse are assured to be helpful.

2. **Gamification of Sustainable Practices**

Engagement management in the metaverse is possible with the help of gamification as an effective instrument. Companies can come up with games or challenges that ensure people engage in environmentally responsible behaviours like recycling e-Resources. Incorporation of point systems, the leaderboard, and virtual incentives means that the users are encouraged to practice sustainable behaviours (Anshari et al. 2022). Policy makers can encourage organization to come up with these interventions as part of the general strategies of sustainability.

3. **Data Analytics as a Tool for Personalization**

It is possible to use the big data to appeal to the people's behaviours and concerns in an effort to bring about the desired sustainability goals. Through observation of user engagement, a businessperson is in a position to provide special alerts that call for energy saving or practices related to sustainability of the environment that concerns the users (Anshari et al. 2022). There is need for policymakers to encourage the emergence of ethical standards for use of data to enhance on transparency and user permission.

4. **Education and Awareness Campaigns**

Education has a significant part in developing behavioural destinies that are sustainable. Companies within the metaverse can organize virtual classes to explain different problems that having an effect on the environment and necessity of the eco-oriented life. Also, the development and application of ground-breaking technologies will improve the learning process by offering participants a virtual look into the consequences of their actions and decisions (Anshari et al. 2022). Policy makers can contribute in such initiatives by financing and encouraging education projects that deal with sustainable principles with regards to the virtual world.

5. **Creating Community-Based Initiatives**

The feeling of togetherness can go a long way in supporting sustainability. There should be another way in which business help people create a community engagement where the users of the metaverse share success and practices of sustainability. This form of communication with the peers can improve the levels of accountability and related motivation amongst the people (Anshari et al. 2022). Policy makers can support promotion of funding in area of community development projects in virtual environment for sustainable living.

6. **Implementing Virtual Environmental Policies**

The responsibility of putting in place the measures that will spell out how sustainability in virtual environments should be affected and lies in the hands of policymakers. Putting into place legislation that requires businesses in the metaverse space to conform can go a long way in realizing behavioural changes. It could involve demanding that Internet properties must operate with net-zero emissions or forcing ethical use of data that should protect the environment (Anshari et al. 2022). Organisations are forced to address new forms of regulation that are developing in these systems.

7. **Partnership with Environmental Organizations**

Ownership of virtual assets by businesses and products in the metaverse can be protected through partnership with environmental organizations which also increases credibility and information exchange. Strategic partnerships result in a concept that suggests the creation of long-lasting digital goods and services, which incorporate biological aspects of ecology into virtual space (Anshari et al. 2022). These relationships may be initiated at a policymaking level through the creation of forums where sector actors can meet, negotiate and share information.

8. **Implementing Virtual Reality for Experiential Learning**

This is because using of VR technology can render significant experiential learning experiences. Some examples are: Businesses can provide humanized scenarios that involve real-life action and visually show the impact of non-sustainable practices so as to invoke user's sympathy and bring about a sustainable shift in his/her behaviour over time (Anshari et al. 2022). It may help when policy makers can provide incentives towards development of VR applications, specifically, in the field of sustainability education, so such technologies can become popular.

References

Anshari M, Syafrudin M, Fitriyani N, Razzaq A (2022) Ethical responsibility and sustainability (ERS) development in a metaverse business model. Sustainability 14(23):15805. https://www.mdpi.com/2071-1050/14/23/15805

Augmented Island Studios (2023) Augmented reality and virtual reality: a game-changer for environmental awareness—augmented island studios. Augmented Island Studios. https://augmentedislandstudios.com/augmented-reality-and-virtual-reality-a-game-changer-for-environmental-awareness/

Chen B, Bao L, Zhang R, Zhang J, Liu F, Wang S, Li M (2024) A multi-strategy computer-assisted efl writing learning system with deep learning incorporated and its effects on learning: a writing feedback perspective. J Educ Comput Res 61(8):60–102. https://doi.org/10.1177/07356331231189294

Cox E (2024) My risk or theirs? An experimental investigation of character-driven and plot-driven engagement as explanations for risk perceptions. All Theses, p 4303. https://open.clemson.edu/all_theses/4303

Engel K, Kilmartin PA, Diegel O (2023) Flexible and multi-material intrinsically conductive polymer devices fabricated via DLP and DIW additive manufacturing techniques. Rapid Prototyping J 29(10):2164–2175. https://doi.org/10.1108/RPJ-02-2023-0037

Gesthuizen R, Tan H, Kidman G (2024) Inspired to build understanding: learning about sustainability and environmental education from within a virtual world. Int Res Geogr Environ Educ 1–18. https://doi.org/10.1080/10382046.2024.2348265

Grabs J, Langen N, Maschkowski G (2016) Understanding role models for change: a multilevel analysis of success factors of grassroots initiatives for sustainable consumption. J Clean Prod 134(A):98–111. https://www.sciencedirect.com/science/article/pii/S0959652615015280

Hsu MH, Chang CM, Yen CH (2011) Exploring the antecedents of trust in virtual communities. Behav Inf Technol 30(5):587–601. https://doi.org/10.1080/0144929X.2010.549513

Immersive Reality: Unleash Augmented, Virtual, and Mixed Reality : Learning Solutions | The Learning Guild (2024) Learningguild.com. https://www.learningguild.com/articles/immersive-reality-unleash-augmented-virtual--and-mixed-reality-/

Iyer S (2023) Examining United Nations VR content production usage of experiential media in climate change storytelling: a qualitative analysis. Malays J Media Stud 25(2):51–66. https://tamilperaivu.um.edu.my/index.php/JPMM/article/view/39442

Klaebe H (2012) Framing local stories for community consumption facilitation and evaluation in narrative driven arts-based projects. In: Biennial International Auto/Biography Association (IABA) conference, 2012-07-17–2012-07-20

Laukkanen T, Xi N, Hallikainen H, Ruusunen N, Hamari J (2022) Virtual technologies in supporting sustainable consumption: from a single-sensory stimulus to a multi-sensory experience. Int J Inf Manage 63:102455. https://doi.org/10.1016/j.ijinfomgt.2021.102455

Law SP-M, Chang MK (2008) Fostering knowledge exchange in online communities: a social capital building approach. In: ICIS 2008 proceedings, p 173. https://aisel.aisnet.org/icis2008/173

Liu J (2023) The influence of narrative transportation on university students' environmental intentions: a serial mediation of empathy with nature and environmental attitudes. J Cleaner Prod 431:139763 https://www.sciencedirect.com/science/article/pii/S0959652623039215

Mosca O, Manunza A, Manca S, Vivanet G, Fornara F (2024) Digital technologies for behavioral change in sustainability domains: a systematic mapping review. Front Psychol 14:12343491234349. https://doi.org/10.3389/fpsyg.2023.1234349

Ouariachi T, Li C-Y, Elving WJL (2020) Gamification approaches for education and engagement on pro-environmental behaviors: searching for best practices. Sustainability 12(11):4565. https://doi.org/10.3390/su12114565

Papaevangelou O, Soultana R, Syndoukas D, Kalogiannidis S (2024) Applying rhetorical methods to teach writing techniques to pre-service teachers. Int J Educ Pract 12(2):406–420. https://archive.conscientiabeam.com/index.php/61/article/download/3694/8018

Parnell C (2022) Designing narrative-based interfaces for collective action: a case study using Amazon, climate change, and consumer behavior. Dartmouth College Undergraduate Theses, p 269. https://digitalcommons.dartmouth.edu/senior_theses/269
Scurati GW, Bertoni M, Graziosi S, Ferrise F (2021) Exploring the use of virtual reality to support environmentally sustainable behavior: a framework to design experiences. Sustainability 13(2):943. https://doi.org/10.3390/su13020943
Spangenberger P, Freytag S-C, Geiger SM (2024) Embodying nature in immersive virtual reality: are multisensory stimuli vital to affect nature connectedness and pro-environmental behaviour? Comput Educ 212:104964. https://doi.org/10.1016/j.compedu.2023.104964
Virtual Reality (2024) How can you use VR to increase environmental awareness? Linkedin.com. https://www.linkedin.com/advice/3/how-can-you-use-vr-increase-environmental-awareness-qsqhc
Wong J (2021) The effects of interactivity on memory relating to presence in virtual environments. Psychological Science Undergraduate Honors Theses. Retrieved from https://scholarworks.uark.edu/psycuht/16
Written by INAUGMENT.com. (2024) Using AR for environmental sustainability. Medium. https://medium.com/@ai4all.linkedin/using-ar-for-environmental-sustainability-dcd48905c1c2
Zhao J, Wang T, Fan X (2015) Patient value co-creation in online health communities: social identity effects on customer knowledge contributions and membership continuance intentions in online health communities. J Serv Manag 26(1):72–96. https://doi.org/10.1108/JOSM-12-2013-0344
Zhao L, Lu Y, Wang B, Chau P, Zhang L (2012) Cultivating the sense of belonging and motivating user participation in virtual communities: a social capital perspective. Int J Inf Manag 32(6):574–588. https://www.sciencedirect.com/science/article/pii/S0268401212000278
Zhao S (2024) Research on the design of a system based on machine learning algorithms for automatic scoring of English writing ability. Scalable Comput Pract Experience 25(5):4005–4012. https://www.semanticscholar.org/paper/2ee28c16da659e4a9a1a1a66117cfee5905e1fdd

Leveraging Sustainability and Business Resilience: Integrating Climate Change Mitigation and Adaptation for Achieving SDGs

Prashant C. Ramteke and Abhishek Prakash Paswan

Abstract Climate change is a significant topic that has to be presented concerning sustainable development and/or business resilience to achieve the general objectives of sustainability. This chapter aims at showing how organizations can build climate resilience, by integrating sustainable ideas, anchored in sustainable development goals (SDGs). Reducing emissions is the goal of avoidance while preparing to address the impacts is the goal of adaptation. These actions should be included with a judgment towards sustainability and it means investing in renewable energy, energy efficiency, and building structures resilience to climate change. In addition, strategies of justice have to be incorporated into business responses so that climate change response is just and fair. The chapter also scrutinizes how new solutions, for example, implementing the Metaverse for effective virtual collaboration and sustainable decision-making can lead to more profound sustainable resilience and development. There are many possibilities for businesses to take action on climate change and enhance its stability even when there are limitations within financial and technological provisions. Businesses shipping goods internationally have got to do much more to ensure they address the worse impacts of climate change and help bring fairness to the climate system. The prioritising of such efforts will help not only in supporting businesses to advance outcomes in climate change adaptation but also guarantee the role of businesses in helping to bring more justice to the climate system. This brings us to the end of the chapter which also points out that further

P. C. Ramteke · A. P. Paswan (✉)
Department of Civil Engineering, Delhi Technological University, New Delhi, India
e-mail: abhishek_mt2k16@dtu.ac.in; appce@mmmut.ac.in

P. C. Ramteke
e-mail: prashant_2k16phdce12@dtu.ac.in

P. C. Ramteke
Department of Civil Engineering, G. B. Pant DSEU Okhla Campus, New Delhi, India

A. P. Paswan
Department of Civil Engineering, Madan Mohan Malaviya University of Technology, Gorakhpur 273010, UP, India

W. Leal Filho et al. (eds.), *Metaverse and Sustainability: Business Resilience Towards Sustainable Development Goals*, World Sustainability Series,
https://doi.org/10.1007/978-3-031-89545-6_6

assessment is needed in order to understand the effectiveness of such schemes and their compatibility with the goals for global sustainability.

Keywords Climate change · Mitigation · Adaptation · Renewable energy · Resilience · Integrated approach · Financial resources · Financial barriers

1 Introduction

Possibly, the most prominent problem of the twenty-first century is climatic change caused by greenhouse gases such as carbon dioxide, methane, and other fumes from activities like fossil fuel burning, deforestation, and manufacturing. This chapter identifies it as a major threat to ecosystems, economies and the well-being of people globally (Abbass et al. 2022; de Souza et al. 2024; Ray Biswas and Rahman 2023). The overwhelming consensus of scientific opinion is that the global climate is warming at a pace unprecedented in Earth's history, with potentially profound impacts on the various natural systems of the planet. The IPCC (Intergovernmental Panel on Climate Change) says that if no aggressive action is undertaken, the average global temperature will stand at 1.5 °C above the pre-industrial level (Allen et al. 2022; Kikstra et al. 2022; Naran et al. 2022). Various consequences due to climate change such as frequent heatwaves, droughts, floods, storms, sea level rise and disruption of ecosystems and biological diversity are other implications of this warming process (Ray Biswas and Rahman 2023; Semenza et al. 2011). Two complementary strategies have emerged as key climate change tactics in response to the increasing urgency of the climate crisis: adaptation and mitigation. Mitigation's objectives are to minimize greenhouse gas (GHG) emissions to eliminate or lessen the impact of climate change by addressing the sources of global warming (Allen et al. 2022). This implies a transition from fossil fuel energy systems to renewable energy sources, increasing energy efficiency, enhancing carbon sink through afforestation and reforestation and utilizing CCS within the atmosphere (Zute et al. 2024). The short-term objectives include the global emission reductions aiming at the extreme outcomes including climate stabilisation which, among other things, the Paris Agreement refers to in terms of limiting the global temperature rise to well below 2 °C above pre-industrial levels (Rogelj et al. 2016).

Adaptation's primary function is to develop or enhance capacities to better cope with current or predicted consequences of climate change effects, while mitigation seeks to eliminate the processes resulting in change in the climate. The term "adaptation" generally encompasses a set of strategies intended to reduce risk and enhance the ability of communities, ecosystems and infrastructure systems to withstand and recover from climate-related risks (Baninla et al. 2022; Ray Biswas and Rahman 2023; Semenza et al. 2011; Tedesco et al. 2024). These include the measures which involve modifying the farming techniques that farmers can practice to mitigate climate change such as changes in rainfall regimes, ecological systems that are relied on to conserve resources from climate hazards include; and improving ways in which the climate information and early warning systems are enhanced enabler mediators farmers to act quickly and adequately in climate risks include. Appreciation of the inter-linkages between social, economic, political and ecological systems as well as the interdependency between human society and the environment is crucial while addressing challenges of climate change mitigation and adaptation (Abbass et al. 2022; Lafontaine et al. 2010). Furthermore, it means that the governments, as well as various business companies, NGOs operating at the civil society level, and common citizens, have to act in unity and synergy at the local, national and global levels to address the climate crisis. It requires innovation, the revolution of paradigms of consumption and behaviour and social reinvention of the connection with nature.

In this context, the purpose of this chapter is five-fold: to describe a range of mitigation and adaptation measures, challenges related to their implementation, and the role of an integrative approach in climate change policies and practices. This chapter also relies on the latest scientific literature, policy discussions, innovations and cases to add to the growing literature knowledge on climate change resilience and sustainability. In this sense, this chapter seeks to make dissemination and knowledge contribution to policymakers, practitioners and stakeholders by using interdisciplinary and intersectoral perspectives and experiences in approaching and applying climate resilience. Finally, its goal is revolutionary for a future that is more sustainable, stable, and just for all of society.

2 Six-Step Guide to Designing Adaptation Initiatives

The process of developing an adaptation initiative can be presented as the six-step model described in detail in Fig. 1, which would provide for a clear identification, planning, and practical application of the adaptation initiative (Lafontaine et al. 2010). In order to suggest how much each of the steps is significant, let us consider it further:

2.1 Defining the Problem

This step of climate change adaptation is all about identifying the specific hazard and risk of a certain area or sector that is prone to climate change effects. Potential sources of uncertainty herein include the likelihood of more frequent intensified natural disasters, changes in sea level, or access to water sources. As a result, stakeholders can collect resources and put priorities for adaptation to focus on the most challenging issues by defining the problem as much as possible.

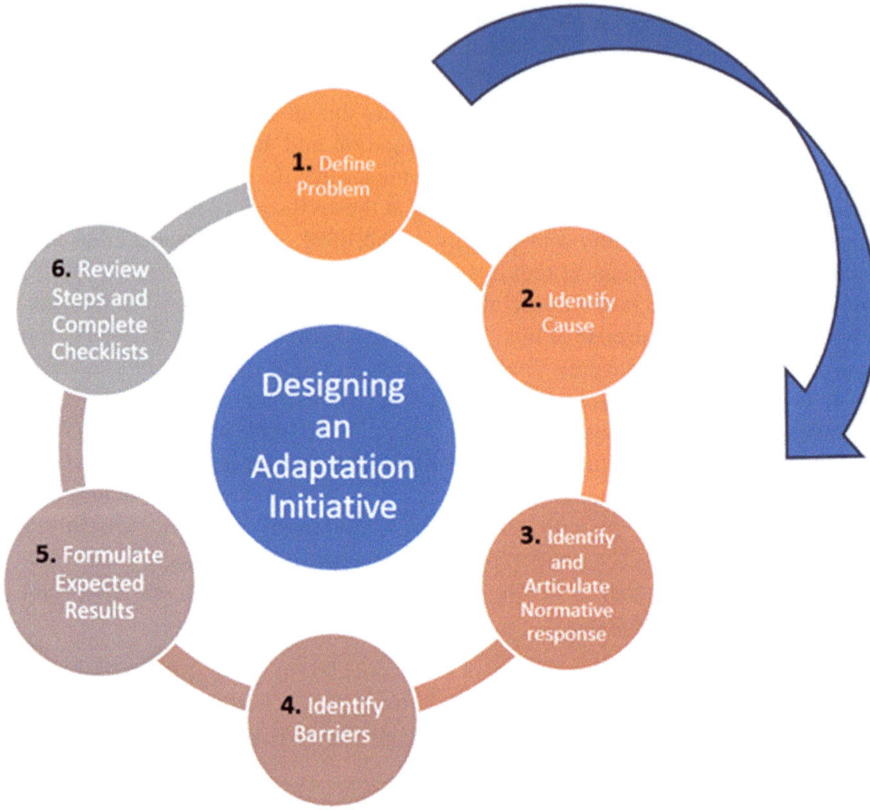

Fig. 1 Six steps for designing an adaptation initiative

2.2 Identifying the Causes of the Problem

After the issue has been identified, stakeholders have to understand the basic causes of the climate-related risks and vulnerabilities. This may require the assessment of fragmentation and density in relation to land use patterns and socioeconomic characteristics as well as estimating risks and using climate change scenarios. Identifying the causes helps the stakeholders to set strategic management strategies development that address the causes and not the effects only.

2.3 Identifying and Articulating the Normative Response

The goals of the adaptation initiative and the intended outcomes are defined and communicated by stakeholders as a result of having considered the problem at hand and its causes. This involves setting of adaptation targets for instance the need to enhance resilience, to reduce vulnerability and to maintain/protection of livelihoods. The adaptation initiative can be directed by evaluating the normative response because the guideline ensures that the approaches taken are within its purpose and mission.

2.4 *Identifying Key Barriers*

The targeted and multiple stakeholders carry out an assessment of the barriers to successful adaptation initiatives. These challenges may include bureaucratic barriers, knowledge limits, limited resource access or other competing needs. It also assists the stakeholders in formulating means of coping with the main challenges towards enhancing the possibility of accomplishment of the adaptation interventions.

2.5 *Designing Project Responses to Overcome Key Barriers*

Based on the latter barriers, stakeholders develop specific adaptation interventions for reducing or erasing them. It may mean implementing CAB clients' adaptation measures, improving infrastructure, or using NBS. Developing project responses ensures that adaptation strategies are targeted to address specific issues and to capitalize on opportunities for enhancing vulnerability.

2.6 *Reviewing the First Five Steps and Completing Checklists*

In the final step of the adaptation planning, stakeholders reflect and evaluate the planning with respect to the identification of all issues. This means having a consultation with the stakeholders, deciding whether the proposed intervention should be realistic, and also doing some final touch up work on the due diligence process. Whenever a checklist is completed, it means that adaptation projects are conducted according to the standards of adaptation planning and implementation and compliance with the regulatory acts and fundamental requirements for financing.

In general, these six steps contain an effective procedure for designing adaptation measures and providing concerned stakeholders with more structured approaches to the identification of the problem, facilitating the process of coming up with the solutions and coordinating the process of their implementation. Conceptualising the complex and evolving nature of climate change adaptation, the process is highly consultative and flexible.

3 Overview of Climate Change and Its Impacts

The level of climate change as well as the current weather patterns are majorly a result of several activities carried out by human beings; this is usually called "climate change" (Awuni et al. 2023; de Souza et al. 2024; Stehr and von Storch 2005). The result of decisions, pursuits, and actions such as burning fossil fuels and cutting down trees raise the global temperature. The temperature rise comes with many issues including heat wave intensification and the ice melting in areas like the North and South Poles (Allen et al. 2022). This brings about the next effects such as the emergence of high levels of sea, which are levels that pose threats to shore regions. In addition to enhancing the occurrences of severe weather events like flooding and storms, climate change is causing changes in the locations of plants and animals so that some species are facing out. Several problems being experienced in the process of growing crops for farmers are culminating in food scarcity and an increase in hunger. Besides, diseases are also spread through the various surrounding conditions and the air and water are also polluted through climate change. The worst effects of climate change are being borne by needy people and countries that lack resources or development. Essential steps are required to prevent the worsening of the situation and to assist communities in handling its consequences.

3.1 Importance of Mitigation and Adaptation in Addressing Climate Change

Climate change response involves two main approaches namely adaptation and mitigation. Avoiding or limiting the amount of greenhouse gases that cause climate change is called mitigation (Allen et al. 2022). By using this strategy, it would be possible to reduce the impacts of climate change's worst effects such as storms and, especially, steadily rising seawater levels, as well as the increase in the overall heat on the planet. To do so, there will be a need to shift from using fuel that pollutes such as coal to renewable sources of energy such as solar and wind energy. This can benefit the environment and also support life to be called sustainable life. On the other hand, adaptation means proactively preparing for the various changes that result from climate change effects. Since we want our communities, environment, as well as infrastructure to survive incidents like heatwaves and floods, whose impacts may be devastating, they require protection. It is most relevant to the least sustainable areas and communities, such as low-income neighbourhoods and coastal cities. For the forces of nature to be useful, we can emulate the example of the great floods by trying to recover forests and wetlands to protect ourselves from natural calamities and disasters (Zute et al. 2024). Adaptation and mitigation are both needed, solely because both are required. While mitigation lets us act on the primary cause by reducing emissions, adaptation prepares us with the tools to handle the ongoing alterations. They offer the highest potential for building a safer and more sustainable world in the future.

The purpose of this chapter is to explore effective solutions for mitigating climate change. It focuses on two strategies: The examination of the measures of preparation for and of protection against its effects makes a splendid read. This paper seeks to explore ways in which climate change can be addressed through, energy conservation, community enhancement, and climate protection. Further, it responds to issues that may hinder the process: insufficient sponsorship and technology are mentioned. However, it goes on from there. It prescribes how these difficulties may be overcome, by collaboration and creation of new and additional solutions. The study emphasizes the importance of the partnership approach and presents examples of integration. On top of that, it sort of informs those in charge of how to make the best decision possible in the climate change issue. The study wants the readers to come together in order to move forward in ensuring our future is safe from climate change and business resilience.

3.2 Climate Change Mitigation

To reduce the negative impact that climate change, different methods are employed as follows. Switching to renewable energy forms such as solar, wind, hydro and geothermal signals another important strategy. It means that it is necessary to create policy among such factors and advocate for these sources' use, despite challenges like investment and infrastructure updating. Also, energy efficiency is important since the probability of emissions is scaled down by using limited energy. This can be done, for instance, by adopting economical energy technologies within branches like industry and transport, and especially within buildings, where often policies, norms and incentives play a leading role. Also, afforestation and reforestation are an essential factor since forests assist in carbon dioxide control steps (Zute et al. 2024). However, there are challenges in the identification of suitable land, Delivering the message to the relevant communities and lastly funding for these undertakings. Moreover, carbon capture and storage carbon capture and storage (CCS) technology is used to capture and store carbon dioxide (CO_2) emissions that are emitted by power plants and industries. Although it faces some challenges during its implementation there are possibilities of integrating it with other further strategies of mitigation.

3.3 Climate Change Adaptation

Implementation of climate change therefore requires systematic strategies of how to approach it to ensure that the impacts are reduced across the identifiable sectors. Infrastructural robustness is important and has to provide for the construction and adaptation of structures in preparation for and handling of disasters and floods. We know that climate change should be an important consideration regarding the organizational development of urban territories. Due to variability in weather conditions and how it affects crops and animal production systems, there is a need to adapt to change. Since these changes are expected to occur, there has to be mechanisms put in place, to assist farmers employ dynamic farming practices to address these changes. Another element is ecosystem-based adaptation which is the protection and restoration of ecosystems for the purpose of climate change, such as coral reefs and wetlands. It helps to build their defence and provide fundamental products such as conserving bio-diversity and controlling floods. Other types of data are also important for adaptation because; early warning systems and climate data provide information to alert people when there is extreme weather. However, problems such as data sharing or lack of proper communication remain to be had, which is why proper funding and capacity development is still needed in this area. In conclusion, adaptation to climate change means the development of an efficient approach that will determine such threats and opportunities in different regions and fields.

4 Integrating Mitigation and Adaptation

What is even more significant is that adaptation and mitigation are to be integrated into a package approach for overall resilience against climatic change. By scaling up action against climate change impact and integrating cost-effective short-term and long-term mitigation measures and strategies for climate change adaptation, communities have the potential to enhance their coping capacity to bear the brunt of change. This integration involves the implementation of measures that also have the effect of preventing exposure to climate risks while also strengthening adjustment mechanisms and decreasing emission levels. Reducing greenhouse gas emissions, for example, enhances energy security and coping with fluctuations in the price of traditional sources of energy. Similarly, green investment raised in the form of Low Impact Development (LID), such as the use of permeable pavements or incorporation of broader green areas in the urban fabric, also suppresses the flood hazard, increases the biological diversity, and decreases the impact of urban heat islands. Other residual issues include the balance between adaptation and mitigation measures, where/when it is possible; and how social justice, equity and human rights impacts are pursued to ensure the benefits of climate action do not accrue disproportionately to the privileged. Lastly, adaptation and mitigation when combined form a better and more effective approach on how to tackle complex issues of climate change delivering lasting and sustainable impacts on human beings and the environment.

4.1 Synergies and Trade-Offs Between Mitigation and Adaptation Strategies

When it comes to climate change, one will observe that there are interactions and especially overlapping between the climate change mitigation and adaptation measures. These complementarities exist when gains from attaining the reduction objectives also contribute to the possibilities or to the needs of adaptation or when the reverse is true. For instance, tree planting can logically be for mitigation through capturing carbon dioxide and for adaptation through providing shade and reducing heat stress, especially with regard to cities and towns (Verkerk et al. 2022). However, a trade-off is understood as a situation when the application of one strategy violates the effectiveness or benefits of the other. For example, building a large-scale dam to develop a generator for hydroelectric power (impact mitigation) may affect riverine habitats along with water-sharing ability with other stations (impact adaptation). It is thus important to understand such complementary relations /substitutabilities/ interactions and competencies in order to develop coherent climate policies that can enhance the benefits of one policy while minimising the negative impacts of another.

4.2 Importance of Holistic Approaches that Consider Both Aspects of Climate Action

In order to combat climate change, a multidimensional approach, which will entail both adaptation and emission reduction measures for climate change, is sufficient. It is possible to integrate these two aspects of climate interventions and come up with elaborate plans that not only reduce emissions of greenhouse gases but also enhance the capacity of communities to withstand the impacts of climate variation. This approach recognises the mutual relationship existing between adaptation and mitigation in accordance with the knowledge that decisions made within either category may affect the other. For example, a shift to renewables enhances the security of energy supply and reduces dependence on the availability and price volatility of fossil fuels as well as emissions reduction (mitigation/adaptation). Likewise, green investment provides adaptation benefits such as the reduction of flood hazard, increase of species richness, and minimization of the effects of urban heat islands. It is possible to identify complementary areas that would allow minimising trade-offs and maximising the benefits of addressing both aspects of climate change at the same time. This will lead to results that are more sustainable for ecosystems and communities and that are also preferred by society.

4.3 Examples of Integrated Climate Policies and Initiatives at National and Local Levels

Unprecedented numbers of climate policies and strategies have been adopted at local and national levels. Thus, using the best strategy laid down in the following instance are beneficial in dealing with Climate change. A comprehensive strategy can work as shown in the following instances.

4.3.1 National Climate Action Plans

A large number of countries have developed National Climate Action Plans that reflect the combination of adaptation and mitigation measures. For instance, the Climate Change Act of the United Kingdom outlines targets for the least partitions of emissions of greenhouse gases and has sections on adaptation discussion that cover water resources and flood protection.

4.3.2 City Climate Action Plans

A large number of cities in the world have developed City Climate Action Plans. These strategies embrace procedures for adaptation and mitigation consistent with the locale. For example, Copenhagen in Denmark now has policies to reduce emissions by supporting public transport and sources of energy such as wind energy. They have also increased the 'hardiness' of cities and constructed additional parks and green areas to better contend with heat waves and rising sea levels.

4.3.3 Ecosystem-Based Adaptation Projects

Various countries have embarked on ecosystem-based adaptation project implementation. These aspirations can therefore be said to involve a combination of nature-based approaches with climate resilience programs. For instance, Draw-down Payment for Ecosystem Services in Costa Rica aims at enhancing the amount of carbon stock through incentivising forest protection and renewal. Other benefits are; the eradication of floods and the protection of wildlife and other flora and fauna species.

4.3.4 Climate Smart Agriculture Programs

Many countries have signed for Climate Smart Agriculture Programs. These programmes combine methods of enhancing the vulnerability of agriculture to climate change with techniques of mitigating emissions of greenhouse gases in agricultural operations. For instance, The Kenya Climate Smart Agriculture Strategy provides guidelines for the implementation of improved methods of irrigation, minimum tillage and trees on farms.

4.3.5 Integrated Water Management Strategies (IWMSs)

IWMSs have been developed as shown with the following examples in many other parts. These approaches incorporate numerous activities intended to reduce water consumption, improve water quality, and increase resilience to floods (Laidlaw and Percival 2024). For instance, the Room for the River program in the Netherlands consists of multiple restoration projects accompanied by flood mitigation processes. In turn, dual-purpose flood plains thus are created as flood mitigation and habitat improvement.

The aforementioned illustrations reveal on how integrated climate policies and initiatives are capable of mitigating the challenges presented by climate change in a very effective manner. These synthesis studies advocate for considerable improvement in dealing with this global issue if the mitigation and adaptation measures developed are formulated to suit specific environments and concerns.

5 Challenges and Opportunities

Thus, in order to counteract climate change a number of tasks have to be solved. These are; Technology, The economic constraints that hinder the financing of clean energy, social issues and the process of raising finances for adaptation in developing countries. Nevertheless, these obstacles present prospects for advancement: On the employment and ideas creation related to a categorical shift towards a sustainable economy, technological advancement provides remedies which can help decrease emissions; nature-based approaches apply economic solutions which have so many benefits; communities help build resilience; and policy leadership help bring change. In order to realise these opportunities, sectors and nations must come together in harmony, and then form coherent co-cooperation to address the complex issues of climate policies with innovative development of an enduring and sustainable world of tomorrow.

5.1 Financial Barriers

The control of climate change might just be hampered by financial factors. For instance, one may be financially constrained from investing in clean energy or retrofitting the building to be more resistant to climate change. That said there are however strategies that can be employed over these hurdles. There are some funds, for instance, the Green Climate Fund that assist developed countries to develop climate projects (Cui et al. 2014). Besides, the imposition of prices on carbon emissions can push companies to decrease pollution levels while financing climate projects. The funds needed to combat climate change can also be obtained through international assistance, climate, bonds and cooperation between governments and companies. Therefore, the mentioned funding strategies will help to fight against climate change and construct a better future for all.

5.2 *Technological Innovations and Research Priorities for Advancing Mitigation and Adaptation Efforts*

The research priorities and the technological improvements are core to adaptation and address to climate change mitigation (Agarwal et al. 2020; Baninla et al. 2022). Studies are being conducted to refine the utilization of renewable sources of electricity for mitigation to include; solar power, wind power, and water power among others. The goal that is sought is reliable, efficient, cheap, and versatile energy storage systems. The viability of carbon capture and storage (CCS) to capture and store carbon emissions from power plants and industrial processes, is also being studied. That is why it is necessary to conduct more research to increase scalability and efficiency. Some of the innovations that are being developed in the adaptation domain include precision farming methods and crops resistant to drought. These aim at minimizing emissions resulting from agricultural practices and enhancing climate change response capabilities (Zute et al. 2024). Moreover, evaluation of climatic risks and timely alerts are determined by early warning systems and climate information systems. Application of ecosystem restoration and green infrastructure are some of the nature-based approaches that are known to be under research at the moment considering the fact that they have benefits for both; adaptation and mitigation. Experiments are underway to investigate how effective they are and how to improve the ways they are delivered. Adaptive and planned development for construction also involves new engineering processes and structures for sheltering structures from climate hazards. To do this, research on integrated planning techniques and climate-resilient materials is required for this. By focusing on investment in research and development in these fields of paramount importance, we can stimulate a technological breakthrough in addressing the concerns caused by climate change. This will, in turn, pave the way for a future that can easily rebound and is sustainable.

5.3 Equity and Justice Considerations in Climate Change Responses

Idealism is part of the climate change intervention, as it ensures efforts towards addressing the impacts of climate change are fair and holistic. As it links to this knowledge, it implies taking into account the issue of climate injustices where the negative impacts of climate change are either not or are hardly is felt by vulnerable and marginalized groups such as Indigenous peoples, low-income, and small populations. Moreover, it involves guaranteeing fairness in the distribution of privileges and assets, structures, and amenities with special attention given to environmentally friendly structures and systems, water, and energy regardless of location or status at the base. This means that further advancement of energy justice requires more focus on measures to increase access to renewable energy solutions for those living in energy poverty. Moreover, it is also necessary to involve numerous stakeholders like native people, and communities located on the front line, while designing and implementing of climate policies to ensure that their voices are heard and considered. Besides, climate justice, which embraces principles of equity and justice for the affected poor and marginalized groups and communities, also requires: acknowledgement and acceptance of past responsibility of climate change and the disparate impact sufferers are experiencing. It is now expected that various adjustment actions must be done in line with the particular conditions of a region sensitive to the concerns of a community and provide priority to the vulnerable ones. In the economic consideration, equity is important in determining the allocation of costs and benefits arising from climate response measures. These concerns provide assistance to those communities and workers who lost jobs due to a decline in fossil fuel use and the promotion of sustainable jobs as a means of boosting employment opportunities that give economic prospects to everyone.

In particular, equity and justice must be addressed in climate responses with an aim of promoting equity and justice. This would ensure that everyone has something to gain from such endeavours and assist the most affected populace to improve their ability to cope with the effects of climate change.

6 Studies Importance and Its Benefits

In the prevailing global conditions where the impacts of climate change are being manifested in more dramatic and unambiguous ways, research on measures to combat the crisis and the ways of adapting to it is possible. In order to manage climate risks and promote sustainable development, there is a pressing need for comprehensive and integrated approaches, which is addressed in this chapter. The significance of this study is highlighted by the following several factors:

- *Taking on critical global challenges*: The matters of concern here are natural ecosystems, human beings, food, and water, today and in future, owing to climate change. Thus this study provides significant recommendations on how to deal with these emergent global challenges and reduce the risks to communities and ecosystems by focusing on, and on.
- *Encouraging decision-making and policy*: Credible information is key when policymakers want to set out strategies and policies that will address the issue of climate change. Based on the intent of the investigation, this study will also give useful advice to undertake local, national and interstate decision-making processes due to the fact that it compiles the latest scientific literature, policies, and case studies.
- *Encouraging integrated methods*: Properly, adaptation and the related mitigation are often entered chiefly as separate disciplines during which their integration is crucial to climate motion. As a result, this study encourages approaches that enhance both climate mitigation and adaptation in relation to climate change impacts. It does this by creating a dependency between these approaches.
- *Increasing Sustainability and Resilience*: Climate resilience and sustainable development are factors related to each other. This study contributes towards the achievement of the broader sustainable development goals by exploring the management of climate change through mitigation and adaptation management. This way it is possible to guarantee that climate change mitigation measures also facilitate economic growth, social integration and conservation of the environment.
- *Encouraging stakeholders and communities*: Consequently, the impact of climate change varies from one community or region to another, meaning that special measures need to be taken. The emphasis of this study therefore rests on the implication of communities and stakeholders in the formulation and implementation of climate change strategies since this affords the local people a chance to give inputs that would inform pragmatic and equitable solutions.
- *Boosting creativity and initiative*: Efficient examples of how and what can be done to mitigate the impact of climate change and to adapt to the changes are shown in this study hence fostering the use of creativity in solving this issue. It enlightens and challenges stakeholders to take action and build on successful models brought by other approaches, thus motivating the implementation of a change.
- *Creating a Knowledge Base to Support Upcoming Research*: This work thus provides a basis for future analysis of climate change and contributes to the constantly growing body of literature on the subject. By pointing out gaps, challenges, and promises in prior scholarly work, it clears the path to future advancement in the field of climatology and policy making.
- *Business Resilience*: Through a view towards sustainability, the study gives information on how the operators can operate more securely against the problems of climate change and weather conditions to ensure long-lasting stability.

- *Alignment with SDGs*: This serves to promote the consideration of and adherence to Sustainable Development Goals within business management activities and where appropriate to SDGs within the climate action, affordable clean energy and responsible consumption and production family of goals.

In other words, this study will address sustainable transition and business resilience in the context of the Metaverse by intertwining climate change mitigation and adaptation. It responds to the increasingly recognised need for an innovative yet rigorous approach to realising the Sustainable Development Goals (SDGs). In that regard, the study is an invaluable tool for fostering cooperation and offering actionable strategy-oriented insight for businesses, communities, researchers as well as policymakers to develop a more resilient sustainable environment cornerstones to exist within the growing digital world.

7 Conclusion

The study has identified significant strategic actions of adaptation and mitigation that are essential in addressing the impacts of climate change. Combating greenhouse gases and the prevention of the factors that cause climate change are the aims of the mitigation measures which incorporate; afforestation, capture and storage of carbon, energy efficiency measures, and the promotion of the use of renewable energy. Adaptation is the process of developing capacities to address the impacts of climate change, and the objectives of adaptation are long-term, as seen in Climate Information Systems, Infrastructure Resilience, Agricultural Adaptation, and Ecosystem-based Adaptation. The problem of climate change does not solve itself, if we want our globe to be livable for future generations, we need to act fast and in unison. The conclusions underline the importance of international cooperation in the fight against climate change and the attainment of a sustainable climate. The problem of climate change requires international cooperation for its solution, and the participation of every nation, company and society. It is beyond us to solve this problem and ensure people's safety and stability for future generations if we do not act as one. In order to adopt big-wattage climate policies, promote innovation and lock in resources for the reduction of emissions, manage the impacts of Climate change and establish a sustainable world all the stakeholders such as government agencies, corporations, civil society and the people are required to work in unity. Climate change is a real threat, and it is incumbent upon all citizens to come together in order to ensure that there is a positive future for future generations to inherit. Moreover, the metaverse offers businesses a great opportunity to embark on a sustainable course. This, allows them to create, cooperate and also connect with the SDGs as well as address the complexities of climate change.

Conflict of Interest In this research, the use of any organization or institution's property was not offered by the authors, and that there are no conflicts of interest as such.

Competing Interests There is no actual or potential conflict of interest to report for each of the authors contributing to this article.

Funding Statement This research did not receive any specific grant from funding agencies in the public, commercial, or not-for-profit sectors.

References

Abbass K, Qasim MZ, Song H, Murshed M, Mahmood H, Younis I (2022) A review of the global climate change impacts, adaptation, and sustainable mitigation measures. Environ Sci Pollut Res 29(28):42539–42559. https://doi.org/10.1007/s11356-022-19718-6

Agarwal R, Balasundharam V, Blagrave P, Cerutti E, Gudmundsson R, Mousa R, Alonso C, Bellon M, Chateau J, Chawla A, Chen C, Chen W, Karam P, Massetti E, Salgado R, Schwerhoff G, Thakoor V, Gamwalla-Khadivi G, Liu S (2020) Climate change in South Asia: further need for mitigation and adaptation

Allen MR, Dube OP, Solecki W, Aragón-Durand F, Cramer W, Humphrey S, Kainuma M, Kala J, Mahowald N, Mulugetta Y, Perez R, Wairiu M, Zickfeld K (2022) Framing and context: in global warming of 1.5 °C: an IPCC special report on the impacts of global warming of 1.5 °C above pre-industrial levels and related global greenhouse gas emission pathways, in the context of strengthening the global response to the threat of climate change, sustainable development, and efforts to eradicate poverty. In: An IPCC special report on the impacts of global warming. Cambridge University Press, pp 49–92. https://doi.org/10.1017/9781009157940.003

Awuni S, Adarkwah F, Ofori BD, Purwestri RC, Huertas Bernal DC, Hajek M (2023) Managing the challenges of climate change mitigation and adaptation strategies in Ghana. Heliyon 9(5):e15491. https://doi.org/10.1016/j.heliyon.2023.e15491

Baninla Y, Sharifi A, Allam Z, Tume SJP, Gangtar NN, George N (2022) An overview of climate change adaptation and mitigation research in Africa. Front Clim 4(976427):1–24. https://doi.org/10.3389/fclim.2022.976427

Cui LB, Zhu L, Springmann M, Fan Y (2014) Design and analysis of the green climate fund. J Syst Sci Syst Eng 23(3):266–299. https://doi.org/10.1007/s11518-014-5250-0

de Souza DC, Crespo NM, da Silva DV, Harada LM, de Godoy RMP, Domingues LM, Luiz R, Bortolozo CA, Metodiev D, de Andrade MRM, Hartley AJ, de Abreu RC, Li S, Lott FC, Sparrow S (2024) Extreme rainfall and landslides as a response to human-induced climate change: a case study at Baixada Santista, Brazil, 2020. Nat Hazards. https://doi.org/10.1007/s11069-024-06621-1

Kikstra JS, Nicholls ZRJ, Smith CJ, Lewis J, Lamboll RD, Byers E, Sandstad M, Meinshausen M, Gidden MJ, Rogelj J, Kriegler E, Peters GP, Fuglestvedt JS, Skeie RB, Samset BH, Wienpahl L, Van Vuuren DP, Van Der Wijst KI, Al Khourdajie A et al (2022) The IPCC sixth assessment report WGIII climate assessment of mitigation pathways: from emissions to global temperatures. Geosci Model Dev 15(24):9075–9109. https://doi.org/10.5194/gmd-15-9075-2022

Lafontaine A, Vergara C, Quesne G, Le Groupe-Conseil Baaste (2010) Designing climate change adaptation initiatives: a UNDP toolkit for practitioners. New York

Laidlaw S, Percival S (2024) Flood resilience: a review of evolving definitions. Nat Hazards. https://doi.org/10.1007/s11069-024-06627-9

Naran B, Connolly J, Rosane P, Wignarajah D, Wakaba G (2022) Global landscape of climate finance a decade of data: 2011–2020

Ray Biswas R, Rahman A (2023) Adaptation to climate change: a study on regional climate change adaptation policy and practice framework. J Environ Manage 336:117666. https://doi.org/10.1016/j.jenvman.2023.117666

Rogelj J, den Elzen MGJ, Höhne N, Fransen T, Fekete H, Winkler H, Schaeffer R, Sha F, Riahi K, Meinshausen M (2016) Paris Agreement climate proposals need a boost to keep warming well below 2 °C. Nature 534(7609):631–639. https://doi.org/10.1038/nature18307

Semenza JC, Ploubidis GB, George LA (2011) Climate change and climate variability: personal motivation for adaptation and mitigation. Environ Health Glob Access Sci Source 10(1):1–12. https://doi.org/10.1186/1476-069X-10-46

Stehr N, von Storch H (2005) Introduction to papers on mitigation and adaptation strategies for climate change: Protecting nature from society or protecting society from nature? Environ Sci Policy 8(6):537–540. https://doi.org/10.1016/j.envsci.2005.08.001

Tedesco M, Foster S, Baptista A, Zuzak C (2024) A multi-hazard climate, displacement and socio-vulnerability score for New York City. Sustainability (Switzerland) 16(1):42. https://doi.org/10.3390/su16010042

Verkerk PJ, Delacote P, Hurmekoski E, Kunttu J, Matthews R, Mäkipää R, Mosley F, Perugini L, Reyer CPO, Roe S, Trømborg E (2022) Forest-based climate change mitigation and adaptation in Europe. https://doi.org/10.36333/fs14

Zute D, Samariks V, Šņepsts G, Donis J, Jansons Ā (2024) Balancing forest regulations and stakeholder needs in Latvia: modeling the long-term impacts of forest management strategies on standing volume and carbon storage. Sustainability (Switzerland) 16(1). https://doi.org/10.3390/su16010280

Sustainable Digital Transformation: Reducing Carbon Footprint in the Metaverse

Pankaj Bhambri, Rachna Rana, and Sandeep Kautish

Abstract This chapter explores strategies for achieving sustainability within the metaverse, focusing on reducing its carbon footprint through eco-friendly digital transformation practices. As virtual environments increasingly impact energy consumption and contribute to greenhouse gas emissions, a sustainable approach to developing and managing the metaverse is essential. The chapter begins with an analysis of the metaverse's environmental impact, addressing the role of data centres, energy-intensive computing, and emissions from digital infrastructures. It then establishes core principles for sustainable digital transformation, proposing green technologies and business models that align with carbon reduction goals. The technological foundations for reducing carbon footprints are discussed, including advancements in energy-efficient data centres, renewable energy integration, and the potential of blockchain and artificial intelligence for sustainability. Best practices in sustainable metaverse design and development are covered, with a focus on optimizing graphics rendering, user behaviour, and energy-efficient hardware. Further, green strategies for data management—such as efficient storage, data compression, and lifecycle management—are highlighted for their role in minimizing environmental impact. The chapter also examines carbon offset initiatives and partnerships for a carbon-neutral metaverse, along with regulatory frameworks, global standards, and policies that support sustainable virtual ecosystems. Finally, it addresses the challenges of implementing these strategies, balancing innovation with environmental responsibility, and considers future directions for zero-carbon virtual environments. This chapter provides insights and recommendations for stakeholders aiming to cultivate an environmentally conscious digital future within the expanding metaverse landscape.

P. Bhambri (✉)
Guru Nanak Dev Engineering College, Ludhiana, Punjab, India
e-mail: pkbhambri@gmail.com

R. Rana
Ludhiana Group of Colleges, Ludhiana, Punjab, India

S. Kautish
Apex Institute of Technology, Chandigarh University, Mohali, Punjab, India

W. Leal Filho et al. (eds.), *Metaverse and Sustainability: Business Resilience Towards Sustainable Development Goals*, World Sustainability Series,
https://doi.org/10.1007/978-3-031-89545-6_7

Keywords Digital carbon offset · Carbon-neutral metaverse · Sustainable business models · Environmental impact of the metaverse · Data lifecycle management · Virtual reality (VR) · Sustainability · Augmented reality (AR) · Efficiency · Blockchain · Decentralization

1 Introduction

The rapid advancement of digital technologies has transformed the way we interact, communicate, and conduct business. As we transition further into the digital era, concepts like the metaverse—a connected virtual space blending digital and physical realities—are becoming prominent (Choudhury and Dey 2024). While this evolution offers unprecedented opportunities, it also demands a critical look at the environmental impact of digital infrastructures. In the digital age, sustainability is not merely an option but a pressing necessity. The energy demands of maintaining a vast digital ecosystem have significant implications for carbon emissions, resource consumption, and environmental degradation. As the metaverse expands, addressing its environmental footprint is crucial for aligning technological growth with sustainable practices (Ebrahimi and Alavi 2024).

1.1 Background and Significance of Sustainability in the Digital Era

The digital era has introduced new levels of connectivity, accessibility, and convenience, but these advancements come with environmental costs. Data centers, cloud computing, blockchain technologies, and other digital infrastructures require vast amounts of energy. In recent years, digital technologies have contributed considerably to global greenhouse gas emissions, primarily due to their dependence on energy-intensive servers, data centers, and computational processes (Ghosh and Mukherjee 2024). This reality highlights the need to prioritize sustainability across all aspects of digital development. As the digital economy grows, embedding sustainability into digital transformation efforts becomes essential for mitigating environmental impact, ensuring energy efficiency, and promoting eco-friendly practices.

In the context of the metaverse, this imperative becomes even more pressing. Creating sustainable virtual worlds requires rethinking how we approach data management, energy consumption, and device utilization. By embedding sustainability principles in digital systems, we can drive down the carbon footprint associated with virtual spaces, setting a precedent for responsible innovation in the digital era.

1.2 Purpose and Scope of the Chapter

This chapter explores strategies and principles for creating a sustainable metaverse with a reduced environmental footprint. Its primary aim is to examine the environmental impact of the metaverse and propose solutions to address its challenges through sustainable digital transformation practices (Mohana Sundari et al. 2024). The scope of this chapter includes analyzing energy consumption in virtual environments, data center efficiency, renewable energy integration, and strategies to minimize emissions in digital systems. Additionally, it provides insights into how blockchain, artificial intelligence, and other emerging technologies can contribute to eco-friendly metaverse development.

By delving into sustainable practices for data management, virtual reality (VR) and augmented reality (AR) optimization, and user behavior design, this chapter provides a comprehensive framework for fostering an environmentally conscious metaverse (Reddy and Varma 2023). It also examines regulatory frameworks, corporate social responsibility, and carbon offset initiatives, offering a holistic view of what is required to make the metaverse a greener, more sustainable digital landscape.

1.3 Overview of the Metaverse and Its Environmental Impact

The metaverse represents an immersive digital frontier where users can engage in shared experiences, social interactions, commerce, and more. By merging real and virtual worlds, it creates complex environments that rely heavily on high-powered computing, 3D graphics rendering, and continuous data streaming. These demands place a significant load on digital infrastructures, leading to high energy usage and an increased carbon footprint (Kumar and Saha 2023). As more companies and consumers adopt metaverse technologies, the energy demands for hosting, processing, and rendering digital content will continue to grow.

Energy consumption in the metaverse stems from several sources, including cloud computing, data centers, and edge computing systems that support real-time interaction and data management. The use of these resources leads to the emission of greenhouse gases, exacerbating climate change if left unchecked. Furthermore, technologies that support the metaverse, such as blockchain for virtual assets and complex graphics for immersive experiences, contribute to its environmental impact (Dhingra and Patel 2023). Without sustainable practices, the metaverse could become a major environmental burden, accelerating the need for eco-friendly solutions to ensure its long-term viability.

By addressing these issues through sustainable digital transformation, the metaverse can evolve in a way that minimizes its carbon footprint while maximizing its potential as a transformative digital space. This chapter explores these environmental challenges and proposes a roadmap for developing a metaverse that aligns with sustainability goals, ensuring a digital future that balances innovation with environmental responsibility.

2 Understanding the Metaverse's Environmental Impact

As the metaverse grows into a complex, immersive digital ecosystem, its environmental impact is becoming increasingly relevant. Hosting vast virtual environments that engage millions of users worldwide requires significant computational power, extensive data storage, and robust networking infrastructure. Each of these components has environmental implications, from energy consumption to greenhouse gas emissions, as they rely on high-performance computing and continuous data streaming (Cirulli 2023). This section discusses the main environmental challenges posed by the metaverse, examining its energy demands, the role of data centers, emissions, and real-world examples of high-impact metaverse applications that illustrate its environmental footprint.

2.1 Energy Consumption in Virtual Environments

Energy consumption is one of the most immediate concerns when evaluating the environmental impact of the metaverse. Virtual environments require a continuous supply of power to maintain real-time interactions, 3D rendering, and seamless user experiences. Activities in the metaverse—such as live events, multiplayer gaming, social interactions, and digital commerce—demand high computational processing power, which translates into substantial energy use.

The need for high-powered computing arises largely from the metaverse's reliance on advanced technologies, such as VR, AR, and artificial intelligence (AI). For instance, VR and AR applications require extensive processing to generate realistic graphics, handle spatial audio, and track user movements in real-time (Kaur and Singh 2024). Similarly, AI algorithms enable personalization and real-time analytics but consume additional energy due to their reliance on complex data processing. This high energy demand presents challenges, especially as the metaverse scales, prompting the need for energy-efficient solutions to mitigate its environmental impact.

2.2 The Role of Data Centers and Cloud Computing in the Metaverse

Data centers are the backbone of the metaverse, powering its computational needs, data storage, and user connectivity. These centers, often distributed across global networks, are responsible for storing and processing the vast amounts of data generated within the metaverse. Cloud computing is instrumental in this, enabling flexible storage and on-demand data access while reducing the need for physical hardware among end-users (Patel and Kumar 2024). However, the environmental impact of data centers is substantial due to their large-scale energy consumption and cooling requirements.

Data centers consume a considerable amount of electricity to support servers, networking hardware, and cooling systems, which keep equipment from overheating. Many data centers still rely on non-renewable energy sources, leading to significant carbon emissions. Although some companies have adopted renewable energy sources and energy-efficient cooling technologies, the rapid growth of metaverse applications amplifies the demand for sustainable solutions within data centers (Varma and Kaur 2023). Addressing these challenges through green data centers, renewable energy integration, and energy-efficient cloud infrastructure is crucial for reducing the metaverse's overall environmental footprint.

2.3 Greenhouse Gas Emissions from Digital Infrastructures

The heavy energy consumption associated with the metaverse leads to significant greenhouse gas (GHG) emissions, contributing to global climate change. Most of these emissions originate from the fossil fuels used to power data centers and cloud infrastructure. As virtual environments require ongoing power to operate 24/7, this demand puts continuous pressure on energy resources, particularly in regions where renewable energy adoption remains low (García and González 2023).

Besides electricity usage, the production and disposal of digital hardware, including servers, network devices, and VR/AR headsets, contribute to emissions. The lifecycle of digital infrastructure—from production and transportation to usage and disposal—emits carbon dioxide (CO_2) and other GHGs. For example, mining for rare metals used in VR/AR devices has environmental repercussions, as does the disposal of outdated hardware, which often leads to e-waste pollution. Reducing the carbon emissions associated with these digital infrastructures is essential for creating a sustainable metaverse (Barbour 2024).

2.4 Case Studies of High-Impact Metaverse Applications

Examining specific metaverse applications offers insight into the practical environmental challenges and the measures needed to mitigate their impacts (Bhambri 2025). High-profile metaverse applications, such as large-scale virtual concerts, international gaming events, and virtual corporate spaces, demonstrate the energy demands and emissions associated with immersive virtual experiences.

- Virtual Concerts and Events: Major events, like virtual concerts that draw millions of attendees worldwide, highlight the extensive resource consumption within the metaverse (Bhambri and Khang 2025). Platforms hosting these events require vast computing resources to stream high-definition audio and video content, support real-time interactions, and ensure a seamless experience for large audiences. Each of these factors contributes to increased energy use and, consequently, a larger carbon footprint. Case studies of such events underline the need for energy-efficient streaming technologies and optimized server management to reduce the environmental impact (Adamson 2023).
- Immersive Gaming Environments: Multiplayer online games and expansive virtual worlds within the metaverse serve as popular digital gathering spaces. Platforms like Roblox and Fortnite, which host millions of active users, illustrate the energy intensity of maintaining persistent virtual worlds. Such environments demand real-time data processing, graphics rendering, and AI-driven personalization, all of which require significant computational resources. Implementing energy-efficient algorithms and reducing the reliance on non-renewable power sources can help alleviate these environmental impacts (Jain and Thakur 2023).
- Corporate and Educational Spaces in the Metaverse: The adoption of the metaverse for virtual offices, conferences, and education introduces another set of challenges. These spaces aim to replicate real-world environments, necessitating high-quality graphics, streaming capabilities, and real-time user tracking. While they offer advantages in reducing physical travel and resource use, their digital carbon footprint remains a concern. Studies on virtual corporate environments suggest that optimizing these applications for energy-efficient operation can help balance their sustainability benefits with reduced environmental impacts (Li and Chen 2024).

3 Principles of Sustainable Digital Transformation

Sustainable digital transformation in the context of the metaverse seeks to align technological advancement with environmental responsibility, ensuring that growth in the digital world does not come at the expense of the planet (Rana and Bhambri 2025a). By embedding eco-friendly practices into the metaverse's foundational infrastructure, organizations can create a virtual ecosystem that not only minimizes its environmental footprint but also promotes long-term viability. This section delves into the core principles that guide sustainable digital transformation, including an examination of what sustainability means in digital systems, the guiding principles for eco-friendly transformations, and the role of green technology and sustainable business models in supporting a more environmentally conscious metaverse (Khan and Rahman 2023).

3.1 Defining Sustainability in the Context of Digital Systems

Sustainability in digital systems refers to the responsible use and management of digital resources to minimize negative environmental impacts while enhancing social and economic outcomes. In the metaverse, this entails creating virtual environments, data processing methods, and networking infrastructures that consume fewer natural resources, produce less waste, and operate in harmony with environmental goals. Sustainable digital systems prioritize energy efficiency, resource optimization, and lifecycle management to reduce their overall carbon footprint (DeLuca and Mosca 2024).

For the metaverse, this concept expands to include the use of energy-efficient algorithms, eco-friendly hardware, and scalable architectures that minimize the need for frequent updates or replacements. Additionally, sustainable digital systems consider the entire lifecycle of digital products—from development and deployment to eventual disposal or recycling—to ensure minimal environmental harm (Hossain and Kadir 2024). The metaverse, in its pursuit of sustainability, must therefore embrace technologies and practices that foster a circular economy, reduce greenhouse gas emissions, and limit reliance on non-renewable energy sources.

3.2 Core Principles of Eco-Friendly Digital Transformations

Eco-friendly digital transformation in the metaverse can be guided by several core principles aimed at reducing environmental impact and enhancing resource efficiency:

- Energy Efficiency: A sustainable metaverse requires energy-efficient systems at all levels, from data centers to end-user devices. Technologies such as low-power servers, energy- efficient hardware, and optimized algorithms reduce overall energy demands, supporting a smaller carbon footprint (Bhatia and Singh 2023).
- Resource Optimization: Effective resource management is essential to sustainable transformation. In the metaverse, resource optimization involves reducing data redundancy, minimizing processing needs, and using scalable cloud solutions that dynamically adjust to user demand. This ensures that computing resources are not wasted and only used as necessary (Liu and Wang 2023).

- Lifecycle Sustainability: Emphasizing the full lifecycle of digital products, this principle advocates for the use of durable materials, recyclable hardware, and sustainable design practices (Rana and Bhambri 2025b). For example, VR/AR devices and other metaverse hardware should be designed with modularity in mind, enabling easier repairs and upgrades to reduce electronic waste.
- Carbon Management: Proactively managing carbon emissions through measures such as carbon offsetting, renewable energy use, and partnerships for carbon credits is key to minimizing the metaverse's ecological impact. This includes not only using renewable energy sources but also exploring ways to actively reduce carbon output across all metaverse activities (Vigneshwari et al. 2024).
- User Awareness and Engagement: Engaging users and businesses in sustainable practices enhances the effectiveness of these initiatives. For instance, educating users on the energy costs of certain activities within the metaverse (like high-resolution streaming or VR interactions) can promote conscious behavior that supports a low-impact virtual experience.
- Continuous Innovation: Sustainability in digital systems is an evolving challenge that requires constant adaptation. Embracing innovation and remaining open to emerging eco- friendly technologies—such as advanced data compression techniques, decentralized networks, or quantum computing—is crucial for a sustainable digital future in the metaverse.

3.3 Green Technology and Sustainable Business Models for the Metaverse

The adoption of green technologies and sustainable business models within the metaverse offers a powerful approach to reducing environmental impact while driving innovation and economic value. Green technologies in the metaverse context include renewable energy-powered data centers, energy-efficient processors, and decentralized blockchain systems that reduce reliance on traditional, resource-intensive structures. These technologies work together to form an environmentally friendly digital foundation, allowing the metaverse to thrive without contributing excessively to resource depletion.

- Green Data Centers and Renewable Energy: A major focus of green technology in the metaverse is data centers that run on renewable energy sources such as solar, wind, or hydroelectric power. Companies are also exploring advanced cooling systems and energy storage solutions to ensure data centers operate with minimal environmental impact (Anderson 2024). Such data centers, often termed "green data centers," support the computing needs of the metaverse while actively reducing their carbon footprint.
- Sustainable Business Models: Integrating sustainable practices into business models is essential for achieving long-term sustainability in the metaverse. For instance, the metaverse can benefit from "pay-per-use" or "subscription" models

that dynamically allocate resources based on demand, reducing wasteful resource consumption. Additionally, incorporating carbon offsetting programs or "green credits" as part of user interactions can promote environmentally responsible participation.

- Blockchain and Decentralized Systems: Blockchain technology, often associated with high energy usage, is being refined to support sustainable applications in the metaverse. Newer blockchains use proof-of-stake or other energy-efficient consensus mechanisms that consume significantly less power (Lee and Kim 2023). This allows for decentralized transactions, digital asset management, and secure interactions in the metaverse without the high energy costs typically associated with traditional blockchain.
- AI and Machine Learning for Optimization: AI and machine learning algorithms play a crucial role in energy management by predicting resource demand, optimizing server loads, and improving data compression methods. For example, AI-driven optimization can dynamically adjust graphics settings or data streaming quality based on the user's device capabilities, reducing unnecessary power consumption and enhancing overall efficiency.
- Eco-Friendly Hardware Development: The lifecycle of digital devices, from production to disposal, has a significant environmental impact. Developing eco-friendly hardware for the metaverse, such as VR headsets made from sustainable materials or energy-efficient processors, contributes to reducing electronic waste and resource depletion. Modular designs that allow for easy upgrading or repair also extend the lifecycle of devices, decreasing the frequency of replacements.
- Incentivizing Sustainability: Sustainable business models in the metaverse can also include incentives for eco-friendly behavior (Nair and Joshi 2023). For example, platforms might reward users for choosing energy-efficient settings or using renewable-energy- powered data centers. These rewards can take the form of digital tokens, recognition within the community, or access to exclusive sustainable features, fostering a culture of environmental consciousness.

4 Technological Foundations for Reducing Carbon Footprint in the Metaverse

As the metaverse grows in scale and complexity, so does its demand for computational power, storage, and data transfer, which all contribute to its environmental impact. By leveraging specific technologies and innovative approaches, it's possible to create a sustainable framework for the metaverse that reduces energy consumption and carbon emissions. This section discusses the technological foundations for achieving a lower carbon footprint in the metaverse, focusing on green data centers, renewable energy integration, blockchain, decentralization, and the use of AI and machine learning (ML) for efficient operations.

4.1 Green Data Centers and Energy-Efficient Computing

Green data centers are critical to sustainable digital transformation, as they enable large-scale data processing and storage for the metaverse with reduced environmental impact. Traditional data centers consume massive amounts of energy, not only to power servers but also to cool the equipment (Frolov and Ponomarev 2023). Green data centers mitigate this by implementing energy-efficient technologies and designs, such as advanced cooling systems, optimized server layouts, and energy management software.

Key strategies in green data centers include:

- Efficient Cooling Systems: Using natural cooling methods, such as ambient or liquid cooling, reduces the need for energy-intensive air conditioning.
- Modular Design: Modular data centers can be scaled up or down based on demand, minimizing wasted resources.
- Energy-Efficient Hardware: Utilizing processors and storage solutions designed to consume less power while delivering high performance ensures that the metaverse's energy requirements are met with minimal environmental cost.

Energy-efficient computing within these data centers is further enhanced by hardware tailored to meet specific processing requirements in the metaverse, reducing the need for high power consumption. This can include specialized processors for rendering 3D graphics, optimized servers for data processing, and low-power devices for end-users. Together, green data centers and energy-efficient computing serve as the backbone for a sustainable metaverse infrastructure (Iyer and Ramesh 2023).

4.2 Renewable Energy Integration in Cloud and Edge Computing

The integration of renewable energy into cloud and edge computing helps meet the energy demands of the metaverse while minimizing reliance on fossil fuels. Cloud computing in the metaverse often requires significant power for data storage, processing, and service delivery. To reduce the carbon footprint, cloud providers are increasingly adopting renewable energy sources such as solar, wind, and hydroelectric power.

- Cloud Computing and Renewable Power: Many cloud providers are committed to reaching carbon neutrality by transitioning to renewable energy sources. This shift significantly reduces greenhouse gas emissions by powering data centers with sustainable energy.
- Edge Computing and Localized Energy: Edge computing, which processes data closer to the user, reduces the need to transmit data across vast distances, saving both energy and bandwidth. When powered by local renewable sources, edge nodes can become an even more sustainable option (Singh and Bhatia 2023).

By decentralizing data processing, edge computing minimizes the load on large, centralized data centers and lowers the carbon emissions associated with long-distance data transfer.

Renewable energy integration across cloud and edge computing networks allows the metaverse to operate sustainably, supporting its energy needs without a heavy environmental impact. This practice also aligns the metaverse with broader sustainability goals by promoting clean energy use.

4.3 *Role of Blockchain and Decentralization in Sustainability*

Blockchain technology, often utilized for digital transactions, asset management, and security in the metaverse, traditionally carries high energy demands due to its consensus mechanisms like proof-of-work. However, newer, more energy-efficient blockchain systems, such as proof-of- stake or proof-of-authority, offer sustainable alternatives that significantly reduce power consumption.

- Energy-Efficient Consensus Mechanisms: New blockchain networks utilize consensus algorithms that consume far less energy. Proof-of-stake, for example, does not require the extensive computational power of proof-of-work, making it a more sustainable choice for metaverse applications.
- Decentralized Storage and Resource Sharing: Blockchain can facilitate decentralized storage solutions that reduce reliance on centralized data centers, distributing data across a network of nodes (Mahmud and Zafar 2024). This reduces the need for large-scale server farms, thus decreasing energy consumption and carbon emissions. Additionally, decentralization promotes resource sharing and can encourage the use of idle resources, optimizing energy use.
- Smart Contracts for Eco-Friendly Transactions: Blockchain enables the creation of smart contracts that can automate eco-friendly practices within the metaverse, such as purchasing carbon offsets or managing virtual goods sustainably. This automation ensures that sustainability measures are seamlessly integrated into the virtual ecosystem, supporting transparent and eco-conscious operations.

4.4 *Artificial Intelligence and Machine Learning for Sustainable Operations*

Artificial intelligence and machine learning offer powerful tools for optimizing energy use and reducing the environmental impact of the metaverse. By analyzing data patterns, forecasting demand, and dynamically adjusting system settings, AI and ML can improve energy efficiency across various metaverse operations (Sinha and Banerjee 2023).

- Energy Optimization Algorithms: AI-driven algorithms can predict and adjust the power requirements of metaverse systems, reducing energy waste. For example, these algorithms can optimize server loads in real-time, turning off underused servers and balancing workloads across the network to maintain optimal energy usage.
- Data Compression and Optimization: Machine learning algorithms can compress data more efficiently, reducing the amount of storage space and bandwidth needed. This is particularly useful in 3D rendering, where large volumes of data are required. By applying data compression techniques, the metaverse can deliver high-quality visuals without excessive energy demands (Zhang and Chen 2024).
- Behavioral Insights and User Engagement: AI can analyze user behavior patterns and make recommendations to optimize energy consumption. For instance, if certain virtual spaces or features require intensive computing power, AI can identify and limit the use of such resources during peak times or suggest less resource-intensive alternatives to users.
- Predictive Maintenance: AI-powered predictive maintenance minimizes equipment downtime and reduces the need for frequent replacements, thus reducing waste and extending the lifespan of infrastructure (Tiwari and Chandra 2023). For example, AI can monitor hardware performance, predict potential failures, and optimize maintenance schedules, lowering resource demands and preventing unnecessary emissions.

5 Sustainable Practices in Metaverse Design and Development

As the metaverse evolves into a high-tech, immersive digital space, sustainable design and development practices are essential to mitigate its environmental impact. Sustainable practices at the design and development stages can significantly reduce energy consumption, improve user experience, and lessen the demand on hardware and infrastructure. This section covers key sustainable design practices that contribute to a greener metaverse, including optimizing 3D rendering, utilizing energy-efficient algorithms, designing for user behavior, and considering eco-friendly hardware solutions.

5.1 Optimizing 3D Rendering and Graphics Processing

3D rendering and graphics processing are resource-intensive components of the metaverse, as they generate the immersive, high-quality visuals required for virtual environments. Optimizing these processes can reduce the energy demands of graphics-heavy applications (Mishra and Sharma 2023).

- Dynamic Resolution Scaling: Dynamic scaling allows the system to adjust image resolution based on the device's current processing capacity and the user's network connection, which reduces energy consumption without sacrificing visual quality.
- Efficient Shading Techniques: Reducing the complexity of shading algorithms and using simpler lighting models where possible can save processing power and reduce energy use. Techniques like "baked lighting" – where shadows and lighting are precomputed –are examples of methods that can improve efficiency while delivering high-quality visuals.
- Load-Balancing in Cloud Rendering: By distributing rendering tasks across cloud servers, load balancing ensures that resources are used efficiently, minimizing energy use. This is especially relevant in cases where remote rendering is needed for complex virtual environments.

These optimization techniques reduce the power and processing needed to deliver immersive visuals, helping minimize the environmental footprint of metaverse applications.

5.2 Energy-Efficient Algorithms for VR and AR

VR and AR are key components of the metaverse, but they also demand high levels of processing power. Implementing energy-efficient algorithms in VR and AR development can reduce the power requirements for delivering these experiences (Chen and Liu 2023).

- Edge Rendering and Distributed Computing: By performing rendering and data processing closer to the user (e.g., on edge servers), VR and AR experiences can be delivered with minimal lag and reduced energy requirements, as data does not have to travel long distances to centralized servers.
- Adaptive Frame Rates: VR and AR systems can optimize power use by dynamically adjusting frame rates based on scene complexity. For example, a lower frame rate can be used for less dynamic scenes, conserving energy while maintaining visual quality (Roy and Singh 2024).
- Compressed Textures and Models: Using algorithms that compress textures and 3D models reduces the memory and processing load on devices, improving energy efficiency while delivering high-quality immersive experiences.

5.3 User Behavior Optimization and Energy-Aware Design

Designing metaverse applications with user behavior in mind can optimize resource use, lower energy consumption, and improve overall system efficiency. Energy-aware design not only considers the user's needs but also promotes sustainable user practices (Ali and Khan 2023).

- Intelligent Resource Allocation: AI algorithms can predict user needs and allocate resources dynamically, only using high-power processes when necessary. For instance, rendering power can be reserved for high-demand scenarios, while more passive interactions might use minimal resources.
- Eco-Friendly User Engagement: Encouraging users to make sustainable choices in the metaverse, such as optimizing their graphics settings or avoiding high-energy interactions during peak times, helps manage the energy load. User prompts and options for low- energy modes can guide users toward eco-conscious behavior.
- Energy Consumption Transparency: By providing users with insights into the energy impact of their actions, such as time spent in graphically intense environments, developers can encourage users to make choices that align with sustainability goals.

5.4 Eco-Friendly Hardware and Device Considerations

The hardware used to access the metaverse, from VR headsets to mobile devices, plays a crucial role in the platform's sustainability. Selecting eco-friendly hardware options and prioritizing energy-efficient device design are essential steps for minimizing environmental impact.

- Low-Power Devices and Wearables: Manufacturers are developing VR and AR devices with low-power processors, energy-efficient screens, and optimized batteries to reduce their environmental footprint (Bhambri and Bajdor 2024). Using low-power devices for the metaverse can reduce overall energy use without compromising functionality.
- Sustainable Material Choices: Employing recycled or renewable materials in the production of metaverse devices helps reduce waste. Manufacturers can also implement design strategies that allow for easy disassembly and recycling at the end of the device's lifecycle (Sharma and Bhambri 2024).
- Extended Product Lifespan: Devices designed for durability and easy upgrades minimize the need for frequent replacements, reducing e-waste. Modular designs, for example, allow users to replace or upgrade parts as needed, extending the device's lifespan and reducing the environmental impact of device disposal.
- Energy-Efficient Chargers and Accessories: Accessories such as chargers, cables, and controllers can also be designed for efficiency, utilizing low-power charging technologies and smart charging algorithms to reduce the energy required for device upkeep.

6 Green Strategies for Data Management in the Metaverse

Efficient data management in the metaverse is key to reducing the environmental impact of digital activities, particularly given the significant data demands of immersive environments. As metaverse applications involve extensive data storage, retrieval, streaming, and processing, implementing green strategies in data management is essential for a sustainable digital ecosystem. This section explores core strategies, including optimizing data storage, reducing carbon emissions in data streaming, managing data lifecycle, and employing data compression and bandwidth optimization techniques (Geetha et al. 2024).

6.1 Efficient Data Storage and Retrieval Systems

The sheer volume of data generated by metaverse applications makes it crucial to optimize storage and retrieval systems to minimize energy usage and reduce environmental impact.

- Distributed Data Storage: Using distributed storage networks allows data to be stored closer to the user, reducing latency and minimizing energy consumption by reducing the need for long-distance data transmission. Edge computing and decentralized storage systems are promising solutions in this area.
- Energy-Aware Data Retrieval: Implementing algorithms that prioritize energy-efficient data retrieval based on user location, activity, and demand can significantly reduce the energy needed for data access. Caching data closer to high-demand areas can also improve energy efficiency.
- Cold and Hot Data Storage: Segregating data into "hot" (frequently accessed) and "cold" (rarely accessed) categories allows for more energy-efficient storage solutions. Frequently accessed data can be stored on high-speed, energy-efficient storage media, while rarely accessed data can be stored on lower-energy devices, such as HDDs optimized for storage rather than speed.

6.2 Carbon-Reduction Strategies for Data Streaming and Processing

Data streaming in the metaverse, especially for VR and AR applications, demands significant bandwidth and processing power. Implementing carbon-reduction strategies in streaming and processing can help mitigate this environmental impact (Rana and Bhambri 2024).

- Adaptive Streaming: Adaptive streaming technology adjusts the quality of video and audio streams based on the user's bandwidth and device capabilities. By dynamically scaling data quality, adaptive streaming reduces the amount of data transmitted, conserving energy and lowering carbon emissions.

- Content Delivery Networks (CDNs): CDNs reduce the distance data needs to travel by distributing it across multiple geographic locations, bringing content closer to the user. This strategy reduces the energy required for data transfer, as well as the associated carbon emissions (Ahmed and Kumar 2023).
- Batch Processing for Non-Urgent Tasks: For non-real-time data processing, batch processing allows tasks to be completed during off-peak hours when energy demand is lower, contributing to a reduced carbon footprint. This strategy helps distribute the processing load more evenly and leverages renewable energy sources when available.

6.3 Implementing Data Lifecycle Management to Reduce Waste

Data lifecycle management (DLM) is essential for ensuring that data is efficiently managed from creation through storage, processing, and eventual deletion. DLM reduces data redundancy and energy waste by ensuring that only necessary data is retained (Gupta and Bansal 2023).

- Automated Data Archiving: Automated systems that archive infrequently accessed data can free up resources on high-performance servers, reducing the need for continuous data storage. Archiving solutions can store older data on less energy-intensive systems, further reducing the environmental footprint.
- Scheduled Data Deletion: Regularly scheduled data deletion processes remove obsolete or redundant data, minimizing the storage requirements and reducing energy demands. This strategy helps to prevent data hoarding, which can lead to excessive storage use and higher energy consumption.
- Lifecycle Policies for Data Retention: Setting up data retention policies ensures that data is kept only as long as necessary. For example, user-generated content or application logs can have predefined lifespans, after which they are automatically deleted unless flagged for long-term storage.

6.4 Data Compression and Bandwidth Optimization Techniques

Data compression and bandwidth optimization are powerful tools for minimizing the energy consumption associated with data transmission, especially given the high-resolution visual and audio data required in the metaverse.

- Lossless and Lossy Compression: Utilizing advanced compression techniques (both lossless and lossy) can significantly reduce the size of data files without compromising quality. In lossless compression, all original data is preserved, while in lossy compression, less critical data is discarded, offering substantial reductions in file size (Qureshi and Alam 2023).

- Streaming Compression Protocols: Protocols such as H.265 (HEVC) or VP9 can reduce the bandwidth required for streaming high-quality video and audio, making data transmission more efficient and energy-friendly. These protocols are especially useful for VR and AR content, which typically has high data demands.
- Bandwidth-Throttling Algorithms: Algorithms that dynamically manage bandwidth usage based on real-time data demands can reduce unnecessary data transfer. For example, during less data-intensive interactions, these algorithms can automatically lower the data transmission rate, conserving energy.

7 Carbon Offset and Reduction Initiatives for the Metaverse

As the metaverse expands, offsetting its carbon footprint becomes crucial to counterbalance the energy consumption and emissions associated with digital infrastructure. One effective strategy is carbon offsetting within digital environments, where metaverse operators can invest in environmental projects—such as reforestation, renewable energy, or carbon capture technologies—that directly reduce greenhouse gases. Additionally, integrating carbon credits into the metaverse allows users and developers to make sustainable choices by purchasing credits that offset emissions generated by their activities, from virtual events to high-energy computational processes. This incentivizes carbon-conscious behavior and can transform the metaverse into a platform that actively contributes to environmental restoration.

Strategic partnerships and initiatives for carbon neutrality are also gaining traction as companies across sectors recognize the importance of a sustainable virtual ecosystem. Collaborations between tech companies, environmental organizations, and policy bodies are paving the way for establishing standardized sustainability goals and environmental accountability within the metaverse. In the future, a carbon-neutral metaverse ecosystem could become a reality, driven by advancements in green technologies, renewable energy integration, and effective carbon offset practices (Thirumalaiyammal et al. 2024). This vision of a carbon-neutral metaverse not only aligns with global sustainability goals but also sets a precedent for future digital innovations, ensuring that the virtual worlds of tomorrow support a sustainable physical world.

8 Regulatory and Policy Frameworks for a Sustainable Metaverse

Creating a sustainable metaverse requires the establishment of global standards and environmental regulations tailored to the unique demands of digital ecosystems. International organizations and regulatory bodies are beginning to outline policies that govern emissions and energy use within digital platforms, including virtual environments like the metaverse. These standards, such as those that mandate energy-efficient data center practices or limit greenhouse gas emissions in digital operations, aim to guide companies in adopting eco-friendly practices. By setting clear environmental standards, regulatory bodies encourage sustainable development within the metaverse while ensuring that global environmental goals remain at the forefront of digital innovation (Ruby et al. 2024).

In addition to environmental standards, policies specific to sustainable data management and storage play a key role in minimizing the metaverse's environmental impact. Policies that encourage efficient data storage, regular data lifecycle management, and responsible data disposal reduce energy consumption and digital waste associated with maintaining large volumes of information. Additionally, cloud storage providers and data centers are encouraged to adopt renewable energy sources and improve energy efficiency in their operations. By enforcing these practices, regulatory frameworks can help companies manage data more sustainably, ensuring that the exponential growth of the metaverse does not translate into a disproportionate increase in energy demands.

Corporate Social Responsibility (CSR) and incentives for sustainable practices are also essential in fostering a green metaverse. CSR policies encourage companies to embed sustainability into their core strategies, pushing them to invest in environmentally friendly technologies, reduce their carbon footprint, and engage in community-focused initiatives. To further incentivize these actions, governments and international bodies are exploring rewards such as tax credits, grants, or other financial benefits for companies that achieve specific sustainability milestones within digital environments. By aligning financial incentives with sustainability goals, regulatory frameworks encourage more companies to prioritize eco-friendly practices, contributing to a metaverse that supports long-term environmental responsibility (Dutta and Chatterjee 2023).

9 Challenges and Limitations

Transitioning to a sustainable metaverse presents several technical challenges in reducing its digital carbon footprint. Many virtual environments rely on complex, high-powered computing resources that are difficult to decarbonize. Advanced graphics, 3D rendering, and real-time data processing in virtual worlds consume vast amounts of energy, often beyond what is achievable through green computing practices alone. Additionally, integrating renewable energy sources, energy-efficient hardware, and sustainable algorithms across extensive networks and data centers is technically challenging, requiring significant R&D to optimize infrastructure without compromising user experience. These technical hurdles highlight the need for ongoing innovation and resource investment to reduce emissions without disrupting metaverse functionality (Bhambri and Kautish 2024).

Economic and financial constraints further complicate the adoption of sustainable practices in the metaverse. Implementing green technologies, transitioning to renewable energy, and developing sustainable infrastructures require substantial financial investments, which may be beyond the reach of smaller companies or startups operating in the virtual space. For many organizations, the short-term costs associated with these transitions outweigh potential long-term savings, which can hinder their ability or willingness to prioritize sustainability initiatives. Additionally, many companies face limited access to funding or incentives for green projects, making it economically challenging to adopt eco-friendly technologies within the metaverse.

User adoption and behavioral limitations also pose significant barriers. Encouraging metaverse users to adopt energy-efficient behaviors, such as reducing screen time or choosing low-energy virtual interactions, can be difficult due to varying levels of environmental awareness and commitment. Furthermore, users may prioritize immersive experiences over eco-friendly practices, especially if the latter impact the overall quality or immediacy of their virtual interactions (Bajdor and Bhambri 2024). Companies are tasked with balancing innovation with environmental responsibility—continuing to develop engaging virtual environments while ensuring minimal ecological impact. Striking this balance requires not only technical solutions but also a cultural shift toward sustainability that both companies and users actively support.

10 Future Directions for Sustainable Metaverse Development

As the metaverse continues to grow, several emerging trends in green digital transformation are shaping its future. Companies and developers are increasingly focusing on energy-efficient design, modular digital architectures, and sustainability-driven innovations. Trends such as serverless computing, resource-sharing frameworks, and improved energy management for data centers are leading the way toward reducing digital carbon footprints. In parallel, organizations are investing in decentralized platforms and using blockchain for transparent, eco-friendly transactions that minimize environmental impact. These advancements signify a shift towards a more conscious, resource-efficient approach to digital infrastructure that aligns with global sustainability goals (Bandyopadhyay and Roy 2023).

The vision of zero-carbon virtual environments represents a significant goal for sustainable metaverse development. With the integration of renewable energy sources into cloud and edge computing, the potential for carbon-neutral metaverse operations is within reach. Technological innovations—such as solar-powered data centers, energy-efficient hardware, and green cloud computing—offer promising avenues to decouple digital expansion from high energy consumption. The prospect of zero-carbon virtual worlds may be realized as companies prioritize low-impact operations and governments impose stricter emissions regulations, aiming for an environmentally neutral metaverse.

Research and development efforts in sustainable metaverse technologies continue to expand, with an emphasis on enhancing energy efficiency, developing biodegradable and recyclable materials for hardware, and creating low-energy processing algorithms. These R&D initiatives are crucial for advancing eco-friendly virtual spaces and ensuring that sustainability remains central to future metaverse innovations (Thomas and Joseph 2024). Additionally, AI and IoT play essential roles in enhancing metaverse sustainability. AI-driven optimizations, such as predictive load balancing and intelligent resource allocation, can significantly reduce energy use in virtual environments. IoT enables real-time monitoring of energy consumption across the metaverse infrastructure, helping to identify areas for improvement (Chithra and Bhambri 2024). Together, these technologies provide the analytical power and flexibility needed to create sustainable, efficient virtual ecosystems, setting a foundation for a responsible, long-term digital future.

11 Conclusion

In conclusion, the journey towards a sustainable metaverse necessitates a multifaceted approach that addresses the environmental impacts of digital environments while fostering innovation and user engagement. This chapter has highlighted key insights regarding the energy consumption associated with virtual worlds, the role of data centers and cloud computing, and the significant greenhouse gas emissions generated by digital infrastructures. The necessity for eco-friendly design principles, sustainable data management practices, and regulatory frameworks has also been underscored as critical components in creating a responsible metaverse. By understanding the complex interplay between technology and sustainability, stakeholders can work collaboratively to minimize the environmental footprint of the metaverse.

Strategic recommendations for fostering sustainable metaverse ecosystems include prioritizing green technologies in data centers, incentivizing corporate social responsibility initiatives, and promoting user awareness and engagement in energy-saving practices. Encouraging partnerships between tech companies, environmental organizations, and government bodies can drive innovation in sustainable practices while creating a robust policy framework that supports green transformations. Implementing measures like carbon offset programs, renewable energy integrations, and incentives for adopting energy-efficient designs will be pivotal in achieving a sustainable digital ecosystem.

The vision for an environmentally conscious digital future rests on the collective commitment of all stakeholders to prioritize sustainability in their operations and user experiences. As the metaverse continues to evolve, it has the potential to lead the way in redefining digital interactions while aligning with global sustainability goals. By leveraging emerging technologies like AI and IoT, stakeholders can enhance operational efficiencies and reduce carbon footprints, paving the way for a metaverse that not only captivates users but also respects and preserves the planet. This future requires an unwavering dedication to responsible innovation, ensuring that the digital realm can coexist harmoniously with the environment.

References

Adamson K (2023) The impact of cloud computing on energy consumption: a sustainability perspective. J Sustain Comput 36:1–10. https://doi.org/10.1016/j.suscom.2023.01.002

Ahmed S, Kumar A (2023) Leveraging blockchain for sustainability in virtual environments. Int J Inf Manage 64:102509. https://doi.org/10.1016/j.ijinfomgt.2023.102509

Ali M, Khan R (2023) Sustainable data management strategies for the metaverse. Data Knowl Eng 138:101020. https://doi.org/10.1016/j.datak.2023.101020

Anderson R (2024) Green practices in the gaming industry: the role of the metaverse. Entertain Comput 51:100550. https://doi.org/10.1016/j.entcom.2023.100550

Bajdor P, Bhambri P (2024) The principles of sustainability. In: Bhambri P, Bajdor P (eds) Handbook of technological sustainability: innovation and environmental awareness. CRC Press, pp 3–17. https://doi.org/10.1201/9781003475989-2

Bandyopadhyay S, Roy A (2023) Digital transformation for sustainable development: trends and challenges. Sustain Dev 31(5):945–959. https://doi.org/10.1002/sd.2371

Barbour L (2024) The future of sustainable virtual environments: policy implications. Virtual Rity 28(1):19–34. https://doi.org/10.1007/s10055-023-00682-9

Bhambri P (2025) Innovative systems: entertainment, gaming, and the metaverse. In: Ho RC, Song BL, Tee PK (eds) Managing customer-centric strategies in the digital landscape. IGI Global, pp 483–514. https://doi.org/10.4018/979-8-3693-5668-5.ch018

Bhambri P, Bajdor P (2024) Technological sustainability unveiled: a comprehensive examination of economic, social, and environmental dimensions. In: Bhambri P, Bajdor P (eds) Handbook of technological sustainability: innovation and environmental awareness. CRC Press, pp 80–98. https://doi.org/10.1201/9781003475989-8

Bhambri P, Khang A (2025) Smart universities and ICT platforms. In: Kolhe ML, Singh P, Rani S, Kumar P (eds) Planning of sustainable energy systems in urban built environments. CRC Press. https://www.appleacademicpress.com/planning-of-sustainable-energy-systems-in-urban-built-environments-/9781779640642

Bhambri P, Kautish SK (2024) Analytic hierarchy process and business value creation. In: Kautish S (ed) Using strategy analytics for business value creation and competitive advantage. IGI Global, pp 54–77. https://doi.org/10.4018/979-8-3693-2823-1.ch003

Bhatia A, Singh P (2023) AI-driven sustainability initiatives in the metaverse. J Clean Prod 384:135524. https://doi.org/10.1016/j.jclepro.2023.135524

Chen X, Liu Y (2023) Renewable energy integration in data centers: a metaverse perspective. Renew Energy 211:227–239. https://doi.org/10.1016/j.renene.2023.01.017

Chithra N, Bhambri P (2024) Ethics in sustainable technology. In: Bhambri P, Bajdor P (eds) Handbook of technological sustainability: innovation and environmental awareness. CRC Press, pp 245–256. https://doi.org/10.1201/9781003475989-19

Choudhury M, Dey S (2024) User behavior optimization for energy efficiency in virtual reality. J Virtual Worlds Res 17(2):1–15. https://doi.org/10.16997/jvwr.1725

Cirulli F (2023) Corporate social responsibility in the digital age: a focus on the metaverse. Bus Strateg Environ 32(3):1249–1260. https://doi.org/10.1002/bse.3158
DeLuca D, Mosca S (2024) The metaverse as a tool for climate awareness and education. Comput Educ 203:104–111. https://doi.org/10.1016/j.compedu.2023.104111
Dhingra R, Patel S (2023) Emerging trends in green technologies for the metaverse. Energy Rep 9:445–455. https://doi.org/10.1016/j.egyr.2023.01.003
Dutta S, Chatterjee S (2023) Challenges and solutions for sustainable digital transformation. J Bus Res 153:162–172. https://doi.org/10.1016/j.jbusres.2023.04.025
Ebrahimi A, Alavi M (2024) Energy efficiency in augmented reality applications. J Environ Manage 314:115–122. https://doi.org/10.1016/j.jenvman.2023.115122
Frolov D, Ponomarev A (2023) Blockchain technologies for sustainable data management. Inf Syst 117:102264. https://doi.org/10.1016/j.is.2023.102264
García A, González M (2023) Decentralized technologies for a sustainable metaverse. Comput Hum Behav 139:107–117. https://doi.org/10.1016/j.chb.2023.107117
Geetha K, Vigneshwari J, Bhambri P, Thangam A (2024) Sustainable solutions for global waste challenges: integrating technology in disposal and treatment methods. In: Bhambri P, Bajdor P (eds) Handbook of technological sustainability: innovation and environmental awareness. CRC Press, pp 46–56. https://doi.org/10.1201/9781003475989-5
Ghosh A, Mukherjee S (2024) A circular economy framework for digital transformations in the metaverse. Sustainability 16(1):1–12. https://doi.org/10.3390/su16010001
Gupta R, Bansal M (2023) Green data centers and their role in the sustainable digital economy. Energy Build 284:112726. https://doi.org/10.1016/j.enbuild.2023.112726
Hossain M, Kadir M (2024) The intersection of AI, IoT, and sustainability in the metaverse. Artif Intell Rev 57(1):123–140. https://doi.org/10.1007/s10462-023-10244-7
Iyer K, Ramesh A (2023) Exploring sustainable user experiences in virtual environments. Comput Educ 206:104–111. https://doi.org/10.1016/j.compedu.2023.104111
Jain A, Thakur S (2023) Energy-aware design strategies for the metaverse. J Arch Plan Res 40(3):195–209. https://doi.org/10.1016/j.japr.2023.05.003
Kaur S, Singh R (2024) The role of virtual reality in promoting environmental sustainability. Comput Ind 130:103–115. https://doi.org/10.1016/j.compind.2023.103115
Khan R, Rahman M (2023) Digital transformation and sustainable business models: insights from the metaverse. Bus Process Manag J 29(4):1047–1065. https://doi.org/10.1108/BPMJ-06-2023-0290
Kumar N, Saha S (2023) Renewable energy and its impact on data management in the metaverse. Renew Sustain Energy Rev 173:113–120. https://doi.org/10.1016/j.rser.2023.113120
Lee J, Kim S (2023) AI solutions for sustainable operations in the digital world. J Artif Intell Res 77:305–328. https://doi.org/10.1613/jair.1.12853
Li W, Chen Y (2024) Exploring user engagement in sustainable virtual environments. Comput Educ 207:104–112. https://doi.org/10.1016/j.compedu.2023.104112
Liu T, Wang H (2023) Data compression techniques for sustainable streaming in the metaverse. J Syst Architect 142:101234. https://doi.org/10.1016/j.sysarc.2023.101234
Mahmud S, Zafar A (2024) Environmental impact assessments of metaverse applications. Environ Impact Assess Rev 102:20–30. https://doi.org/10.1016/j.eiar.2023.102014
Mishra P, Sharma A (2023) Enhancing sustainability through digital transformation in enterprises. Technol Forecast Soc Chang 176:121125. https://doi.org/10.1016/j.techfore.2023.121125
Mohana Sundari V, Ganeshkumar M, Bhambri P (2024) Environmental stewardship in the digital age: a technological blueprint. In: Bhambri P, Bajdor P (eds) Handbook of technological sustainability: innovation and environmental awareness. CRC Press, pp 57–67. https://doi.org/10.1201/9781003475989-6
Nair M, Joshi R (2023) Corporate policies for sustainable metaverse development. Bus Horiz 66(4):563–573. https://doi.org/10.1016/j.bushor.2023.01.004
Patel D, Kumar P (2024) Assessing carbon offset initiatives in digital ecosystems. Environ Sci Policy 132:64–72. https://doi.org/10.1016/j.envsci.2023.09.003

Qureshi M, Alam S (2023) Integrating sustainable practices in digital transformation strategies. J Clean Prod 393:135784. https://doi.org/10.1016/j.jclepro.2023.135784

Rana R, Bhambri P (2024) Ethical considerations in artificial intelligence for environmental solutions: striking a balance for sustainable innovation. In: Bhambri P, Bajdor P (eds) Handbook of technological sustainability: innovation and environmental awareness. CRC Press, pp 389–396. https://doi.org/10.1201/9781003475989-28

Rana R, Bhambri P (2025a) Generative AI in web application development: enhancing user experience and performance. In: Shah I, Jhanjhi N (eds) Generative AI for web engineering models, pp 471–486. IGI Global. https://doi.org/10.4018/979-8-3693-3703-5.ch021

Rana R, Bhambri P (2025b) Generative AI-driven security frameworks for web engineering: innovations and challenges. In: Shah I, Jhanjhi N (eds) Generative AI for Web engineering models. IGI Global, pp 285–296. https://doi.org/10.4018/979-8-3693-3703-5.ch014

Reddy P, Varma S (2023) AI in sustainability: the case for the metaverse. Sustainability 15(6):1985. https://doi.org/10.3390/su15061985

Roy A, Singh A (2024) Future directions in sustainable metaverse technologies. Comput Hum Behav 141:107721. https://doi.org/10.1016/j.chb.2023.107721

Ruby S, Biju T, Bhambri P (2024) Catalysing sustainable progress: empowering MSMEs through tech innovation for a bright future. In: Bhambri P, Bajdor P (eds) Handbook of technological sustainability: innovation and environmental awareness (pp. 374–388). CRC Press. https://doi.org/10.1201/9781003475989-27

Sharma R, Bhambri P (2024) Digital duplicity and the disintegration of trust: a quantitative inquiry into the impact of deep fakes on media sustainability and societal equilibrium. In: Bhambri P, Bajdor P (eds) Handbook of technological sustainability: innovation and environmental awareness, pp 273–291. CRC Press. https://doi.org/10.1201/9781003475989-21

Singh R, Bhatia A (2023) A comprehensive review of green IT practices in the metaverse. Environ Manage 72(2):290–304. https://doi.org/10.1007/s00267-023-01783-5

Sinha R, Banerjee S (2023) The role of IoT in achieving sustainability goals in digital platforms. J Ambient Intell Humaniz Comput 14(8):4283–4298. https://doi.org/10.1007/s12652-023-04556-4

Thirumalaiyammal B, Steffi PF, Bhambri P (2024) Green horizons: navigating environmental challenges through technological innovation. In: Bhambri P, Bajdor P (eds) Handbook of technological sustainability: innovation and environmental awareness. CRC Press, pp 292–304. https://doi.org/10.1201/9781003475989-22

Thomas A, Joseph M (2024) Sustainable practices in digital marketing within the metaverse. J Bus Res 158:182–190. https://doi.org/10.1016/j.jbusres.2023.12.015

Tiwari R, Chandra M (2023) Analyzing the economic implications of sustainable practices in the metaverse. Technol Forecast Soc Chang 180:121623. https://doi.org/10.1016/j.techfore.2023.121623

Varma S, Kaur N (2023) Digital identity and its implications for sustainability in the metaverse. Comput Secur 136:103847. https://doi.org/10.1016/j.cose.2023.103847

Vigneshwari J, Senthamizh Pavai P, Maria Suganthi L, Bhambri P (2024) Eco-ethics in the digital age: tackling environmental challenges through technology. In: Bhambri P, Bajdor P (eds) Handbook of technological sustainability: innovation and environmental awareness. CRC Press, pp 201–213. https://doi.org/10.1201/9781003475989-16

Zhang L, Chen H (2024) The convergence of AI, IoT, and sustainability in the metaverse. Artif Intell 310:103867. https://doi.org/10.1016/j.artint.2023.103867

Blockchain Powered Sustainability: A Qualitative Research Agenda

Kumar Saurabh, Neelam Rani, Debasisha Mishra, and M. Ramkumar

Abstract Blockchain as technology intends to provide significant sustainable solutions to influence human, social, economic, and environmental factors. The research proposes mapping sustainability theories-based themes areas to understand the current usage of blockchain technologies (based on characteristics) to drive sustainability. The paper aims to investigate the role of blockchain technology in propelling sustainable ecosystems using the proposed blockchain-led sustainability framework. A systematic review of blockchain sustainability studies is discussed using the Theory, Context, Characteristics and Methodology (TCCM) approach. 19 global expert interviews were conducted with relevant blockchain sustainability practitioners and user personas from industry and research. MAXQDA qualitative data analysis software has been used to analyse the interviews and opinions data thematically. Research findings can help managers envisage sustainable organisations to instrument and balance human rights, consensus, moral reasoning, and societal boundary contexts. The research proposes the role of sustainable digital practices in exercising novel methods, codes of conduct, policies, standards, compliance, law, and regulation requirements using blockchain technologies.

Keywords Blockchain · Smart contracts · Sustainability · Decentralisation · TCCM · MAXQDA

K. Saurabh (✉) · D. Mishra
Strategic Management, Indian Institute of Management, Shillong, Meghalaya, India
e-mail: kumar.phd20@iimshillong.ac.in

D. Mishra
e-mail: debasisha@iimshillong.ac.in

N. Rani
Finance and Control, Indian Institute of Management, Shillong, Meghalaya, India
e-mail: neelam@iimshillong.ac.in

M. Ramkumar
Operations Management, Indian Institute of Management, Raipur, Chhattisgarh, India
e-mail: mramkumar@iimraipur.ac.in

W. Leal Filho et al. (eds.), *Metaverse and Sustainability: Business Resilience Towards Sustainable Development Goals*, World Sustainability Series,
https://doi.org/10.1007/978-3-031-89545-6_8

1 Introduction

Today, several metrics, such as social responsibility and sustainability indicators, are being deployed to provide a robust and transparent ecosystem. Hence, a publicly available data-based scoring model design can represent an organisation's commitment to sustainable practices (Aronow and Fitzpatrick 2016). In this regard, the United Nations has also declared the importance of sustainable development goals (SDGs). It facilitates the channelisation of efforts against the global non-availability of a fair (transparent), inclusive, responsible and accountable system to ensure an end to poverty and prosperity for all (Van Vuuren et al. 2015). Sustainable development emphasises critical values of transparency, trust, tokenisation and decentralisation. Blockchain elements have been recognised as a fundamental technology implementation paradigm for fulfilling sustainability targets.

Blockchain technology supports traceability, accountability, authentication, contracting, and tokenisation characteristics, which is pivotal in achieving the sustainability goals of digital transformation (DT) (Kouhizadeh et al. 2021). Successful sustainable development requires diverse groups to work together. Blockchain develops trust via a cryptographic process through the secure exchange of information without a middleman linking diverse groups. Henceforth, when blockchain becomes the DT carrier, blockchain becomes the tool and accelerator for DT (Hughes et al. 2019). Moreover, sustainable efforts require an unbiased, trust-based consensus system (Hastig and Sodhi 2020). It is evident from previous studies that blockchain led sustainability to deliver a trust mechanism built on a consensus formed by the group of nodes, where the network authenticates transactions and approves their addition to the blockchain (Risius and Spohrer 2017). The primary goal of this qualitative research is to respond to the following two questions.

RQ1. What is the role of blockchain technology in driving sustainability in any prospective organisation?

RQ2. How are the sustainability theories and blockchain technology characteristics sequenced to envisage technology-led sustainable ecosystems?

While limited literature on blockchain-led DT is available with limited use cases for specific industries, research studies on how blockchain-enabled DT can lead to better sustainable development are scarce. This question motivated us to research the need for blockchain technology as a sustainable development (SD) lever. Section 1 paper works on a literature survey and targets to cluster established sustainability theories-based themes to blockchain characteristics. The Sect. 2 study contributes, by its rich content, to understanding the use of blockchain in sustainability and allows researchers and practitioners to work in a structured way using Moher et al.'s (2009) PRISMA work. Section 3 presents the data collection method and analysis of the research. Research work focuses on collecting interview data from 19 respondents across geographies. MAXQDA software was used to analyse the interview data qualitatively and presented it using Word Cloud, Code Matrix Browser, Code

Frequency, and Code Maps. Blockchain-led sustainability framework is conceptualised using thematic codes and interview transcripts. Finally, in Sect. 4, the TCCM framework is leveraged to weave theory, concept, development, evidence, research and management implications and discuss the study's limitations (Paul et al. 2021). We understand the role of blockchain in sustainability and how users leverage the applications, their substitutes, values, and their influence on humanity.

The research proposes mapping sustainability theories-based theme areas (using MAXQDA thematic analysis) to understand the current usage of blockchain technologies (based on characteristics) in sustainable development. The paper provides levers to examine theoretically sequenced sustainability factors with practical blockchain implementations. It also covers how research scholars and managers perceive the value and relationship between sustainability factors and blockchain using the TCCM framework.

2 Sustainability Theories Mapped with Blockchain Solutions

Sustainable development barriers look for collaborative methods to understand the real problem driven by technology but backed by established sustainability theories (McElroy et al. 2008). This section aims to understand how blockchain technologies can help devise a sustainability journey backed by relevant, sustainable theories-based themes validated in the following sections using MAXQDA.

2.1 Utilitarian/Deontological Business Ethics

Sustainable digital ethics require baselining business ethics using digital methods, toolkits, and guidelines (Christensen and Kohls 2003). Digital businesses must integrate with seminal ethical theories to practice the code of conduct, virtues, goods, and duties under various consequences and lay down the guidelines for sustainability programs. It assists in managing sustainability matters and offers the instruments to handle operational and ethical business issues (Ferrell et al. 1989). In this regard, two main sustainability theories can facilitate this aim powered by blockchain, as detailed below:

2.1.1 Triple Bottom Line (TBL)

Companies have widely utilised the TBL approach to gauge the benefits attained through the adoption of sustainable solutions (Treiblmaier 2019). TBL considers factors beyond the organisation's bottom-line profits and impacts the social system, environment, and economics (primarily considers this whole world planet) (Schinckus 2020; Mangla et al. 2021). Also, the TBL approach helps measure sustainability, strategies and enterprise-specific sustainability indicators (Norman and MacDonald 2004). Some studies also suggest that the TBL environmental approach measures the business impact on the ecosystem based on resource consumption. Furthermore, it monitors the consumption levels of organisations and society and audits natural resources based on green policies. Likewise, the TBL economic approach ascertains the stability of businesses and how it contributes to its community's financial health, risk mitigation, and economic growth (Elkington 1998).

Considering the importance of the TBL approach in accomplishing sustainable developmental goals, blockchain solutions can further enhance the effectiveness of the TBL economic approach following the decentralised ecosystem. In this regard, benefits provided by blockchain in the form of smart contracts, trust, traceability and transparency can help fulfil SDGs of natural resources with financial inclusion.

2.1.2 Corporate Social Responsibility (CSR)

An organisation can be socially accountable by practising a self-regulation-based business strategy and being conscious of society, environment, and economics by positively impacting society (Daly 1990). When performed in a volunteering programme, CSR practices, society benefits, givebacks, and philanthropy help foster community habits and form stronger bonds (Norman and MacDonald 2004).

The role of technology can be utilised by following the theories of 'utilitarian and deontological ethics' and mapping it with CSR theory for successfully achieving sustainable transformation. It could be executed in documentation, authentication of information using digital signatures and transparent tracking of the resources in a decentralised way. Such audit trailing, monitoring, and decentralised management provide greater power for enhancing brand image, commitment toward human rights, and capturing corruption-based frauds.

2.2 Customer/User Privacy, Integrity and Safety

Past studies have provided substantial evidence for integrating user privacy, integrity and safety rights with sustainable development theories (Upadhyay et al. 2021). To fulfil the necessities for good living conditions for human beings, following the principles of equity and inclusivity can be best established by adopting blockchains (Upadhyay et al. 2021). In this regard, the two sustainability theories that are powered by blockchain can help achieve these aims.

2.2.1 Equity and Human Rights

The social systems have undergone massive transformations considering the requirements associated with existing rights while paving the way for new rights (Venkatesh et al. 2020). Older human rights refer to civil, political, economic, social and cultural rights. These rights are based on the Universal Declaration of Human Rights (UDHR), 1948, and refer to rights about life, equality, democratic participation in political activities, and overall well-being (Sen 1992). Considering the new technological advancements, data privacy, protection, and integrity rights (for instance, GDPR compliance) must also be considered (Venkatesh et al. 2020; Hughes et al. 2019). It involves collecting, using and disseminating the data by any agency or authority (Schinckus 2020).

Here, blockchain builds a system tuned to origin (older roots) and can be well deployed to fulfil privacy, security, and integrity requirements (Venkatesh et al. 2020; Hughes et al. 2019). It establishes accountability through digital authentication, identity, and equity-based tokenisation, with incentivisation based on proof of effort and wages. Hence, it ensures human rights adherence and avoids labour abuses (Schinckus 2020). There are many use cases in different parts of the world where this technology ensures fair political systems and provides better health care (Leng et al. 2020; Salmerón-Manzano and Manzano-Agugliaro 2019). Sustainability and traceability trade-offs capture the essence of blockchain adoption. The degree of distrust is proportional to avoiding sustainable solution implementations of blockchain (Biswas et al. 2023). Smart contracts can be written sustainably so that all the participants (humans) of the network (society) are facilitated with equal opportunities for rendering services (mining resources) and asking for services (mining requests) (Andryukhin 2018).

2.2.2 Behavioural Systems

In the past, the role of behaviour sustainability has been well acknowledged in various global forums such as G8-Summit, UN, and World Economic Forum (Vemuri and Costanza 2006). It works on personal and community-based principles and mutual action toward society (Mulder et al. 2006). Blockchain works on decentralised autonomous organisations (DAO). Using consensus algorithms helps balance the limited resources and consumption-based infrastructure, organisation structure, design, and sustainability needs (Sayeed et al. 2020).

2.3 *Organisational Design and Governance Processes*

Sustainable organisational design and governance processes refer to the community of systems, stakeholders, structures, styles and how they communicate to achieve sustainable development goals with a shared vision. In this context, the two main sustainability theories that help accomplish this are powered by blockchain (Clohessy and Acton 2019; Atzori 2017).

2.3.1 Institutional Theory (Factors)

DiMaggio and Powell (1983) explained this using the institutional isomorphism theory, namely 'coercive, normative and mimetic isomorphism. Scholars termed the theory as game rules (pressures) set by humans, society, state or institutions that make organisations or corporates successful, stable and overall competitive (Porta et al. 1998; Claessens and Fan 2002).

Blockchains fully support institution management as they maintain trust according to the requirements of participants (stakeholders) (Treiblmaier 2019). All nodes maintain data integrity to ensure data protection. Agreements are key to institutional factors as smart contracts handle them well without intermediaries or agencies (Tan et al. 2020). Blockchain-enabled workflows ensure fairness and inclusiveness. It works on consensus algorithms and data collected using social structure integration, reproduces many patterns and helps to remove disparity and discrimination (Murray et al. 2019; Andryukhin 2018).

2.3.2 Transition Theory

Transition theory is based on a change and transformation journey using multiple perspectives. It looks for transitions for new reformed strategic actions (Geels et al. 2004). These transitions can happen in socio-technical macro settings or niche settings for various innovation and experimentation scenarios (Avelino et al. 2016). Transition is successful in any organisation backed by robust 'governance' guidelines (Loorbach and Rotmans 2006). It can include organisational structures, normative rules, incumbents' intentions, stakeholder analysis, bias, and power rules (Kouhizadeh et al. 2021).

Blockchains maintain standards of data in decentralised structures. The blockchain system transparently aligns with a process that ensures proper and consensus-based techniques are followed for the desired outcome, appreciating transition at a community or organisation level (Davidson et al. 2016). Blockchain technology-led algorithms drive an organisation's governance and influence sustainability areas of ethics and policy making (Chatterjee et al. 2024). Governance is achieved by blockchain design-level accountability that can be audited for regulatory and compliance requirements. Automation agents can perform tasks based on immutable distributed ledgers for better governance (Allen and Berg 2020).

3 Methodology

3.1 Review Method

We conducted a PRISMA statement complied scientific method-based literature review to produce transparent, fair and inclusive results. Steps were adapted based on specific attributes from similar prior work suggested in the literature (Moher et al. 2009). The literature search was conducted during the Nov 2023–Feb 2024 timeline in two cycles without any filters on the timeframe. The review used recognised databases like Wiley, Web of Science, Scopus, Google Scholar, and Science Direct. Further review is conducted using the reference of the shortlisted article to fortify the search process.

3.2 Article Selection

We conducted four steps based on a bibliographic search- identification, screening, eligibility and inclusion (Fig. 1). We performed a bibliographic trail search using references other than a keyword-based search. We screened and analysed the paper's titles and abstracts and excluded articles based on irrelevant search results. As authors, we devised eligibility criteria to include (e.g., peer-reviewed publications, proceedings articles, book chapters, analyst reports) and exclude (e.g., non-English papers, generic articles, technical blockchain software articles) research papers for our focused study. Initially, abstracts were analysed to cluster the paper based on themes, and finally, specific full-text-required articles were retrieved. Irrelevant full-text research papers are eliminated in this phase to get the targeted review articles. We included 112 papers related to our anticipated work themes according to eligibility. The papers were searched based on various criterion keywords like (1) blockchain sustainability (2) blockchain in green programs (3) blockchain sustainable models, (4) blockchain social development, and (5) blockchain in corporate social responsibility. Finally, 25 articles were included for qualitative synthesis. We used the thematic content analysis (MAXQDA) method to cluster the articles based on the inclusion criteria.

The final 25 articles shortlisted for our study were considered the body of knowledge adapted from the blockchain-led sustainability literature. Table 1 depicts the bibliographic journal sources from which the papers are identified with their frequencies (with percentage). Table 2 presents the industry-wise information of the study with author details. Table 3 depicts the authors' commonly used methods in studying blockchain-led sustainability.

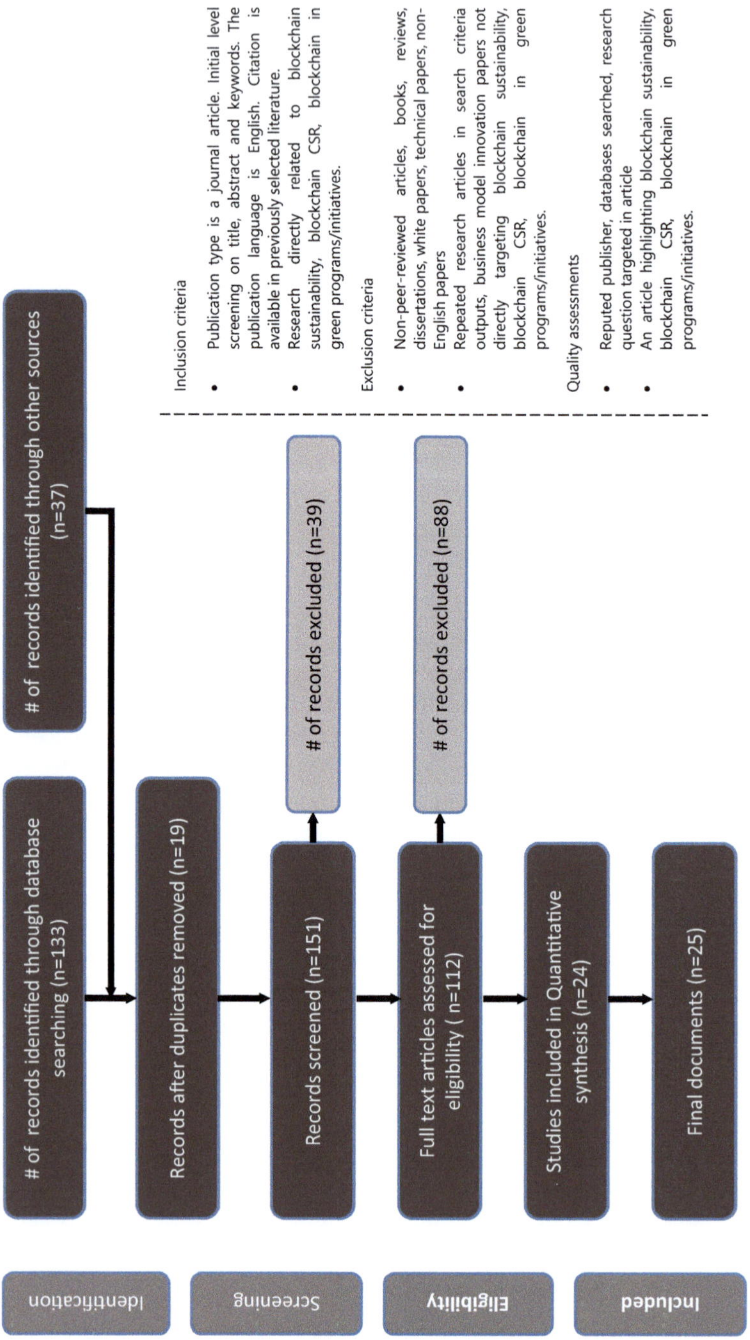

Fig. 1 PRISMA flow diagram. *Source* Moher et al. (2009)

Table 1 Bibliographic sources on blockchain led sustainability literature

Bibliographic source	Articles considered for review	Total (%)
Journal of Cleaner Production	4	16
Journal of Business Research	2	08
Technological Forecasting and Social Change	2	08
Transportation Research Part E: Logistics and Transportation Review	2	08
International Journal of Information Management	2	08
International Journal of Electrical Power & Energy Systems	2	08
Information Processing & Management	1	04
Computers in Industry	1	04
Sustainable Production and Consumption	1	04
Computers in Human Behaviour	1	04
International Journal of Production Economics	1	04
Cities	1	04
Procedia Computer Science	1	04
Futures	1	04
Computer Standards & Interfaces	1	04
Robotics and Computer-Integrated Manufacturing	1	04
Journal of Systems Architecture	1	04
Total	25	100.00

Source Author's own compilation

Table 2 Industry-wise literature context

Industry	Authors	Percentage
Generic	de Villiers et al. (2021), Dursun and Üstündağ (2021), Böckel et al. (2020), Devine et al. (2021), Drljevic et al. (2020), Upadhyay et al. (2021)	24
Logistics-supply chain	Ali et al. (2021), Kshetri (2021), Kouhizadeh et al. (2021), Sunmola (2021), Venkatesh et al. (2020)	20
Real estate-construction	Mora et al. (2021), Bagloee et al. (2021), Kumar et al. (2021), Figueiredo et al. (2022)	16
Power and energy	Wang et al. (2022), Tsao et al. (2021)	08
Government	Choi and Luo (2019), Schulz et al. (2020)	08
Food industry	Tsolakis et al. (2021), Mangla et al. (2021)	08
Agriculture	Dey and Shekhawat (2021), Li et al. (2020)	08
Aviation	Di Vaio and Varriale (2020)	04
Fashion	Wang et al. (2020)	04

Source Author's own compilation

Table 3 Commonly applied methods

Method	Author(s)	Percentage (%)
Literature survey	Böckel et al. (2020), Wang et al. (2020), Mora et al. (2021), Ali et al. (2021), Kouhizadeh et al. (2021), Di Vaio and Varriale, (2020), Bagloee et al. (2021), Devine et al. (2021), Drljevic et al. (2020), Venkatesh et al. (2020), Figueiredo et al. (2022)	36.67
Conceptual framework/ model	de Villiers et al. (2021), Dursun and Üstündağ (2021), Sunmola (2021), Tsolakis et al. (2021), Venkatesh et al. (2020), Kumar et al. (2021)	20.00
Case study approach/use case method	Tsolakis et al. (2021), Schulz et al. (2020), Di Vaio and Varriale (2020), Dey and Shekhawat (2021)	16.67
Theoretical/mathematical model	Choi and Luo (2019), Tsao et al. (2021), Li et al. (2020)	10.00
Proposition development	Kshetri (2021)	3.33
Thematic analysis	Devine et al. (2021), Upadhyay et al. (2021)	6.67
Simulation and modelling	Mangla et al. (2021)	3.33
Expert interview and validation	Wang et al. (2020)	3.33

Source Author's own compilation

4 Findings

Software-assisted qualitative analysis method—Computer-assisted Qualitative Data Analysis Software (CAQDAS) is implemented to evaluate blockchain influences when adopted as a digital solution for sustainability in an organisation. Technical explanations can be better visualised qualitatively for social and ethical aspects-based intelligence (Holloway and Galvin 2016). The study broadly observed the sights of blockchain practitioners, blockchain potential adopters, sustainability solutions advisors and ESG experts. It supports comprehending their opinions on how individuals and organisations wanted the digital transformation to be carried out sustainably.

4.1 Data Collection: Data Analysis

The research followed a systematic two-way communication-based informal data collection approach to conduct the in-depth interviews (Mason 2002). Virtual face-to-face interviews were conducted via MS Teams/Zoom call. Using qualitative methods, 19 interviews were scheduled to gauge the sustainable dimensions and their perspectives. The research considered purposive sampling based on 19 interviewee experts to fulfil the study criteria. The pool of experts comprises blockchain chain experts,

C-level executives, sustainability researchers, and experts from governance, risks and compliance teams who have performed exemplary in their field. A similar methodology has been followed in previous studies (Etikan et al. 2016). Research uncovered aspects not touched on in the earlier studies, as interview analysis of blockchain-led sustainability is one of its kind. Researchers contacted participants by establishing personal connections and references. Fourteen (14) interactions/discussions were scheduled between Mar 04, 2023, and Mar 31, 2023; five (5) interactions/discussions were scheduled between Nov 12, 2023, and Dec 16, 2023. All the interviewees represent a global footprint comprising 9 countries and 10 from India. Research practised the snowball effect with the specific purpose of exercising the sampling method. It helped to meet the two goals while conducting the interview (1) research subject (2) research questions. It ensured that the interviewees were equipped with the relevant information and became more participative in the research.

The method of our research interview process is open-ended semi-structured discussions. Each interview is available as transcribed data shared and validated with the interviewee. The analysis aims to produce the themes and codes during the talks to develop the broad categories. We analysed the common themes to link one individual's views with the other. We developed notes of each interview in relation to the topics discussed during the interview. It is developed in the form of memos to generate the categories of data to record ideas and map theories. The interview transcript data were again analysed, and this time, the objective was to remove the worthless information first and note down all the important headings or keywords used during the interviews. The categories surveyed in the previous phase under different headings are analysed and grouped under the umbrella or big headings. The objective is to reduce the categories and finalise the broader one, covering the various subtopics/headings underneath. Two co-authors were requested to develop the thematic category systems with headings that independently identified thematic research categories compared with each other, and necessary alterations were made to safeguard against possible biases. We used coloured coding across the transcripts to distinguish all the transcripts using the MAXQDA software. We used various qualitative features of the software to develop word clouds, code maps, frequency matrices, etc., and to visualise the qualitative data as much as possible. When writing the research stage's conclusions, we referred back to the transcripts as and when required for clarity. Interview data were transcribed as the first step before the start of the content analysis. Finally, 4 main themes were developed after the rigorous review of the transcripts. Two researchers initially worked on the themes, and two worked as reviewers before finalising and importing the themes into the MAXQDA qualitative analysis software. Axial coding as a thematic analysis method was leveraged to understand subthemes' roles and associated relationships. The iterative process was followed to validate the themes and subthemes based on the data extracted by the interviewees during the discussions (Heung et al. 2011). We rejected any source of prejudice by eliminating the likelihood of biasedness to intensify data reliability (Fischer 2009). Transcripts were shared with the interviewees after the interviews before they were used for content analysis.

4.2 Interviewee Details

Table 4 describes the interviewee details, confirming the diverse persona of the experts and demographic information before we present the MAXQDA qualitative analysis output.

4.3 Word Cloud

MAXQDA software provides a word cloud feature to depict the word frequency and percentage across the transcripts. Figure 2 describes a word frequency-based cloud formed using the interview transcripts. It depicts the frequently used words used by the interviewees when responding to the questions (Sinclair and Cardew-Hall 2008).

Figure 2 is the testimony of the blockchain sustainability-based highest frequency words revealed after thoroughly analysing the interview transcripts. Each transcript adds the commonly used words in its frequency from transcripts. The size of the word is proportional to the frequency of the word. The greater the word frequency, the larger the size of the word. It is evident that words like blockchain, the lexical combination of the words sustainability, human, strategy, maturity, compliance, audit, model, organisation, behavioural, business, and institutional, appear bigger because of their occurrence, repetition, and preferential words used in the interview transcripts.

4.4 Themes and Subthemes Study

Transcript data were analysed to evaluate the possibilities of driving sustainability programs and solutions using blockchain technology. Research equated the relationship of the interview transcripts using the unique coding process. The coding process is a phased approach to analysing the transcripts (Singh and Dey 2020) (Fig. 3).

As the first step, code labels were tagged to the areas of interest of the data provided through the interviews during the coding process of developing the themes. The code labelling process is iterative and goes for multiple changes after each interview as the researcher seizes the interviewee's experience. The changes were performed on the data continuously till theoretical foundations were attained.

The developed themes during the process worked like the fundamental unit of qualitative research. The analogy presented the interviewee's intention, viewpoint, and choices while exercising blockchain to pursue sustainable interests. All the themes developed during the process are based on data analysis, field comprehension and interview transcript interpretation (Braun and Clarke 2006; Holstein and Gubrium 1997).

Themes(T)/sub-themes (ST) are presented in Table 5 with the associated word frequencies used by the interviewees during the discussions. Research presented 4

Table 4 Interviewee profiles

S. No	Interviewee code	Gender	Age	Country	Date of interview	Experience in years	Role	Industry
1	Int01	M	39	India	02-Nov-23	14	Researcher	Education
2	Int02	F	49	India	03-Nov-23	17	CTO	Telecom
3	Int03	F	53	Canada	09-Nov-23	26	CEO-start-up	IT
4	Int04	M	36	Qatar	13-Nov-23	11	Lead-sales	Business
5	Int05	M	47	India	14-Nov-23	21	Researcher	Comp Sci
6	Int06	M	38	India	16-Nov-23	12	Gazetted officer	Government
7	Int07	F	43	USA	17-Nov-23	19	Procurement Mgr	Utilities
8	Int08	M	39	UK	20-Nov-23	15	Lead developer	IT
9	Int09	M	46	Ireland	21-Nov-23	23	Product manager	Healthcare
10	Int10	M	41	India	22-Nov-23	16	Chief architect	Retail
11	Int11	F	50	USA	23-Nov-23	24	Head-CSR	Manufacturing
12	Int12	M	48	India	24-Nov-23	22	Head-innovation	Logistics
13	Int13	F	55	India	30-Nov-23	31	Delivery manager	IT
14	Int14	M	49	Germany	31-Jan-24	26	CIO	Automobile
15	Int15	M	38	France	12-Feb-24	21	CISO	Bank
16	Int16	M	33	India	15-Feb-24	9	Compliance Mngr	Finance
17	Int17	F	37	India	20-Feb-24	12	Sustain. lead	Construction
18	Int18	M	42	Malaysia	22-Feb-24	18	Head-audit	Aviation
19	Int19	F	40	India	27-Feb-24	17	Lead-CSR	Government

Fig. 2 Blockchain sustainability word cloud

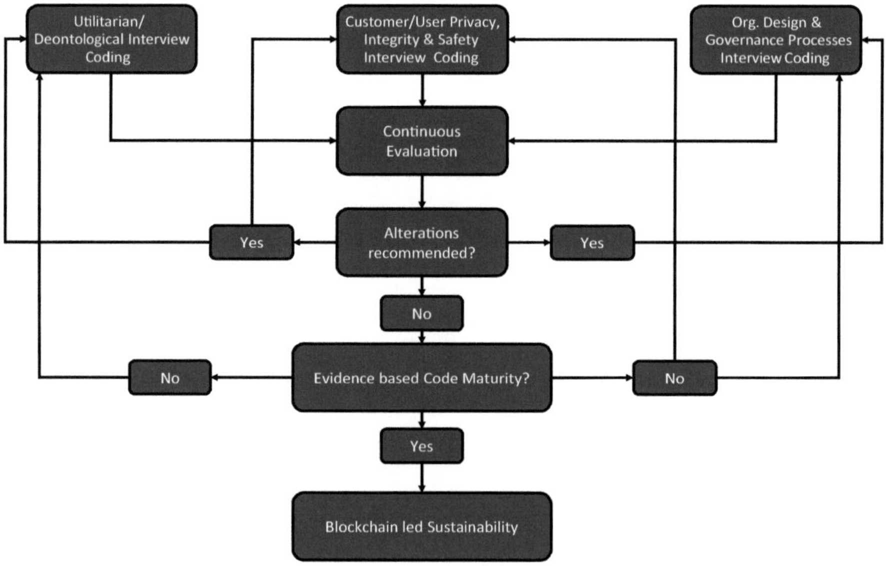

Fig. 3 Coding process

themes and 6 subthemes in the below table. The most pervasive subthemes indicated in all the interviews are 'Customer/User Privacy, Integrity & Safety', 'Organizational Design & Governance Processes', 'subtheme-Transition Factors' and 'subtheme-Behavioural Systems'.

MAXQDA salient feature- code matrix browser is presented in Fig. 4 to visualise the themes/subthemes based on data analysis. The code matrix browser helped

Table 5 Themes (T)/sub-themes frequency analysis

Themes/Subthemes	Frequency	Percentage
Blockchain led sustainable behaviour	46	15.18
Organisational design and governance processes	38	12.54
Customer/user privacy, integrity and safety	37	12.21
Organizational design and governance processes\transition factors	34	11.22
Customer/user privacy, integrity and safety\behavioural systems	30	9.90
Customer/user privacy, integrity and safety\equity and human rights	29	9.57
Utilitarian/deontological business ethics\triple bottom line	26	8.58
Organizational design and governance processes\institutional factors	24	7.92
Utilitarian/deontological business ethics\corporate social responsibility	22	7.26
Utilitarian/deontological business ethics	17	5.61

analyse all the transcripts in their entirety as it would be challenging to comprehend relationships, validate the codes, compare them, and weave the bigger picture without it. Figure 4 provides a snapshot of all the themes and subthemes for each document-based frequency code. Frequencies are shown using the square shape. The square size is proportional to the repetition of the codes across the transcript. It helps to visualise code distribution among the interviewees based on the size of the square. Various insights can be formulated using the size of the theme/subtheme available in the code matrix browser. For instance, int 16, the interviewee comes from the finance field as a compliance manager, and his concerns were related to equity and human rights 'and 'Behavioural Systems' as depicted by the bigger size square.

MAXQDA code maps (Fig. 5) can be used to visualise the outline of the themes(T) /subthemes (ST). It helps to depict the relationship based on mapping codes and frequency distribution among the transcripts. It is impossible to comprehend all the hidden relationships till it is not mapped using the MAXQDA utility. The code maps do not visualise the directions as they help present and comprehend the codes relationship. Figure 5 reveals that blockchain-led sustainability is a complex phenomenon influenced by multiple themes(T)/subthemes (ST).

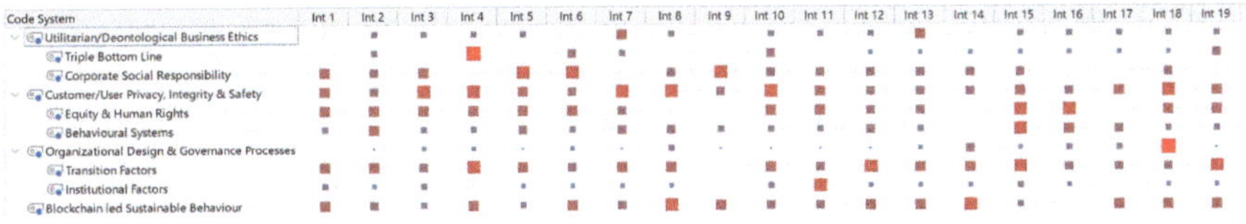

Fig. 4 Code matrix browser

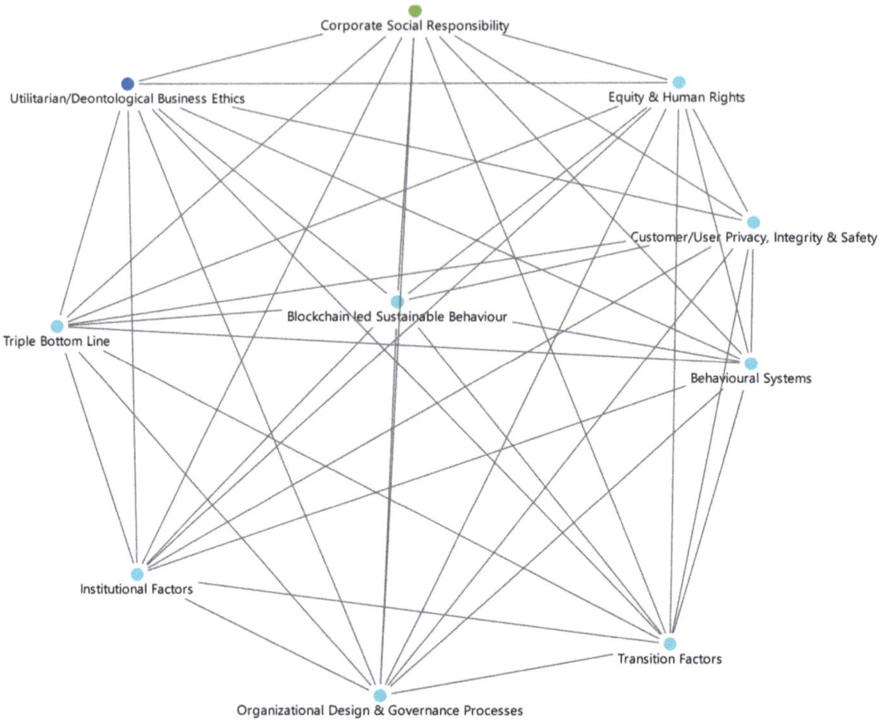

Fig. 5 Code maps

5 Discussion

Using the thematic analysis approach, we worked beforehand on 4 central themes (T1, T2, T3 and T4) with 6 subthemes after the detailed discussion during the interview. The themes are developed based on thorough literature survey results and validated using thematic analysis using MAXQDA.

Research uses blockchain technologies based on the 12 questions to weave the sustainability narrative. The discussions were undertaken not strictly adhering to the questions, but they were used more to captivate the plot of the subject, open up the interviewee, and share their experience. The paper presents the 4 final themes analysed based on the interview transcripts and discussions. The study encapsulates themes and mapped subthemes and excerpts from the interview's transcript data in the section below.

1. Utilitarian/Deontological Business Ethics (T1)
2. Customer/User Privacy, Integrity and Safety (T2)
3. Organizational Design and Governance Processes (T3)
4. Blockchain led Sustainable Behaviour (T4).

5.1 T1-Utilitarian/Deontological Business Ethics

Utilitarian/Deontological Business Ethics is the first theme anticipated by the researchers based on the existing literature on sustainable strategy. Researchers for business ethics expected two themes. Interviews are intricated following sub-thematic codes for T1.

- T1-ST1- Triple Bottom Line
- T1-ST2- Corporate Social Responsibility.

T1-ST1 and T1-ST2 are described and evaluated in the following section, authenticated by valuable interviewee transcripts.

5.1.1 T1-ST1-Triple Bottom Line

This was an exciting discussion theme, where half of the respondents expressed positive views on the importance of the Triple Bottom Line propelled by the blockchain system. They unarguably positioned the need for sustainable blockchain systems that at least lack today at a required level. Some excerpts of the transcriptions are given below, as discussed during the interview.

> Sustainability is a big concept. So far; I have not seen technology like blockchain perfect for balancing social ecosystem in a so-called 'sustainable way. Interviewee—**Int05**

> Decentralisation and immutability are the core of blockchain and the heart of sustainability; blockchain enables trust and transparency, which are the needs of the hour. As a company, we bank on it to serve triple bottom line areas like social and environmental impact as discussed with you. Interviewee—**Int14**

> I am practising compliance management for my organisation. We have created many MOOCs to make our employee base aware of the ethics code of conduct. However, after discussing with you the Utilitarian/Deontological (somewhat difficult to pronounce) issues, I will request my learning team to look once again at the requirements. Interviewee—**Int16**

5.1.2 T1-ST2-Corporate Social Responsibility

The corporate social responsibility-based theme recommended that traditional ethics concepts be based on societal, community, individual, and organisational theories to give moral reasoning to values that deliver happiness, human welfare, etc. The discussions very well give importance to rights, duties and virtues for blockchain CSR activities. Most respondents indicated that interpreting such ethical cognition to technology (like blockchain) is a phenomenal chore. Some excerpts of the transcriptions given below validate the discussion.

> In my opinion, ethical theories are critical, but I doubt that you know this (except compliance teams) till you have not told me. It opened my eyes. I will try emulating something in our coding practices. In my opinion, these theories are vital to thinking of sustainable alternatives about the CSR. Interviewee—**Int10**

> It is easy to say and preach, but practising CSR in a real field is difficult. When a customer faces a critical outage, optimisation targets, a shorter timeline, moral reasoning, virtues, and well-being take a hit. I am not against this but emphasising that customers also must understand the ethics like providers and should not ask unrealistic outcomes. Interviewee—**Int02**

> As per my profile, my job is to bring transparency in CSR activities when the government itself is involved in the great initiative. I hope blockchain can give me relief if my superiors are ready to adopt it. Interviewee—**Int19**

5.2 T2-Customer/User Privacy, Integrity and Safety

Customer/User Privacy, Integrity, and Safety is the second theme anticipated by the researchers based on the existing literature on sustainable strategy and data collected. Researchers developed this theme to a certain extent based on the existing literature on sustainable user experience for the sustainability journey. Interviews are intricated following sub-thematic codes for T2.

- T2-ST1-Equity and Human Rights
- T2-ST2-Behavioural Systems.

5.2.1 T2-ST1-Equity and Human Rights

This was one of the hottest topics of the interview, and it took a long time to probe equity and human rights issues. Blockchain systems depend on publicly available data (but it is immutable and decentralised), and if it is compromised, it will affect humans, their rights, the social ecosystem, its privacy, security and integrity systems, etc. While working on blockchain-led sustainable systems, we debated humanity theory factors and how they affect equity and human rights, behavioural aspects like knowledge, Values, Attitudes, well-being, self-worth, etc. The concerns are very well covered in the transcript given below.

> As a researcher, I can assure you that blockchain is best for ensuring the privacy and security of humans while using the systems. However, we should learn such sustainable techniques of ethical coding, even architecting solutions that support sustainability. Interviewee—**Int01**

> This is the basis of trust. What can be more sustainable than this? I believe blockchain characteristics can achieve this, but I have asked all my providers to demonstrate this as POC. So far, no response. Interviewee—**Int12**

> I am confident that our solutions help achieve equality and practice for human rights as they are powered by blockchain. I have multiple stakeholders, contractors, and subcontractors to maintain business. It was challenging to see what was happening on the ground. Now, using blockchain, I get a cockpit view of different processes getting executed in my organisation, where we can ensure no violation of human rights, unbiased and inclusive system. Interviewee—**Int03**

5.2.2 T2-ST2-Behavioural Systems

All respondents agreed that blockchain systems should be equipped with behavioural characteristics like reliability and safety to sustain humanity. Efficient data handling is the key to enabling reliability and safety in behavioural systems. Respondents appreciated the theories and discussed that it is understood that blockchain systems can be used to implement business models and channels for society, environment, and economic factors. The transcript below demonstrates the importance of sustainable behavioural systems.

> Blockchain as a technology is reliable. The most important thing is how to design it. Fundamental characteristics enable reliability and consensus to establish trust before taking action. The supply chain has some excellent use cases to emulate, as I use them in my workplace. Interviewee—**Int07**

> I represent the construction industry. Of late, I have heard many things about blockchain as a sustainability leader. It is good now that we discuss the managerial way of achieving sustainable systems outside the book of knowledge. It is also important for us to comply with many government norms. Interviewee—**Int17**

5.3 T3-Organizational Design and Governance Processes

Organisational design and governance process is the third theme anticipated by the researchers based on the existing literature on sustainable strategy. Researchers developed this theme to a certain extent based on the existing literature and thematic analysis of sustainable organisational design and governance processes. Interviews are intricated following sub-thematic codes for T3.

- T3-ST1-Transition Factors
- T3-ST2-Institutional Factors.

5.3.1 T3-ST1-Transition Factors

Mostly, all respondents talked about the significance of transition factors in blockchain operating processes in the sustainability journey. They emphasised that blockchain systems should be context-driven, well-established, workflow-based systems with proper audit features. Most respondents agreed that a blockchain system helps structure, refer, and work on smart contracts, provide accountability to specific dilemmas and moral reasoning, and sustain conflict management. Our transcripts also prove the exact requirements given below during the interview.

> There is no doubt that blockchain characteristics are good for governance. I have introduced it in my organisation for some of the use cases in logistics for asset tracking as well as managing the external vendors invoicing and many workflow-related automation jobs. Interviewee—**Int12**

I am using it for auditing purposes and compliance auditing. It is implemented using cloud technologies and helps us to maintain the log of transactions that is one of the compliance requirements. Interviewee—**Int16**

Some service providers discussed the blockchain-based log management system in our aviation industry. We have conducted the pilot, and the results were satisfactory. We are building on it. Interviewee—**Int18**

5.3.2 T3-ST2-Institutional Factors

Respondents accepted that institutional factors are essential criteria for achieving a sustainable system. Researchers developed this theme to a certain extent based on the existing literature on sustainable community-based organisation design. It was discussed that blockchain supports the idea that size, structure, style, shared values, and technology play a vital role in the organisation's design. The importance of this is highlighted in the transcripts given below.

We have implemented a decentralised autonomous organisation (DAO), which is one of the key concepts of blockchain in organisation design. We are getting good outcomes on that. External IT giant is helping us achieve this. Interviewee—**Int09**

Take the example of DAO; I am not surprised how this approach will change all the theories of sustainable organisation design. It will bring a revolution to running an organisation just on smart contracts. Certainly, it will take time to take to reality, but it is not far. Interviewee—**Int09**

My opinion is different; I accept organisation design influences sustainability, but I doubt getting it translated using technology. I want to see how this works. Decentralisation will not help every time. It depends on the design and requirement. Interviewee—**Int06**

It admires to open my eyes; I never thought technology could play such a role to bring unbiased, fair, inclusive and what you say 'Immutable' transparent trust-based social system (with a pun intended). Interviewee—**Int04**

5.4 T4 Blockchain-Led Sustainable Behaviour

Blockchain-led sustainable behaviour is the last theme anticipated by the researchers based on the existing literature on sustainable strategy and data collected. This theme was developed by researchers based on the existing literature on sustainability theories for the sustainable journey. Principally, it is achieved using discussed thematic analysis, evaluation and assessment of sustainability theories factors. All respondents avowed the need for sustainable behaviour to be powered by blockchain to perform SDG activities. Some excerpts of the transcriptions are available to baseline the argument.

I do not have any second opinion that we should not use blockchain for sustainability. It requires a systematic approach as it has all the principles to choose what is good for human beings or better, I can say for our entire planet. Interviewee—**Int05**

As a head of the information security department, I see many perspectives of using blockchain in sustainable ecosystems. It is not about bringing security to financial organisations but contributing to UN SDG programs' overall agenda. Interviewee—**Int15**

With this interview, I got a new perspective on sustainability through the lens of technology. I never thought we could expect my organisation's digital journey using blockchain-led sustainable theories. Thanks for arranging this. Interviewee—**Int12**

5.5 Blockchain-Led Sustainable Behaviour Framework

A literature survey, interview transcripts, and MAXQDA software helped us code the developed themes and subthemes of blockchain-led sustainability. The dimensions of utilitarian/deontological, customer/user privacy integrity and safety and organisational design and governance processes emerged as the elements to represent a blockchain-led sustainability framework (Fig. 6). The research proposes that the first component- relevant sustainability theories of the triple bottom line and corporate social responsibility steer utilitarian and deontological ethics. The second component, customer/user privacy integrity and safety, is driven by the sustainability theory of equity and human rights and behavioural systems. The third component of the framework is organisational design and governance processes influenced by organisational sustainability theories of transition and institutional factor theories. Finally, the other elements, utilitarian/deontological, customer/user privacy integrity and safety, organisational design and governance processes, enable blockchain technologies to influence sustainable behaviour.

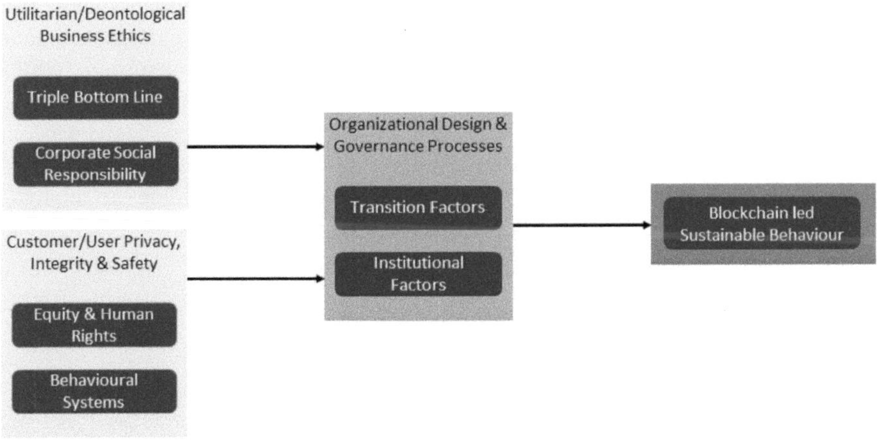

Fig. 6 Blockchain-led sustainable behaviour framework

6 Discussion

The theory, context, characteristics and methodology (TCCM) framework is used to discuss and consolidate the research results that create the discussion's foundation (Paul et al. 2021; Paul and Rosado-Serrano 2019).

6.1 Theory

Authors have mainly focussed on blockchain-driven sustainability use cases in different settings. Based on the literature survey, a few papers (9.52% only) banked on the established theory to build context. Our literature survey indicates that only two papers (Tsolakis et al. 2021; Mangla et al. 2021) out of 25 papers find the theoretical genesis. Tsolakis et al. (2021) use the Principal-Agent Theory, Resource-Based View Theory, Network Theory, and Transaction Cost Analysis Theory to propose a blockchain SDG-based food (fish industry) supply chain. The paper suggests data asymmetry is the biggest problem in achieving SDG targets. The authors present our design principles for the supply chain using blockchain technologies. Similarly, Mangla et al.'s (2021) work uses systems theory to investigate the societal impacts of blockchain technology on agriculture, the food industry and rural development. The research presents issues related to the food supply chain and proposes using blockchain for transparent and end-to-end tracking. It is important to note that we have not found established sustainability theories (like utilitarian/deontological business ethics, triple bottom line (TBL), corporate social responsibility (CSR), equity and human rights, institutional theory and transition theory) while using blockchain sustainability as keywords during the literature review. However, our paper contributes to mapping related literature and sustainability theories based on blockchain characteristics. MAXQDA software is used to validate the thematic analysis of the theories using interview transcripts.

6.2 Context

This review observed that the blockchain sustainability development use cases are driven in multiple contexts. 6 out of 25 (24.0%) papers targeted the generic industry-based market segmentation for sustainability requirements (Table 3). Logistics-Supply Chain is the most apparent and common industry where multiple use cases and contextualised requirements are researched (5 out of 25 papers, 20.0%, Table 3). We also found specific studies in the context of blockchain and sustainable supply chain management in developing countries assessing standards like environmental accounting, product quality and social impact. The real estate construction industry also shows the promising adoption of blockchain technologies for sustainable development. The research investigates various options for smart cities where blockchain can be used for sustainable service delivery, resource management and city administration (Mora et al. 2021; Bagloee et al. 2021). Social aspects of sustainable development related to the food industry are also investigated in two papers (country context: Turkey, Thailand) to design principles and an integrated technology implementation framework. The area investigated was to reduce food fraud, enhance welfare programs, increase proximity to the food market and food security, access and social acceptability using blockchain (Tsolakis et al. 2021; Mangla et al. 2021). Sustainability issues related to the power and energy industry (country context: Vietnam) are also highlighted (in two papers) to increase profitability and customer satisfaction (Wang et al. 2022; Tsao et al. 2021). Applications like digital identity, climate finance, and social inclusion in the government sector have been well-researched in two papers (country context: India and Netherlands).

6.3 Characteristics

Most of the papers highlighted the importance of blockchain technology usage for UN SDG requirements. Researchers indicated how blockchain-based solutions could become the catalyst for socially sustainable business. Past studies delivered a blockchain model to ensure social and economic coexistence with new trust relationships among stakeholders using smart contracts. Many papers highlighted the literature survey for urban use cases about governance, logistics and transportation (smart cities). Investigations provide context-aware, secure, and auditable (using blockchain) information-sharing platforms for sustainable visibility in heterogeneous scenarios. However, the literature shows wide-ranging unawareness of sustainability characteristics of TBL, CSR, human rights and equity, transition, institutional culture and climate. Demographics (age, exp, gender, location, etc.) as control variables can reveal the fresh outlook for the adoption and diffusion of technology missing in the literature.

6.4 Methodology

Quality Research in any area requires proper balance in terms of applied methods like literature review, qualitative methods—case method, thematic analysis, cluster analysis, parametric and non-parametric studies, path and mediation analysis. A literature survey is the most preferred study method, with a frequency percentage of 36.67 (Table 3). The author's developed/compiled conceptual framework/model contributes heavily to survey literature (20.0, Table 3). The case study approach is the obvious choice for investigating the impact of blockchain technologies on sustainability (16.67, Table 3). It is essential to note that quantitative analysis is missing as the outcome of our study. Factor, path, mediation, partial least square, parametric and non-parametric studies are also not available to maintain the balance of qualitative and quantitative studies. Similarly, the absence of interview analysis in the literature indicates the non-availability of blockchain-led sustainability program experts. Methodological rigour can be applied in the sample identification, data collection and quantitative techniques to strengthen the anticipated research areas.

7 Research Implications

Blockchain-led sustainability development is influencing the broader avenues. It is executed on many relationships, from sustainability theories to blockchain characteristics and sustainable performance. It is an excellent opportunity to research utilitarian/deontological business ethics theory factors (TBL, CSR), humanity and behavioural theory factors (equity and human Rights, behavioural) and organisational theory factors (institutional, transition) to strengthen the proposed work. Proposed work and interview analysis overlay a compact route to envisage the next-generation blockchain-led digital agency for laying down ethics, agreements, regulations, and guidelines to safeguard sustainable perils. Expert interview analysis gives a substantial prospect to research the guidelines for blockchain-led sustainability' standards benchmarking agency' to achieve successful sustainability. Figure 7 outlines the future research opportunities following the TCCM approach based on the study conducted in the paper. Although we have extensively studied established sustainability theories tested empirically in the last 50–60 years in numerous circumstances, a comprehensive analysis can be achieved using theme-based constructs and sustainability testing. Appropriate statistical/ mathematical modelling can be applied and tested to the hypothesis developed based on themes (T) / subthemes (ST) and their relations. Research outcomes can be examined for small and medium enterprises, considering industries at micro and macro levels. Interviews and thematic analysis suggest the need for blockchain-led fair and inclusive systems, sustainable software architectural principles, sustainable social engineering design considerations, and theoretical sustainable development models to be brought to the blockchain-learning forums at the early stages of education.

8 Managerial Implications

Enabling sustainability in an enterprise by experts, community/society leaders, or activists entails assignation energies. The manager's job becomes more complex to get the right mix of ethics, human rights and behavioural and institutional transition factors to manage and design sustainable programs and policies. It requires holistic workforce participation, and managers must understand the subject's sensitivity, such as human rights and behavioural consequences that may bring down the whole

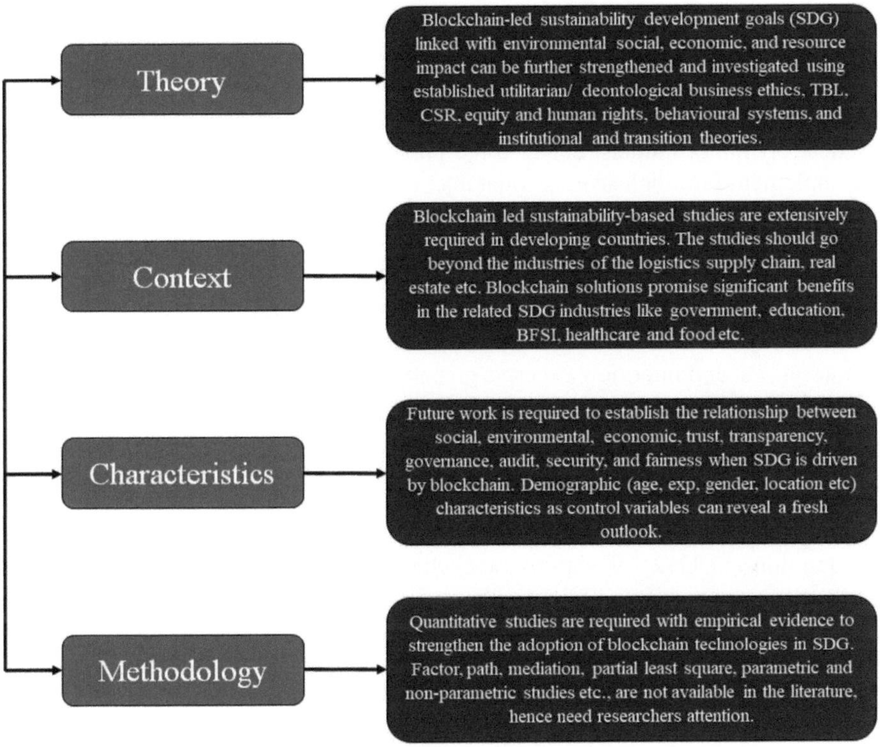

Fig. 7 TCCM approach based on future research opportunities

sustainable initiative if not appropriately handled. Blockchain-led proposed work can help in policymaking, assessing, evaluating, controlling, and timely auditing, among other things, which are the reasons to expect a sustainable digital blockchain program. Sustainability looks for leadership sponsorship and advocacy blockchain; our findings support it. We can choose the sustainability brand ambassador based on a person's work and self-sufficiency, promoting the correct social and communication characteristics standards while scheming and designing blockchain solutions for sustainability. Implementing blockchain-led DT sustainability can be best done using administrative practices. First, administrators are accountable for making the material and assets related to sustainability available to personnel using prevalent networks. Based on the proposed themes and subthemes, developed research acumens can be transformed into practices to produce a suitable blockchain-led digital sustainable learning program.

Findings from the paper can help administrators develop apparatuses and ecosystems for various circumstances related to sustainable development. Practitioners can use it as a sustainable device to work out morality, human rights, governance and organisational theory. Organisational systems, strategies, compliances, laws and whistleblowing codes can be established for pertinent services and products. The modelling of sustainability use cases can also be anticipated as the simulation of the proposed work can give good insights into various factors influencing the conclusion. Blockchain led to sustainable digital transformation, which is a combination of theory and practices. However, the study can become the reference point to practice practical operating models, enhance the public experience, enforce human rights, and implement fair, inclusive, accountable, responsible, and auditable propositions for a sustainable journey.

Trust, transparency, open architecture, and resiliency are the wide-ranging influences of blockchain technology on SDGs. Equality, infrastructure, production/ consumption and climate change are the fine-grained topics of SDG principles under blockchain discussions. Similar issues were discussed among the practitioners during the interviews, and interview excerpts give multiple glimpses of the specific SDGs. Affordable and cleaner energy (SDG7) was the topic of choice by many practitioners across the board. Blockchain technology ensures the practice of building resilient infrastructure for inclusive usage (SDG 9) and various industrialisation use cases. Blockchain promises the innovation for sustainable development to provide inclusive growth and immutable infrastructure (SDG9). Responsible production and consumption (SDG12) of various blockchain offerings can support the logistics supply chain, as confirmed by interviewees during the interviews. Many traces of literature and interview excerpts confirmed blockchain technology's role in climate action (SDG13). P2P energy trading, certificates as renewable tokens, and blockchain carbon credits were some discussion points while discussing the SDG 7 outputs of blockchain development. The use cases of the circular economy were adjudged as game-changers to create a more robust and resilient infrastructure common for public consumption but with accountable participation and sourcing. Greenhouse emissions climate change dashboards demand traceability, transparency and accountability, which can accomplished by blockchain technology to ensure reliable, accurate and available data for future use and analysis. The highlights of the interview analysis were the mentions of climate action and account (CA2) particular interest group (SIG) and CASIG during the interviews to foster multi-stakeholder open-source communities.

9 Conclusion

Research work influences the five dimensions of blockchain-led sustainability. Blockchain-led sustainability study using the interview method is one of the first pieces of research for decision-making through a systematic analysis of various factors. Second, research becomes a rich source of surveyed literature (using PRISMA), mapping technology (blockchain) and the theoretical side of the sustainability studies, which are established with deeper foundations and qualitative evidence on sustainability (thematic analysis using MAXQDA). Thirdly, the study delivers the gamut of impact features concerning business ethics (utilitarian/ deontological, TBL, CSR), humanity and behavioural theory factors (equity and human Rights, behavioural systems), and organisational theory factors (institutional, transition) to propose a blockchain led sustainability framework. The study is augmented by qualitative analysis of in-depth expert interviews using MAXQDA thematic analysis with transcript quotes from diverse personas from various parts of geography representing different verticals. Finally, the findings are detailed using the TCCM methodology to chart the future agenda. The paper theorises virtuous acumens to scholars and managers to appreciate the influencing factors of blockchain when preferred as a sustainability vehicle.

Blockchain-led sustainability promotes inclusivity, immutability, autonomy, democracy, and tokenisation of DAO services, products, and works of art. Hence, it opens the technology doors for creating the instruments of sustainable development goals (SDGs) and ethical practices to be addressed as a research gap in the coming time. Blockchain-led sustainability practices call for transformational actions to establish trust levels powered by smart contracts-based algorithmic limitations. The smart contract-led organisational operations need a true research shift to transform sustainability practices, reporting methods, traceability, product information storage, and updating and collecting methods. Blockchain sustainability being autonomous, supported by anonymous technology contributors, mainly in the community space, leveraging open source fixed the scope of the paper. Blockchain qualitative evaluation of adoption and diffusion in the enterprise space for specific industries to use as a sustainability tool can generate future thematic (deduced/induced) topics of research interests while interviewing prospective practitioners or relative personas from targeted verticals. Blockchain sustainability literature is scarce and relatively new in academic literature. The current blockchain sustainability literature is available to showcase the blockchain sustainability communication channels and their governance between the member participants and consumers at a certain level. The research lenses are required to consider blockchain sustainability as a new subject in engineering and management to research the more significant perspectives of virtual meta-autonomous blockchain-led sustainability management practices. The research is qualitative and restricted to specific vertical industries and geographies; it necessitates a broader reach and characteristic facades. It would be good if we test the proposed work using empirical methods. The qualitative method of interviews can affect the author and interviewee during the discussion stages, expecting additional wide-ranging persona selection and other empirical procedures. It would be great to realise the result of the research as a practical framework for industry and get it deployed in SMEs, large organisations, social systems, and societies willing to use technology (blockchain) for sustainability.

References

Ali MH, Chung L, Kumar A, Zailani S, Tan KH (2021) A sustainable blockchain framework for the halal food supply chain: lessons from Malaysia. Technol Forecast Soc Chang 170:120870

Allen DW, Berg C (2020) Blockchain governance: what we can learn from the economics of corporate governance. J Br Blockchain Assoc 3:1–10

Andryukhin AA (2018) Methods of protecting decentralized autonomous organizations from crashes and attacks. Труды института системного программирования РАН 30(3):149–164

Aronow S, Fitzpatrick P (2016) Introducing a score for CSR and sustainability performance to the gartner supply chain top 25. ID: G00293155 Gartner

Atzori M (2017) Blockchain technology and decentralised governance: is the state still necessary? J GovAnce Regul 6:44–57

Avelino F, Grin J, Pel B, Jhagroe S (2016) The politics of sustainability transitions. J Environ Planning Policy Manage 18:557–567

Bagloee SA, Heshmati M, Dia H, Ghaderi H, Pettit C, Asadi M (2021) Blockchain: the operating system of smart cities. Cities 112:103104

Biswas D, Jalali H, Ansaripoor AH, De Giovanni P (2023) Traceability vs. sustainability in supply chains: the implications of blockchain. Eur J Oper Res 305:128–147

Böckel A, Nuzum AK, Weissbrod I (2020) Blockchain for the circular economy: analysis of the research-practice gap. Sustain Prod Consum 25:525–539

Braun V, Clarke V (2006) Using thematic analysis in psychology. Qual Res Psychol 3:77–101

Chatterjee S, Chaudhuri R, Vrontis D, Ajith Kumar VV (2024) Adoption of blockchain technology in organisations: from morality, ethics and sustainability perspectives. J Inf Commun Ethics Soc 22:38–57

Choi TM, Luo S (2019) Data quality challenges for sustainable fashion supply chain operations in emerging markets: roles of blockchain, government sponsors and environment taxes. Transp Res Part E Logist Transp Rev 131:139–152

Christensen SL, Kohls J (2003) Ethical decision making in times of organisational crisis: a framework for analysis. Bus Soc 42:328–358

Claessens S, Fan JP (2002) Corporate governance in Asia: a survey. Int Rev Financ 3:71–103

Clohessy T, Acton T (2019) Investigating the influence of organisational factors on blockchain adoption: an innovation theory perspective. Ind Manag Data Syst 119:1457–1491

Daly H (1990) Toward some operational principles of sustainable development. Ecol Econ 2:1–6

Davidson S, De Filippi P, Potts J (2016) Economics of blockchain. SSRN 2744751

de Villiers C, Kuruppu S, Dissanayake D (2021) A (new) role for business–promoting the United Nations' sustainable development goals through the internet-of-things and blockchain technology. J Bus Res 131:598–609

Devine A, Jabbar A, Kimmitt J, Apostolidis C (2021) Conceptualising a social business blockchain: the coexistence of social and economic logics. Technol Forecast Soc Chang 172:120997

Dey K, Shekhawat U (2021) Blockchain for sustainable e-agriculture: literature review, architecture for data management, and implications. J Clean Prod 316:128254

Di Vaio A, Varriale L (2020) Blockchain technology in supply chain management for sustainable performance: evidence from the airport industry. Int J Inf Manage 52:102014

DiMaggio PJ, Powell WW (1983) The iron cage revisited: institutional isomorphism and collective rationality in organisational fields. Am Sociol Rev 48:147–160

Drljevic N, Aranda DA, Stantchev V (2020) Perspectives on risks and standards that affect the requirements engineering of blockchain technology. Comput Stand Interfaces 69:103409

Dursun T, Üstündağ BB (2021) A novel framework for policy based on-chain governance of blockchain networks. Inf Process Manage 58:102556

Elkington J (1998) Cannibals with forks. New Society Publishers, Gabriola Island, British Columbia

Etikan I, Musa SA, Alkassim S (2016) Comparison of convenience sampling and purposive sampling. Am J Theor Appl Stat 5:1–4

Ferrell OC, Gresham LG, Fraedrich J (1989) A synthesis of ethical decision models for marketing. J Macro Mark 9:55–64
Figueiredo K, Hammad AW, Haddad A, Tam VW (2022) Assessing the usability of blockchain for sustainability: extending key themes to the construction industry. J Clean Prod 343:131047
Fischer CT (2009) Bracketing in qualitative research: conceptual and practical Matters. Psychother Res 19:583–590
Geels FW, Elzen B, Green K (2004) General introduction: system innovation and transitions to sustainability. Edward Elgar Publishing, pp 1–16
Hastig GM, Sodhi MS (2020) Blockchain for supply chain traceability: business requirements and critical success factors. Prod Oper Manag 29:935–954
Heung VC, Kucukusta D, Song H (2011) Medical tourism development in Hong Kong: an assessment of the barriers. Tour Manage 32:995–1005
Holloway I, Galvin K (2016) Qualitative research in nursing and healthcare. Wiley
Holstein JA, Gubrium JF (1997) Active interviewing. In: Silverman D (ed) Qualitative research: theory, method and practice, vol 5. Sage, London, pp 113–129
Hughes L, Dwivedi YK, Misra SK, Rana NP, Raghavan V, Akella V (2019) Blockchain research, practice and policy: applications, benefits, limitations, emerging research themes and research agenda. Int J Inf Manage 49:114–129
Kouhizadeh M, Saberi S, Sarkis J (2021) Blockchain technology and the sustainable supply chain: theoretically exploring adoption barriers. Int J Prod Econ 231:107831
Kshetri N (2021) Blockchain and sustainable supply chain management in developing countries. Int J Inf Manage 60:102376
Kumar P, Gupta GP, Tripathi R (2021) TP2SF: a trustworthy privacy-preserving secured framework for sustainable smart cities by leveraging blockchain and machine learning. J Syst Architect 115:101954
Leng J, Ruan G, Jiang P, Xu K, Liu Q, Zhou X, Liu C (2020) Blockchain-empowered sustainable manufacturing and product lifecycle management in industry 4.0: a survey. Renew Sustain Energy Rev 132:110112
Li X, Wang D, Li M (2020) Convenience analysis of sustainable E-agriculture based on blockchain technology. J Clean Prod 271:122503
Loorbach D, Rotmans J (2006) Managing transitions for sustainable development. Understanding industrial transformation. Springer, Dordrecht, pp 187–206
Mangla SK, Kazancoglu Y, Ekinci E, Liu M, Özbiltekin M, Sezer MD (2021) Using system dynamics to analyse the societal impacts of blockchain technology in milk supply chains. Transp Res Part E Logist Transp Rev 149:102289
Mora H, Morales-Morales MR, Pujol-López FA, Mollá-Sirvent R (2021) Social cryptocurrencies as model for enhancing sustainable development. Kybernetes 50(10):2883–2916
Mason J (2002) Qualitative researching. Sage Publications, London
McElroy MW, Jorna RJ, van Engelen J (2008) Sustainability quotients and the social footprint. Corp Soc Responsib Environ Manag 15:223–234
Moher D, Liberati A, Tetzlaff J, Altman DG, Prisma Group (2009) Reprint- preferred reporting items for systematic reviews and meta-analyses: the PRISMA statement. Phys Ther 89:873–880
Mulder K, Costanza R, Erickson J (2006) The Contribution of built, human, social and natural capital to quality of life in intentional and unintentional communities. Ecol Econ 59:13–23
Murray A, Kuban S, Josefy M, Anderson J (2019) Contracting in the smart era: the implications of blockchain and decentralised autonomous organisations for contracting and corporate governance. Acad Manag Perspect 35:622–641
Norman W, MacDonald C (2004) Getting to the bottom of triple bottom line. Bus Ethics Q 14:243–262
Paul J, Merchant A, Dwivedi YK, Rose G (2021) Writing an impactful review article: what do we know and what do we need to know. J Bus Res 133:337–340
Paul J, Rosado-Serrano A (2019) Gradual internationalisation vs born-global/international new venture models: a review and research agenda. Int Mark Rev 36:830–858

Porta RL, Lopez-de-Silanes F, Shleifer A, Vishny RW (1998) Law and finance. J Polit Econ 106:1113–1155

Risius M, Spohrer K (2017) A blockchain research framework. Bus Inf Syst Eng 59:385–409

Salmerón-Manzano E, Manzano-Agugliaro F (2019) The role of smart contracts in sustainability: worldwide research trends. Sustainability 11(11):3049

Sayeed S, Marco-Gisbert H, Caira T (2020) Smart contract: attacks and protections. IEEE Access. 8:24416–24427

Schinckus C (2020) The good, the bad and the ugly: an overview of the sustainability of blockchain technology. Energy Res Soc Sci 69:101614

Schulz KA, Gstrein OJ, Zwitter AJ (2020) Exploring the governance and implementation of sustainable development initiatives through blockchain technology. Futures 122:102611

Sen A (1992) The political economy of targeting. World Bank, Washington, DC

Sinclair J, Cardew-Hall M (2008) The folksonomy tag cloud: when is it useful? J Inf Sci 34(1):5–29

Singh H, Dey AK (2020) Listen to my story: contribution of patients to their healthcare through effective communication with doctors. Health Serv Manage Res 34:178–192

Sunmola FT (2021) Context-aware blockchain-based sustainable supply chain visibility management. Procedia Comput Sci 180:887–892

Tan A, Gligor D, Ngah A (2020) Applying blockchain for Halal food traceability. Int J Logist Res Appl 25(6):1–18

Treiblmaier H (2019) Combining blockchain technology and the physical internet to achieve triple bottom line sustainability: a comprehensive research agenda for modern logistics and supply chain management. Logistics 3:10

Tsao YC, Thanh VV, Wu Q (2021) Sustainable microgrid design considering blockchain technology for real-time price-based demand response programs. Int J Electr Power Energy Syst 125:106418

Tsolakis N, Niedenzu D, Simonetto M, Dora M, Kumar M (2021) Supply network design to address United Nations sustainable development goals: a case study of blockchain implementation in Thai fish industry. J Bus Res 131:495–519

Upadhyay A, Mukhuty S, Kumar V, Kazancoglu Y (2021) Blockchain technology and the circular economy: Implications for sustainability and social responsibility. J Clean Prod 293:126130

Van Vuuren DP, Kok M, Lucas PL, Prins AG, Alkemade R, van den Berg M, Stehfest E (2015) Pathways to achieve a set of ambitious global sustainability objectives by 2050: explorations using the IMAGE integrated assessment model. Technol Forecast Soc Chang 98:303–323

Vemuri A, Costanza R (2006) The role of human, social, built, and natural capital in explaining life satisfaction at the country level: toward a national well-being index (NWI). Ecol Econ 58:119–133

Venkatesh VG, Kang K, Wang B, Zhong RY, Zhang A (2020) System architecture for blockchain based transparency of supply chain social sustainability. Robot Comput Integr Manuf 63:101896

Wang B, Luo W, Zhang A, Tian Z, Li Z (2020) Blockchain-enabled circular supply chain management: a system architecture for fast fashion. Comput Ind 123:103324

Wang T, Hua H, Wei Z, Cao J (2022) Challenges of blockchain in new generation energy systems and future outlooks. Int J Electr Power Energy Syst 135:107499

Examining Metaverse-Driven Medical Education Systems: A Triple Bottom Line Assessment and Alignment with Sustainable Development Goals

Vartul Shrivastava and Shekhar Shukla

Abstract The traditional medical education system is expensive from an institutional standpoint. In general, it also remains insufficient to cater to certain practical learning needs of medical students in a scalable and cost-effective manner. In clinical training, the educational value derived from immersive experiences is of paramount importance. The recent advancements in the metaverse technology can offer promising solutions to address these posed challenges, by introducing anatomy lessons with immersion, enhancing economic benefits, and delivering experiential learning to foster a skilled workforce. The metaverse technology can be a sustainable alternative in the scenario of a medical education system as it offers a reduced carbon footprint, and long-term usability and holds alignment with Sustainable Development Goals (SDGs). In this chapter, we provide a convergent understanding of the metaverse-driven medical education (MDME) ecosystem from a sustainability perspective using the Triple Bottom Line approach (TBL). This work aims to foster and consolidate towards a holistic premise on the integration of metaverse and medical education, to render sustainable solutions in alignment with sustainable development goals.

Keywords Metaverse · Sustainable development goals (SDGs) · Medical education · Scalability · Triple bottom line (TBL)

1 Introduction

Medical education has been in a position of paramount importance in society since it holds the very foundation that shapes the minds and skills of future medical professionals (Shafei 2023). They are entrusted with the responsibility to provide health

V. Shrivastava (✉) · S. Shukla
Indian Institute of Management Indore, Indore, India
e-mail: vartul.shrivastava@gmail.com

W. Leal Filho et al. (eds.), *Metaverse and Sustainability: Business Resilience Towards Sustainable Development Goals*, World Sustainability Series,
https://doi.org/10.1007/978-3-031-89545-6_9

care services and drive advancements in medical research. Therefore, this imperatively necessitates rigorous clinical training of medical professionals in order to render high standards of academic rigour (Malone-Lee 2008). However, the traditional medical education landscape comes with many barriers, which might decrease the effectiveness and accessibility as intended (Hodson 2020).

The high costs associated with the maintenance of medical schools, hospitals and training facilities pose financial constraints on the institutions, government and the students (Johnston 2018). Due to the resource-intensive nature of medical apprenticeship, which includes cadaver surgical training (CST), clinical rotations, and simulation labs, the scalability and accessibility of these programs are significantly limited, particularly for individuals carrying out training in remote and underserved facilities (Miyano et al. 2023; Shah 2023). Additionally, the theory-centric curricula may hinder the integration of practical skills, which can furnish major gaps in experiential learning (Baird 2010; Irwin and Ryan 2013). Exposure to hands-on learning positively enhances medical students' decision-making ability in real-world cases (Gubler 2024). These challenges necessitate sustainable and effective solutions to delineate experiential learning to those in need while ensuring long-term benefits and improved outcomes in medical education.

The emergence of metaverse technology can offer some promising solutions towards these issues that the medical fraternity has been facing. The metaverse, distinguished by its features such as immersiveness and ability to render interactive virtual environments, can help medical institutions in the dissemination of educational initiatives (Riches et al. 2024). Metaverse helps in enacting interactive spaces, which can simulate clinical procedures and introduce hyper-spatiotemporal collaboration (Kataria et al. 2023), furnishing novel methodologies to conduct training. It can aid in the transformation of medical anatomy lessons in engaging 3D environment through the use of Augmented Reality (AR) and Virtual Reality (VR), allowing for visualization of virtual cadavers resembling real-life training setups (Ebrahimzadeh and Safa 2024).

These transformations can further provide the generation of simulation-based assessments within the metaverse-driven medical training, that can help in the differentiation of real-time feedback to the students through adaptive hands-on learning pathways in a safer and controlled environment (Mystakidis and Lympouridis 2024). However, these initiatives might not replace the experiential and practical values that are attained through actual real-life simulations (Mah 2023); however, they can enhance reasoning through an ideal convergence established by medical practitioners, which could employ metaverse training in coherence with traditional methods. Moreover, questions remain on the extent of depth covered by these virtual setups, as opposed to the details contained in actual practical simulations; which cannot just be replaced in some aspects of medical training. However, coupling the metaverse in an idyllic manner can complement the traditional methodological frame of medical education (Mah 2023).Therefore, through the holistic adoption of a metaverse in medical education, a cohesive and sustainable ecosystem can be provided, that is capable of handling crucial aspects of sustainability.

This chapter applies the Triple-Bottom Line approach to bring forward a comprehensive understanding towards the economic, social, and environmental aspects of metaverse-driven medical education (MDME). We also shed light on metaverse applicative alignments (MAAs) with the sustainable development goals (SDGs) to yield holistic reasoning in the adoption of metaverse within medical education systems. We outline benefits and trade-offs, to trace recommendations, insights, and the latest developments in this field. We believe this approach will help the reader to build up a balanced view, considering the impacts of metaverse adoption in medical training, and how it can impact economic, social, and environmental dimensions with the ambition of promoting a more sustainable, and effective medical education.

The remainder of the chapter is organized as follows: Sect. 2 outlines front-end and back-end technologies that are rooted in the foundational architecture of the metaverse. Section 3 discussed the alignment of metaverse artefacts with the SDGs from extant literature. Subsequently, Sect. 4 comprehensively explains TBL dimensions interplay in MDME. Section 5 elucidates on challenges that are associated with MDME. Section 6 offers the key takeaways from the chapter and discusses critical prospects that can be examined in a similar research direction. Lastly, we acknowledge the references that contributed to shaping the reasoning behind the analysis presented. Having outlined the structure of this chapter, we now explore how metaverse technology is positioned to address these issues through a technological lens.

2 Overview of Metaverse Technology

The terminological ideation of 'metaverse' was articulated by a prominent science fiction writer, 'Neal Stephenson', in his novel named 'Snow Crash' which was published in 1992 (Evans et al. 2022). The term metaverse is an amalgamation of the prefix 'meta' implying transcendence, and the term 'universe', which signifies fusion; as a result, referring metaverse as a parallel digital twin which depicts a virtual environment closely associated with the physical world (Nayak et al. 2024). The foundational architecture of the metaverse (refer to Fig. 1) can be understood by categorizing its services into two broad categories: back-end metaverse services (BeMS) and front-end metaverse services (FeMS).

To erect the technological infrastructure and support system in the metaverse, services that render connectivity, intelligence, security and scalability are paramount which can be categorized within BeMS. This tier chiefly includes the use of HPC (High-Performance Computing) clusters, which aid in rendering 3D graphics while processing large volumes of data streams concurrently (Alawneh and Hamou-Lhadj 2022). As the metaverse is a digital twin, it necessitates the use of spatial intelligence and respective content-aware mechanisms through the use of AI-based agents, that help in undertaking contextual decisions and actions accordingly (Durante et al. 2024). Moreover, to enact high-speed Ultra-Reliable Low-latency communication, the use of edge computing and 5G services is crucial. This can help in providing

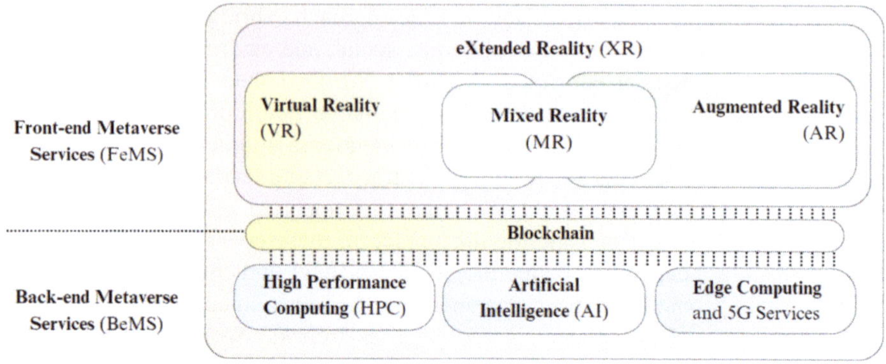

Fig. 1 Front-end and Back-end technologies of the metaverse

idyllic user engagement with minimal lag which is needed for immersion (Sarah et al. 2023). Further, the integration of Blockchain in BeMS advocates for transparency among transactions, for securing digital ownership and data integrity (Pineda et al. 2024; Zhu et al. 2023).

The aforementioned back-end services are integrated with FeMS to emulate Metaverse for user engagement. FeMS is entrusted to expose the emulated virtual environment to the user base through immersive technologies such as Virtual Reality (VR) which utilizes head-mounted displays (HMDs) to engage users completely through visuals (Ren et al. 2023), Augmented Reality (AR) which tweaks real-world experience through the super-imposition of digital artefacts (Singh and Ahmad 2024), and Mixed Reality (MR) that elucidates the inclusion of VR and AR into single strata; by allowing the coexistence of physical and digital objects concurrently (Carmigniani and Furht 2011). To represent VR, AR and MR as a spectrum of immersive technologies, eXtended Reality (XR) acts as a comprehensive framework to support the metaverse's functioning (Le Noury et al. 2022). Together, these frontend services create a rich, engaging, and interactive interface for users in the metaverse, enabling them to experience the metaverse pleasantly, by blurring the boundaries between the physical and virtual worlds. With an understanding of metaverse technology in place, let us now examine how it aligns with global sustainability efforts, particularly the United Nations Sustainable Development Goals (SDGs).

3 Metaverse Alignment with Sustainable Development Goals

In view of metaverse's technological artefacts, it can be systematically aligned with SDGs because of its ability to emulate virtual spaces; thereby helping to address prominent challenges such as provision of quality education, the promotion of gender equality, and raising awareness about climate change, among others. In alignment with the extant literature (Jauhiainen et al. 2022; Joseph 2023; Kulkarni et al. 2024; Naved and Gupta 2023), Table 1 highlights metaverse applicative alignments (MAAs) with SDGs; aiming to showcase practical viability of metaverse in fostering development goals across varied verticals.

With the alignment of the metaverse and SDGs outlined, we now turn to the critical analysis of metaverse-driven medical education through the lens of the Triple Bottom Line framework.

4 Analysing Metaverse-Driven Medical Education (MDME): TBL-Based Analysis

In recent years, the intersection of metaverse technologies and medical education has garnered increasing attention, as these innovations hold the potential to revolutionize traditional learning paradigms. Few studies have explored this potential intersection of metaverse and medical education. For instance, a study found that metaverse-based cadaveric dissection courses for East African surgeons were beneficial, though further validation is needed to establish its effectiveness on a global scale (Narang et al. 2023). Another study emphasized the evolution of medical education through virtual and augmented reality, identifying challenges related to technology integration and accessibility (Lewis et al. 2024). Also in the extant current technological landscape, there exist notable platforms such as Osso VR, Touch Surgery, HoloAznatomy, Anatomage Table, Proximie among many others; that are delineating medical education through VR and AR by realistic surgical simulations, 3D anatomical visualizations, haptic feedback, and remote collaboration. As medical education adapts to these immersive platforms, it becomes essential to evaluate their broader impact, not only in terms of technological advancement but also through the lens of sustainable development.

For benchmarking the tenets of an ecosystem, the traditional myopic approach is to analyse economic dimensions, which often neglect the broader sustainable implications. It is vital to consider the social and environmental aspects of an ecosystem, along with economic aspects; to render a holistic approach to assessing long-term viability and to foresee the trade-offs. The Triple-Bottom Line (TBL) approach advocates for such a philosophy (Ahmad et al. 2023), where it emphasizes probing into economic, social and environmental factors, while not solely focusing on tenets of financial viability (Naz and Magda 2023).

TBL as a framework was conceptualized by John Elkington in the mid-1990s, he introduced a 3-P formula which includes Profit, People and Planet; aiming to furnish social equity, community well-being and ecological balance (Srivastava et al. 2022). To comprehensively understand the impact of metaverse-driven medical education systems, we will analyse this intersection through the TBL framework (refer to

Table 1 Exploring the synergy between SDGs and metaverse characteristics

SDGs	Metaverse applicative alignments (MAAs)
SDG-I: No poverty	Through immersion, metaverse can delineate education, professional training and financial opportunities for marginalized strata of society
SDG-II: Zero hunger	Addressing concerns over food security by providing immersive training to farmers
SDG-III: Good health and well-being	By utilizing immersive technologies within the metaverse to provide health education, telemedicine and virtual consultations
SDG-IV: Quality education	By enhancing global collaboration in the virtual immersive settings, metaverse can delineate education without explicit need for full-fledged infrastructure
SDG-V: Gender equality	Through virtual spaces, it can yield inclusion by empowering women and spectrum of gender identities through impartial access to resources
SDG-VI: Clean water and sanitation	Metaverse might not have a direct impact on SDG-VI, but by impacting SDG-IV it can influence SDG-VI by offering virtual training on water management and effective sanitation methodologies
SDG-VII: Affordable and clean energy	Impact on SDG-VII is rooted in the premise of metaverse. Through hyper-spatiotemporality, it surpasses constraints related to demographical delimitations; hence rendering sustainability
SDG-VIII: Decent work and economic growth	By employing technologies such as digital twins within the metaverse, businesses can enhance productivity which can create novel employment opportunities
SDG-IX: Industry, innovation, and infrastructure	To leverage the metaverse in the long term, a scalable and sustainable infrastructure is needed which can be rendered through enterprise-scalable innovations
SDG-X: Reduced inequalities	By transcending demographical delimitations, metaverse adoption in various verticals can drastically reduce differences by empowering underserved strata of society
SDG-XI: Sustainable cities and communities	The conceptualization of virtual cities can sustainably support urbanization. The current resource-intensive nature of the metaverse needs to be advanced further accordingly

(continued)

Table 1 (continued)

SDGs	Metaverse applicative alignments (MAAs)
SDG-XII: Responsible consumption and production	The core tenet of a metaverse which is AR can offer virtual try-ons, and 3D object rendering among many others; which can reduce waste and promote consumption sustainably
SDG-XIII: Climate action	Metaverse can inherently comply with environmental stewardship through simulation of climate change adversities resulting in enhanced awareness
SDG-XIV and SDG-XV: Life below water and life on land	While metaverse might not directly impact SDGs XIV and XV. Through direct influence on SDG-XIV and SDG-19, climate change effects on fauna below water and land can be visualized to raise consciousness
SDG-XVI: Peace, justice, and strong institutions	By leveraging immersive spaces within the metaverse, international diplomatic dialogue can be fostered. This can create more transparent and accountable governing practices leading to resilient enterprises
SDG-XVII: Partnerships for the goals	The decentralized nature offered by blockchain in the metaverse can help promote trust among the stakeholders

Fig. 2). This will enable us to identify its benefits and recognize the trade-offs comprehensively. We begin by assessing the economic sustainability of metaverse-driven medical education, focusing on its financial viability and long-term cost efficiencies.

4.1 Economic Sustainability in MDME

Economic sustainability in MDMEs refer to system's ability to maintain financial viability while being able to deliver high quality in the long run. This involves evaluation of cost efficiencies, optimization of resource allocation, and encouragement of novel revenue streams so that system can sustainably support itself while being adaptive to the uncertainties that can emerge in the changing economic landscape.

The metaverse can deliver cost efficiency, optimized resource allocation, and novel revenue streams in the medical education paradigm, which aligns with SDG-VIII. Also, by enacting virtual hospitals and clinical workflows; MDMEs can minimize the necessity for expensive physical infrastructure, enabling sustainable financial instruments in clinical training. Subsequently, the use of immersive artefacts such as VR for medical training contributes to SDG-IX by promoting advancements in healthcare while reducing operational costs, rendering the system economically viable in the long run.

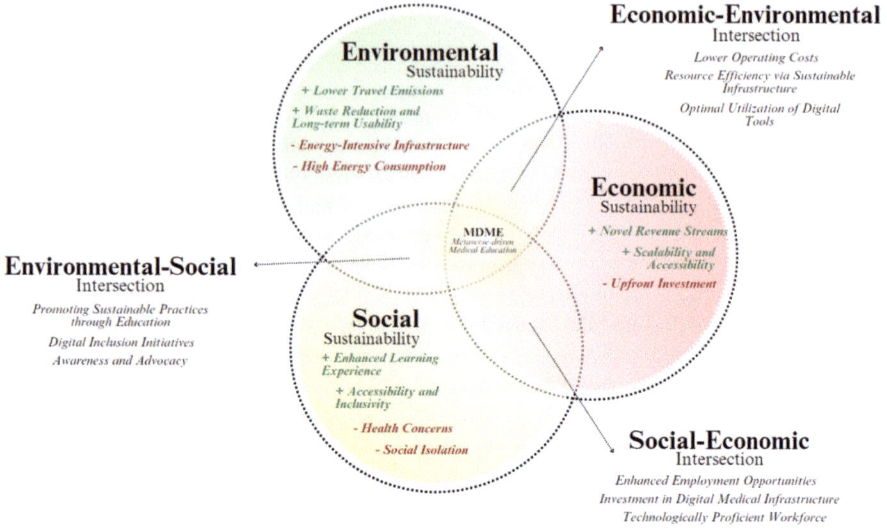

Fig. 2 Triple-Bottom Line analysis of MDME systems

As the metaverse facilitates the creation of virtual hospitals and clinical workflows, it can provide diagnosis and treatment of numerous conditions; hence, optimizing healthcare delivery while reducing operational costs (Curtis and Brolan 2023). This approach brings a reduction in the need for a physical infrastructure, which can help in saving costs on infrastructure and maintenance of medical facilities. Virtual environments can be employed for training purposes and education, which could reach large pool of students without too many constraints which come with the physical travel and accommodation (Bhatia and Joshi 2023). This reach not only increases accessibility, but ensures that the institutions can attract diverse range of students, which can help in strengthening financial viability.

Moreover, use of VR systems such as MAGES offers cost-effective and widely accessible medical training solutions (Papagiannakis et al. 2018). By providing a fail-safe, realistic environment; it can help in enhancing skills through the use of VR simulations in an affordable and scalable manner. This kind of system helps transform traditional medical training, into a much more economical process. Similar insights are also supported by the Virtual Reality in Medicine and Surgery (VRiMS) initiative; this initiative showcased that metaverse platforms can facilitate global health education programmes. This initiative allowed over 1000 delegates from over 61 countries, to participate in the surgical training courses without the need for physical presence; leading to a reduction of the associated travel and accommodation costs (Narang et al. 2023). Additionally, VR training systems can help in simulating clinical procedures and environments which can lower the need for costly physical materials (Tan et al. 2022). The economic benefits of MDME also lie in enhancing the experience of surgical training, as it can offer close to real simulations (Kataria et al. 2023). For instance, in the practice of performing orthopaedic surgery; VR-based clinical training has shown optimal performance in certain procedures such as total knee arthroplasty (TKA) and total hip arthroplasty (THR) (Mah 2023). Beyond the economic aspects, social sustainability plays a pivotal role in shaping inclusive and equitable medical education, which is discussed next.

4.2 Social Sustainability in MDME

Social sustainability in MDME examines prospects concerning to inclusiveness, accessibility, and social equity. In light of metaverse-adopted clinical training, direct interaction can be established among the vested individuals; which can eliminate barriers towards race, gender, and physical disabilities (Kumari et al. 2024). Hence, sets a benchmark for an inclusive learning environment, which could address diverse student diasporas who are coming from various cultures and demographics across the globe. For instance, commence of virtual worlds, such as Second Life; have been successful in emulating the paradigm of Continuing Medical Education (CME), whose results indicated that individuals who were part of CME had enhanced confidence. Thus, showcasing a better learning curve as compared to the traditional mode of teaching (Wiecha et al. 2010). Moreover, in addition to offering inclusiveness, MDME might minimize the gap in the existing digital divide which exists in between various socio-economic groups. As metaverse reduces the need for a physical structure and other costs, it can help in lowering tuition costs; which in turn enhances the affordability of clinical training while also being compliant with social inclusivity.

Coming to SDGs with respect to social dimensions of MDME, it directly supports SDG-IV by making medical education more accessible on a global frontier, eliminating barriers posed by geography, and reducing the costs of traditional education. Through inclusive virtual environments, MDME addresses SDG-V and SDG-X, enabling diverse student populations, regardless of gender, race, or physical disabilities, by dispensing off high-quality medical training. This fosters social equity and ensures that a wider range of socio-economic groups can access essential clinical education without the constraints of traditional systems. Metaverse-driven platforms can offer continuous professional development and can deliver learning opportunities for medical trainees. Such platforms would not only provide access to the latest information, virtual conferences, and certification programs; but will also allow professionals to keep up with medical developments, regardless of their geographical location.

The literature shows that similar to any other virtual world, Second Life has the potential to enhance medical education, through the facilitation of hands-on training, simulations, and worldwide collaboration among medical professionals; for enhanced access towards clinical training (Duan et al. 2021). Besides, virtual environments navigate opportunities for students and professionals, to foster collaboration and networking across the globe. This aids in enriching the educational experience through real-time engagement and also delineating feedback from peers and instructors resulting in a strong and assistive network (Wang et al. 2022). AI-driven learning platforms in the metaverse can be designed to offer a much more personalized education experience for each student, based on one's needs and learning style. Such type of approach assists in self-directed learning and reinforces independent thinking with problem-solving skills; enabling students to have better educational outcomes, and greater social sustainability within the learning environment (Chen et al. 2024).

Subsequently, training on social determinants of health can be emphasized in the MDME setups. This could help in mitigating various factors that bear health outcomes. This can be tackled through the use of virtual simulations to showcase patient scenarios, which might lead to cultural affinity, and may help in the development of empathy in medical trainees (Rahman Khan et al. 2023). In addition to social and economic considerations, environmental sustainability is essential to understanding the ecological impact of metaverse-driven medical education.

4.3 Environmental Sustainability in MDME

Medical education integrated within the technical premise of metaverse can provide several environmental benefits. The practitioners utilizing this adoption can thereby considerably reduce their reliance on physical resources, which can lower carbon emissions that are imperative with conventional training techniques (Yue and Qian 2024). The resource-consumptive nature of operation theatres and other facilities that utilize a lot of energy in operation; thus, by shifting to VR and AR simulation in the metaverse, such resource-intensive practices could be considerably reduced. Perhaps, the most important ecological advantage of metaverse adoption is embodied in the reduction of physical travel (Florido-Benítez 2024). Traditional medical conferences and training require a huge amount of travel, which in itself translates to big volumes of carbon emissions. For example, exposure to carbon emissions due to international travel during medical conferences is significantly high. For instance, at the American Thoracic Society conference in San Diego, air travel to and from the venue was purported to have resulted in 10,779 tonnes of CO_2 emissions from an estimated 15,000 participants. This is equivalent to the average annual carbon footprint of 11,000 people living in India or 110,000 people living in Chad (Roberts and Godlee 2007).

MDME aligns with SDG-XII by minimizing the use of physical resources, such as travel and disposable medical materials, in training environments. Virtual simulations allow for repeated practice without the environmental costs associated with physical training methods. Furthermore, MDME contributes to SDG-XIII by reducing the carbon emissions linked to medical conferences and international travel. However, the energy-intensive infrastructure that supports the metaverse presents challenges to its environmental impact, which necessitates more research to fully assess its sustainability.

Thus, digitalising educational platforms can contribute to sustainability, by reducing the ecological load of healthcare education activities. For example, the use of VR simulations offers a way to repeat specific practices without limitation and also discards the use of disposable materials that are unavoidable in cadaver labs and operating rooms (Mah 2023) Besides this, paper-bound textbooks, and other related materials that are printed are extensively used in training methods, which have a direct effect on the generation of paper waste (Tracana and Carvalho 2012). Thereby, adhering to the metaverse can help in saving the available biological resources which can further yield lesser biomedical waste. Also, metaverse usage can lead to reduced usage of traditional training facilities, which ensures optimized use rendering lower environmental impact due to healthcare activities. This transition will further support the goals of environmental sustainability but will also echo the broader agenda of digitalization and modernization in the verticals of medical practices. While there are undeniable advantages, there are also key challenges that must be considered in fully realizing the potential of metaverse-driven medical education.

5 Challenges in MDME Highlighted Through TBL

We examined complexities of MDMEs through the lens of TBL, which encompasses economic, social and environmental aspects; which are elucidated as follows:

Economic Challenges in MDME: To setup immersive training in MDMEs, it will demand significant upfront cost in assembly of high-performance computing and high-speed connectivity hardware. Many institutions especially belonging in the developing regions, can struggle to set up and maintain this cost-intensive infrastructure (Koohang et al. 2023). Also, the metaverse is an evolving technological premise; thus, the supporting technical setup within these institutions has to be looked for advancements periodically. This can pose financial distress in low-income regional institutions, which might create the economic divide subsequently (Mancuso et al. 2023).

To overcome these economic challenges, underserved institutions can explore partnerships, government subsidies, and shared infrastructure models while leveraging cloud-based solutions and scalable technologies that reduce the need for heavy upfront investments; thereby making MDME more accessible and affordable, especially in the developing regions.

Social Challenges in MDME: While there are benefits, there are also concerns that need to be addressed about the mental and physical health of individuals exposed to such virtual environments. For example, the metaverse can induce cybersickness, a form of motion sickness that could affect motor skills adversely (Mohamed and Al-Barazie 2023). Dizziness, fatigue, and headaches may also be induced if followed by a prolonged period of engagement. Also, certain psychological issues can be formed through the unrestful use of VR, leaving the affected person confused and conflicted with real life.

Moreover, concerns over factors such as privacy, the safety of patients, also on technological literacy challenges, are some of the important and pressing issues that need to be addressed. The major risks include the exposure of personal data, which can be revealed through virtual attacks against users, and via the manipulation of virtual environments (Fernández-Parrado and Perandones-González 2023). Social behaviour and communication skills can also get altered, since people can project themselves differently through avatars which can lead to complex criminal situations. Thus, MDME might disable social connections which can lead to instances of virtual crimes, resulting in real-world maladaptation.

Thus, to overcome the aforementioned social challenges in MDME, it becomes crucial to implement guidelines, and safety protocols alongside user-education programs. To ensure that MDME can leverage the well-being and privacy of individuals whilst utilizing virtual environments.

Environmental Challenges in MDME: The current energy-intensive architecture that supports the traditional metaverse is not in a position to lead to ecological balance, as it consequently leads to an increase in greenhouse emissions (Kshetri and Dwivedi 2023). Data centres, which are crucial in the functioning of the metaverse, consume close to 1.3% of the world's total electricity production; they critically affect the environment of the earth (Samuel et al. 2022). Subsequently, it is also believed that the metaverse can lower emissions by 10 gigatonnes of carbon dioxide equivalent by the year 2050 (Zhao and You 2023), there remains substantial concern that redirects criticality in the examination of MDMEs.

However, to mitigate the carbon footprint that poses an issue in the current setup of the metaverse, the adoption of green technologies can be incorporated to reduce ecological impact. The use of renewable energy sources which include solar, wind and hydropower can suffice the energy requirements of servers (Ma 2023), which is central to MDME. Beyond renewable energy, the use of energy-efficient servers is critical as well. The incorporation of ARM (Advanced RISC Machines) processors and liquid cooling can enhance energy efficiency in data centres (Xu and Chang 2017). Adding to this, technological directions that prioritize modularity, recyclability and material efficiency can be adapted to MDME's hardware practices. This can render lower e-waste and promote a greener circular economy (Chen 2023). Having discussed both the potential benefits and challenges, we now conclude with future directions for research and implementation in metaverse-driven medical education.

6 Conclusion and Future Directions

MDME systems showcase affirmative potential to change the way medical education is dispensed and catered towards future physicians. More importantly, the immersive and interactive features within the metaverse could improve upon the learning opportunities; by offering simulations and promoting international collaboration between teachers and students. The main advantages are expected in the verticals of academics, which can aid students in clinical practice. MDMEs can increase access towards medical education, particularly for individuals who are located in remote or resource-poor locations. Many tools can make learning more interactive and efficient through the three-dimensional visualization of complex anatomical structures, performing virtual surgeries, and collaborating in real-time while working on problem-solving exercises.

The economic benefit associated with metaverse integration is high because it offers a reduction in physical infrastructure while being compliant with scalability. Other than economic benefits, the metaverse also promotes inclusivity and accessibility by marching beyond traditional thresholds. As it can render quality education without resource constraints and geographical differences. Environmentally, metaverse could help mitigate ecological footprint since it diminishes the need to travel, rendering its candidacy as a sustainable tool in the delivery of education. In our assessment of MAAs with the SDGs in sync with the TBL framework, we found that SDG-VIII and SDG-IX are critical for economic sustainability in MDME. SDG-IV, SDG-V, and SDG-X were instrumental in evaluating the social impact, emphasizing inclusivity and equitable access to medical training. For environmental sustainability, SDG-XII and SDG-XIII alignments helped in addressing how metaverse adoption could reduce the ecological footprint.

However, the adoption of metaverse into medical education is not without its challenges. These premises require an enormous technological infrastructure (reliable high-speed internet and high-performance computer hardware) which is discouraging mostly to the institutions coming from under-developed countries. Another crucial aspect is the financial investment, which will be required in deploying and maintaining such systems at a large scale. Subsequently, due to the extensive use of technology, concerns about security and privacy are also raised. The huge amount of data that will be recorded in the metaverse, can lead to the possibility of data breaches. Other concerns are concerning social equity and accessibility. The advent of metaverse adoption by institutions and students may cultivate already existing disparities in medical education. Also, excessive use of simulations at the cost of real clinical experience can be detrimental as well. Environmentally, the current landscape of the metaverse might not contribute towards the broader goal of sustainability. As the current technological premise of the metaverse needs conventional data centres, which have extensive energy requirements, leading to an overhaul in greenhouse emissions.

Looking ahead, there remain several practical avenues that can be incorporated which can render MDME more sustainably. The inclusion of green technologies to build sustainable and scalable infrastructure can alleviate inequalities targeting SDG-X and can deliver immersive education resources to underserved institutions aligning with SDG-IV. Collaboration between medical institutions, technologists, and regulatory bodies will help in enacting ethical standards for data usage which will ensure MDME's stakeholders' safety and privacy; which can strengthen SDG-XVI. Also, concurrent research that sheds light on the long-term effects of using immersive technologies needs to be studied, as mitigating risks such as cybersickness and fatigue is essential to render MDME holistically; consequently, supporting

SDG-III. To target SDG-XVII, future efforts that can promote global partnership through resource-sharing and knowledge exchange can foster MDME setup without exacerbating the digital divide among institutions. Thus, the following future research questions can be explored to strengthen the premise of MDME from technical and societal perspectives:

F.R.Q.1 What educational frameworks can be introduced to enhance the technical literacy of educators and students in MDME, without afflicting mandatory approaches in clinical training?
F.R.Q.2 Which adaptive learning algorithms can be integrated into MDME to cater to personalized learning curves for medical students?
F.R.Q.3 To reduce racial division and gender discrimination, how large-scale collaborative premises can be built for MDME?

Future research in coherence with these probes can enhance the applicability of MDMEs across the globe. Thus, in our study, we use the TBL framework to comprehensively assess tenets of sustainability in MDME systems. Though the metaverse provides ample opportunities by offering immersive simulations and global connectivity in the verticals of medical education, it also brings associated challenges that need to be addressed. Such issues can be tackled by holistic efforts at all levels, which may succinctly embed coherent use of metaverse in clinical training; consequently, paving the way for more sustainable and inclusive medical education in future and beyond.

References

Ahmad S, Wong KY, Butt SI (2023) Status of sustainable manufacturing practices: literature review and trends of triple bottom-line-based sustainability assessment methodologies. Environ Sci Pollut Res 30(15):43068–43095

Alawneh L, Hamou-Lhadj A (2022) Locating and categorizing inefficient communication patterns in HPC systems using inter-process communication traces. J Syst Softw 194:111494. https://doi.org/10.1016/j.jss.2022.111494

Baird J (2010) The theory–practice gap. Assess Educ Princ Policy Pract 17(2):113–116. https://doi.org/10.1080/09695941003732357

Bhatia B, Joshi S (2023) Applications of metaverse in the healthcare industry. Int Conf Innov Data Commun Technol Appl (ICIDCA) 2023:344–350. https://doi.org/10.1109/ICIDCA56705.2023.10099897

Carmigniani J, Furht B (2011) Augmented reality: an overview. In: Furht B (ed) Handbook of augmented reality. Springer, New York, pp 3–46. https://doi.org/10.1007/978-1-4614-0064-6_1

Chen G, Jin Y, Chen P (2024) Development of a platform for state online education services: design concept based on meta-universe. Educ Inf Technol. https://doi.org/10.1007/s10639-024-12792-y

Chen Y (2023) The metaverse's inspiration for sustainable business: restructuring economic logic, capital, assets, organization, and industry. In: Walker T, Turtle HJ, Kooli M, Nikbakht E (eds) Fintech and sustainability: how financial technologies can help address today's environmental

and societal challenges. Springer Nature Switzerland, Cham, pp 129–145. https://doi.org/10.1007/978-3-031-40647-8_7
Curtis C, Brolan CE (2023) Health care in the metaverse. Med J Aust 218(1):46–46. https://doi.org/10.5694/mja2.51793
Duan H, Li J, Fan S, Lin Z, Wu X, Cai W (2021) Metaverse for social good: a university campus prototype. In: Proceedings of the 29th ACM international conference on multimedia, pp 153–161. https://doi.org/10.1145/3474085.3479238
Durante Z, Huang Q, Wake N, Gong R, Park JS, Sarkar B, Taori R, Noda Y, Terzopoulos D, Choi Y, Ikeuchi K, Vo H, Fei-Fei L, Gao J (2024) Agent AI: surveying the horizons of multimodal interaction (arXiv:2401.03568). https://doi.org/10.48550/arXiv.2401.03568
Ebrahimzadeh F, Safa R (2024) Unlocking the potential of metaverse in innovative and immersive digital health (arXiv:2406.07114). https://doi.org/10.48550/arXiv.2406.07114
Evans L, Frith J, Saker M (2022) The roots of the metaverse. In: From microverse to metaverse: modelling the future through today's virtual worlds. Emerald Publishing Limited, pp 15–24. https://www.emerald.com/insight/content/doi/10.1108/978-1-80455-021-220221002/full/html
Fernández-Parrado M, Perandones-González H (2023) A new universe in dermatology: from metaverse to dermoverse. J Eur Acad Dermatol Venereol 37(4):e470–e471. https://doi.org/10.1111/jdv.18525
Florido-Benítez L (2024) Metaverse cannot be an extra marketing immersive tool to increase sales in tourism cities. Int J Tour Cities. https://doi.org/10.1108/IJTC-01-2024-0001
Gubler KD (2024) The impact of experiential learning, operation bushmaster, on medical student decision-making compared to directed learning. MilY Med, usac394. https://doi.org/10.1093/milmed/usac394
Hodson N (2020) Landscapes of practice in medical education. Med Educ 54(6):504–509. https://doi.org/10.1111/medu.14061
Irwin RL, Ryan TD (2013) Get real: using engagement with practice to advance theory transfer and production. Sport Manag Rev 16(1):12–16. https://doi.org/10.1016/j.smr.2011.12.007
Jauhiainen JS, Krohn C, Junnila J (2022) Metaverse and sustainability: systematic review of scientific publications until 2022 and beyond. Sustainability 15(1):346. https://doi.org/10.3390/su15010346
Johnston SC (2018) The US training system for physicians-need for deeper analysis. JAMA 320(10):982–983. https://doi.org/10.1001/jama.2018.12879
Joseph A (2023) An exploratory study on metaverse and SDGs. In: El Khoury R, Alareeni B (eds) How the metaverse will reshape business and sustainability. Springer Nature, pp 83–93. https://doi.org/10.1007/978-981-99-5126-0_9
Kataria S, Kedia AK, Ravindran V (2023) Metaverse: evolving role in healthcare delivery and implications. J R Coll Physicians Edinb 53(3):186–191. https://doi.org/10.1177/14782715231189900
Koohang A, Nord JH, Ooi K-B, Tan GW-H, Al-Emran M, Aw EC-X, Baabdullah AM, Buhalis D, Cham T-H, Dennis C, Dutot V, Dwivedi YK, Hughes L, Mogaji E, Pandey N, Phau I, Raman R, Sharma A, Sigala M, Wong L-W et al (2023) Shaping the metaverse into reality: a holistic multidisciplinary understanding of opportunities, challenges, and avenues for future investigation. J Comput Inf Syst 63(3):735–765. https://doi.org/10.1080/08874417.2023.2165197
Kshetri N, Dwivedi YK (2023) Pollution-reducing and pollution-generating effects of the metaverse. Int J Inf Manage 69:102620. https://doi.org/10.1016/j.ijinfomgt.2023.102620
Kulkarni P, Gaddi A, Ramona B, Popescu V, Hiremath V, Shetty SK, Kagawade S (2024) A sustainable approach on metaverse in education: a SCOPUS-based bibliometric analysis to test TMIOS (a-sustainable-approach-on-metaverse-in-education) [Chapter]. IGI Global. https://doi.org/10.4018/979-8-3693-2845-3.ch011
Kumari VS, Ponmaniraj S, Selvi GV, Kaur S (2024) The future of immersive experience: exploring metaverse application development technologies and tools. In: Technology innovation pillars for industry 4.0. CRC Press.

Le Noury P, Polman R, Maloney M, Gorman A (2022) A Narrative review of the current state of extended reality technology and how it can be utilised in sport. Sports Med 52(7):1473–1489. https://doi.org/10.1007/s40279-022-01669-0

Lewis KO, Popov V, Fatima SS (2024) From static web to metaverse: reinventing medical education in the post-pandemic era. Ann Med 56(1):2305694. https://doi.org/10.1080/07853890.2024.2305694

Ma Z (2023) Energy metaverse: the conceptual framework with a review of the state-of-the-art methods and technologies. Energy Inform 6(1):42. https://doi.org/10.1186/s42162-023-00297-w

Mah E (Ted) (2023) Metaverse, AR, machine learning & AI in Orthopaedics? J Orthop Surg 31(1):10225536231165362. https://doi.org/10.1177/10225536231165362

Malone-Lee J (2008) A defence of rigorous method in clinical science. Neurourol Urodyn 27(6):462–465. https://doi.org/10.1002/nau.20550

Mancuso I, Messeni Petruzzelli A, Panniello U (2023) Digital business model innovation in metaverse: how to approach virtual economy opportunities. Inf Process Manage 60(5):103457. https://doi.org/10.1016/j.ipm.2023.103457

Miyano G, Takahashi M, Suzuki T, Iida H, Abe E, Kato H, Yoshida S, Lane GJ, Ichimura K, Sakamoto K (2023) Remote cadaveric minimally invasive surgical training. Front Pediatr 11:1255882

Mohamed A, Al-Barazie R (2023) Metaverse technology in health and family: assessing impacts, identifying guidelines, and examining perceptions. In Selvaraj H, Chmaj G, Zydek D (eds) Advances in systems engineering. Springer Nature Switzerland, pp 523–536. https://doi.org/10.1007/978-3-031-40579-2_49

Mystakidis S, Lympouridis V (2024) Designing simulations in the metaverse: a blueprint for experiential immersive learning experiences. In: Geroimenko V (ed) Augmented and virtual reality in the metaverse. Springer Nature Switzerland, pp 65–79. https://doi.org/10.1007/978-3-031-57746-8_4

Narang K, Please H, Dhanda J (2023) 940 the metaverse as a potential solution for the delivery of global health education. Br J Surg 110(Supplement_7):znad258.393. https://doi.org/10.1093/bjs/znad258.393

Naved M, Gupta AK (2023) Metaverse as a tool for the achievement of SGDs: challenges, opportunities, and applications. In: El Khoury R, Alareeni B (eds) How the metaverse will reshape business and sustainability. Springer Nature, pp 105–129. https://doi.org/10.1007/978-981-99-5126-0_11

Nayak A, Patnaik A, Satpathy I, Patnaik BCM, Baral SK (2024) Metaverse: the transformation of customer engagement with brands in the digital age. In: Green metaverse for greener economies. CRC Press

Naz F, Magda R (2023) Triple bottom line in sustainability and populism. In: Chacko Chennattuserry J, Deshpande M, Hong P (eds) Encyclopedia of new populism and responses in the 21st century. Springer Nature, pp 1–5. https://doi.org/10.1007/978-981-16-9859-0_10-1

Papagiannakis G, Lydatakis N, Kateros S, Georgiou S, Zikas P (2018) Transforming medical education and training with VR using M.A.G.E.S. SIGGRAPH Asia 2018 Posters, 1–2. https://doi.org/10.1145/3283289.3283291

Pineda M, Jabba D, Nieto-Bernal W (2024) Blockchain architectures for the digital economy: trends and opportunities. Sustainability 16(1):1. https://doi.org/10.3390/su16010442

Rahman Khan MN, Ahmmed F, Al Zabir MK, Lippert KJ (2023) Virtual patient in medical education. In: 2023 7th international symposium on multidisciplinary studies and innovative technologies (ISMSIT), pp 1–6. https://doi.org/10.1109/ISMSIT58785.2023.10304938

Ren Z, Fu X, Dong K, Lai Y, Zhang J (2023) Advanced study of optical imaging systems for virtual reality head-mounted displays. Photonics 10(5):555. https://www.mdpi.com/2304-6732/10/5/555

Riches S, Williams G, Saidel S (2024) The metaverse as a therapeutic space: opportunities for public health benefits and clinician-industry partnerships. Int J Hum Comput Interact 1:1–4. https://doi.org/10.1080/10447318.2024.2357902

Roberts I, Godlee F (2007) Reducing the carbon footprint of medical conferences. BMJ 334(7589):324–325. https://doi.org/10.1136/bmj.39125.468171.80

Samuel PK, Moses PM, Davies S, Wekesa C (2022) Analysis of energy utilization metrics as a measure of energy efficiency in data centres: case study of Wananchi Group (Kenya) limited data centre. IEEE PES/IAS PowerAfrica 2022:1–5. https://doi.org/10.1109/PowerAfrica53997.2022.9905342

Sarah A, Nencioni G, Khan MdMI (2023) Resource allocation in multi-access edge computing for 5G-and-beyond networks. Comput Netw 227:109720. https://doi.org/10.1016/j.comnet.2023.109720

Shafei I (2023) Medical education. In: Loh E, Long PW (eds) Textbook of medical administration and leadership, pp 413–427. Springer Nature. https://doi.org/10.1007/978-981-99-5211-3_19

Shah R (2023) Does cadaveric surgery have a role in modernizing anatomical education and sparking surgical interest in medical students? Am J Surg 226(4):559–561. https://doi.org/10.1016/j.amjsurg.2023.05.019

Singh G, Ahmad F (2024) An interactive augmented reality framework to enhance the user experience and operational skills in electronics laboratories. Smart Learn Environ 11(1):5. https://doi.org/10.1186/s40561-023-00287-1

Srivastava AK, Dixit S, Srivastava AA (2022) Criticism of Triple Bottom Line: TBL (with special reference to sustainability). Corp Reput Rev 25(1):50–61. https://doi.org/10.1057/s41299-021-00111-x

Tan TF, Li Y, Lim JS, Gunasekeran DV, Teo ZL, Ng WY, Ting DSW (2022) Metaverse and virtual health care in ophthalmology: opportunities and challenges. Asia Pac J Ophthalmol 11(3):237–246. https://doi.org/10.1097/APO.0000000000000537

Tracana RB, Carvalho GS (2012) Ecosystems, pollution, and use of resources in textbooks of 14 countries: an ecocentric emphasis. Int Sch Res Not 2012(1):419782. https://doi.org/10.5402/2012/419782

Wang G, Badal A, Jia X, Maltz JS, Mueller K, Myers KJ, Niu C, Vannier M, Yan P, Yu Z, Zeng R (2022) Development of metaverse for intelligent healthcare. Nat Mach Intell 4(11):922–929. https://doi.org/10.1038/s42256-022-00549-6

Wiecha J, Heyden R, Sternthal E, Merialdi M (2010) Learning in a virtual world: experience with using second life for medical education. J Med Internet Res 12(1):e1337. https://doi.org/10.2196/jmir.1337

Xu SS-D, Chang T-C (2017) A feasible architecture for ARM-based microserver systems considering energy efficiency. IEEE Access 5:4611–4620. https://doi.org/10.1109/ACCESS.2017.2657658

Yue D, Qian Y (2024) Clinical skills training of metaverse-based learning: a systematic review. Int J Acad Res ProgIve Educ Dev 13(3):950–964

Zhao N, You F (2023) The growing metaverse sector can reduce greenhouse gas emissions by 10 Gt CO2e in the united states by 2050. Energy Environ Sci 16(6):2382–2397. https://doi.org/10.1039/D3EE00081H

Zhu P, Hu J, Li X, Zhu Q (2023) Using blockchain technology to enhance the traceability of original achievements. IEEE Trans Eng Manag 70(5):1693–1707. https://doi.org/10.1109/TEM.2021.3066090

Learning in the Digital Frontier: How the Metaverse Transforms Education

Jasdeep Kaur and Kuldeep Kumar

Abstract The Metaverse is the next frontier of digital interaction, a collective virtual shared space formed by merging enhanced physical reality with persistent virtual environments, encompassing all virtual worlds, augmented reality, artificial intelligence (AI), blockchain, machine learning, and the internet. With the rapid advancement of technology, the Metaverse presents unparalleled prospects for revolutionizing the field of education. It offers immersive, interactive learning experiences that transcend geographical barriers and traditional limitations. This study looks at how the Metaverse can improve accessibility, engagement, and individualized learning to transform the way that education is delivered. While admitting issues with accessibility, digital literacy, privacy, security, and cost, the article also discusses the major advantages, such as enhanced learning outcomes and worldwide collaboration. The article concludes by examining the potential for integrating the Metaverse into education going forward, with a focus on creative pedagogies and chances for lifelong learning. This study attempts to shed light on the significant influence of the Metaverse on education through a thorough examination and offer suggestions for educators, decision-makers, and interested parties on how best to utilize its potential.

Keywords Metaverse · Education · Virtual reality (VR) · Immersive learning · Virtual classrooms · Artificial intelligence (AI)

J. Kaur (✉)
Department of Commerce, S.G.T.B. Khalsa College, University of Delhi, New Delhi, India
e-mail: jasdeep@sgtbkhalsa.du.ac.in

K. Kumar
Department of Physics, S.G.T.B. Khalsa College, University of Delhi, New Delhi, India
e-mail: kuldeep@sgtbkhalsa.du.ac.in

W. Leal Filho et al. (eds.), *Metaverse and Sustainability: Business Resilience Towards Sustainable Development Goals*, World Sustainability Series,
https://doi.org/10.1007/978-3-031-89545-6_10

1 The Metaverse: An Overview

The Metaverse is a collective virtual environment that combines the digital and physical worlds to offer more than just traditional virtual reality. It offers interactions with others, immersive experiences, and business opportunities. Fast-moving technology and improved connectivity are enabling it to advance quickly, with the potential to completely change the way people work, socialize, and engage with the digital environment. A network of linked virtual worlds that offer a variety of activities, such as attending virtual events and working remotely or shopping for digital items, is referred to as the "metaverse." It can be seen as an advanced, three-dimensional cyberspace where users can interact with various applications and services without difficulty. The metaverse makes the virtual world feel more substantial by enabling more immersive and engaging interactions than the present internet, which primarily consists of people navigating around webpages. Virtual reality (VR), augmented reality (AR), and cutting-edge internet and semiconductor technologies are all used in the fast-evolving metaverse, a digital world that offers consumers lifelike experiences in both personal and professional settings (What Is the Metaverse and Where Will It Lead Next? 2022). Although the metaverse is still in its early phases of development, Facebook's 2021 rebranding as Meta and $10 billion investment in its development attracted a lot of attention and investment. This action brought attention to the metaverse's expanding significance and promise, which in turn encouraged billion-dollar investments in the field from other major tech companies including Microsoft, Google, Sony, and Nvidia. In addition to investing $1 million in a Center for Responsible AI at IIT Madras and Wadhwani AI for agricultural AI solutions, Google is also improving its models for Indic languages through a variety of data collection initiatives (Google Grants $1 Million to IIT Madras for AI Research, 2022). The big tech companies are now involved in shaping up this new technology and are focusing on a number of metaverse-related topics for smooth virtual interactions (Lavania 2025).

A number of key components help the metaverse to function as an engaging and interactive virtual world. Virtual reality (VR) and augmented reality (AR) are the two main technologies that help to create completely virtual environments or integrate the digital content with the real world. This results in engaging and authentic experiences. In order to ensure the smooth operation of these virtual realities, we depend on robust computer systems, such as fast networks and cloud computing. These systems offer the necessary processing power and uninterrupted connectivity to uphold the intricate environments of the metaverse. Another important factor is the user-generated content and virtual worlds that allow both users and developers to shape and build the metaverse with diverse personalized experiences. Users have the ability to effortlessly navigate and engage with virtual realms through the use of tools such as AR glasses and VR headsets, allowing entry into the metaverse. In the end, the metaverse's sustainability relies on strong privacy, security, and governance frameworks for it to be a secure, fair, and valuable digital space. The metaverse is a vast, interconnected digital space that integrates these elements to create an experience nearly identical to the real world.

2 Components of the Metaverse

A complex network of interconnected technologies, the metaverse produces persistent, immersive, and interactive digital experiences. The following section outlays the essential elements that support this digital world:

- **Virtual Reality (VR)**: Virtual reality is an essential component since it makes digital environments more engaging, realistic, and immersive. It is a key technology that constructs the metaverse, along with extended reality (XR) and AI. Virtual reality improves the visual experiences in number of industries, including real estate, healthcare, and education, by submerging people in lifelike circumstances. This promotes more learning and engagement. Additionally, it facilitates social engagement within the metaverse, promoting user cooperation and communication within a common digital environment. As a result, VR is essential for building immersive virtual worlds and laying the technological foundation required for the metaverse (Tukur et al. 2024).
- **Augmented Reality (AR)**: According to Jadhav (2013), augmented reality is a simulation of the real world that is supplemented with computer-generated sensory input, such as images, music, video, or GPS data. It can be used in many different streams, including designing, gaming, education, and retail. This kind of reality combines digital visual elements with the actual world, in contrast to virtual reality, which is entirely digital. Interactive video learning resources that encourage creativity and provide students with a variety of educational resources are made possible by the use of Augmented Reality and the Metaverse (Zahra et al. 2024).
- **Extended Reality (XR)**: A wide range of technologies, such as mixed reality, virtual reality, and augmented reality, are together referred to as "extended reality". A real-time fusion of virtual and real-world settings where things coexist is called mixed reality. Our understanding of reality can be expanded or altered by these technology advancements that help create engaging content. These technologies are becoming more and more useful in different sectors of the economy, such as healthcare, education, entertainment, and collaborative work environments. It is possible that future developments change how people work, learn, and interact with the outside world.
- **Blockchain**: Blockchain is a like decentralized digital ledger that stores transactions in a more secure and transparent way through the use of several computers. Each transaction is linked to the other, creating a "block chain" that cannot therefore be forged or altered in any way. Blockchain provides the digital assets and business conducted in the Metaverse with security and credibility. Through the establishment of the virtual currencies, including NFTs and smart contracts within the Metaverse, it allows for safe ownership and exchange of digital products and services. Blockchain makes it easier to create decentralised applications and platforms where users have a say in what happens to them through decentralisation because the latter fosters trust and transparency. In the metaverse, virtual product ownership and exchange are done using blockchain technology as the backbone. Given that many things, including art, virtual land, and in-game items, are relatively easy to buy and sell, it alters commerce (Shukla and Taneja 2024).
- **Artificial Intelligence (AI)**: The Metaverse relies heavily on Artificial Intelligence and computer vision as it seeks to improve immersive experiences and help design virtual realities (The et al. 2023). The intelligent interactions and personalisation of user experiences through Artificial Intelligence boosts the Metaverse

significantly. AI-supported voice interfaces enable real conversation and immediate assistance, and AI-provided content accelerates the creation of complex and diverse virtual environments.

- **Internet of Things (IoT)**: The Internet of Things (IOT) is crucial for improving the Metaverse since it lays a foundation between the real and virtual worlds. Smart sensors and wearables belong to the (IoT) that collect data in real time from the physical environment and transfer into the Metaverse. Because of this integration, virtual worlds can replicate real-life situations and interactions in a much more realistic and accurate manner. Homes for example, smart home equipment can interact with the digital home equipment and manipulate virtual environments to establish a conjoined environment. As a rule, through these possibilities, IoT expands the Metaverse by incorporating devices and data from outside this environment into various regulated domains.
- **Cloud Computing**: The Metaverse is highly dependent on cloud computing as it brings necessary resources for complex and resource-hungry environments. Cloud services offer the networking, storage, and computing capabilities required for accommodating several users simultaneously and hosting vast virtual worlds. This technology ensures that the interactive features and quality images and videos can flow, thus enhancing the smoothness of the overall experience. Furthermore, the use of cloud computing contributes to real-time processing and handling of data for updates and Metaverse operations. Thus, to increase Metaverse's reach, it is crucial to further investigate its potential for various use cases, sizes, and performance while providing engaging user experiences for people worldwide through the use of cloud assets.

3 Potential Benefits of using Metaverse in Education

The increasing use of artificial intelligence in the education sector is helping improve various aspects, including academic performance, optimal utilization of resources, and the provision of personalized learning (Gaurav 2023; Ahmed et al. 2024). The inclusion of artificial intelligence in education is crucial to ensure that students are educated students for a world in which the use of AI will be incorporated seamlessly in their daily lives (Pullen and Bien 2023). AI tools enhance student learning by personalizing experiences and streamlining instructional strategies through knowledge tracing and spaced repetition (Roy 2023). The use of metaverse which is an AI-driven immersive virtual environment has the potential to revolutionize education by providing immense benefits. The education system is improved and stimulated, becoming more open-sourced and distinctive, by establishing a learning-centered atmosphere that permits students to participate in meaningful learning activities outside of the classroom. It has the potential to completely alter the way the learning and teaching are done, particularly when it comes to connectedness, inclusivity, and openness. If academic institutions and IT companies continue to search for and invest this technology, the metaverse will continue to be an exciting future that has the ability to transform education and offer special opportunities for both students and teachers. The following points illustrate how the Metaverse can help achieve this:

- **Immersive Learning Experiences**: The feeling of presence in a learning environment obtained with virtual and augmented reality, coupled with the natural and appealing simulations, are changing the concept of learning. Through the creation of immersive learning environments, virtual reality enables students to engage in interactive topic exploration (Ramasamy 2024). The studies also emphasizes that VR can enhance the practical thinking process by providing students with a more realistic way to interact with concepts that are often challenging to grasp (Poupard et al. 2024). For instance, when students are provided with realistic assignments which are active in nature, it is quite effective in enhancing recall and understanding of complex topics (Pramanik 2024). One such example is the use of virtual reality in training medical students whereby students can practice surgery in a safe environment that is free from any real harm.
- **Individualized Learning**: Personalized learning routes based on students' strengths, limitations, and learning styles can be created via artificial intelligence (AI) and machine learning operating within the metaverse. Adaptive learning environments are provided by platforms such as DreamBox Learning and Squirrel AI, which use data analytics to modify the pace and content to meet the needs of individual students. This results in a more efficient and tailored education. In India BYJU'S has announced BYJU'S WIZ that uses generative learning to give students individualized and successful learning experiences (Malik 2023).
- **Enhancing Accessibility**: With the removal of regional constraints and support for a diverse student body, the Metaverse dramatically improves accessibility. Through virtual classrooms, students from underserved or rural places can receive high-quality instruction and materials, filling in any gaps that traditional systems would overlook. Additionally, studies show that the Metaverse can increase accessibility for people with disabilities by providing individualized virtual experiences that can be customized to meet a range of needs, resulting in a more welcoming learning environment (Zhang and Chemnad 2024)
- **Transcending Geographical Barriers**: The Metaverse opens unique opportunities in learning and collaboration on an international level due to geographic barriers' elimination. This way, people, who are in different locations, share the same physically undefined space and have an opportunity to engage in similar practices and processes by benefitting from the Metaverse. The study shows it also enhances social cohesion as individuals can easily work together and relate with one another. The Metaverse is also used for training and education, enabling users to obtain information and sources that may be physically inaccessible (Mkedder and Das 2024).
- **Collaborative Learning Environments**: The Metaverse activates enhanced engagement through realistic and experimental learning scenarios. It makes group assignments, observation, working while on a project and engaging in interdisciplinary tasks conception beneficial by providing students with more opportunities

to get involved, interact, and excel in their study endeavors. Research shows that the Metaverse boosts students' engagement and interaction levels since many of the academic settings are not very conducive to bringing out the community feeling in a group (Esguerra et al. 2024).

4 Challenges and Considerations in the Metaverse for Education

The Metaverse brings significant advantages to the education sector—with its focus on experiential learning, personalized content, and global alignment. However, if we want to ensure that it will be both effective and long-lasting, there are also some challenges and considerations with its application.

- **Digital Divide**: The disparities in either the possession or access of technology or the internet amongst the demographical sexually, geographically, or economically are known as the 'digital divide.' This divide often presents or can be defined through inequality in opportunities in civil participation, work, or education. Many people who do not possess necessary equipment or stable Internet connection feel rather handicapped in the modern world that has recently become more and more technologically advanced. Thus, the Metaverse is based on innovative technology and a high-speed internet connection, which can only increase the feeling of isolation among those who cannot afford such tools (Wang 2023).
- **Cost of Technology**: One of the biggest obstacles is the high cost of the technology needed to reach the metaverse. Immersion learning requires top-tier hardware, such as VR headsets, powerful PCs, and reliable high-speed internet. The expense of technology is exacerbating the digital gap and running the risk of establishing a two-tiered educational system where only the wealthiest can take full advantage of the metaverse's possibilities. Though the metaverse offers many benefits, institutions wishing to invest in this changing digital landscape should carefully examine the upfront expenses and ongoing challenges.
- **Technical Issues**: Technical issues are a major barrier preventing the metaverse from being widely adopted. System compatibility becomes crucial because as difficult as it is to make sure that all of the devices, apps, and virtual platforms can operate together harmoniously. Additionally, reliable and high-speed internet connection is crucial in order to prevent any interruptions and to provide an exciting experience. Also, there is the concern that such technology has to be widespread, but sadly, it is possible that not all places, especially developing or rural ones, have this connectivity. These technical challenges need to be addressed to ensure that equity and efficiency are achieved when accessing the metaverse.
- **Privacy and Security Concern**: While the concept of the metaverse offers new ideas for learning, it raises concerns of privacy and security immensely. This is particularly important since students' records, including academic credentials and personal information, should be kept secure. Security measures, including access

constrains and data encryption, should be implemented on this information. Other risks affecting the virtual environment include cyber threats such as harassment and hacking. Web technologies, augmented reality, and virtual reality can lead to entirely new cybersecurity challenges. Thus, it is crucial that the users are protected from dangers like identity theft and loss of property in cyberspace. (Al-Emran et al. 2024).

- **Pedagogical Challenges**: It is highly challenging to design effective teaching strategies for the metaverse. Specific competencies and experience are required to generate knowledge meaningful and engaging for education, aligned with learning outcomes in this new digital environment. Another challenge is also combining the application of metaverse experience with today's curriculum without compromising the quality of education. They also have to contemplate how best to preserve the attraction of the technology while being mindful of proper teaching and learning.
- **Ethical and Social Considerations**: When it comes to Metaverse education moral and social factors constitute great concern. One weakness of shift to virtual classes and other virtual settings may be reduction of physical contact that is important in the development of interpersonal skills. In addition, students' mental and physical states might be affected with negative consequences due to digital dependency resulting from the richness of the Metaverse. Privacy is also an issue originated from the processing of huge quantities of individual data in these contexts. However, social-inequalities will only be enhanced in the Metaverse as poorer learners are denied opportunities for probe technology that expands learning inequalities to a higher level. This is because to ensure that the Metaverse enhances education without negatively affecting equity and well-being of the learners these problems needs to be addressed.
- **Evaluation and Assessment**: It can be seen that there are a number of challenges in the process of evaluation and assessment in context of metaverse education; particularly, while developing the measures for evaluation and assessment there are some barriers in the equitable application of the same. There have been recent discussions regarding the creativity of new measures in the context of performance in virtual environments where traditional measures can be nonrepresentative of the depth of the interactions and feedback that learners receive in such environments. The additional complexity arises in a need to ensure that these assessments have to enhance the conventional teaching systems and practices. It also becomes challenging to develop a standard set of assessments that can be fairly adapted to a variety of platforms while ensuring that all students are assessed fairly.

5 Metaverse Applications in Indian Education

Metaverse has presented a roster of innovative, immersive, and interactive learning experiences that go beyond conventional classrooms, and they are slowly changing Indian education. IIT Madras project is a good example where a virtual campus was created through which students can participate in fully academic activities, have lessons and make friends. Besides eradicating geographical barriers, the virtual campus is beneficial in elaborating and engaging and enhancing the students' understanding and aids in recalling complex concepts.

The integration of AR and VR in vocational training in India has opened a new era of change through programs like TCS iON. This EdTech innovator has introduced targeted programmes that equip learners with what they require to design VR and AR experiences in cooperation with academic institutions. In vocational training, where they create environments that the learner can practice intricate activities which require a lot of risk such as welding, operating machines, and performing surgery, these technologies are particularly transformative. For example, as mentioned before, VR simulations are applied to teach procedural and security measures to employees of logistics and automobile manufacturing firms. This helps immensely in raising efficiency and reducing the propensity for errors during a worker's working hours. Similarly, the implementation of AR in medical industry allows students practice on real operations and other procedures but not practical on human beings. These applications ensure that students are better prepared in case of any challenge they meet in their day to day life as well as enhance the learning process by minimizing the theory-practice gap. It is apparent that the possibilities for integrating VR and AR technology in vocational education in India are great, which opens the door to a workforce that is more competent and talented as they grow.

Additionally, Moreover, the technological advancement in the learning process has also been embraced by the Maharashtra State Board of Technical Education (MSBTE) which is now using AR/VR platforms as a learning advancement tool, which greatly improves the student's learning experience. One of the major initiatives is virtual history tours, which are virtual tours of historical sites planned by MSBTE in collaboration with historians and archaeologists. These virtual surroundings enable students to learn about the architectural styles, as well as the culture and historical backgrounds of such historic temples, forts, and monuments that they can virtually tour. Through geographical simulations involving virtual reality, MSBTE has endeavored to expose students to the Himalayan Mountain range, the Western Ghats and the coastal plains among other terrains. These simulations offer an engaging approach to learning key geographic concepts such as geography, climate, and ecosystems. MSBTE has also revamped science labs to recreate role play scenarios to replace old techniques such as frogs to dissect or chemical compounds to study. It is an effective creative strategy to promote practical learning but in a safe and accessible online environment. The commitment shown by MSBTE for integrating AR and VR speaks volumes about their modern and innovative approach that helps uplift the educational standards, making it more engaging, interactive, and more appropriate for the new generation.

Moreover, BYJU'S, an Indian EdTech company, has mastered the use of AR in its instructional design. This enables students to work on the 3D models and simulations which enhance the understanding of topics that may otherwise remain abstract. This creative method brings an aesthetic touch and interactivity to education which make learning far more effective and efficient. This enhances comprehension at the same time as interest and results in a deeper analysis of the given areas. Adding to it, BYJU'S is making lessons similar to usual TV shows and presenting them in the form of Augmented Reality based learning, addressing different learning styles and helping the students comprehend and focus on difficult concepts much better.

From these presented cases and examples, it can be seen how Metaverse is introducing new prospects of education in India, as well as enhancing the quality, availability, and interest to education at various subjects and areas of the country. Such technologies create opportunities to level the playing field and prepare students for the demands of the workforce of the twenty-first century.

6 The Future of Education in the Metaverse

Learning may be fully engaged in a rich, dynamic, and individualized setting thanks to the Metaverse, which has the potential to transform education and skill development. Geographical restrictions are removed, and chances for lifelong learning are made available to people of all ages in a variety of domains, such as professional development, technical proficiency, and interpersonal skills. The enhancement of new competencies can be attracted to the Metaverse through virtual exercises, and self-learning is an efficient way to follow up on competency growth. Educators should embrace metaverse technology, make training investments where appropriate, and guarantee open access for all students, regardless of their position as digitally disadvantaged or limited users, in order to maximize these advantages.

Policymakers should encourage the development of requisite infrastructure and provide potential funding to help offset start-up costs in order to improve digital connection. However, they also need to set up policies and procedures in the Metaverse to guarantee the security and quality of the instruction provided. It is also recommended that researchers carry out an empirical investigation to determine the effectiveness of Metaverse learning in comparison to traditional learning methods; examine and determine the optimal practices that can be adhered to during the implementation of a Metaverse; and assess future trend analysis to predict modifications. Through effective management of these obstacles, stakeholders can leverage the Metaverse as a means of transforming the educational process and offering each learner a relevant virtual experience.

7 Conclusion

There are a plethora of educational options in the Metaverse. It facilitates the execution of lifelike simulations and hands-on learning to improve comprehension and retention of complex ideas. By providing classroom and social platform opportunities for students to engage with peers and teachers from across the globe, a global learning community can be established. The diverse learning opportunities in the Metaverse allow students to move at their own speed while still addressing their unique needs and interests, which improves the effectiveness and efficiency of education for all.

However, the transition to Metaverse-based education is not without its challenges. Digital literacy, privacy, security, accessibility, and affordability are important concerns. Digital gaps may already be there, but they could widen if certain communities lack access to essential technologies like VR headsets and fast internet. Addressing the financial obstacles related to the adoption of cutting-edge technologies is equally as important as ensuring privacy and security in virtual environments. To fully utilize the Metaverse's educational potential, these issues must be resolved.

This study has looked into a number of the Metaverse's components and its possible uses in education, particularly in Indian educational institutions. Innovative pedagogies and chances for lifelong learning can be provided by the incorporation of Metaverse technologies. In the future, incorporating the Metaverse into education could revolutionize the way that knowledge is imparted and experienced. Teachers may design dynamic, interesting, and accessible learning environments by utilizing the Metaverse's unique features and overcoming its current challenges. To fully realize the promise of the Metaverse, educators, legislators, and other stakeholders must work together to remove obstacles, and incorporate practical plans.

To sum up, the Metaverse presents a paradigm-shifting chance for education that might improve student performance, encourage international cooperation, and promote lifelong learning. The study highlights the necessity for continuous efforts to address issues and maximize the advantages of this cutting-edge technology and offers insights into how the Metaverse might be used into instructional procedures. By doing this, we can open the door to a future in education that is inclusive and more engaging.

References

Ahmed ZE, Hashim AHA, Saeed RA, Saeed MMA (2024) AI-enhanced education: bridging educational disparities. In: Ahmed ZE, Hassan AA, Saeed RA (eds) Advances in educational technologies and instructional design. IGI Global, pp 88–107. https://doi.org/10.4018/979-8-3693-2728-9.ch004

Al-Emran M, Al-Sharafi MA, Foroughi B, Iranmanesh M, Alsharida RA, Al-Qaysi N, Ali N (2024) Evaluating the barriers affecting cybersecurity behavior in the metaverse using PLS-SEM and fuzzy sets (Fsqca). Comput Hum Behav 159:108315. https://doi.org/10.1016/j.chb.2024.108315

BYJU'S launches BYJU'S WIZ, a suite of cutting-edge AI models for hyper-personalized learning (n.d.) Retrieved 30 July 2024 from BYJUS website https://byjus.com/press/byjus-launches-byjus-wiz-a-suite-of-cutting-edge-ai-models-for-hyper-personalized-learning/

Chinta SV, Wang Z, Yin Z, Hoang N, Gonzalez M, Quy TL, Zhang W (2024) Fairaied: Navigating fairness, bias, and ethics in educational AI applications. https://doi.org/10.48550/arXiv.2407.18745

Esguerra JR, Figueroa Jr, RB, Rodriguez F, Fukuda A, Anzai Y (2024) Collaborative online international learning: a reflexive thematic analysis of narratives from the metaverse. Int J Inf Technol Gov Educ Bus 6(1):100–108. https://doi.org/10.32664/ijitgeb.v6i1.140

Gaurav (2023) AI in education: changing the learning landscape. Retrieved 30 July 2024, from Embibe Exams website: https://www.embibe.com/exams/artificial-intelligence-ai-in-education/

Google grants $1 million to IIT Madras for AI research (2022). Retrieved 30 July 2024, from INDIAai website: https://indiaai.gov.in/news/google-grants-1-million-to-iit-madras-for-ai-research
How AI will displace teachers: what we can expect in the future (2023) Retrieved 31 July 2024, from 21K School India website: https://www.21kschool.com/in/blog/how-ai-will-displace-teachers/
How to leverage AI tools in content creation (n.d.) Retrieved 31 July 2024, from Edulyte website: https://www.edulyte.com/educate/ai-content-generation/
IIT-Delhi and Wipro launches CoE on generative AI (2023) Retrieved 30 July 2024, from INDIAai website: https://indiaai.gov.in/news/iit-delhi-and-wipro-launches-coe-on-generative-ai
Jadhav R (Director) (2013) An introduction to augmented reality with mobile system introduction to augmented reality with mobile system. Retrieved from https://typeset.io/papers/an-introduction-to-augmented-reality-with-mobile-system-3zsflu5ce7
Lawton G (n.d.) What is generative AI? Everything you need to know. Retrieved 24 July 2024, from Enterprise AI website: https://www.techtarget.com/searchenterpriseai/definition/generative-AI
Malik P (2023) BYJU'S introduces generative AI models for hyper-personalised learning. Retrieved 30 July 2024 from https://yourstory.com/2023/06/byjus-wiz-generative-ai-models-hyper-personalised-learning-edtech
Meet Khanmigo: Khan Academy's AI-powered teaching assistant & tutor (n.d.) Retrieved 30 July 2024 from https://khanmigo.ai/
Mkedder N, Das M (2024) Metaverse integration challenges: an in-depth ISM and MICMAC analysis. J Retail Consum Serv 77:103684. https://doi.org/10.1016/j.jretconser.2023.103684
Poupard M, Larrue F, Sauzéon H, Tricot A (2024) A systematic review of immersive technologies for education: learning performance, cognitive load and intrinsic motivation. Br J Educ Technol 56(1):5–41. https://doi.org/10.1111/bjet.13503
Pramanik S (2024) Immersive innovations: exploring the use of virtual and augmented reality in educational institutions. In: Augmented reality and the future of education technology, pp 66–85. IGI Global
Pullen M, Bien H (2023) Benefits of ai in education. Retrieved 31 July 2024 from https://www.jamf.com/blog/pros-cons-ai-in-education/
Ramasamy T (2024) AI in higher education. Int J Sci Res Eng Manag 8(6). https://ijsrem.com/download/ai-in-higher-education/
Roy D (2023) Enabling education: Dev Roy from BYJU'S explains how generative AI tools can create better learners and teachers—CNBC TV18. Retrieved 30 July 2024, from CNBCTV18 website: https://www.cnbctv18.com/views/enabling-education-dev-roy-from-byjus-explains-how-generative-ai-tools-can-create-better-learners-and-teachers-16639001.htm
Shukla P, Taneja B (2024) Navigating the metaverse in business and commerce: opportunities, challenges, and ethical consideration in the virtual world. In: Research, innovation, and industry impacts of the metaverse, pp 183–196
Teacher training expert—Khanmigo: Khan academy India (2024) Retrieved 30 July 2024, from DevInfo.in website: https://devinfo.in/job/khan-academy-delhi-ncr-lucknow-19-teacher-training-expert-khanmigo-khan-academy-india/
The T, Pham Q, Pham X, Do-Duy T, Reddy Gadekallu T (2023) AI and computer vision technologies for metaverse. In: Hoang DT, Nguyen DN, Nguyen CT, Hossain E, Niyato D (eds) Metaverse communication and computing networks, 1st ed. Wiley, pp 85–124. https://doi.org/10.1002/9781394160013.ch5
Tukur M, Schneider J, Househ M, Dokoro AH, Ismail UI, Dawaki M, Agus M (2024) The Metaverse digital environments: a scoping review of the techniques, technologies, and applications. J King Saud Univ Comput Inf Sci 36(2):101967. https://doi.org/10.1016/j.jksuci.2024.101967
Wang Q (2023) On development of metaverse and digital ecology safety. Front Comput Intell Syst 5(3):111–116. https://doi.org/10.54097/fcis.v5i3.14008

Zahra A, Dewi IP, Hendriyani Y, Novaliendry D (2024) Pengembangan media augmented reality pada pembelajaran videografi di smk negeri 2 padang panjang. Voteteknika (Vocational Teknik Elektronika dan Informatika) 12(2):184–195. https://doi.org/10.24036/votete knika.v12i2.127459

Zhang CY, Chemnad K (2024) Is the metaverse accessible? An expert opinion. Nafath 9(25). https:// doi.org/10.54455/MCN2507

Pixels and Profits: A Deep Dive into Virtual Economies

Rihan Banerjee, Vansh Sharma, Deepika Dhingra, and Anjali Munde

Abstract A virtual economy, existing in the digital world, exchanges virtual goods and services with cryptocurrency acting as its mediator (i.e. currency). Virtual economies have transcended mere entertainment, evolving into complex ecosystems with real-world value. This chapter delves into the historical foundations of these economies, while critically examining existing shortcomings and loopholes. Connecting the major nodes of the entire system of Technological Leverage-Decentralized markets-mega profits, to achieve sustainable growth, building trust and long-term involvement of consumers. The chapter will explore the technology's inherent benefits, highlighting its potential to revolutionize economic interaction. Furthermore, the chapter will analyse the psychological factors influencing user acceptance and full participation within virtual economies. Employing a comprehensive psychographic analysis to gain a deeper understanding of user needs and primary usage drivers. Finally, the chapter will engage in a crucial debate: is a centralized regulatory body necessary, or can inclusivity and open trade practices pave the way for a more equitable system? Finding a correlation between the lack of involvement of the government that may-may not be the root for potential frauds-capital focused influencing-creating profits, along with the location of these trades i.e. where are they performed and the marketplaces that exist. Through this comprehensive analysis, the chapter aims to contribute to the construction of a virtual economic framework that fosters both prosperity and fairness.

R. Banerjee · V. Sharma · D. Dhingra (✉)
School of Management, BML Munjal University, Kapriwas, India
e-mail: mithudhingra@gmail.com

A. Munde
Business Analytics, Southampton Malaysia Business School, University of Southampton, Gelang Patah, Malaysia

W. Leal Filho et al. (eds.), *Metaverse and Sustainability: Business Resilience Towards Sustainable Development Goals*, World Sustainability Series,
https://doi.org/10.1007/978-3-031-89545-6_11

1 Introduction

A deeper dive into the metaverse, exploring the virtual economy, and the wealth-creating functions of the metaverse. In the synthetic world, the prospect economy is to navigate economic activities and find the nodes to generate wealth.

The excitement and interest in exploring and working towards a virtual world has increased over the past few years. The rapid development of technology and the changing generation greatly aid this factor. The idea of the metaverse boomed since the company Facebook (now meta) changed, investing heavily into the metaverse in recent years. First started in 1992, in a science fiction novel "Snow Crash" by Neal Stephenson. The novel describes life in a metaverse with the duality of leisure and work—the ignition points to several forthcoming games and an attempt to develop an entire economy.

A virtual world is a world run by avatars representing real-world humans and its economy is the transaction between the avatars trading virtual items and services. In the earlier models of virtual worlds, the primary focus of virtual items was on game-related assets, including swords, skins, and special powers. These items were designed to enhance gameplay and could be acquired by players through purchases made with real-world currency, allowing for a more personalized and powerful in-game experience. For example, in games like Minecraft, Minecraft coins can be bought for real-world money and these Minecraft coins allow the users to purchase skins and several items in the game. Entropia Universe, another example of the same, that functions in the virtual world, has allowed users to earn money and in some cases an opportunity to make a living out of it. In the Entropia universe the currency 'Project Entropia dollars' has a fixed exchange rate of 10:1 with the US Dollar (10 PED = 1 US$) (Entropia Universe-Deposits n.d.) via which the transactions are conducted within the game. Several other games have been created in the same domain such as World of Warcraft, Everquest, Second Life, etc. Yet most of these innovations and developments occur in the realm of gaming. Meta (formally Facebook) has been one of the major companies backing the development of the Metaverse, which has expressed the change in the dynamics of this industry and how it has the potential to change the everyday interactions one has.

With gaming as its history, the acceptability and adaptability of the virtual world remain a major question mark. Mark Zuckerberg (CEO of Meta) has emphasized the importance of AR/VR changes in the future of human interaction by saying that future interaction will involve people attending meetings as avatars and digital holograms (Martins 2023). This comes at a high pace with the development of new technology. The adoption rate has increased since 2022 yet when looking at the long-term uses, a sustainability model will have to work in place as well and the ethicality of the ecosystem will have to be questioned for the benefit of the consumers.

With an entire generation who have better adaptability to virtual reality comes as a great boon for the same, yet several questions need to be addressed while we venture into the use of the metaverse and the virtual economy that resides in this virtual world. Aspects like security, purpose, and market availability need to be investigated and built trust through the process.

Several elements of virtual worlds like Blockchain technology and NFT have a wide range of usability that need to be investigated and experimented with along the road map of the development of a virtual economy. Blockchain is one of the major core competencies of a virtual economy. As the world advances, there's an increasing emphasis on data security, control, and most importantly, decentralization—goals effectively met by blockchain technology. NFTs are another major node for the same, as the primary purpose of NFTs as identity codes, which are tradable. This comes in as the authentication factor for digital art and digital real estate, they serve a major role in identity services.

Finding a monetary value in this new ecosystem remains a major focus for several companies. Brands like Dell have had to close their operations in the virtual world due to a lack of participation, lower profits, and operational costs. Along with exploring the value of the virtual economy, it is essential to define its purpose clearly. Until now the primary focus of a virtual world has been entertainment and gaming. As we progress, the apparent use of the virtual world becomes versatile thereby having to introspect the main purpose of the virtual economy in the metaverse.

The metaverse can result in one of the major milestones in revolutionizing society as it holds the power to change how interactions occur in the world. This is an extremely collaborative process which involves a series of developments and research.

2 Understanding the Metaverse

2.1 History and Evolution

2.1.1 Key Components: Technological Advancements and How Can We Leverage It

The development of the Metaverse has been marked by several key milestones and technological advancements. The advent of high-speed internet, powerful computing hardware, and sophisticated software has enabled the creation of increasingly immersive and interactive virtual experiences. The technical leverage in virtual economies is the use of emerging technologies that may lead to efficiency in a decentralized market which will also boost productivity and enhance innovation these technologies function as a bridge that connects virtual economies with the latest tech like by providing enhanced security, increased stability and provide accessibility.

Blockchain technology is one of the core competencies of virtual reality and the virtual economy. Its main characteristics of data privacy, quality, sharing, interoperability, and integrity are some of the driving forces for its use, and a lot of different applications like in banking or any part of the economy (Gadekallu et al. 2022).

Blockchain is the foundational technology behind cryptocurrencies. Blockchain provides infrastructure for virtual assets and decentralized finance ensuring safety security and transparency in the transactions. It works on a no intermediaries model which automatically leads to decreased transaction cost, increased trust between participants, and increased speed (Catalini and Gans 2020). Blockchain technology is changing how we do business by allowing to make transactions directly with each other, without needing banks or middlemen. This not only cuts costs but also opens new chances for people all around the world to join and make money. Because blockchain is built to be safe and fair, it is a key part of the metaverse, which is like a digital universe where users can share, play, and trade. As the metaverse grows, blockchain will be important for handling digital assets, keeping user information private, and building trust among everyone involved. To encourage more people to join this exciting space, it is vital to offer them benefits, such as those provided by decentralized finance (DeFi), which can make their experiences richer and more rewarding.

The concept is applied in Decentraland and The Sandbox, using blockchain for ownership and monetization of virtual real estate, users can buy, develop, and sell digital properties within these virtual worlds. Cryptocurrencies, such as MANA in Decentraland, facilitate transactions within the virtual space. Powered by the Ethereum blockchain, Decentraland functions in a separate metaverse with its native token, MANA, which is used to transact business within its worlds (Ashmore 2023). With the advent of 5G, better connectivity has been unlocked making it more immersive for consumers, and at the same time cloud computing has given the edge in managing vast data sets (Exploring the fusion of cloud computing and metaverse 2024). AI plays a key role in enhancing user experience, optimizing business processes, and predicting market trends. As AI can analyze large data sets at a very high pace, simultaneously minimizing potential errors, thereby producing greater insights into consumer behavior, it leverages its descriptive and predictive edge to optimize the result.

A major node for the application of blockchain is the NFTs, unique digital assets verified on the blockchain, which are used to represent ownership of virtual items, artworks, and collectibles, creating new economic opportunities within the Metaverse, another major building block of the metaverse is its social and economic forum (Wang et al. 2021; Dionisio et al. 2013). For further automation, smart contracts are used to automate transactions and create agreements without intermediaries. They operate on blockchain technology. These contracts are self-executed based on terms and conditions written in its code that ensure transparency in transactions and function as proof of transactions and their conditions. Smart contracts reduce the risk of fraud and ensure compliance with the terms of the agreement, which is crucial in decentralized environments where trust is often lacking (Gatteschi et al. 2018). By removing intermediaries like banks and legal institutions smart contracts are one of the ways to optimize transactions leading to increased efficiency and maximizing profitability in virtual economies. The Metaverse encompasses complex social and economic ecosystems that mirror and extend beyond the physical world. Social interactions in the Metaverse are facilitated by avatars, digital representations of users

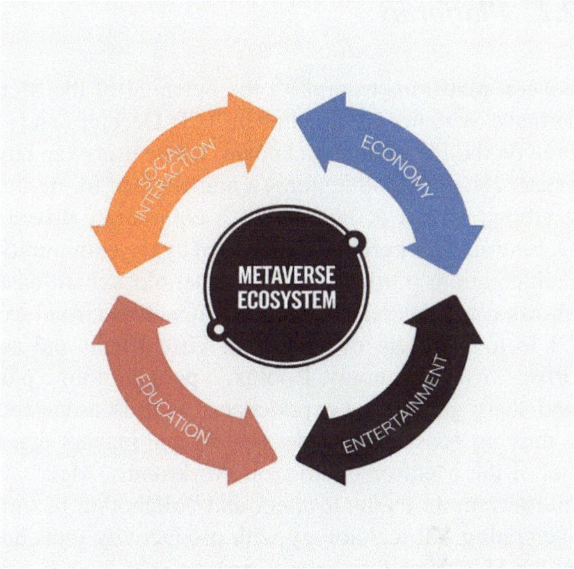

Fig. 1 The application of metaverse

that can communicate, collaborate, and engage in various activities (Dionisio et al. 2013). These interactions occur in virtual spaces, such as virtual worlds, social VR platforms, and online communities.

Virtual reality (VR) and Augmented Reality (AR) have been major cornerstones of the development of the metaverse. An immersive experience is one of its primary goals therefore also, making it one of the most sought-after technologies to work on (How does the metaverse work 2023; Interactive and Interactive 2023). AR on the other hand overlays digital elements onto the real world, enhancing the user's perception of their physical surroundings. This has a wide range of applications, it has its branches spread out to gaming, education, entertainment, etc. as shown in Fig. 1 (Azuma 1997; Weidner 2024).

For example, VR can provide an immersive training atmosphere, for instance stimulation-based training like the pilot training or to an extent surgical stimulation in the medical field can be made possible with the development of VR. In the very important medical field, VR has given a new edge that not only helps the studies and the development of medical sciences but also in surgery (News-Medical 2021). The American Board of Internal Medicine (ABIM) asserted the importance of this type of training before undertaking many particularly high-risk procedures (Rogers 2024). AR, on the other hand, is used in applications like navigation, where digital directions are superimposed onto the physical world, or in retail, where customers can visualize products in their homes before making a purchase.

2.2 Platforms

Several platforms exemplify the potential of the Metaverse, offering users diverse virtual experiences. Launched in 2020, Decentraland is A decentralized virtual world built on the Ethereum blockchain, where users can buy, sell, and develop virtual real estate. Decentraland features a marketplace for digital goods and services, creating a vibrant virtual economy. It's a community-driven virtual reality where land in Decentraland is permanently owned by the community, giving them full control over their creations (Ordano n.d.). Another blockchain-based virtual world, The Sandbox allows users to create, own, and monetize virtual experiences. The Sandbox uses NFTs to represent ownership of virtual land and assets, fostering a community-driven virtual economy. Roblox, a popular online platform, enables users to create and share games and experiences, with its user-generated content model creating a thriving ecosystem of developers and players that highlights the creative potential of the Metaverse. Horizon Workrooms, Meta's virtual collaboration platform, allows remote teams to meet and collaborate in immersive virtual environments, integrating VR technology with productivity tools and demonstrating the potential of the Metaverse for professional use.

2.3 Market State

The virtual economy is in a state of rapid growth and evolution. The current market size of the metaverse is estimated at 94.1 Billion dollars (from 65.2 billion dollars in the year 2022) (Metaverse market size 2022–2032 2024) and the surge of the global virtual production market (valued at USD 2.97 billion in 2023, the market is projected to grow from USD 3.37 billion in 2024 to USD 10.07 billion by 2032, exhibiting a CAGR of 14.7% during the forecast period.) (Insights 2024). As more users and businesses enter the metaverse, the demand for virtual goods and services continues to rise. With an estimated 600 million users in 2023 with Roblox having 55 million users in the virtual reality domain. Yet it's not just restricted to gaming as about 60% of gamers have used the Metaverse for non-gaming activities such as socializing and shopping and Around 52% of internet users around the world join the Metaverse for work (Hryziuk 2024). This has led to the emergence of new markets and opportunities for economic activity, such as virtual real estate, digital art, and online education. According to McKinsey, corporations, venture capitalists, and private equity firms have already poured over $120 billion into the Metaverse in the first five months of 2022—more than double the $57 billion invested throughout all of 2021 (Chandrasekar n.d.). However, the market is also characterized by volatility, particularly in the value of cryptocurrencies and virtual assets, which can fluctuate significantly based on market demand and external factors.

2.4 Business Applications

The metaverse has provided a new platform for the business to work on. When companies operate in the Metaverse, they can use fewer physical resources and minimize environmental impact. The possibility of replacing the current form of meetings i.e. offline, with a prospected virtual meet means less travel and a smaller carbon footprint. This advantageous transformation is not restricted and has its implications in factors. BMW has entered the virtual world for optimizing its operations along with Nvidia's Omniverse. With Omniverse, the BMW team can consolidate data into powerful models, integrate their specialized software tools, and facilitate live collaboration among multiple users across different locations. The BMW team aims to incorporate a robot in a limited floor space. The team improvises, allowing logistics and production planners to envision and determine the optimal placement (Geyer 2023).

Considering the remarkable opportunities within the virtual economy, it is imperative to acknowledge the associated challenges. The acquisition of Metaverse technology entails significant costs, presenting a barrier for small enterprises and individuals seeking to invest in the requisite infrastructure and expertise. Furthermore, the complex nature of mastering this technology may dissuade individuals from wholeheartedly embracing the virtual economy.

2.5 SWOT Analysis

To get a holistic understanding of the virtual reality, a SWOT analysis was performed,

3 Fairtrade, Ethical Consumption, and Sustainability

3.1 Sustainability

For the Metaverse to turn into a suitable alternative to the current ways of interaction it is important to define a certain model or a structure that will help it grow. At the same time, the structure should be sustainable, for a long period. To begin the process, we must first define what we understand by sustainability in the metaverse. Sustainability refers to the ability to meet present needs without compromising the ability of future generations to meet their own 120 and 240 billion kilowatt-hours per year. Environmental impact, promote social equity and ensure long-term economic viability. Therefore, the study takes place under 3 keywords economic, environmental, and social sustainability (Transforming our world 2030).

The age demographics of the virtual world show that most (about 80%) of the participants in the virtual economy are under the age of 16 (Howarth 2023). This serves as a major catalyst for the adoption of metaverse in the upcoming years. Since this generation is more adjusted to virtual trade, it serves as an incentive for several companies to try to set up stores in the metaverse. The structure needed for economic sustainability needs continuous development in the metaverse in terms of the market providing new product opportunities and sale venues for the involved firms to participate in (Giovanni 2023). At the same time, firms must be willing to participate more in the market with new products and applications of metaverse components like the use of NFTs. The core factor pushing the development and sustainability of the virtual economy relies on the activities and trade happening inside the metaverse.

Environmental sustainability is essential to the successful development of the metaverse (Sustainable Innovation Technology-Google Sustainability n.d.). The running costs of the metaverse are extremely high. A single NFT transaction can emit about 100 $KgCO_2$ to mint which is about a 1–2 h flight, e.g. London-Frankfurt (Akten 2023) but after several amounts of research and development, new technical insights, in an event known as The Merge, the Ethereum (NFT trades) energy consumption has been reduced by ~ 99.95% (The Merge n.d.). Similarly, crypto mining is a major part of the metaverse, which is an extremely energy-consuming process. Crypto mining takes approximately about 120 and 240 billion kilowatt-hours per year which is more than the combined usage of all the world's data centers (Energy 2024). Adding to it, the running energy costs of the data centers and transactions increase largely. The environmental damage is not only restricted to that of the energy, it expands to manufacturing environmental damage. The production of the equipment required to be a part of this ecosystem, for example, VR headsets, chips, computers, servers, etc. all have production costs at the same time causing production wastes. Meta (formally Facebook, owner of Oculus VR headsets) has taken several steps toward a more sustainable model, with an attempt to reduce the GHG emissions to a net zero by 2030 (Palmieri and Changigi 2020). On the upside with the development of the metaverse and the virtual space, it holds potential for several activities. A virtual space can be utilized for marketing, sales, holding interaction, gaming, etc. giving an environmental edge as it is less resource-intensive, and to perform these activities, the amount of waste generated is also reduced. With the reduced movement due to the metaverse, the carbon footprints are also cut for every participant in this space. For example, virtual fashion shows and digital clothing collections can significantly reduce the environmental footprint of the fashion industry. Brands can showcase their designs in virtual environments, reducing the need for physical samples, runway shows, and travel. Digital fashion items, represented as NFTs, can be bought and worn by avatars in the Metaverse, promoting sustainable consumption and reducing textile waste.

When the interaction between humans converts into that of avatars interacting with each other, it becomes crucial to introduce social norms. As humans, several attributes make our identity which include physical appearance, race, gender, emotions, culture, social status, persona, etc. When we enter a virtual space we tend to lose some of these defining factors as we are being represented by a "pseudo" version of our very own self. The positive aspect remains that of lesser discrimination, prejudice, racism, etc. With its leverage of accessibility, it can also be a great tool to reach previously inaccessible locations for purposes such as education (Giovanni 2023). Yet some of the problems remain constant compared to the current world with as cyberbullying, addiction, mental health concerns, etc. Since the metaverse is a new and unfamiliar environment for many less tech-savvy individuals, they are more vulnerable to scams due to their limited technical knowledge, including their security and privacy-related scams. With the advent of social media, we were introduced to new ways of interacting with our peers, and the metaverse similarly can foster global connectivity. Better communication helps us connect areas that used to be cut off. This opens doors to many great things, like education, science breakthroughs, and healthcare services.

Sustainability use cases in the metaverse

Regulatory	Governments and various organizations that make rules are beginning to explore how the metaverse could aid sustainability efforts. With the new ways of collecting data, it risks the consumer's privacy invasion, manipulation, and misinformation at the same time The metaverse could allow corporations to monitor and monetize users on an unprecedented scale (Rosenberg 2022). The regulatory factor is closely related to the fair-trade practices within the metaverse
Market volatility and psychology	The cryptocurrency market has seen several market fluctuations over the years and has been a victim of several pump-and-dump schemes which are often controlled in nature, outsiders might be tempted to take part in pump-and-dump schemes due to their overestimation of their market timing skills or the attraction of unusually high short-term profits
Securities	With an entirely new market, new market services, products, and assets become available. along with this comes its security issues and challenges. At the same time, the requirement of a new legal order made to this new world's requirements and market mentality is crucial

3.2 *Fairtrade*

Building trust and getting more people into the Metaverse calls for Fairtrade principles. Fairtrade focuses on making sure producers in developing areas get fair pay and decent working conditions. In the Metaverse, these ideas help ensure that virtual goods are made ethically and that creators earn a fair wage for their hard work (What is Fairtrade n.d.).

3.3 *Ethical Consumption*

Just like a coin has two sides, the Metaverse has its ups and downs. Ethical consumption means making choices that reflect our values—like buying eco-friendly or ethically made stuff. In the Metaverse, this looks like supporting creators who prioritize sustainability and fairness.

Ethical Responsibility and Sustainability

Ethical responsibility is highly intertwined with sustainability. When it comes to business or carrying economic activities inside the metaverse, businesses must consider the ethical implications of their operations when it comes to environmental impact, data privacy, and social equity. Until some time ago only a few factors could be recorded when it comes to information about a consumer. However, with the advent of the metaverse and the virtual economy, information venues have increased in number. "The metaverse will be data collection on steroids" as described in an article by Mastercard (Zohar 2015). The volume of collected data has increased significantly. The data collected within the metaverse now consist of detailed personal Insights such as intimate psychological information, emotions, and behavioural patterns which raises concern regarding unethical usage, (Zohar 2015). For references blockchain which is one of the most common methods to exchange or create virtual assets. Must be employed in ways that minimize energy and consumption and remove environmental impact like carbon emissions. Additionally, it needs to be noted that companies make sure that their virtual goods and assets are accessible to all users irrespective of their socio-economic background or geographic location. As data is the fuel for the metaverse, businesses must navigate the complex ethical challenges surrounding its collection, storage and use (Foley et al. 2019). Implementing ethical data practices is essential to protect user privacy and keep information safe. This means being open about how data is used and taking strong measures to guard it against or leaks. In today's digital world, personal and financial details are often connected to our online identities, making it crucial to secure them properly. In places like the metaverse, where people interact and share a lot of information, having strong rules and protections for data is very important. By focusing on privacy and safety, we can help build trust and ensure everyone feels secure while using technology.

The metaverse, a growing digital space where people interact and trade, faces many risks because it lacks proper oversight, unlike regular banks and financial systems that are monitored for safety. Users can fall victim to scams, ransomware attacks, and other harmful activities without any protective measures in place (Auer and Claessens 2018). To make the metaverse a safer place, it is important to create rules that follow government standards. These rules would help protect users, stop illegal actions, and improve security overall (Catalini and Gans 2020). With a solid regulatory framework, we can reduce many of the dangers that come with decentralized platforms and make the digital world a more trustworthy space for everyone.

Fraud in virtual economies	The decentralized nature of virtual economies provides opportunities for fraud, including pump-and-dump schemes and phishing attacks in virtual asset markets (Foley et al. 2019)
Capital manipulation	Speculative trading inflates asset prices, creating bubbles that eventually burst, causing losses to investors. Whale traders manipulate prices due to market liquidity control (Gandal et al. 2018)
Unregulated markets and systemic risks	The lack of regulatory bodies in virtual economies leads to systemic risks like fraud, market instability, and cybersecurity breaches. These markets lack consumer protection and regulatory frameworks (Zohar 2015)
Anonymity in transactions	Anonymity, while providing privacy, also enables illegal activities like market manipulation, tax evasion, and money laundering without fear of detection (Auer and Claessens 2018)

4 Business Resilience in the Metaverse Era

Business resilience is how well a company can adapt and recover from disruptions in its industry. In the growing world of the metaverse, being resilient is especially important as businesses face constant changes and new challenges. To thrive, companies need to keep innovating and finding new ways to operate. By exploring different business models and offering digital goods and services, they can create new income sources while staying relevant. The metaverse also allows for remote work, which gives employees more flexibility and helps businesses keep running smoothly during tough times. Investing in new technologies and promoting sustainable practices are key steps for businesses to secure their future. This includes protecting data, encouraging creative thinking, and working closely with everyone involved, like customers and partners. Effective leadership plays a big part, as leaders must set a vision and ensure that the company operates in a fair and responsible way. Overall, building resilience in the metaverse means being ready for change and creating a strong, sustainable future.

Leaders must prioritize sustainability and resilience, integrating these principles into business strategies and decision-making processes. This involves setting ambitious sustainability goals, monitoring progress, and reporting on performance. Corporate governance frameworks should include policies and practices that support data security, ethical conduct, and stakeholder engagement.

5 Economic Outlook of the Metaverse

Mankind has shown several developments in handling how its money flows. Economics is a social science examining how goods and services are produced, distributed, and consumed (Hayes 2024). Humans have always taken part in trade, with the barter system in 6000 BC to the current economy and at the same time set regulations for them. At the same time, these markets where trade occurs have had a governing body. With the advent of virtual currency and the virtual world, certain aspects of the current-day economy might see transformation. Along with all the advantages of virtual currency and economy, a major advantage is its cryptocurrency. This can be seen in live action in countries such as Nigeria. Nigeria happens to be one of the biggest users of cryptocurrency, with a constant surge in its adoption in the country (Mills 2024). A 2020 online survey by the data platform Statista found that 32% of Nigerian participants used cryptocurrencies, the highest proportion of any country in the world (BBC News 2021). In 2021, Nigeria saw a 2,467.2% increase in their number of users within a year as well as 10.34% (In 2022) of Nigerians are involved in cryptocurrency (Cryptocurrency ownership Nigeria 2022 n.d.). One of the leading factors behind this happens to be the unstable economy of the country. Over the past few years, several reforms have been made by the government that have caused severe economic stress to the citizens. As it happens in the country of Nigeria,

Moyo Sodipo co-founder of Nigeria-based cryptocurrency exchange Busha, said in a statement "People are constantly looking for opportunities to hedge against the devaluation of the naira and the persistent economic decline since COVID," (Reuters 2023). As Nigeria faces these issues concerning the value of currency, this happens to be a concerning and important discussion regarding the world economy. With the ideas of the metaverse, come the ideas of free Markets and regulations which bring along several geopolitical changes and change the dynamics of international relations.

6 Virtual Economy in the Metaverse

6.1 Virtual Goods and Services

The metaverse provides an extremely interconnected system for the virtual economy to act. With an extremely well-connected system where digital goods, services, and currencies are created, exchanged, and monetized it involves, various components, including virtual real estate, digital assets, in-game currencies, and decentralized finance (DeFi) platforms.

6.1.1 Virtual Real Estate

Virtual real estate is an aspect of the digital world involving digital plots of land or spaces within the Metaverse that users can buy, sell, and develop. Platforms like Decentraland and The Sandbox allow users to own virtual land and build customized environments, such as virtual homes, businesses, and entertainment venues (Decentraland n.d.). Inside the metaverse, Virtual real estate is often represented as NFTs, providing verifiable ownership and the ability to trade these assets on blockchain marketplaces (Nakamoto n.d.).

Virtual real estate offers unique opportunities for economic activity and investment. For example, businesses can establish virtual storefronts, host events, and create branded experiences within the Metaverse. Virtual landowners can monetize their properties by renting or selling digital spaces, creating a new revenue stream that does not rely on physical resources, for example, Walmart's virtual store for sales.

6.1.2 Digital Assets and NFTs

Since the history of the virtual world has been gaming, this aspect of it doesn't fade away. Digital assets in the Metaverse include items like avatars, outfits, accessories, virtual art pieces and collectibles.

NFTs are special tokens verified by blockchain that show ownership of these digital goodies. They confirm what's real and unique and let people trade items that hold value. NFTs are super popular in areas like art, music, gaming, and fashion. Artists can make and sell their works as NFT, musicians can release special tracks, and gamers can buy and swap in-game items. Owning and trading these digital treasures opens new economic doors and creates opportunities for creative jobs.

6.1.3 In-Game Currencies and Virtual Services

In-game currencies are tokens used within games or virtual worlds to buy things. Players can earn them or purchase them to get virtual goods and services—creating a loop of play, for example, Minecraft coins at Mincraft, Bucks in Fortnite, Robux in Roblox, and MANA in Decentraland.

The possibilities of virtual services are immense, they range from designing spaces to planning virtual real estate development and digital fashion design to virtual event planning and content creation. For instance, the possible role of a virtual architects can design custom virtual spaces, while digital artists can create personalized avatars and accessories.

6.2 Cryptocurrencies in the Virtual Economy

6.2.1 Role of Cryptocurrencies

Cryptocurrencies play a major role in the virtual economy, they serve an important aspect of it, keeping transactions safe and clear. The element of crypto, its Blockchain technology helps manage digital assets by proving ownership while letting people exchange directly with each other. Many virtual worlds have adopted cryptocurrencies as its primary source of transaction.

For example, platforms like Decentraland and Sandbox uses tokens Mana and Sandbox which make it easier for users to buy and sell many things.

6.2.2 Decentralized Finance (DeFi) and Virtual Economies

Decentralized finance, or DeFi, is a revolutionary idea that changes how we think about money. Instead of relying on traditional banks and financial institutions, DeFi allows users to manage their money directly through technology called smart contracts and decentralized platforms. This means people can lend, borrow, trade, and invest without needing a middleman. When combined with the Metaverse, which is a virtual world where people interact, DeFi offers amazing new ways to access financial services and control digital assets safely and easily (Catalini and Gans 2016). For example, users can earn interest by staking their cryptocurrencies or make extra money through yield farming. This blend of DeFi and the Metaverse opens doors for more people to join the financial system, especially those who may not have easy access to banks. It paves the way for economic empowerment, making financial tools available to everyone.

7 Conclusion

Despite the idea of the metaverse and a virtual world being in the air for a long time, the usability factor remains missing and at the same time, market sentiment for this field has been erratic. We might just be on the interface of another economic revolution happening with the help of the metaverse and the virtual economy. The main idea is to make it work and make it work for a long period, the key to it lies in its regulations and its practices, at the same time concentrating on the factors that promote sustainability i.e. continuous development. This continuous development refers to the development of its business model, products, etc. While the metaverse grows internally, there are external factors that come into play like geopolitics and world economy.

With its possible effects seen in many aspects of the world, and at the same time we realize its nascent form, it is crucial to move forward in little steps not rush to make it a profitable source for consumers to exploit. There needs to be more research in the field in the upcoming years which will help diagnose several factors of the metaverse and give a better view of the possible new world.

References

Akten M (2023) The unreasonable ecological cost of #CryptoArt. Part 1|Medium. Medium. https://memoakten.medium.com/the-unreasonable-ecological-cost-of-cryptoart-2221d3eb2053

Ashmore D (2023) 7 top metaverse coins. Forbes Advisor INDIA. https://www.forbes.com/advisor/in/investing/cryptocurrency/top-metaverse-coins/

Auer R, Claessens S (2018) Regulating cryptocurrencies: assessing market reactions. https://papers.ssrn.com/sol3/papers.cfm?abstract_id=3288097

Azuma RT, Hughes Research Laboratories (1997) A survey of augmented reality. Presence: Teleoperators Virtual Environ 6(4):355–385. https://www.cs.unc.edu/~azuma/ARpresence.pdf

BBC News (2021) Cryptocurrencies: why Nigeria is a global leader in Bitcoin trade. https://www.bbc.com/news/world-africa-56169917

Catalini C, Gans JS (2016) Some simple economics of the blockchain. SSRN Electron J. https://doi.org/10.2139/ssrn.2874598

Catalini C, Gans JS (2020) Some simple economics of the blockchain. Commun ACM 63(7):80–90. https://doi.org/10.1145/3359552)

Chandrasekar D (n.d.) How much is metaverse worth? Centific. [Online] https://www.centific.com/how-much-metaverse-worth#:~:text=McKinsey%20reports%20that%20corporations%2C%20venture,invested%20in%20all%20of%202021

Cryptocurrency ownership Nigeria 2022 (n.d.) Triple-A—Triple-A. https://www.triple-a.io/cryptocurrency-ownership-data/cryptocurrency-ownership-nigeria-2022

De Giovanni P (2023) Sustainability of the metaverse: a transition to Industry 5.0. Sustainability 15(7):6079. https://doi.org/10.3390/su15076079

Decentraland. Decentraland. [Online] https://decentraland.org/

Dionisio JDN, Burns WG III, Gilbert R (2013) 3D Virtual worlds and the metaverse. ACM Comput Surv 45(3):1–38. https://doi.org/10.1145/2480741.2480751)

Dionisio JDN, Burns WG III, Gilbert R (2013) 3D Virtual worlds and the metaverse. ACM Comput Surv 45(3):1–38. https://doi.org/10.1145/2480741.2480751

Energy J (2024) Crypto energy consumption and crypto energy explained. Just Energy. https://justenergy.com/blog/crypto-energy-consumption-crypto-energy/#:~:text=Of%20course%2C%20crypto%20is%20more,is%20a%20big%20energy%20user

Entropia Universe-Deposits (n.d.) https://account.entropiauniverse.com/account/deposits/

Exploring the fusion of cloud computing and metaverse (2024) CloudThat resources. https://www.cloudthat.com/resources/blog/exploring-the-fusion-of-cloud-computing-and-metaverse#:~:text=Data%20Storage%20and%20Management%3A&text=Cloud%20platforms%20provide%20scalable%20and,immersive%20experience%20within%20virtual%20worlds

Foley S, Karlsen JR, Putniņš TJ (2019) Sex, Drugs, and bitcoin: how much illegal activity is financed through cryptocurrencies? Rev Fin Stud 32(5):1798–1853. https://doi.org/10.1093/rfs/hhz015

Foley S, Karlsen JR, Putniņš TJ (2019) Sex, drugs, and bitcoin: how much illegal activity is financed through cryptocurrencies? Rev Financ Stud 32(5):1798–1853

Gadekallu TR, Huynh-The T, Wang W, Yenduri G, Ranaweera P, Pham Q, Benevides DCD, Liyanage M (2022) Blockchain for the metaverse: a review. arXiv (Cornell University). https://doi.org/10.48550/arxiv.2203.09738

Gandal N, Hamrick J, Moore T, Oberman T (2018) Price manipulation in the bitcoin ecosystem. J Monet Econ 95:86–96. https://doi.org/10.1016/j.jmoneco.2017.12.004

Gatteschi V, Lamberti F, Demartini C, Pranteda C, Santamaría V (2018) Blockchain and smart contracts for insurance: Is the technology mature enough? Future Internet 10(2):20. https://doi.org/10.3390/fi10020020)

Geyer M (2023) BMW group starts global rollout of NVIDIA Omniverse|NVIDIA Blog. NVIDIA Blog. https://blogs.nvidia.com/blog/bmw-group-nvidia-omniverse/

Hayes A (2024) Economics defined with types, indicators, and systems. Investopedia. https://www.investopedia.com/terms/e/economics.asp

How does the metaverse work? (2023) Default. https://connect.comptia.org/blog/how-does-the-metaverse-work#:~:text=The%20primary%20goal%20of%20the,%2C%20socialization%2C%20learning%20and%20shopping

Howarth J (2023) 75+ Metaverse Statistics (New 2024 data). Exploding Topics. https://explodingtopics.com/blog/metaverse-stats

Hryziuk C (2024) Exploring the Future: 20 Metaverse Stats for 2024. Embryo. https://embryo.com/blog/metaverse-stats-2024/#:~:text=General%20Stats,million%20active%20players%20every%20day.

Insights, Fortune Business. Virtual Production Market Size, Share & Industry Analysis, By Component (Technology/System, and Services), By Type (Pre-production, Production, and Post-production), By Application (TV series, Commercial Ads, Movies, E-sports, and Others), and Regional Fo. Fortune Business insights. [Online] 2024. https://www.fortunebusinessinsights.com/virtual-production-market-107105

Interactive S, Interactive S (2023) The history of virtual reality (VR). SPREE Interactive. https://jointhespree.com/the-history-of-virtual-reality/#:~:text=The%20first%20proper%20virtual%20reality,in%20front%20of%20the%20user

Martins L (2023) Mark Zuckerberg can't quit the metaverse. The Verge. https://www.theverge.com/23894188/mark-zuckerberg-metaverse-connect-2023-ai-vr

Metaverse market size 2022–2032|Statista (2024) Statista. https://www.statista.com/statistics/1295784/metaverse-market-size/#:~:text=In%202023%2C%20it%20was%20estimated,and%20digital%20worlds%20come%20together

Mills L (2024) Crypto in Nigeria—surge in adoption and regulatory shifts. Crypto Council for Innovation. https://cryptoforinnovation.org/crypto-in-nigeria-surge-in-adoption-and-regulatory-shifts/

Nakamoto S (n.d.) Bitcoin: a peer-to-peer electronic cash system. https://bitcoin.org/bitcoin.pdf

News-Medical (2021) Applications of virtual reality in medicine. https://www.news-medical.net/health/Applications-of-Virtual-Reality-in-Medicine.aspx

Ordano E, Meilich A, Jardi Y, Araoz M, Decentraland (n.d.) Decentraland. In: Decentraland, pp 4–14. https://decentraland.org/whitepaper.pdf

Palmieri E, Changigi (2020) The next decade: how Facebook is stepping up the fight against climate change. Engineering at Meta. https://engineering.fb.com/2020/09/14/data-center-engineering/net-zero-carbon/

Reuters (2023) Nigeria crypto usage growing further, report says. Reuters. [Online] 19 September 2023. https://www.reuters.com/technology/nigeria-crypto-usage-growing-further-report-says-2023-09-19/

Rogers S (2024) The use of virtual reality in the medical field. ProMed: your source for healthcare education. https://blog.promedcert.com/the-use-of-virtual-reality-in-the-medical-field/

Rosenberg L (2022) Regulation of the metaverse: a roadmap. Int J Ind Organ. https://doi.org/10.1145/3546607.3546611

Sustainable Innovation and Technology—Google Sustainability (n.d.) Sustainability. https://sustainability.google/

The Merge|Ethereum.org (n.d.) ethereum.org. https://ethereum.org/en/roadmap/merge/#what-is-the-merge

Transforming our world: the 2030 Agenda for Sustainable Development|Department of Economic and Social Affairs (n.d.) https://sdgs.un.org/2030agenda

Wang Q, Li R, Wang Q, Chen S (2021) Non-Fungible Token (NFT): overview, evaluation, opportunities and challenges. arXiv (Cornell University). https://doi.org/10.48550/arxiv.2105.07447

Weidner JB (2024) Why google glass failed. Investopedia. https://www.investopedia.com/articles/investing/052115/how-why-google-glass-failed.asp#:~:text=Google%20Glass%20was%20launched%20in,the%20market%20again%20in%202023

What is Fairtrade? (n.d.) Fairtrade International. https://www.fairtrade.net/about/what-is-fairtrade

Zohar A (2015) Bitcoin: under the hood. Commun ACM 58(9):104–113

Zohar A (2015) Bitcoin. Commun ACM 58(9):104–113. https://doi.org/10.1145/2701411

Sustainability of the Metaverse for Indigenous Tribes: Opportunities, Challenges, and Pathways Forward

Neetu Mishra, Astha Joshi, Anil Vashisht, Vijay Prakash Gupta, Neelesh Kumar Mishra, and Parul Asati

Abstract This research delves into the relationship between indigenous communities and the metaverse, specifically looking at how their social, economic, and cultural activities might be preserved in virtual spaces. The goal is to find ways to honour Indigenous peoples' traditions and sovereignty while also integrating them into the metaverse in a way that benefits everyone. We will also look for ways to promote and preserve indigenous history. Helping indigenous communities make the most of the metaverse's possibilities and avoid its pitfalls is the overarching goal of this research. To guarantee that virtual environments positively aid in the preservation and development of indigenous communities and cultures, it promotes inclusive and sustainable practices. This chapter provides a comprehensive review of digital technologies available to indigenous communities, including blockchain for digital sovereignty, augmented and virtual reality (AR/VR) for cultural preservation, and artificial intelligence (AI) for economic development. Tribe development in the metaverse can become an example of sustainability by paying attention to these details allowing for technological progress while also ensuring social responsibility, economic viability, and environmental protection. As augmented and virtual reality (AR/VR) and other digital technologies collide with physically persistent virtual space, a new collective virtual shared place called the metaverse is born. Many new lines of inquiry can be opened up by investigating how this new digital environment can aid in the long-term prosperity of Indigenous people.

Keywords Metaverse · Sustainable development · Digital technologies · Culture · Economic development

---

N. Mishra (✉) · A. Joshi · A. Vashisht · P. Asati
Amity Business School, Amity University, Gwalior, Madhya Pradesh, India
e-mail: onemishraneetu@gmail.com

V. P. Gupta
Institute of Business Management, G.L.A. University, Mathura, India

N. K. Mishra
XLRI—Xavier School of Management, Jamshedpur, Jharkhand, India

W. Leal Filho et al. (eds.), *Metaverse and Sustainability: Business Resilience Towards Sustainable Development Goals*, World Sustainability Series,
https://doi.org/10.1007/978-3-031-89545-6_12

1 Introduction

1.1 Indigenous Community

Indigenous which means Original or ancestral) of people having historical, cultural, and spiritual connections with the land where they live (icki Grieves 2008). That includes being recognized for what they truly are, Indigenous people who have always had a role in and added value to their natural environment. Indigenous are similarly forced into the margins of society from more dominant or colonizing cultures (Cervantes 2011).

Indigenous people number more than 4.76 million, or some 6.2% of the world's population, and are based in approximately ninety countries (Rowan et al. 2021). There are over 5000 individual ethnic groups in the world; all of whom have their own unique history and customs. India's Indigenous minority officially known as "Scheduled Tribes" account for roughly 104 million or 8.6% of the population in all with 705 ethnic groups (Parakkal 2023; Prasad n.d.) recognized within this demography, many more can be called as having indigenous communities but have yet to be acknowledged. Now, this just adds one more level of complexity to an already complicated demographic. The northeastern states and the central tribal belt are where most of them live in India.

1.2 Sustainability

Sustainability has got to be one of the most overused buzzwords in recent years. It seems there is nothing that cannot be described as sustainable: all can either lead to it or follow a hyphen after. Sustainable cities, economies, resource management business, and livelihoods sustainable development. Sustainability has turned, par excellence, what is called a 'boundary term' (Mishra et al. 2025a) i.e., where science meets politics and politics meets science. The global political economy, and the 'boundary work' around sustainability collective efforts to construct epistemic communities (Mishra et al. 2025b) that share a common appreciation of why environmental needs must be twinned with economic development concerns has now also become an international project. A case can be made that much of what networks and alliances have brought into being, the institutions, organizations; projects, etc. put in place to resolve/finance/address sustainability has happened so not because unsustainability. It is at this intractable intersection of science and politics that boundary work unfolds, where frequently polysemic, politically consequential words are vital to processes of policy formation and implementation.

In today's modern era, technology has surrounded everyone and at the same time technology has given a new form to the economy (Asati et al. 2024), we know this new form by the name Metaverse. Sustainability, in particular sustainable development, is perhaps a concept that resulted from broader interdisciplinary discussions. These

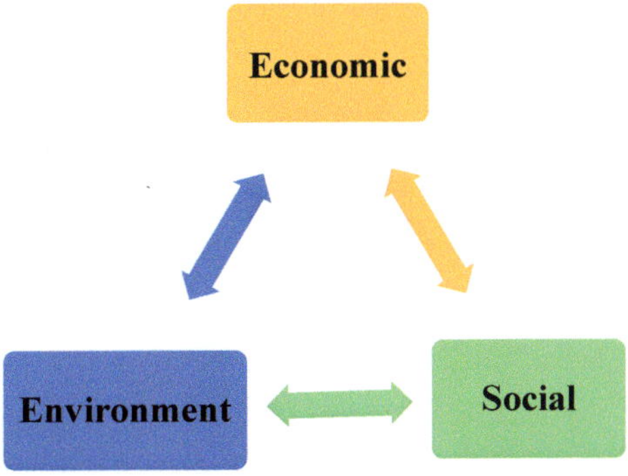

Fig. 1 Dimensions of sustainability

issues have also been widely debated at the scholarly and policy level globally since the 1980s, particularly leading up to before 1992 United Nations Conference on Environment and Development (UNCED) in Rio de Janeiro Brazil.

"***Development that is sustainable ensures that current requirements are satisfied without jeopardizing future generations' capacity to do the same***" (Brundtland 1985).

Sustainability refers to the capability of processes to be continued indefinitely without exhausting resources for future generations. Most often, sustainable development can be defined as "development that meets the needs of the present without compromising the ability of future generations to meet their own needs.

The three pillars are often seen to be economic, environmental, and social sustainability (Purvis et al. 2019; Boström 2012). Social sustainability looks at the well-being of people and healthy communities, while environmental sustainability takes care of nature and conserving ecosystems as well as natural resources. The three pillars are interrelated, and to achieve sustainability it is required that all of the three measure Dimensions(Pillars) focus on (Purvis et al. 2019). However, the figure above explains how sustainability is associated with three dimensions (Fig. 1).

1.3 Metaverse

The metaverse combines physical reality with the digital world by using virtual technologies, including Virtual Reality (VR) Augmented Reality (AR) Mixed Reality (MR) accessible via computers (Efe 2022; Mystakidis 2022; Lee et al. 2021). The concept is to form a sandbox or stable region where avatars link up across different experiences without losing the self in data transition.

The concept of the metaverse has its roots in science fiction and actual technological developments, dating back to the 1960s (Ball 2022; Knox 2022). The concept of a metaverse was first used in 1992 with the novel Snow Crash by author Neal Stephenson to describe an interconnected virtual world inhabited by avatars (Zakarneh et al. 2024). Over the years, especially with smartphones and the internet

Fig. 2 Types of metaverse

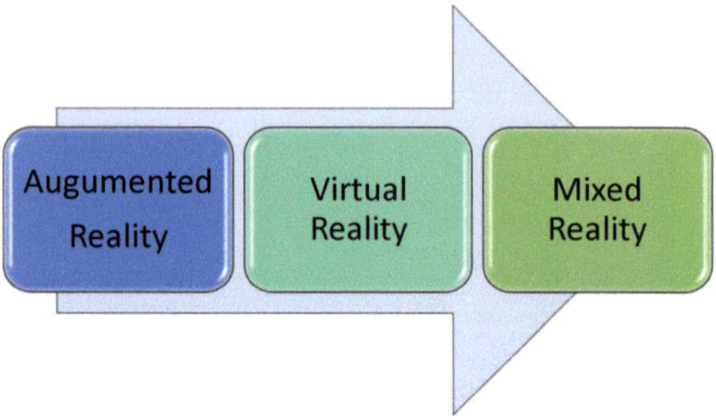

bettering the lives of target groups (read all above 25-year-old) we have seen a growing sense of metaverse as to how it makes possible what other platforms could only promise or boast. We are beginning to see this in the real world, as industries find practical applications of the metaverse. The e-commerce industry is already seeing rising sales of virtual goods interacting with physical products, and experts expect this to increase. Furthermore, they also use it for training purposes and digital twins the process of creating a virtual model to mimic real-world objects while generating data. Meta verse are of three types which are as follows (Fig. 2).

Augmented Reality: Augmented Reality is a technology in which users can enhance their real-world experiences by overlaying computer-generated images on reality (Aggarwal and Singhal 2019; Mishra et al. 2024b; Tiwari et al. 2016). This approach allows for a more interesting recreation of the cultural heritage places and objects by introducing an interactive layer between the user and the environment. **Example**: The Old Borneo Traditions Museum's augmented reality tour is one of several recent initiatives that showcase Indigenous traditions and arts, including rattan weaving and the Pua Kumbu textile heritage. By utilizing augmented reality, individuals may engage with these cultural elements more dynamically, deepening their comprehension and admiration of Indigenous artistry.

Virtual Reality: Virtual reality (VR) is a technology where users can have this experience of immersive worlds in real-time, that remain realistic through creating an artificial environment (Mishra et al. 2024c). Virtual reality is defined as an immersive, interactive experience of a 3D computer-generated world in which the user has some degree of control over how they interact (Gandhi and Patel 2018; Halarnkar et al. 2012). A common example that gets lumped into specialized hardware are headsets and gloves generally used to interact with environments for lifelike sensory. **Example**: Arapaho Elicitation Northern Arapaho VR, the prioritizing of Indigenous endangered languages It makes the learning experience better, and keeps dying languages alive by allowing those whose native language is endangered to share their knowledge in digital world replicas of significant cultural sites. For Indigenous language revitalization, virtual reality has been hailed as a powerful tool. For instance, there have been studies that show ways in which VR can serve the purpose of being used to learn and practise languages a vital asset when it comes to maintaining one's cultural identity.

Mixed Reality: Together, the act of integrating digital and physical worlds became mixed reality—coexisting with presence in multiple realities at once (Benyon 2012). This includes aspects of both virtual reality (placing users in an entirely digital environment) and mixed-reality systems where objects appear to inhabit the real world when viewed through a VR headset. That's because mixed reality (MR) enables us to combine the presence and interaction of digital content with actual objects from a physical world achieving more interactive, immersive user experiences. **Examples**: Mixed reality technology goes a long way to support the sharing of Indigenous knowledge and legacy, as it unveils hidden histories and alternate narratives. This technology also offers a unique platform that suggests the natural heritage of Indigenous people in science disciplines, amplifies their presence within educational environments, and caters to being active while preserving traditions.

1.4 Catalogue of Digital Technologies

Indigenous communities' ability to use the Metaverse in the future depends on their ability to include a range of digital technologies that can help with cultural preservation while also achieving environmental objectives. By enabling environmentally responsible activities and providing immersive experiences, these technologies also empower Indigenous people, giving them back ownership of their traditional narratives.

1.5 Digital Technologies for Culture Preservation

In the Metaverse, digital technology is helping Indigenous communities keep their traditions alive. For example, efforts like creating Indigenous Cultural Embassies online give Indigenous communities a voice and a place to showcase their history and culture to an international audience. Also, by creating art and cultural materials based on Non-Fungible Tokens (NFTs,) we can help keep traditions alive while making money through online sales.

The development of AR and VR is a major step forward in digital technology; it is changing many different businesses and providing people with new kinds of experiences. Despite their primary differences, both augmented reality and virtual reality serve to unite the digital and physical realms. Providing a thorough review of the effects of AR and VR technologies across many industries, this paper details their salient characteristics, uses, benefits, and drawbacks.

1.6 Metaverse in Dimensions of Sustainability

In terms of sustainability, which includes ecological, monetary, and social considerations, the Metaverse offers a tangled web of possibilities and threats. Although it could encourage eco-friendly habits, issues with power usage and necessary hardware must also be resolved. The design and implementation of the Metaverse have a significant impact on the sustainability outcomes it produces as it develops (De Giovanni 2023; Vlăduțescu and Stănescu 2023).

1.7 Cultural Heritage and Its Relevance to Modern Society

Cultural heritage broadly refers to the tangible and intangible movable elements that have been transmitted from one generation to another, which are conserved by the current or future actors with a view of sharing them for such purposes. Material legacy includes buildings, monuments, and landscapes but also literature, art, and artifacts. It includes intangible cultural assets like folklore, traditions, languages, and knowledge. Every single one of these pieces of our cultural inheritance is intricately intertwined to keep the wisdom, traditions, and practices alive well into the future. This model explains how the 3 Dimensions are connected with the metaverse and associated with emerging society.

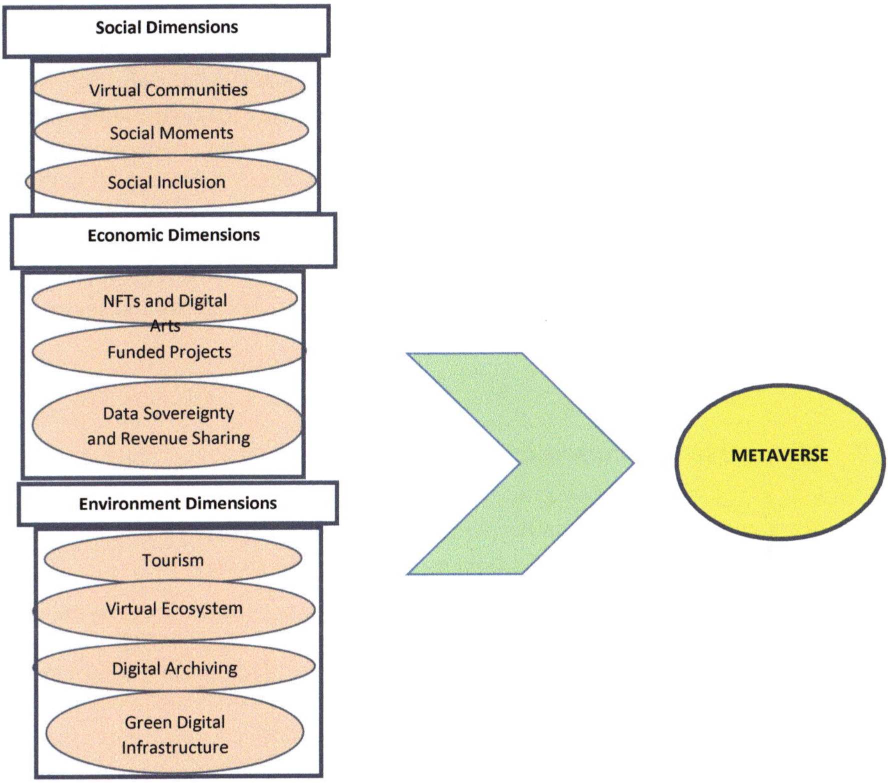

1.8 *Metaverse in Social Dimensions (Allam et al. 2022)*

By facilitating community connections, raising cultural awareness, and offering venues for social engagement and interaction, the metaverse may enhance Indigenous culture and history in a socially significant manner.

Virtual Communities: The metaverse could be host to accepted virtual cultural centers created by indigenous groups as knowledge-sharing and participating sites. These centers become an avenue for people to connect and foster relations wherever they are located. The metaverse enables global Indigenous networking by allowing Indigenous people to communicate from afar. They are able to share stories, gain knowledge, and contribute by working together to cultivate their culture.

Social Moments: The metaverse offers an alternative for Indigenous groups to advocate against climate change, land grabbing, and cultural erosion amidst other social and environmental issues. Online campaigns, educational events, and protests through demonstrations around the world can be employed to activate support for Indigenous interests. Community leaders and activists in Indigenous groups can use the metaverse to ensure that their cultural heritage endures, both within the virtual space itself as well as outside this new dimension. And the best way to do it is by approaching as many minds as possible for their cultural and social rights.

Social Inclusion: With the right design, a metaverse can create virtual footprints that let disabled or remote people in on cultural happenings they might never touch back in meatspace. Which ensures that everyone is involved in Aboriginal cultural events, promoting social inclusion. The metaverse makes for a compelling experience for relieved Indigenous youth to learn about and engage with their culture in an age when you could get caught up on wearing a business suit on your screen. It provides a platform for them to discuss their origins with everyone from around the world hence making them feel proud and justified.

1.9 Metaverse in Economic Dimensions

There are a number of opportunities to monetize Indigenous peoples' culture and legacy in the Metaverse. Some important ways it can help are as follows:

Digital Art and NFTs: NFTs or Non-Fungible Tokens, are digital pieces by Indigenous creators that can be minted and sold (Kale et al. 2022; Wang et al. 2021a, b). This exemplifies an additional revenue stream of tokens that still represent traditional art, stories, or symbols but allow the original owners and creators to retain ownership and control of cultural production (Mishra et al. 2024c). Cultural Authenticity: Indigenous artists maintain cultural expressions safe from exploitation and counterfeiting by ensuring their NFTs are authentic.

Funded Projects: The Metaverse can serve as a platform for crowdfunding projects that aim to promote or preserve Indigenous history through online exhibits, digital archives, or other cultural endeavors (Malerba 2023; Rubino 2023; Xiong 2021).

Data Sovereignty and Revenue Sharing: Data Sovereignty in the Metaverse allows our Indigenous communities to be masters of our own data (Paganopoulos 2022). This will ensure that their cultural data, whether shared or utilized, is monetized and kept within close quarters. To make sure communities are economically recognized for their cultural contributions, revenue-sharing plans could be provided related to just how Indigenous social resources support the Metaverse (Rothe et al. 2024; Wheelock 2022).

Metaverse in Environmental Dimension: The Metaverse can play a significant role in promoting Indigenous culture and heritage in an environmentally sustainable way (Mystakidis 2022). Here are some key environmental benefits and contributions:

Tourism: In place of conventional tourism, in the Metaverse tourists will be able to digitally visit Indigenous villages and landscapes, as well as cultural places (Allam et al. 2022). It conserves fragile natural habitats from the destructive effects of mass tourism and also cuts back on carbon emissions associated with travel. If you encourage people to visit virtually, the contribution and impacts will be much less than if they all fly into the location while still providing many educational as well as cultural benefits in tourism (Dayoub et al. 2024; Kouroupi and Metaxas 2023; Su et al. 2023).

Digital Archiving: The Metaverse could hold traditional wisdom, languages, and artifacts from indigenous cultures in digital storage. At the same time, this method would store and release our living history for generations to come while reducing the strain on other vital resources. Cultural rituals and the sort can be held in the metaverse, which would lower if not remove the need for large gatherings that might strain local ecosystems (eg. food sources, transport systems waste management) (Naruta-Moya 2015; Senier 2014).

Virtual Ecosystem: Interactive virtual landscapes can exhibit indigenous knowledge about local ecosystems and biodiversity, helping people understand the necessity of sustainable living and ecological balance (Wang et al. 2021a, b; Adade Williams et al. 2020).

Green Digital Infrastructure: This is in perfect harmony with the Indigenous cardinal virtue of stewardess for our Earth, and will not only support economic development through sustainable digital infrastructures that are energy efficient and environmentally friendly within enterprise operating environments but also well beyond (Mishra et al. 2023). Hence, Indigenous peoples can participate in the digital economy by selling NFTs virtual goods in Metaverses. By doing this, they are able to generate income without relying on resource-intensive businesses and thus reduce their environmental footprint.

1.10 Relation Between Metaverse in Indigenous Culture and Heritage (Hutson 2024)

In the metaverse, we have a once in an epoch chance to uplift and immortalize Indigenous societal stories. Tribes around the world can share their stories as tribes have indigenous values and interact in virtual experiences. Making sure not just cultural sensitivities but also authenticity which is a bigger risk of course the benefits are humungous especially to be able to represent culture in such a powerful way.

S. No.	Opportunities for cultural preservation	Description
01	Preserving and enhancing (Ngulube 2002)	One way that Indigenous Communities might directly engage with this new anthropological imagination is by building online galleries where artifacts, stories, and traditional practices can be experienced. Biskaabiiyaang is part of a new wave of virtual reality projects which are preserving Anishinaabe storytelling traditions for future generations. The Metaverse has an Indigenous Cultural Embassy that aims to "acknowledge and augment Aboriginal cultural, spiritual and political processes
02	Educational initiatives (McKinley 2005)	Metaverse educational apps allow people, especially adolescents to engage deeply with cultural heritage. One of those initiatives is Indigital, which teaches Indigenous languages and history to young Australians through virtual reality. And that even appealed to his heritage, so it enhances learning and stands as a connecting bridge between youth identity
03	Collaborating across cultures (Prest 2020)	The metaverse provides a meeting point for different Indigenous communities which also enables members to meet online and exchange relevant information while working together on joint projects. For example, programs such as Global Collaboration Village allow visitors to walk through a life-like Amazon rainforest environment where Indigenous storytelling informs an understanding of the continent across cultures while raising consciousness about environmental threats
04	Difficulties with being authentic	The same is true even if there are opportunities to make sure that the kind of Indigenous cultural representations we will inevitably see in the metaverse will be authentic, respectful, and crafted with serious reference to existing traditions. However, because issues of cultural appropriation and representation can be extremely complex it remains crucial to debate with the locals whenever necessary. Developers of virtual experience need to intentionally find rural Indigenous communities where the primary language is English or French, rather than an Indigenous one

(continued)

(continued)

S. No.	Opportunities for cultural preservation	Description
05	Technological accessibility (Brown 2007)	The basic and simple difficulty with creating metaverse applications is a struggle many Indigenous groups face when looking at technology. Inequalities in internet and digital tool access could make it difficult for many Indigenous groups to join metaverse efforts. For Indigenous peoples to equitably participate in the benefits of a metaverse, these barriers need to be dismantled

1.11 Exploring the Application of Metaverse

To know about the Metaverse, the author focuses only on the article that is connected with the metaverse, and in Scopus the author found 101 such papers in which the author found out that different authors have used different Metaphors in different ways and in different forms. After that, using the VoS Viewer application and doing the bibliometric analysis (Yu et al. 2020) the author created a cluster by combining the keywords of all the authors and extracting what metaphors could be said in different areas, after seeing the cluster the author which makes us realize that Metaphor is not a new word, Metaphor is an already old medicinal word but it was not yet associated with new technology (Mishra et al. 2024b) (Fig. 3).

1.12 Opportunities for Indigenous Communities to Use Technology

There are tremendous opportunities available to Indigenous communities through the adoption of technology in areas such as healthcare, education, governance, and cultural preservation. Using digital tools, indigenous communities can become stronger or more self-reliant and gain access to services that are important to them. Daisy Chains and The Human Frontier. These technological dynamics together have allowed people to be more expressive with themselves, make economies richer in quality jobs, and lead well-governed governments being replaced by those that many might perceive as offering citizens an increase of greater freedom. Following are the key areas where the opportunities have been seen there and more Indigenous communities were focusing more on opportunities and these are as follows:

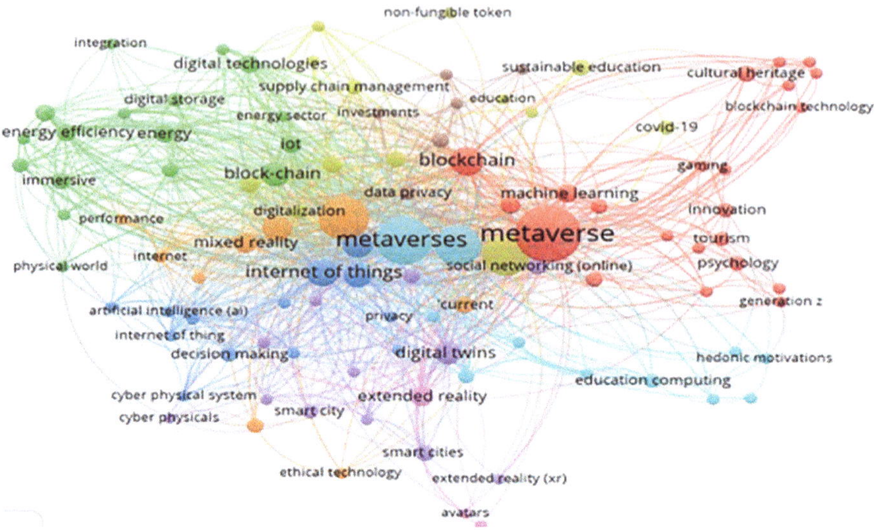

Fig. 3 Cluster keywords

Cultural Preservation: Utilizing augmented and virtual reality (AR/VR) technology makes it even more accessible for the rich cultural traditions of Indigenous communities powering this bevy to reach a wider audience through ever-existing storytelling. Cultural Identity & Continuity digitization fosters a strong sense of cultural identity and continuity for Indigenous peoples as its immersive, experiential means allow users to engage with Indigenous heritage in innovative ways.

Education Empowerment: Indigenous adolescents must-have digital tools available if they want top opportunities in education. Courses such as computer science, web design, and digital literacy can give people the necessary skills to communicate using media. With the expansion of technology in classrooms, academic success for Aboriginal students has as a direct result enabled more bonded with their languages and cultures.

Healthcare Access: In rural Indigenous communities, access to healthcare can be greatly enhanced through the use of technology. Thanks to telehealth technologies, patients can consult with doctors remotely, saving them the trouble and expense of long-distance travel. Reducing travel-related stress and improving the quality of treatment offered to Indigenous peoples, is especially good for managing maternal health and chronic diseases.

Economic Opportunities: E-commerce and other forms of Internet business are one way in which indigenous peoples can leverage their ability to build a future for themselves economically during the digital era. We can open the entryway of advanced aptitudes to one side vocations all things considered from neighborhood specialties and innovation, with more individuals becoming economically autonomous.

Governance and Political Engagement: Digital platforms can increase participation in local governance and accountability. Additionally, they allow community members to voice their concerns and needs directly to governing bodies, which strengthens self-determination and civic engagement. Technology facilitates better governance by improving communication and transparency within Indigenous communities.

Bridging the Digital Device: Improving Indigenous communities' digital infrastructure can help close the digital gap and provide equal access to data and tools. Communities can benefit from increased access to the Internet in many ways, including better educational and economic prospects (Rao 2005).

1.13 Challenges for Indigenous Communities in Using Technology

Indigenous folks have a magnitude of very unique barriers that they must combat in order to use, adopt, and adapt technology. That makes it more difficult for them to participate in the digital economy and access essential services. These problems are worsened by a lack of infrastructure, economic disparity, cultural divides, and overall digital illiteracy (Helsper 2021). Solving this problem is crucial for the empowerment of Indigenous people, making their voices heard in technology advancements.

Limited Infrastructure: The accessibility of technology has been one challenge for Indigenous communities, but the biggest hurdle is whether they have proper infrastructure to support that use. This is because Indigenous communities live in remote or isolated regions which makes it difficult for them to connect with the larger digital ecosystem and access critical elements of an internet connection, such as broadband connectivity (Gondia et al. 2022). While they may be spared from the challenges connected with internet access, this deficit inhibits their socio-economic growth.

Economic Barriers: Economic issues challenge other Indigenous communities' access to digital tools and tech as well. This is even more applicable to economically backward areas, where the cost of devices, internet, and technical know-how could be astronomically high (Baines et al. 2020). This economic barrier restricts their ability to leverage technology for personal as well as professional benefit.

Digital Literacy: The lack of digital literacy in Indigenous populations poses a myriad of problems when it comes to accessing technology. How public participation and process look different due to things like historical oppression, and access over the years with formal education systems creates a problem where communities may have difficulty receiving online resources and information (Dobson and Willinsky 2009).

Cultural Awareness and its Value: Despite the solutions being available, there are gaps between technological solutions and having them back to Indigenous communities as they will not be applicable due to many natures of these appreciations as its availabilities may not consider the cultural values where those people exist. Rather than imposing outside frameworks that may not resonate with Indigenous worldviews, the design of digital tools should be responsive to and cognizant of cultural protocols in order to privilege indigenous ways of knowledge and action (Dutta 2019; Christen 2012).

Marginalization from Technological Advancements: When it comes to talking about the creation and distribution of technology, indigenous communities are usually left out in the dust. By and large, technology has been developed for the most part by indigenous peoples are certainly not representative of this group of people as a whole (Gundemeda 2007). This gap is facilitated by choices made without their input in the matter of constructing rights and culture, leaving them behind.

Risk of Culture Erosion: The fast-moving Technology makes indigenous communities fear that their culture is been washed away. One concern is that, as digital technologies continue to infiltrate their communities, these new mediums will further diminish or destroy traditional practices and languages. As far as interactions with popular digital content go, it is even more critical across younger generations who may not have their cultural legacy top of mind for them (Johnston 2014).

Historical and Institutional Barriers: Historical injustices and contemporary institutional inequities hinder a fair relationship between Indigenous peoples and technology (Grant and Söderbergh 2020). This difficulty is compounded by the fact that many of these communities are at best ignored and often punished, being ruled by systems that deny their sovereignty over information. To create a more inclusive technical ecosystem, there needed to be adjustments to the existing historical and institutional limitations.

2 Discussion

Pathways for Indigenous Communities: Historical injustices and contemporary institutional inequities hinder a fair relationship between Indigenous peoples and technology. This difficulty is compounded by the fact that many of these communities are at best ignored and often punished, being ruled by systems that deny their sovereignty over information (Batty et al. 2016; Popp and Woratschek 2016). To create a more inclusive technical ecosystem, there needed to be adjustments to the existing historical and institutional limitations. Following are the pathways that are really applied by the Government and NGOs then definitely the Indigenous communities are able to stay with the non-indigenous community people.

Education and Digital Literacy program: The most important routes ahead for Aboriginal people are education and training in digital literacy. One such initiative, the Empowering Indigenous Peoples via Digital Skills Project aims to equip people with basic digital skills needed for them to be able to participate in the digital economy (Audretsch et al. 2024). Culturally appropriate training will be more likely to reinforce traditional values and traditions promoting community buy-in and this will help the Indigenous community show their skills all over the world.

Technical Infrastructure: In order to make digital tools more accessible, it is crucial to improve the technical infrastructure. The Indigenous Connectivity and Technology Division and the National Tribal Library Broadband Council are two groups that can work together to build strong broadband networks in places that are currently under-connected (Shankar et al. 2024). Communities are able to make better use of digital resources and services when investments are made in sustainable technical infrastructure (Mishra et al. 2024a).

Technology Solutions That Are Culturally Appropriate: It is crucial to provide technological solutions while honoring and incorporating Indigenous knowledge systems and cultures. Culturally sensitive technology that speaks to community needs is fostered through initiatives like the Techno Indigenous project, which co-designs technology with Indigenous involvement (Mishra et al. 2024a, b, c, d). This method makes sure that technology can only help strengthen cultural identities, not weaken them.

Collaborating with Tech Companies: The use of new technological solutions and the allocation of resources needed can be hastened by entering partnerships with tech companies (Lei 2000). Indigenous Programming has partnered up with Kyndryl to inspire more Indigenous people into STEM careers. These partnerships can provide technical assistance and training, leading to economic empowerment (Zharchenko 2024).

Government Support and Funding: Policymakers and funders must unite to advance Indigenous peoples' access to technology. Separately, to enhance broadband access among tribes, the Biden-Harris administration created initiatives such as the Tribal Broadband Connectivity Program (Khare et al. 2022). However, communities themselves can create the revenue needed to invest in tech-focused projects if they search for it and advocate for leverage.

Community Engagement and Leadership Development: One of the best ways by which technology adoption can be driven is through advancing community engagement utilizing local leadership. Empowering community leaders to lead digital initiatives helps meet the local demand and increase participation on-ground (Adomako and Nguyen 2023). The mechanisms and norms associated with technology adoption depend on the belonging behaviors it is designed to reinforce or dissuade communities.

Utilization of Traditional Culture: These solutions can be innovative if traditional knowledge accompanies current technologies to meet the needs of a specific community. For instance, the practical knowledge of animal management possessed by many Indigenous Peoples can be supported by technical tools like GIS mapping and drones to strengthen conservation efforts (Moretto et al. 2019). Only then collaboration towards traditionalism and sustainability could be a long-term perspective.

2.1 Metaverse and Indigenous Communities

This is also a crucial new area of telling and retelling that is being done by indigenous tribes in the metaverse. Through innovative tech, these projects provide ground-breaking audio and visual experiences that bring Indigenous stories to more people. This provides its platforms and programs to further culture in new environments, blending traditional storytelling with evolving virtual spaces (Agnihotri et al. 2023). Following are the examples and projects that are now a day continuously working in this community to set the culture and heritage among the world.

Global Collaboration Village: In its work on Global Collaboration Village, the World Economic Forum has partnered with Accenture and Microsoft to bring those tales of Indigenous peoples, particularly from the Amazon. Discover the Amazon from a new perspective with this metaverse project, featuring stories of Indigenous elders sharing their culture and history as well as the obstacles they face today. As well as empowering Indigenous people further in the digital sphere, it helps highlight the cultural and environmental value of Amazonia and perhaps will make more people care about protecting this ecosystem.

Indigenous Protocol and Artificial Intelligence Working Group: The IPAI (Indigenous Protocol and Artificial Intelligence) Working Group focuses on Indigenous protocol studies in the context of AI, VR, and metaverse. The intention behind it is to ensure that Indigenous ethics and cultural narratives are accurately depicted in virtual/AI technology development by coalescing with age-old traditional experiences of the Native people (Abou-Foul et al. 2023). These virtual worlds are most often used to promote Indigenous worldviews, Indigenous practices, and/or tales through the functioning of immersive technologies.

XR Tribal Arts by Xarpie Labs: Indian XR (Extended Reality) firm which has started working with indigenous communities in India to safeguard and digitize Indian traditional arts. Unified with AR, VR artists are capable of developing experiences that can be used to become one such as experiencing tribal art within a digital realm. Warli art is one of the great artistic traditions in Maharashtra, and now they are working on a virtual reality experience around it (Dhanshyam and Srivastava 2023). Visitors to these virtual realms can discover computerized re-enactments of Warli craftsmanship as it additionally gives an understanding of the narrative and history behind each plan utilized in their images. This metaverse can expand by infusing the efforts into any type of tribal art and culture, making it possible to interactively explore that on various levels.

Virtual Indigenous Storytelling Platforms by Start-Ups: Mythoverse Studios is another Indian company working on immersive storytelling systems, with a focus on tribal tales and Tagbin is also involved in using XR to recreate historical events from the perspective of India (Indian culture)(Sjödin et al. 2023). Now there is a project that weaves through the oral tales that brings in digital innovation to tell these old stories for an audience who would never have come across them otherwise, Mythoverse and it too projects our myths into the metaverse the ancient lore of this land gets new life. Through this interplay with these platforms we can take India's rich oral tribal heritage into the digital world; and make people come across the stories, settings, and characters that bear witness to our indigeneity.

Adivasi and Tribal Museum in the Metaverse: These days more and more people have been thinking in the direction of digitization of Adivasi, Tribal Museums as a part of the Digital India Plan. In India, cultural institutions are seeking ways to transport traditional tribal artifacts and festivities into a wholly digital experience take the example of Tripura's Tribal Research and Cultural Institute (Bhatti et al. 2023). In essence, the aim is to enable home-grown Gond groups to produce their culture and share it with society around the globe.

3 Conclusion

The metaverse can be a tool for Indigenous communities to become economically empowered environmental stewards, while not only preserving but also sharing their deep cultural history. Immersive technologies like virtual reality (VR) and augmented reality (AR) can help indigenous communities preserve their language, culture, and traditional knowledge (Tjønndal 2020). For the environment and sacred sites, this is an alternative to mass physical tourism as Indigenous communities will have chances to tell their cultural stories in high fidelity within a metaverse. Where Indigenous artists and creators have the option to hock their wears in NFTs or virtual marketplaces on this new digital platform that allows for sustainable revenue. Yet the integration of these technologies must be sustainable and address issues such as cultural appropriation stealing Indigenous knowledge and closing the digital divide across all spectrums. Indigenous groups must also have full control over their digital stories and belongings to uphold cultural identity and data sovereignty. By combining tradition, technology, and sustainable living in the classical worlds with a complete virtual community. Allowing the land to pass to future generations, promotes resilience and cultural pride in both terms of environmental protection. Keeping all this in mind, the Government and some such Organizations along with Private Organizations are and will be working on many such projects so that the tribal or Indigenous community can be brought in front of everyone in the whole world and the young generation are able to understand our culture and heritage through which we are originated and respect that fruitful up brings and methods and styles which today we make it in modern way and use in a modern way.

References

Abou-Foul M, Ruiz-Alba JL, López-Tenorio PJ (2023) The impact of artificial intelligence capabilities on servitization: The moderating role of absorptive capacity-a dynamic capabilities perspective. J Bus Res 157:113609. https://doi.org/10.1016/j.jbusres.2022.113609
Adade Williams P, Sikutshwa L, Shackleton S (2020) Acknowledging indigenous and local knowledge to facilitate collaboration in landscape approaches—lessons from a systematic review. Land 9:331
Adomako S, Nguyen NP (2023) Digitalization, inter-organizational collaboration, and technology transfer. J Technol Transf. https://doi.org/10.1007/s10961-023-10031-z
Aggarwal R, Singhal A (2019) Augmented reality and its effect on our life. In: 2019 9th international conference on cloud computing, data science & engineering (Confluence). IEEE, pp 510–515
Agnihotri A, Bhattacharya S, Yannopoulou N, Thrassou A (2023) Foreign market entry modes for servitization under diverse macroenvironmental conditions: taxonomy and propositions. IMR 40:561–584. https://doi.org/10.1108/IMR-09-2021-0287
Allam Z, Sharifi A, Bibri SE, Jones DS, Krogstie J (2022) The metaverse as a virtual form of smart cities: opportunities and challenges for environmental, economic, and social sustainability in urban futures. Smart Cities 5:771–801
Asati P, Raghuwanshi S, Hasan A, Zeffer A (2024) The role of electric vehicles in the agriculture industry using iot: turning electricity into food. In: Singh K, Kolar P (eds) Digital agricultural ecosystem. Wiley, pp 369–379. https://doi.org/10.1002/9781394242962.ch20
Audretsch DB, Aronica M, Belitski M, Piacentino D (2024) Natural selection or strategic adaptation? Entrepreneurial digital technologies and survival of the species. J Technol Transf. https://doi.org/10.1007/s10961-024-10065-x
Baines T, Ziaee Bigdeli A, Sousa R, Schroeder A (2020) Framing the servitization transformation process: a model to understand and facilitate the servitization journey. Int J Prod Econ 221:107463. https://doi.org/10.1016/j.ijpe.2019.07.036
Ball M (2022) The metaverse: and how it will revolutionize everything. Liveright Publishing
Batty RJ, Cuskelly G, Toohey K (2016) Community sport events and CSR sponsorship: examining the impacts of a public health agenda. J Sport Soc Issues 40:545–564. https://doi.org/10.1177/0193723516673189
Benyon D (2012) Presence in blended spaces. Interact Comput 24:219–226
Bhatti SH, Rashid M, Arslan A, Tarba S, Liu Y (2023) Servitized SMEs' performance and the influences of sustainable procurement, packaging, and distribution: the mediating role of eco-innovation. Technovation 127:102831. https://doi.org/10.1016/j.technovation.2023.102831
Boström M (2012) A missing pillar? Challenges in theorizing and practicing social sustainability: introduction to the special issue. Sustain Sci Pract Policy 8:3–14. https://doi.org/10.1080/15487733.2012.11908080
Brown D (2007) Te Ahu Hiko: digital cultural heritage and indigenous objects, people, and environments
Brundtland GH (1985) World commission on environment and development. Environ Policy Law 14:26–30
Cervantes JM (2011) What is indigenous about being indigenous? The Mestiza/o experience. In: Latina/o healing practices. Routledge, pp 37–62
Christen KA (2012) Does information really want to be free? Indigenous knowledge systems and the question of openness. Int J Commun 6:24
Dayoub B, Yang P, Omran S, Zhang Q, Dayoub A (2024) The role of metaverse in silk road's tourism: a qualitative study within china's belt and road initiative (BRI) context. Int Rev Spat Plan Sustain Dev 12:63–78
De Giovanni P (2023) Sustainability of the metaverse: a transition to Industry 5.0. Sustainability 15:6079

Dhanshyam M, Srivastava SK (2023) Decision framework for efficient risk mitigation in BOT highway infrastructure service projects. J Constr Eng Manage 149:04023081. https://doi.org/10.1061/JCEMD4.COENG-13014

Dobson T, Willinsky J (2009) Digital literacy. The Cambridge handbook of literacy, vol 10, pp 286–312

Dutta U (2019) Digital preservation of indigenous culture and narratives from the global south: in search of an approach. Humanities 8:68

Efe A (2022) Taking virtual reality and augmented reality to the next level: artificial intelligence with mixed reality. Kamu Yönetimi Ve Teknoloji Dergisi 4:141–165

Gandhi RD, Patel DS (2018) Virtual reality–opportunities and challenges. Virtual Reality 5:2714–2724

Gondia A, Ezzeldin M, El-Dakhakhni W (2022) Dynamic networks for resilience-driven management of infrastructure projects. Autom Constr 136:104149. https://doi.org/10.1016/j.autcon.2022.104149

Grant P, Söderbergh C (2020) Minority and Indigenous trends 2020-focus on technology. Minority Rights Group

Gundemeda N (2007) Information technology and marginalization: a sociological understanding of India. nationalizing crises: the political economy of public policy in contemporary India 311

Halarnkar P, Shah S, Shah H, Shah H, Shah A (2012) A review on virtual reality. Int J Comput Sci Issues (IJCSI) 9:325

Helsper E (2021) The digital disconnect: the social causes and consequences of digital inequalities

Hutson J (2024) Combining large language models and immersive technologies to represent cultural heritage in the metaverse context. In: Geroimenko V (ed) Augmented and virtual reality in the metaverse, springer series on cultural computing. Springer Nature Switzerland, Cham, pp 265–281. https://doi.org/10.1007/978-3-031-57746-8_14

icki Grieves VG (2008) Aboriginal spirituality aboriginal spirituality: a baseline for: a baseline for indigenous knowledges development in Australia. Can J Nativ Stud 28:363–398

Johnston AM (2014) Is the sacred for sale: tourism and Indigenous peoples. Routledge

Kale MVRK, Singh C, Khilari DS (2022) Non-fungible tokens (NFT's): the future of digital collectibles. Available at SSRN 4166708

Khare VK, Verma P, Mishra N (2022) A study of the influence of the skills development programme and its components on the growth in income of herbal tribe vendors in Patalkot, Madhya Pradesh. Int J Agric Rural Econ Res 10

Knox J (2022) The metaverse, or the serious business of tech frontiers. Postdigit Sci Educ 4:207–215. https://doi.org/10.1007/s42438-022-00300-9

Kouroupi N, Metaxas T (2023) Can the metaverse and its associated digital tools and technologies provide an opportunity for destinations to address the vulnerability of Overtourism? Tour Hosp 4:355–373

Lee LH, Braud T, Zhou P, Wang L, Xu D, Lin Z, Kumar A, Bermejo C, Hui P (2021) All one needs to know about metaverse: a complete survey on technological singularity, virtual ecosystem, and research agenda. arXiv preprint arXiv:2110.05352

Lei DT (2000) Industry evolution and competence development: the imperatives of technological convergence. IJTM 19:699. https://doi.org/10.1504/IJTM.2000.002848

Malerba S (2023) Exploring the potential of the metaverse for value creation: an analysis of opportunities, challenges, and societal impact, with a focus on the Chinese context

McKinley E (2005) Locating the global: culture, language and science education for indigenous students. Int J Sci Educ 27:227–241. https://doi.org/10.1080/0950069042000325861

Mishra N, Vashisht A, Raghuwanshi S (2024) Transforming Indian agriculture: unleashing the potential of digital agriculture using efficiency analysis. Digit Agric Ecosyst: Revolut Adv Agric 123–143

Mishra NK, Chakraborty A, Singh S, Ranjan P (2023) Efficiency analysis of engineering colleges in India: decomposition into parallel sub-processes systems. Socioecon Plann Sci 89:101708

Mishra NK, Pande Sharma P, Chaudhary SK (2024a) Redefining agile supply chain practices in the disruptive era: a case study identifying vital dimensions and factors. J Glob Oper Strat Sourc

Mishra NK, Mishra N, Sharma PP (2025a) Unraveling the relationship between corporate governance and green innovation: a systematic literature review. Manag Res Rev 48(5):825–845

Mishra NK, Raj A, Mukherjee I, Chakraborty A (2025b) Blockchain adoption in the supply chain management based on technology–organization–environment framework: evidence from developing and developed economies. J Bus Ind Market

Mishra NK, Raj A, Jeyaraj A, Gupta R (2024) Antecedents and outcomes of blockchain technology adoption: meta-analysis. J Comput Inf Syst 64:342–359. https://doi.org/10.1080/08874417.2023.2205370

Mishra NK, Sahoo S, Agarwal S, Sharma PP, Ilahi F (2024) Impact of institutional pressures and security on blockchain technology adoption and organization performance: an empirical study. J Technol Transf. https://doi.org/10.1007/s10961-024-10098-2

Moretto A, Grassi L, Caniato F, Giorgino M, Ronchi S (2019) Supply chain finance: from traditional to supply chain credit rating. J Purch Supply Manag 25:197–217. https://doi.org/10.1016/j.pursup.2018.06.004

Mystakidis S (2022) Metaverse. Encyclopedia 2:486–497

Naruta-Moya A (2015) Indigenous digital archive. Retrieved 13 May 2016

Ngulube P (2002) Managing and preserving indigenous knowledge in the knowledge management era: challenges and opportunities for information professionals. Inf Dev 18:95–102. https://doi.org/10.1177/026666602400842486

Paganopoulos M (2022) The evolution of Representation from cave painting to terraforming: metaverse worlds and the new dialectics of space. Anthropol AI Futur Hum Soc

Parakkal N (2023) Ruptures in global development: restorying'inclusive education for all'with indigenous/Adivasi Youth in Casteist-Colonial India (PhD Thesis). University of Michigan

Popp B, Woratschek H (2016) Introducing branded communities in sport for building strong brand relations in social media. Sport Manag Rev 19:183–197. https://doi.org/10.1016/j.smr.2015.06.001

Prasad BD (n.d) Assessment of Implementation of PESA in Schedule V Areas

Prest A (2020) Cross-cultural understanding: the role of rural school–community music education partnerships. Res Stud Music Educ 42:208–230. https://doi.org/10.1177/1321103X18804280

Purvis B, Mao Y, Robinson D (2019) Three pillars of sustainability: in search of conceptual origins. Sustain Sci 14:681–695. https://doi.org/10.1007/s11625-018-0627-5

Rao SS (2005) Bridging digital divide: efforts in India. Telematics Inform 22:361–375

Rothe D, Boas I, Farbotko C, Kitara T (2024) Digital Tuvalu: state sovereignty in a world of climate loss. Int Aff 100:1491–1509

Rowan CJ, Eskander MA, Seabright E, Rodriguez DE, Linares EC, Gutierrez RQ, Adrian JC, Cummings D, Beheim B, Tolstrup K (2021) Very low prevalence and incidence of atrial fibrillation among Bolivian forager-farmers. Ann Glob Health 87

Rubino F (2023) Museums on-chain? A designerly contribution in the development of blockchain-based digital strategies in cultural institutions

Senier S (2014) Digitizing indigenous history: trends and challenges. J Vic Cult 19:396–402

Shankar U, Kaul V, Gupta V, Mishra NK (2024) Workplace ostracism and counterproductive work behaviour: parallel mediation of felt accountability and emotional exhaustion. J Manag Dev

Sjödin D, Parida V, Kohtamäki M (2023) Artificial intelligence enabling circular business model innovation in digital servitization: conceptualizing dynamic capabilities, AI capacities, business models and effects. Technol Forecast Soc Chang 197:122903. https://doi.org/10.1016/j.techfore.2023.122903

Su P-Y, Hsiao P-W, Fan K-K (2023) Investigating the relationship between users' behavioral intentions and learning effects of VR system for sustainable tourism development. Sustainability 15:7277

Tiwari V, Tiwari VP, Chudasama D, Bala K (2016) Augmented reality and its technologies. Int Res J Eng Technol (IRJET) 3:1416–1420

Tjønndal A (2020) "What's next? Calling beer-drinking a sport?!": virtual resistance to considering eSport as sport. SBM 11:72–88. https://doi.org/10.1108/SBM-10-2019-0085

Vlăduțescu Ș, Stănescu GC (2023) Environmental sustainability of metaverse: perspectives from Romanian developers. Sustainability 15:11704

Wang N, Fang M, Beauchamp M, Jia Z, Zhou Z (2021) An indigenous knowledge-based sustainable landscape for mountain villages: the Jiabang rice terraces of Guizhou, China. Habitat Int 111:102360

Wang Q, Li R, Wang Q, Chen S (2021) Non-fungible token (NFT): overview, evaluation, opportunities and challenges

Wheelock R (2022) Indigenous spectrum sovereignty: creating a place in virtual space? Indig Policy J 33

Xiong S (2021) Digital solutions supporting citizen participation in urban environments

Yu Y, Li Y, Zhang Z, Gu Z, Zhong H, Zha Q, Yang L, Zhu C, Chen E (2020) A bibliometric analysis using VOSviewer of publications on COVID-19. Ann Transl Med 8

Zakarneh B, Annamalai N, Alquqa EK, Mohamed KM, Al Salhi NR (2024) Virtual reality and alternate realities in Neal Stephenson's—Snow Crash‖. World 14

Zharchenko A (2024) Sustainable development as opportunity for business (based on Cardinal Plumbing Heating & Air case) (PhD Thesis). Private Higher Educational Establishment-Institute Ukrainian-American

Carbon Footprint Reduction in Virtual Worlds: Innovations and Implications

Alok Singh Chauhan, Mega Novita, and Bhavna

Abstract Reducing carbon footprints seeks to minimize the impact of climate change by encouraging sustainable behaviors in all spheres of society and reducing greenhouse gas emissions. Unexpected environmental challenges have arisen as a result of the spread of virtual worlds, with energy-intensive data centers becoming a major contributor to carbon emissions. This study looks into cutting-edge tactics and how they might affect virtual worlds' carbon impact. Key advancements like hardware optimization, software efficiency enhancements, and the integration of renewable energy sources are reviewed through an extensive analysis of literature and industry practices. Furthermore, the potential for cutting emissions from upcoming technologies like decentralized hosting and quantum computing is investigated. Exploring emerging technologies is crucial for advancing carbon footprint reduction efforts in virtual environments. Technologies such as artificial intelligence (AI), blockchain, and virtual reality (VR) offer promising solutions for optimizing energy usage, enhancing resource efficiency, and promoting sustainable practices. The viability, efficacy, and overall environmental impact of these solutions are examined in relation to their implications. The study also emphasizes how crucial it is for stakeholders to work together and how helpful legislative frameworks are in order to promote sustainable virtual world practices. By shedding light on these advancements and their implications, this work contributes to the urgent global endeavour to address environmental sustainability in digital ecosystems.

Keywords Carbon footprint reduction · Virtual worlds · Sustainability · Innovations · Environmental impact · Emerging technologies

A. S. Chauhan (✉)
School of Computing Science and Engineering, Galgotias University, Greater Noida, Uttar Pradesh, India
e-mail: alokchauhan.1983@gmail.com

M. Novita
Universitas PGRI Semarang, Central Java, Indonesia

Bhavna
D. Y. Patil Vidyapeeth, Pune, India

W. Leal Filho et al. (eds.), *Metaverse and Sustainability: Business Resilience Towards Sustainable Development Goals*, World Sustainability Series,
https://doi.org/10.1007/978-3-031-89545-6_13

1 Introduction

The obvious and immediate risks posed by rising global temperatures, harsh weather, and the destruction of natural ecosystems have made addressing climate change more urgent in recent years. Reducing carbon footprints in all spheres of society is essential to addressing these issues and lessening the negative consequences of greenhouse gas emissions. However, a new frontier has emerged: digital ecosystems, specifically virtual worlds. Traditionally, the focus has been on industrial operations, transportation, and energy generation.

1.1 Overview of Climate Change and the Need for Carbon Footprint Reduction

One of the most important issues confronting humanity today is climate change, which is mostly caused by the buildup of greenhouse gases (GHGs) in the atmosphere. The consequences of industrial processes, energy generation, deforestation, and other human activities that have increased since the Industrial Revolution are these gases, chiefly carbon dioxide (CO_2). Rising worldwide temperatures, the melting of polar ice caps, an increase in the frequency and intensity of extreme weather events, and ecological disruptions are just a few of the significant and far-reaching effects of climate change. As a result, it is now urgently necessary for all spheres of society to reduce their carbon footprints.

Climate change is a major concern facing humanity today, primarily due to the accumulation of greenhouse gases (GHGs) in the atmosphere. These gases, primarily carbon dioxide (CO_2), are byproducts of industrial operations, energy production, deforestation, and other human activities that have increased since the Industrial Revolution. A few of the major and far-reaching implications of climate change are rising global temperatures, the melting of polar ice caps, a rise in the frequency and intensity of extreme weather events, and biological changes. Consequently, lowering carbon footprints is now an imperative for all sectors of society.

1.2 Emergence of Virtual Worlds as a New Frontier in Sustainability Efforts

Virtual reality (VR) environments, online gaming platforms, and the emerging metaverse are examples of virtual worlds that have permeated contemporary digital life. Through immersive experiences provided by these platforms, users can interact with one another, work, play, and create in incredibly detailed computer-generated settings. These virtual worlds have grown at an exponential rate due to technological developments, more access to the internet, and a global move towards digital interactions.

Nonetheless, there are substantial environmental consequences associated with this quick expansion. The energy-intensive infrastructure that underpins virtual worlds, especially data centers, is needed to run. These data centers, which contain the networking hardware and servers required to run virtual environments, frequently run on fossil fuels, which raises the carbon dioxide emissions from the atmosphere. Virtual worlds are predicted to have a larger carbon footprint as they continue to expand in size and complexity, which will present a new obstacle for sustainability initiatives.

It is critical to investigate methods for lowering virtual worlds' carbon footprint due to the negative effects they have on the environment. Examining the methods and tools that can improve the sustainability and energy efficiency of virtual environments is part of this. Virtual worlds are becoming a major source of carbon emissions worldwide, which opens up new challenges in the larger fight against climate change and calls for creative thinking and cooperative strategies.

1.3 Related Work

Studies on carbon footprints are essential for comprehending how different activities and technological advancements affect greenhouse gas emissions. Numerous studies have compared approaches, evaluated effects, and investigated technological approaches to lower carbon footprints.

Comparative Studies of Carbon Footprints

In order to demonstrate the various effects of various energy sources, Pata and Kumar (2021) examined the impact of coal and hydropower usage on greenhouse gas emissions in China and India. In their comparison of carbon footprints obtained from databases for life cycle assessment (LCA) and ecologically extended input–output analysis (EEIOA), Steubing et al. (2022) drew attention to differences between the ecoinvent and EXIOBASE databases. Their results highlight how crucial it is to select suitable approaches in order to accurately measure carbon footprint. In order to provide a complete picture of regional carbon emissions, Wu et al. (2021) expanded the definition of carbon footprint to include primary and intermediate inputs. Their method makes it easier to comprehend how diverse industries affect emissions in different geographical areas.

Advancements in Carbon Footprint Studies

Chen et al. (2022) examined current developments in urban ecosystem carbon footprint research. They talked about how different approaches are applied and the difficulties in precisely calculating the carbon footprint in urban settings. Hu and Yin

(2021) carried out a scientometric analysis of international studies on carbon footprints, pinpointing important patterns and fields of study. This overview outlines the development and evolution of the field of carbon footprint research and offers suggestions for future lines of inquiry. Chen et al. (2021) evaluate how carbon footprint research has progressed, examining its changes and the main areas of attention over time. Their analysis demonstrates a move away from a purely tracking approach to one that is more integrated and takes into account a wider range of environmental and social factors.

Technological Innovations and Strategies

In their exploration of technologies and viewpoints for achieving carbon neutrality, Wang et al. (2021) covered a range of technological approaches and solutions that may be useful in achieving carbon neutrality objectives. The technical aspects of using renewable energy sources to lower CO_2 emissions were the main emphasis of Razmjoo et al.'s (2021) study. The significance of dependable renewable energy sources in attaining energy sustainability is shown by their analysis.

In their review of carbon capture technology, Wilberforce et al. (2021) talked about developments and how they might reduce emissions. One of the most important technologies for lowering atmospheric CO_2 levels is carbon capture. Hertwich (2021) provided an illustration of the wider effects of economic activity on emissions by discussing how more investments in the manufacture of materials lead to a higher carbon footprint. In their study of global trends in climate change mitigation technologies, Probst et al. (2021) paid particular attention to the development and application of these advances. A thorough analysis of the methods and tools available for lowering carbon footprints is given by Lobus et al. (2023). In their work, the importance of energy efficiency initiatives, carbon capture developments, and renewable energy technology is discussed as critical approaches to lowering world emissions.

Regional and Sectoral Impacts

Hu et al. (2021) investigated important industries for their production, consumption, and network-based carbon, water, and energy footprints in the region. Their analysis shows the effects particular to certain sectors and reveals key contributors to regional footprints.

With a focus on the Yellow River Basin, Yuan et al. (2022) examined embodied carbon transfer and carbon footprint at the provincial level in China. Their study sheds light on the differences in carbon footprints between different regions.

Global and Policy Perspectives

A thorough bibliometric survey of research on carbon neutrality was carried out by Zhang et al. (2023), who also provided an overview of current patterns and future prospects.

Yu et al. (2021) examined the problems with carbon leakage under regionally specific climate policies, emphasizing obstacles and possible fixes. Using satellite data, Jiang et al. (2022) evaluated the carbon neutrality status of the major CO_2

emitters worldwide, offering a global perspective on the progress made towards carbon neutrality. Wang et al. (2022) mapped carbon neutrality research hotspots using bibliometric analysis. They used CiteSpace to identify important subjects that are at the centre of current research, including renewable energy, carbon capture and storage (CCS), and policy frameworks. An analysis of the negative effects of human activity on the environment, specifically the increase in atmospheric carbon dioxide, is provided by Kabir et al. (2023). Their research demonstrates how the uses of fossil fuels, deforestation, and industrialisation have exacerbated climate change.

Emerging Technologies and Future Outlook

In order to manage peak CO_2 emissions and achieve carbon neutrality, Yang et al. (2022) assessed current global issues as well as future prospects. Their research offers an outlook on the approaches required to deal with these difficulties. Xu et al. (2023) highlighted novel methods for carbon capture while talking about recent developments in microalgae's ability to fix CO_2 and its potential to contribute to carbon neutrality.

Renewable Energy and Sustainability

Olabi and Abdelkareem (2022) examined how various renewable technologies support sustainability while examining the importance of renewable energy in the fight against climate change. An overview of renewable energy technologies and their development was given by Deshmukh et al. (2023), emphasizing their significance in meeting energy demands and tackling environmental issues.

Bioenergy and Carbon Sequestration

The potential of bioenergy to lessen the effects of climate change and its role in attaining environmental sustainability were investigated by Rather et al. (2022). In their evaluation of carbon biosequestration techniques, Nayak et al. (2022) emphasized how these strategies may help achieve environmental sustainability and carbon neutrality.

Energy Transition and Policy

In their analysis of the shift to energy security and carbon neutrality, Rabbi et al. (2022) covered the function of various energy sources and legislative frameworks in reaching carbon neutrality.

In their analysis of the impact of novel energy sources in reaching carbon neutrality, Zou et al. (2021) concentrated on the ways in which creative energy solutions lower carbon emissions.

Technological Innovations in Carbon Management

Chen (2021) talked about the idea of carbon neutrality and how it will affect sustainability initiatives in the future, highlighting the necessity for creative thinking and cutting-edge technology. Sun et al.'s (2023) research highlighted the wider social and economic advantages of switching to sustainable energy sources by examining how green energy resources might advance energy justice and shared prosperity. In their bibliometric analysis of solar energy research, Obaideen et al. (2023) looked at trends, applications, and the contribution of solar energy to the achievement of sustainable development objectives.

Synthesis of Findings

The literature analysis indicates a thorough comprehension of the technical advancements, policy viewpoints, and carbon footprint assessment required to reach carbon neutrality. Important topics include the necessity of coordinated policy initiatives, the significance of precise carbon footprint measurement, and the role of diverse technologies in lowering emissions. Technologies for capturing carbon dioxide, microalgae CO_2 fixation, and renewable energy sources have made significant strides in the mitigation of climate change. Research is still being conducted, but it is clear that stakeholders must keep coming up with new ideas and working together to solve the complex problems associated with environmental sustainability and carbon neutrality. Future studies should concentrate on improving measurement techniques, investigating cutting-edge technology, and creating sensible legislative frameworks to aid in international efforts to reduce carbon footprints.

1.4 Purpose and Scope of the Study

The goal of this research is to look into and evaluate innovative approaches and technologies that try to lessen virtual worlds' carbon footprint. Through an emphasis on significant developments including software efficiency improvements, hardware optimization, and the incorporation of renewable energy sources, this research aims to find workable solutions that may be applied in digital ecosystems. The study also looks at how new technologies like blockchain, decentralized hosting, quantum computing, artificial intelligence (AI), and virtual reality (VR) may affect the amount of energy used and carbon emissions released in virtual environments.

This study's scope includes both present methods and potential future developments, offering a thorough analysis of the technological, financial, and legal aspects of reducing carbon footprints in virtual settings. By conducting a thorough analysis of current research, industry standards, and case studies, this study hopes to provide insightful information to practitioners and legislators who are trying to solve the environmental issues that virtual worlds bring.

1.5 Key Research Questions and Objectives

Underlying this investigation are the following major research questions:

1. In virtual worlds, what are the main sources of carbon emissions, and how do they differ from traditional sectors?

2. What technological advancements and approaches are being used right now to lessen virtual worlds' carbon footprint?
3. How might cutting-edge technologies like artificial intelligence (AI), blockchain, and quantum computing improve virtual environments' energy efficiency?
4. How can users, politicians, platform operators, and technology developers all contribute to the promotion of sustainability in virtual worlds?
5. How might sustainable practices be implemented in virtual worlds, and what are the possible obstacles and challenges that may arise?

This study aims to:

- Examine how virtual worlds affect the environment and determine the main causes of their carbon footprint.
- Assess how well-suited the present technology and tactics are for cutting emissions and energy use in virtual settings.
- Examine how new technology can help lower virtual worlds' carbon footprint even more.
- Examine how legislative frameworks and cooperation might promote sustainable practices in digital ecosystems.
- Make suggestions for upcoming studies and legislative initiatives that will aid in the long-term expansion of virtual worlds.

2 The Environmental Impact of Virtual Worlds

2.1 Growth of Virtual Worlds and Their Infrastructure

Expansion of Online Gaming, Virtual Reality, and Metaverses

Virtual reality (VR), online gaming, and the rise of metaverses have all contributed to a dramatic shift in digital entertainment and social interaction during the past ten years. Originally designed as straightforward multiplayer games, online gaming platforms have developed into intricate virtual worlds with millions of players interacting at once. These encounters have been further improved by virtual reality (VR), which provides fully immersive surroundings that obfuscate the distinction between the real and virtual worlds.

The idea of the metaverse, a shared, persistent virtual environment where people can do anything from work and socialize to purchase and have fun, is becoming increasingly popular. Prominent technology firms are making significant investments in the creation of metaverses, seeing them as the next phase on the internet. Millions of people are anticipated to utilize these virtual worlds at once, necessitating enormous amounts of processing power and data storage.

The infrastructure needed to support these digital platforms has grown along with their popularity and usage. This infrastructure allows for the real-time processing, storing, and distribution of enormous volumes of data. It consists of data centers, networking devices, and cloud computing resources. As a result, the demand for digital infrastructure has grown significantly due to the emergence of virtual worlds, which has increased energy consumption and the environmental effect of those systems.

Energy Demands of Data Centers and Digital Infrastructure

The foundation of virtual worlds are data centers, which offer the processing capacity required to manage user interactions, display immersive environments, and store enormous volumes of data. Thousands of servers, networking equipment, and storage systems are housed in these facilities, and for all of them to function properly, they need constant electricity and cooling. The energy needs of the data centers that power virtual worlds rise in direct proportion to their complexity and user base growth.

Since data centers frequently rely on non-renewable energy sources, such coal and natural gas, to meet their power needs, their energy consumption is a major concern. One of the main environmental issues with virtual worlds is data centers because of this reliance, which results in significant carbon emissions. Data centers' carbon footprint is further increased by the energy-intensive cooling systems needed to keep them working at ideal temperatures.

Beyond data centers, the digital infrastructure supporting virtual worlds consists of global networks of servers, communication hubs, and cloud services. It takes a lot of energy to keep these networks up and running, guarantee low-latency connections, and deliver flawless user experiences. The demand on digital infrastructure will only increase as more people interact with virtual worlds and as these settings become more data-intensive, which will have an adverse effect on the environment and increase energy consumption.

2.2 *Carbon Emissions from Digital Ecosystems*

Quantifying the Carbon Footprint of Virtual Worlds

Because the digital infrastructure supporting virtual worlds is so diverse and interconnected, quantifying their carbon footprint is a challenging undertaking. A virtual world's carbon footprint includes the emissions from data centers, network operations, end-user devices, and the hardware's lifespan effects.

As was previously said, data centers have a significant impact on virtual worlds' carbon footprint. These facilities use a lot of energy, and when that energy comes from fossil fuels, it emits a lot of CO_2. According to estimates, data centers use 1% or so of the world's electricity, and as demand for digital services grows, this percentage is predicted to rise. Virtual worlds can have a significant overall carbon footprint when you factor in the emissions from end-user devices and network operations.

Studies aiming at measuring the carbon footprint of particular virtual environments, like well-known online games or virtual reality platforms, have demonstrated that these kinds of digital ecosystems can have a big effect on the environment. The need for sustainable practices in the gaming industry is important, as some studies have calculated that the carbon footprint of large-scale online gaming platforms is comparable to that of big cities.

Comparison with Traditional Sectors

In contrast to conventional industries like manufacturing, transportation, and agriculture, digital ecosystems have comparatively smaller carbon footprints. Nonetheless, the exponential expansion of virtual worlds and the growing digitization of society imply that this impact will persist. The emissions from virtual worlds are related to the energy consumption of digital infrastructure, which may be reduced by technical advancements and environmentally friendly practices, in contrast to traditional sectors where emissions are frequently connected to physical activities.

The analogy between traditional industries and virtual worlds emphasizes how critical it is to address the environmental effects of digital ecosystems as soon as possible. Even if the carbon footprint of virtual worlds is currently smaller than that of sectors like energy production or transportation, it is impossible to discount the possibility of exponentially increasing emissions as virtual environments evolve. In the era of digital transformation, the digital industry has the chance to lead by example by adopting sustainable practices that will lower carbon emissions and encourage environmental stewardship.

3 Innovative Strategies for Carbon Footprint Reduction

Effective methods to lessen virtual worlds' carbon footprint are critical as their effects on the environment become more obvious. This section examines a number of cutting-edge strategies for lowering emissions and energy use related to digital ecosystems. These tactics include software efficiency improvements, hardware optimization, and the use of renewable energy sources.

3.1 Hardware Optimization

Energy-Efficient Server Technologies

Using energy-efficient server technologies is one of the best strategies to lower virtual worlds' carbon footprints. The workhorses of data centers are the servers, which handle and store the enormous volumes of data needed to run virtual environments. A large portion of the energy used by traditional servers is lost as a result of inefficient processing and power conversion.

The creation and implementation of energy-efficient servers have taken precedence in order to address this. Modern CPUs, such as those with energy-efficient architectures that use less power while yet providing excellent performance, are used in these servers. Furthermore, servers can lower their energy consumption during times of low demand by adjusting their power consumption based on workloads thanks to technologies like dynamic voltage and frequency scaling (DVFS).

The replacement of conventional hard disc drives (HDDs) with solid-state drives (SSDs) is another significant development in server technology. SSDs help data centers save energy overall since they are not just speedier but also use less power. Additionally, data centers may replace or upgrade individual components without completely revamping the system thanks to developing server designs that prioritize modularity and upgradability, which lowers energy consumption and electronic waste.

Advances in Cooling Systems and Resource Management

In order to keep servers and other data centre equipment working at their ideal temperatures, cooling systems are necessary. However, energy-intensive conventional cooling techniques like air conditioning greatly increase the carbon footprint of data centers. Significant advancements in cooling systems that lessen their impact on the environment and increase energy efficiency have been made to solve this.

One such invention is liquid cooling, which transfers heat more effectively than air-based systems by using liquid coolants to directly absorb heat from server components. By using this technique, the amount of energy needed for cooling is decreased while simultaneously enabling servers to run at higher densities, making the most use of the available space and resources.

Using immersion cooling, which involves immersing servers in a liquid that conducts heat but not electricity, is another effective strategy. This technology can significantly lower the energy consumption linked to conventional cooling technologies while providing even higher cooling efficiency.

Another important factor in maximizing data centre operations is resource management. By distributing the workload across servers, effective resource management algorithms can minimize energy waste and the need for surplus capacity. Virtualization technologies further improve resource efficiency and lower total energy consumption by enabling the operation of several virtual computers on a single physical server.

3.2 Software Efficiency Enhancements

Optimized Coding Practices

For virtual worlds to have a lower carbon footprint, software efficiency is just as important as hardware optimization. Code that is not well optimized might cause needless computational overhead, which raises energy usage. Consequently, using optimized coding techniques is crucial to developing software that is low energy and reduces the environmental effect of digital ecosystems.

Code that is both functionally sound and resource-efficient is written using optimized coding techniques. Using fewer computationally demanding processes, simplifying algorithms, and avoiding repetitive code execution are some examples of this. Software can be developed with an efficiency-focused approach, which saves energy by utilizing less computing resources to do the same tasks.

Using profiling tools to examine program performance and pinpoint locations where energy consumption can be minimized is a useful strategy for optimized coding. With the use of these tools, developers can identify wasteful code segments and improve their efficiency. Adopting best practices like code reuse and modular programming can also result in more effective software development processes, which can save even more energy.

Efficient Algorithms for Reduced Energy Consumption

The core of software operations are algorithms, which dictate how effectively actions are carried out in virtual environments. One important tactic for lowering the carbon footprint of virtual worlds is the development and application of energy-efficient algorithms.

The goal of energy-efficient algorithms is to reduce resource consumption and computational load without sacrificing functionality. Algorithms that give priority to efficient data processing and storage, for instance, can drastically lower the energy needed to handle big datasets—which are typical in virtual environments. Algorithms that minimize latency and optimize network traffic can also result in more effective communication between servers and end-user devices, which lowers the system's overall energy usage.

There are encouraging opportunities to create algorithms that are energy-efficient through machine learning and artificial intelligence (AI). By utilizing real-time data, AI-driven optimization algorithms can dynamically modify resource allocation, guaranteeing that energy is only utilized when and where it is required. These adaptive algorithms have the potential to save a substantial amount of energy, especially in large-scale virtual environments with highly variable processing demands.

3.3 Integration of Renewable Energy Sources

Adoption of Renewable Energy in Data Centers

Integrating renewable energy sources into data centre operations is one of the most effective ways to lower the carbon footprint of virtual worlds. Among the biggest energy users in the digital economy are data centers, thus switching to renewable energy can dramatically lower their carbon emissions.

An environmentally friendly substitute for fossil fuels is hydroelectric electricity, wind, and solar energy. These energy sources are already being used by a lot of data centers to power their operations. For example, to produce electricity directly from renewable sources, some data centers have erected wind turbines or solar panels on their premises. Others are collaborating with suppliers of renewable energy to get clean electricity from outside grids for their buildings.

Utilizing renewable energy not only helps data centers minimize their carbon footprint, but it also makes virtual worlds more sustainable overall. The use of renewable energy will be essential to ensuring that the growing demand for digital services does not come at the expense of the environment.

Case Studies of Sustainable Digital Infrastructure

A number of prominent technology businesses have established standards for environmentally friendly digital infrastructure by utilizing renewable energy sources and creative approaches. As an illustration:

- **Google**: Google has pledged to use only renewable energy sources to power its data centers. In order to minimize overall energy use and maximize the use of renewable energy, the company has invested in solar and wind farms and employs artificial intelligence (AI) to optimize energy usage in its data centers.
- **Microsoft**: With an aim to become carbon zero by 2030, Microsoft has also made notable progress in sustainability. The business has created cutting-edge cooling systems and made investments in renewable energy projects to lower the amount of water and energy used in its data centers.
- **Facebook (Meta)**: Facebook uses energy-efficient technologies and renewable energy sources in their well-planned data centers, which prioritize sustainability. The company's data centre in Luleå, Sweden, minimizes the need for additional energy-intensive cooling by using the naturally cold air in the area to cool servers. The facility is fully powered by hydroelectric energy.

4 Emerging Technologies and Their Potential Impact

Emerging technologies present fresh chances to further lessen virtual worlds' carbon impact as the digital landscape changes. This section examines three main topics, highlighting their possible effects on energy use and sustainability in virtual environments: decentralized hosting, quantum computing, and the functions of AI, blockchain, and VR.

4.1 Decentralized Hosting

Concept and Benefits of Decentralized Hosting

Instead of depending on centralized data centers, decentralized hosting refers to the dissemination of data and computational resources across a network of interconnected nodes. In order to host and manage digital environments, this method makes use of a peer-to-peer (P2P) architecture, in which numerous users or devices contribute resources. By distributing the workload among several smaller nodes, decentralized hosting eliminates the need for concentrated infrastructure, in contrast to traditional centralized hosting, which depends on massive data centers.

The ability of decentralized hosting to increase energy efficiency is its main advantage. Decentralized hosting can lessen the burden on individual data centers, resulting in lower energy usage, by distributing the computing load over a large network. Decentralized networks can also reduce latency and minimize data transmission distances by dynamically assigning jobs to nodes that are closer to the end users, hence optimizing resource utilization. An infrastructure that is more resilient and energy-efficient is a result of this distributed strategy, which also increases system resilience and lowers the possibility of single points of failure.

Decentralized hosting also fits in nicely with the increasing interest in edge computing, which is about processing data closer to the source (such end-user devices) instead of processing it in centralized data centers that are far away. Because of this close proximity, less data must be transmitted, which lowers the energy required for network operations.

Potential for Reducing Energy Consumption

Because decentralized hosting can optimize resource utilization and lessen dependency on energy-intensive data centers, it has the potential to lower energy usage. Decentralized hosting leverages underutilize d resources, such as idle computer capacity on consumer devices that would otherwise be wasted by dispersing tasks. Because more nodes may be added to the network as needed, this distributed approach also enables superior scalability by eliminating the need for big, energy-intensive data centers.

Decentralized hosting may also result in increased regional energy use, which makes it easier to power with renewable energy sources. To ensure the cleanest possible energy use, nodes in areas with a surplus of renewable energy can be given priority when it comes to hosting jobs. This strategy helps the wider shift to sustainable energy practices while also lowering the overall carbon footprint.

4.2 *Quantum Computing*

Overview of Quantum Computing Technology

By utilizing the ideas of quantum physics, quantum computing offers a major improvement in processing capacity by enabling calculations to be carried out at speeds that are not possible for traditional computers. Due to the fact that quantum bits, or qubits, can exist in numerous states at once, quantum computers are able to handle enormous amounts of data concurrently. Because of this feature, quantum computing is especially well-suited to tackle difficult issues involving massive amounts of computational power, like large-scale simulations, cryptography, and optimization.

Although it is still in its infancy, quantum computing has the power to completely transform a number of sectors, including virtual worlds and digital ecosystems. With its potential to process data faster and more efficiently, quantum computing holds great promise for lowering the carbon footprint of virtual environments. This might result in significant energy savings associated with data processing.

Implications for Energy Efficiency in Virtual Worlds

The implications of quantum computing for virtual world energy efficiency are significant. Traditional data centers will face increasing processing demands as virtual environments grow more intricate and data-intensive. By enabling quicker and more effective data processing and lowering the need for energy-intensive activities, quantum computing may be able to ease these pressures.

For instance, resource allocation in virtual worlds could be streamlined by quantum algorithms built for optimization, reducing the amount of energy needed to run and maintain these settings. Quantum computing may also make simulations and modeling more effective, enabling programmers to design virtual environments that are both energy-efficient and rich in detail.

The ability of quantum computing to lower the energy costs involved with data encryption and security—two essential elements of virtual worlds—is another noteworthy benefit. Digital ecosystems may save energy if quantum-based encryption approaches prove to be less computationally demanding and more secure than existing methods.

Although quantum computing is still in its early stages, it has the potential to drastically alter virtual worlds' energy environments. As technology advances, it may turn into a vital instrument in the endeavor to build sustainable digital landscapes.

4.3 AI, Blockchain, and VR

Role of AI in Optimizing Energy Usage

Virtual worlds are not an exception to the rule that artificial intelligence (AI) is becoming an indispensable tool for optimizing energy usage across industries. Large volumes of data may be analyzed in real time by AI algorithms, which can spot trends and inefficiencies that human operators would miss. Because of this feature, AI is a very useful technology for lowering virtual environments' energy usage.

Artificial Intelligence (AI) can be applied in virtual worlds to forecast user behavior, optimize server operations, and control network traffic. AI-driven load balancing, for example, can guarantee that servers are only powered on when necessary, saving energy during times of low demand. AI is also capable of optimizing cooling systems by anticipating temperature changes and modifying cooling power appropriately, which results in additional energy savings.

AI can also improve the effectiveness of developing and designing virtual worlds. Artificial Intelligence has the potential to decrease the computing burden needed to construct and manage virtual environments by automating parts of the development process, such as content generation and rendering. As a result of this decrease in computing demand, less energy is consumed, which strengthens the foundation of the digital ecosystem.

Blockchain for Sustainable Digital Ecosystems

Sustainable digital ecosystems may benefit from the decentralized and transparent features of blockchain technology. But the energy-intensive nature of blockchain, especially in proof-of-work (PoW) consensus procedures, has sparked questions about how it may affect the environment. In order to combat this, the industry is investigating less computationally demanding, more energy-efficient alternatives like proof-of-stake (PoS) and other consensus algorithms.

Blockchain technology can be used to build more transparent and sustainable ecosystems in virtual worlds. Blockchain-based platforms have the potential to facilitate decentralized transactions and digital asset ownership, hence mitigating the necessity for centralized infrastructure and the corresponding energy usage. Blockchain technology can also be used to monitor and confirm the usage of renewable energy in digital ecosystems, guaranteeing that the energy utilize d to power virtual worlds comes from sustainable sources.

Furthermore, in virtual worlds, incentives for sustainable behavior can be established by using the transparency and immutability of blockchain technology. For instance, adopting energy-efficient behaviors or making contributions to the platform's sustainability might earn users and developers tokens based on the blockchain.

VR as a Tool for Promoting Sustainable Practices

A special platform for encouraging environmentally conscious behavior and increasing public knowledge of environmental issues is provided by virtual reality (VR). VR can replicate the effects of pollution, climate change, and other environmental issues through immersive experiences, giving people a firsthand look at the results of unsustainable behavior. These encounters can serve as effective instructional and inspirational resources, inspiring people and institutions to embrace more environmentally friendly behaviors.

Virtual reality has the potential to improve education, but it can also reduce the demand for physical resources, which makes it more sustainable. VR, for instance, can reduce transportation-related carbon emissions by substituting virtual experiences for some forms of travel and in-person meetings. Similar advantages can be obtained by holding conferences, training sessions, and virtual workspaces in virtual reality (VR), which reduces the need for energy-intensive activities in the real world.

Moreover, before putting sustainable ideas into practice in the real world, VR can be utilized to develop and evaluate them there. This strategy contributes to overall sustainability by lowering the chance of costly errors and enabling more effective utilization of available resources.

5 Collaboration and Policy Frameworks

Technological developments alone won't be enough to achieve significant reductions in the carbon footprint of virtual worlds; concerted efforts from stakeholders and supportive policy frameworks are also necessary. The significance of stakeholder collaboration is discussed in this section, along with the function of legal and regulatory frameworks in advancing digital sustainability.

5.1 Importance of Stakeholder Collaboration

Roles of Technology Developers, Platform Operators, and Users

Difficulty reduction in the environmental effect of virtual worlds is a joint duty among users, platform operators, and technology developers. Every group is essential to advancing sustainability initiatives and making sure that virtual environments develop in a way that respects the environment.

- **Technology developers**: Developing the systems and tools that enable virtual worlds, developers are at the vanguard of innovation. Their responsibilities include creating algorithms that maximize resource utilization, devising hardware and software solutions that are energy-efficient, and integrating renewable energy sources into digital infrastructures. The environmental effect of virtual worlds can be considerably decreased by technology developers by giving sustainability top priority during the design and development phases.
- **Platform Operators**: Digital ecosystems are daily managed by operators of virtual world platforms, which include social media networks, online gaming corporations, and metaverse platforms. They have to make sure that every aspect of their platforms—from user interactions to server operations—are optimized for energy efficiency. Platform owners can also establish sustainability guidelines for their ecosystems, which will incentivize users and developers to follow green behaviors.
- **Users**: There is a part that virtual world users can play in encouraging sustainability. Users can reduce wasteful energy use (e.g., by logging out of inactive sessions), choose platforms that prioritize environmental responsibility, and support projects that encourage green habits to help ensure the general sustainability of digital environments. Users can also encourage more sustainable practices in the communities they participate in, which can increase awareness and spur group action.

To achieve the common objective of lowering virtual worlds' carbon footprint, cooperation amongst various stakeholders is crucial. Users can offer input and support for these projects, and technology developers can collaborate with platform operators to include sustainable solutions into their businesses. In order to establish and preserve sustainable digital ecosystems, all stakeholders must work together, and our cooperative strategy guarantees this.

Case Studies of Successful Collaborations

The potential for group action to promote sustainability in virtual worlds has already been shown by a number of fruitful stakeholder collaborations. These case studies demonstrate how concerted efforts can result in notable advantages for the environment.

- **The Green Gaming Alliance**: A combination of gaming corporations, developers, and environmental organizations, the Green Gaming Alliance aims to lessen the gaming industry's negative environmental effects. The partnership has lowered the carbon footprint of online gaming platforms by exchanging best practices, creating energy-efficient technologies, and increasing player awareness. For instance, a number of member businesses have encouraged the use of renewable energy in their data centers and installed energy-efficient server technology.
- **The Sustainable Digital Infrastructure Alliance (SDIA)**: A network of organizations called the SDIA is tasked with improving the sustainability of digital infrastructure. Through the collaboration of technology businesses, data centre operators, and legislators, the SDIA has enabled the digital ecosystem to adopt energy-efficient practices more widely. Additionally, the alliance has been a major force in promoting legislative reforms that assist the development of energy-efficient data centre standards and incentives for the use of renewable energy sources.
- **Partnership for AI-Driven Sustainability**: In this project, a number of top IT firms have teamed up with academic institutions to provide AI-driven solutions for virtual world energy optimization. The collaboration aims to develop algorithms that dynamically allocate resources, minimizing energy usage while preserving optimal efficiency. Due to this partnership, AI-based technologies have been used across a number of virtual platforms, saving a substantial amount of energy.

These case studies highlight how crucial cooperation is to attaining digital sustainability. Stakeholders can lessen the environmental impact of virtual worlds by collaborating to develop creative solutions that make use of their combined knowledge and resources.

5.2 *Legislative and Regulatory Frameworks*

Existing Policies for Digital Sustainability

- A number of legislative and regulatory frameworks aiming at fostering sustainability have been developed as a result of the increased understanding of the environmental impact of digital ecosystems. These regulations offer organizations direction and financial incentives to implement environmentally responsible activities and lower their carbon footprints.

- **The European Green Digital Coalition**: This European Commission-initiated program aims to quicken the shift to a sustainable digital economy. The coalition unites tech firms that are dedicated to cutting their carbon footprints and adopting energy-saving techniques. As part of the coalition's objective to become carbon neutral by 2050, members are expected to report on their environmental impact and follow sustainability guidelines.
- **The United States Clean Energy Data Center Initiative**: Data centers are encouraged by this program to increase energy efficiency and use renewable energy sources. For data centers that invest in green technologies, like energy-efficient servers and cooling systems, the project offers subsidies and tax benefits. To further encourage accountability and openness, the program also supports the creation of standards for monitoring and disclosing data centre energy consumption.
- **China's Green Internet Policy**: China has put laws into place to lessen the effects of its quickly expanding digital economy on the environment. Internet companies are required by the Green Internet Policy to use renewable energy sources and energy-saving technologies. The strategy also specifies how data centers and other digital infrastructure's carbon emissions would be tracked and regulated.

The government's role in supporting digital sustainability is exemplified by these current policies. These frameworks ensure that the digital economy develops in an ecologically responsible way by establishing clear criteria and offering incentives for green practices.

Recommendations for Policy Development

More policy development is required to promote the sustainability of virtual worlds even further. The following suggestions highlight important areas on which legislators should concentrate their efforts:

- **Incentivize Renewable Energy Adoption**: Governments must to provide data centers and digital platforms with more financial incentives to switch to renewable energy sources. For the installation of solar panels, wind turbines, and other renewable energy technologies, this could include tax credits, grants, and subsidies.
- **Establish Energy Efficiency Standards**: Energy-efficient guidelines for digital infrastructure, such as servers, cooling systems, and software, should be created and implemented by policymakers. These guidelines ought to be updated frequently to take into account new developments in technology and could be based on industry leaders' best practices.
- **Promote Transparency in Environmental Reporting**: Digital platforms and data centers should be required by law to submit periodic reports on their energy use and carbon emissions. Better monitoring of the industry's environmental

impact will be possible because to this transparency, which will also motivate businesses to switch to more sustainable methods.

- **Support Research and Development**: Funding should be set aside by policy-makers for the investigation and creation of novel technologies that lessen virtual worlds' carbon footprint. This could include funding for educational institutions, research facilities, and commercial businesses engaged in the development of hardware, algorithms, and renewable energy integration.

Incentives for Adopting Sustainable Practices

Apart from governmental regulations, incentives are essential in motivating companies to embrace environmentally friendly practices. These incentives can come in many different ways, such as cash payouts, public recognition, and access to resources that help with environmental projects.

- **Financial Incentives**: Companies may be encouraged to invest in energy-efficient technologies and renewable energy sources by tax cuts, grants, and subsidies. A data centre that installs solar panels, for instance, may be eligible for a sizable tax credit, which would mitigate the initial outlay and increase the viability of switching to renewable energy sources.
- **Recognition Programs**: Sustainability awards and certifications can help businesses get public acknowledgement for their work, which can improve their brand and motivate more investment in environmentally friendly operations. Organizations might be encouraged to prioritize environmental responsibility through programs such as the Energy Star accreditation for data centers or the sustainability awards for digital platforms.
- **Access to Resources**: To assist businesses in their sustainability initiatives, governments and industry associations can provide resources including training courses, best practice manuals, and access to green technology solutions. With the aid of these tools, organizations can make sure they are prepared to make decisions that will not negatively impact the environment while navigating the challenges of implementing sustainable practices.

6 Technological and Economic Barriers

Reducing virtual worlds' carbon footprint requires overcoming a number of technical and financial obstacles. While recognizing and removing these obstacles is essential to achieving environmental goals in digital ecosystems, they can impede the adoption of sustainable practices and solutions.

6.1 Technical Limitations and High Costs

Technical Limitations and High Costs

1. **Technical Limitations**

   - **Infrastructure Constraints**: Data centers and server farms are examples of existing digital infrastructure that was not initially created with energy efficiency in mind. It might be difficult and technically demanding to update or modify these buildings to accept new, more sustainable technologies.
   - **Scalability Issues**: Realizing state-of-the-art technology like quantum computing and sophisticated AI-driven systems necessitates substantial hardware and software developments. It's possible that current technology isn't yet developed or scalable enough to provide the projected energy efficiencies.
   - **Integration Complexity**: Compatibility difficulties are frequently encountered when integrating new technology with preexisting systems. For example, implementing new energy-efficient protocols or switching to decentralized hosting necessitates significant changes to the platforms and operating procedures already in use.
   - **Limited Research and Development**: The field of virtual world sustainability technology is currently developing. Insufficient research and development funding might hinder the creation of novel solutions and cause delays in their execution.

2. **High Costs**

   - **Initial Investment**: Adopting innovative, energy-efficient technologies or putting renewable energy solutions into practice can have significant upfront expenditures. This covers the cost of replacing infrastructure, implementing renewable energy systems, and buying new hardware.
   - **Operational Costs**: Even while some technologies might save money in the long run, their deployment and upkeep can come with significant operational costs. For instance, there are continuous costs associated with maintaining sophisticated cooling systems or infrastructure for quantum computing.
   - **Research and Development Expenses**: It costs a lot of money to engage in R&D to create and improve sustainable solutions. Startups and smaller businesses may find it difficult to set aside the money required for innovation.
   - **Economic Uncertainty**: Investments in green technologies may not be financially viable due to shifting energy prices, governmental regulations, and general economic situations. Businesses may decide not to pursue sustainability measures as a result of this uncertainty.

Resistance to Change in Industry Practices

1. **Cultural and Organizational Barriers**

   - **Inertia and Tradition**: Many organizations have long-standing, deeply embedded procedures and practices. Employees or management who are used

to traditional procedures may be resistant to adopting new, more sustainable practices.
- **Risk Aversion**: Perceived risks, such as possible operations interruptions, unproven technology, or uncertain returns on investment, may make companies reluctant to invest in new technologies or techniques.
- **Lack of Awareness**: It's possible that some stakeholders are ignorant of the environmental effects of their existing activities or do not completely comprehend the long-term advantages of implementing sustainable practices.

2. Economic and Competitive Pressures
    - **Cost Competitiveness**: Companies may put cost-cutting tactics ahead of sustainability in highly competitive marketplaces, particularly if the immediate financial benefits of green technologies are not immediately apparent.
    - **Market Differentiation**: If other companies don't have the same level of commitment to environmentally friendly methods, organizations could find it difficult to differentiate them through sustainability. A "race to the bottom" may result from this, in which immediate financial gain takes precedence over long-term environmental objectives.
    - **Regulatory Uncertainty**: Regulations pertaining to sustainability that have ambiguities or inconsistencies may cause uncertainty and discourage the adoption of new practices. Organizations could be uncertain about how to adhere to changing regulations or worry about possible legal ramifications.

6.2 Opportunities for Innovation and Leadership

Business Models for Sustainable Virtual Worlds

1. **Green Business Models**
    - **Energy-as-a-Service (EaaS)**: Under this approach, energy management is contracted out to outside companies that specialize in integrating renewable energy sources and maximizing energy consumption. Businesses can take advantage of cost savings and skilled management by implementing EaaS instead of having to shoulder the entire expense of energy infrastructure upgrades.
    - **Circular Economy Models**: Circular economy models concentrate on extending the lifecycle of products and materials, with an emphasis on resource efficiency and waste reduction. This can entail developing hardware that is easier to recycle or repurpose and putting in place software that uses fewer resources when it comes to virtual worlds.
    - **Subscription-Based Services**: The long-term cost distribution of sustainable technologies can be facilitated by implementing subscription models for software and services. This strategy can lessen the initial financial load and help organizations budget for green initiatives.

2. **Sustainable Revenue Streams**

   - **Carbon Offset Programs**: Users of virtual world platforms may be able to purchase carbon offsets, which can be used to finance greenhouse gas emission reduction initiatives. This strategy satisfies growing consumer demand for ecologically friendly items while giving users a way to support sustainability initiatives.
   - **Green Certifications and Eco-Labels**: Businesses can set themselves apart in the market by using eco-labels and certifications when they reach sustainability milestones. These certifications can give businesses a competitive edge and draw in environmentally conscientious customers.

Opportunities for Leadership in Green Technology

1. **Innovation Leadership**

   - **Pioneering Sustainable Technologies**: Businesses that invest in and create innovative green technology have the chance to rise to the top of their respective industries. These businesses have the power to change industry norms and procedures by establishing sustainability standards and proving the practicality of creative solutions.
   - **Collaborative R&D**: Collaborating with educational establishments, research groups, and additional businesses helps expedite the creation of novel sustainable technology. Research and development collaborations can combine knowledge and resources to produce faster and more efficient progress.

2. **Policy and Advocacy Leadership**

   - **Advocating for Stronger Regulations**: Businesses and groups who actively support stricter environmental laws and policies have the power to significantly influence how digital sustainability develops in the future. Through active participation in business associations and policymakers, these leaders may facilitate systemic change and advance best practices.
   - **Educational Initiatives**: Taking the lead in educating stakeholders about the advantages of sustainable technologies and practices can help to promote an environmentally conscious culture. In order to spread awareness and exchange expertise, this can involve holding workshops, publishing research, and taking part in industry forums.

3. **Strategic Partnerships**

   - **Forming Alliances**: Sustainability initiatives can be strengthened through strategic alliances with other businesses, trade associations, and non-profits. These partnerships can share resources, encourage cooperative projects, and boost the effectiveness of sustainability programs.
   - **Community Engagement**: Creating a sense of shared responsibility and increasing support for sustainable practices can be achieved through interacting with the larger community, which includes users and environmental

organizations. Green technology adoption can gain momentum and spread thanks to community-driven projects and efforts.

7 Case Studies and Industry Practices

Analyzing real-world instances and business practices offers insightful information about practical methods for lowering virtual worlds' carbon footprint. In this part, noteworthy case studies are examined and best practices that can serve as industry models are highlighted.

7.1 Case Study 1: Sustainable Practices in Virtual Gaming

Overview: This case study looks at a well-known online gaming company that has successfully reduced its environmental impact through the use of sustainable practices. The business has come a long way in incorporating eco-friendly methods and technology into its daily operations.

Key Initiatives

- **Energy-Efficient Data Centers**: The business has switched to energy-efficient servers and cooling systems in its data centers. The organization has achieved significant energy savings by optimizing server use and implementing cutting-edge cooling solutions.
- **Renewable Energy Integration**: The business has pledged to run its operations entirely on renewable energy. This effectively lowers its carbon footprint by sourcing electricity from hydroelectric, solar, and wind projects.
- **Green Game Design**: The business has included sustainability into the process of creating new games. This involves encouraging in-game activities that spread knowledge of environmental issues and streamlining code to use less energy.
- **User Engagement**: The business has started programs to get customers involved in environmental initiatives. This involves encouraging players to support environmental causes through game-related campaigns and providing in-game rewards for eco-friendly behavior.

Outcomes

- **Reduced Carbon Emissions**: As a consequence of the company's efforts, carbon emissions have significantly decreased, in line with its sustainability objectives.
- **Enhanced Brand Image**: By putting sustainability first, the business has drawn in eco-aware customers and enhanced its reputation.
- **Positive User Response**: The company's sustainability measures have received excellent feedback from users, which has increased engagement and loyalty.

7.2 Case Study 2: Green Data Centers in Virtual Reality Platforms

Overview: In order to enable its immersive digital experiences, the virtual reality (VR) platform covered in this case study has implemented green practices in its data centers.

Key Initiatives

- **Energy-Efficient Infrastructure**: The VR platform has made investments in low-power servers and cutting-edge cooling systems as well as energy-efficient data centre infrastructure. Considerable energy savings have resulted from this expenditure.
- **Smart Resource Management**: The platform optimizes energy use based on demand by utilizing intelligent resource management strategies including load balancing and dynamic scaling. This method enhances overall efficiency and lowers energy waste.
- **Green Building Certifications**: The data centers' dedication to environmentally friendly design practices is evidenced by their attainment of green building certifications like LEED (Leadership in Energy and Environmental Design).
- **Partnerships for Renewable Energy**: To guarantee that a significant amount of its energy comes from sustainable sources, the platform has worked with providers of renewable energy.

Outcomes

- **Energy Savings**: Significant energy and cost savings have resulted from the use of energy-efficient technologies and astute management techniques.
- **Environmental Impact**: The platform's carbon footprint has substantially decreased as a result of the switch to renewable energy sources, in line with its environmental goals.
- **Industry Recognition**: The VR platform's dedication to eco-friendly methods has been acknowledged by environmental organizations and industry peers.

7.3 Industry Best Practices for Carbon Footprint Reduction

1. **Adopting Energy-Efficient Technologies**
    - **Upgrade to Energy-Efficient Hardware**: To lower energy usage and operating expenses, make investments in networking hardware, cooling systems, and servers that use less energy.
    - **Implement Smart Cooling Solutions**: Reduce the amount of energy needed for cooling in data centers by implementing cutting-edge cooling technologies like free cooling and liquid cooling. Reduce the amount of energy needed for

cooling in data centers by implementing cutting-edge cooling technologies like free cooling and liquid cooling.

2. **Integrating Renewable Energy**
   - **Source Renewable Energy**: Switch to renewable energy sources to cut down on carbon emissions and lessen dependency on fossil fuels, such as solar, wind, and hydroelectric power.
   - **Purchase Green Energy Certificates**: By acquiring green energy certificates, you can lessen the environmental effect of using non-renewable energy sources and support renewable energy projects.
3. **Optimizing Software and Infrastructure**
   - **Enhance Software Efficiency**: Create and implement software with energy-saving algorithms and optimized code. Energy consumption can be significantly decreased with the help of efficient program design.
   - **Employ Virtualization**: Reduce the number of physical servers needed and increase server utilization with virtualization technologies to cut down on energy usage and related expenses.
4. **Promoting Sustainable Practices**
   - **Educate and Engage Stakeholders:** Bring up the topic of sustainability with partners, staff, and users. In order to encourage the adoption of green practices, offer resources and training.
   - **Set Sustainability Goals:** To monitor progress and promote ongoing development in the reduction of the carbon footprint, clearly define your sustainability goals and KPIs.
5. **Supporting Industry Collaboration**
   - **Participate in Green Alliances**: Join groups and coalitions in the sector that are devoted to advancing sustainability, and communicate best practices with colleagues.
   - **Collaborate on Research and Development**: Take part in joint R&D projects to develop environmentally friendly products and technologies. To promote innovation, collaborate with other organizations and research institutions.
6. **Leveraging Policy and Regulation**
   - **Advocate for Sustainable Policies**: Encourage and support laws and rules that improve the sustainability of the environment in digital ecosystems.
   - **Comply with Standards**: Follow industry guidelines and sustainability-related certifications, such as ENERGY STAR for energy efficiency or ISO 14001 for environmental management.

8 Future Directions and Implications

As we progress towards more environmentally friendly virtual environments, it is critical to take into account the changing environment and its consequences for both practice and research. In addition to examining long-term effects on the world's carbon footprint, this section also highlights research gaps and potential future directions for virtual worlds.

8.1 Evolution of Sustainable Virtual Worlds

Predictions for the Future of Virtual Worlds

1. **Advancements in Technology**
    - **Increased Integration of Green Technologies**: It is anticipated that in the future, virtual worlds will incorporate more and more green technologies, such as advanced cooling systems, hardware with low energy consumption, and renewable energy sources. These inventions will be widely adopted as technology advances and become more accessible and affordable.
    - **Enhanced AI and Machine Learning**: The optimization of energy usage and resource management in virtual environments will be greatly aided by AI and machine learning technology. More sophisticated algorithms will make data processing, predictive maintenance, and dynamic energy management possible.
2. **Growth of Immersive Experiences**
    - **Expansion of Virtual Reality (VR) and Augmented Reality (AR)**: It is projected that the adoption of VR and AR, which provide more engaging and immersive experiences, would increase dramatically. The need to build sustainable infrastructure to meet these technologies' growing demands will become more pressing as they proliferate.
    - **Development of Metaverses**: It is very conceivable that the idea of metaverses—virtual worlds that are connected—will come to pass. Creative thinking and concerted action are needed to manage the environmental impact of these vast digital ecosystems.
3. **Shift Towards Decentralization**
    - **Decentralized Hosting and Blockchain**: Adoption of blockchain and decentralized hosting could revolutionize the administration and operation of virtual worlds. These technologies may present fresh possibilities for cutting energy use and improving sustainability by promoting transparency and distributed resource management.

Long-Term Impacts on Global Carbon Footprint

1. **Reduction in Overall Emissions**
    - **Increased Efficiency**: Virtual worlds have the potential to significantly lower overall carbon emissions as they adopt more energy-efficient practices and technologies. The adoption of renewable energy sources and infrastructure optimization is expected to mitigate the carbon footprint of digital ecosystems.
2. **Potential for Carbon Offsetting**
    - **Carbon Credits and Offsets**: The expansion of virtual worlds might present new chances to offset carbon emissions, such funding initiatives aimed at lowering greenhouse gas emissions. Users of digital platforms might have the opportunity to acquire carbon credits or take part in offset programs, supporting international efforts to reduce carbon emissions.
3. **Impact on Traditional Sectors**
    - **Reduced Resource Consumption**: The amount of resources used and the environmental effect of physical infrastructure may be reduced by moving some operations to virtual settings. Virtual gatherings and events, for instance, may lessen the need for travel and the emissions that go along with it.

8.2 Implications for Research and Practice

Gaps in Current Research

1. **Lack of Comprehensive Metrics**
    - **Measurement and Reporting**: Standardized criteria and procedures are required in order to precisely quantify and disclose the carbon footprint of virtual worlds. Comprehensive frameworks for evaluating the environmental impact of digital infrastructure and activities are lacking in current research.
2. **Limited Exploration of Emerging Technologies**
    - **New Technologies**: While some study has been done on cutting edge technologies such as decentralized hosting and quantum computing, more research is required to fully understand how these innovations might reduce carbon footprints and what that means for virtual worlds.
3. **Integration Challenges**
    - **Practical Implementation**: Theoretical aspects of sustainability technologies are frequently the focus of research, with less attention paid to the difficulties of actual implementation. Research on the practical challenges of incorporating green technologies into current virtual world infrastructures is needed.

Directions for Future Studies

1. **Development of Standardized Metrics**
   - **Frameworks and Guidelines**: The main goal of future study should be to provide uniform measures and reporting standards for the carbon footprint of virtual worlds. This will make evaluations and comparisons between various platforms and technologies more realistic.
2. **Exploration of Sustainable Technologies**
   - **Innovative Solutions**: Examine how cutting-edge technologies like decentralized hosting, blockchain, and quantum computing might improve sustainability in virtual environments. The best ways to incorporate these technologies and their long-term effects on energy use and carbon emissions should be investigated in research.
3. **Case Studies and Practical Applications**
   - **Real-World Implementation**: Perform research and case studies on how sustainable technologies and practices are being applied in real-world virtual environments. This entails looking at effective cases, recognizing difficulties, and creating integrated best practices.
4. **Policy and Regulation Research**
   - **Impact of Policies**: Examine how proposed and current regulations affect virtual world sustainability. In addition to identifying areas for policy improvement, research should examine how regulatory frameworks might encourage the adoption of green technologies and behaviors.
5. **User Behavior and Engagement**
   - **Behavioral Studies**: Examine the ways in which user behavior affects virtual world sustainability. Research ought to look into how to encourage people to adopt environmentally beneficial habits and how their participation can help achieve sustainability objectives as a whole.

9 Conclusion

It is becoming more and more important to manage the environmental impact of virtual worlds as they develop and grow. This part provides a summary of the study's main conclusions, emphasizes how digital ecosystems contribute to environmental sustainability, and makes some last recommendations for the future.

Summary of Key Findings

1. **Environmental Impact of Virtual Worlds**

- **Growth and Infrastructure**: Virtual reality (VR), online gaming, and metaverses are examples of virtual worlds that are rapidly expanding and resulting in higher energy requirements and carbon emissions from data centers and digital infrastructure.
- **Carbon Emissions**: Because data centers and digital operations require a lot of energy, virtual worlds have a major impact on global carbon emissions. Targeted reduction measures are required, as shown by the quantification and comparison of these emissions with traditional sectors.

2. **Innovative Strategies for Reduction**

- **Hardware Optimization**: Reducing the carbon footprint of virtual worlds requires the use of resource management techniques, sophisticated cooling systems, and energy-efficient server technology.
- **Software Efficiency**: Enhanced overall efficiency and reduced energy usage can be achieved through the implementation of efficient algorithms and optimised coding methods.
- **Renewable Energy Integration**: Reducing environmental effect requires data centers to use renewable energy sources and to build sustainable digital infrastructure.

3. **Emerging Technologies**

- **Decentralized Hosting and Blockchain**: By improving transparency and distributing resource management, these technologies have the potential to lower energy usage and promote sustainability.
- **Quantum Computing**: Even though it's still in its infancy, quantum computing has the potential to improve virtual world energy efficiency by enabling more sophisticated and effective data processing.
- **AI and VR**: While VR can encourage sustainable practices and increase public awareness of environmental issues, AI can optimize energy usage.

4. **Barriers and Opportunities**

- **Technical and Economic Barriers**: High prices, technological restrictions, and industry opposition to change are some of the difficulties. To overcome these obstacles, we need to be innovative, collaborative, and committed.
- **Opportunities for Leadership**: Sustainable business models, like energy-as-a-service and circular economy strategies, offer chances for green technology innovation and leadership.

5. **Case Studies and Best Practices**

- **Successful Implementations**: Case studies of businesses using green data centers and virtual gaming to promote sustainable practices show successful tactics and provide role models for others in the sector.
- **Best Practices**: Adopting energy-efficient technologies, incorporating renewable energy, and encouraging sustainable practices among stakeholders are important industry activities for reducing carbon footprints.

6. **Future Directions**
    - **Technological Advancements**: Long-term sustainability objectives depend on the advancement of green technology and their incorporation into virtual environments.
    - **Research Gaps**: To improve knowledge and adoption of sustainable practices, standardized measures, research into cutting-edge technologies, and useful case studies are required.
    - **Policy and Regulation**: The adoption of green practices and technologies will be aided by efficient laws and regulations, and systemic change will be sparked by cooperative efforts.

Contribution to Environmental Sustainability in Digital Ecosystems

The following ways that this work advances our knowledge of environmental sustainability in digital ecosystems:

- **Providing Insights**: Providing thorough research, analysis, and insights on virtual worlds' carbon footprints as well as creative mitigation techniques.
- **Highlighting Best Practices**: Showcasing effective case studies and best practices from the industry that can help other businesses implement sustainable practices.
- **Identifying Future Directions**: describing the gaps in knowledge and potential paths forward for improving sustainability in virtual environments, in order to support and steer current initiatives.

Final Thoughts on the Path Forward

Sustainability must continue to be a top priority as virtual worlds expand and become more integrated into daily life. Crucial actions to take in order to proceed are:

- **Investing in Innovation**: Support the development and adoption of new technologies and practices that reduce carbon footprints and enhance environmental sustainability.
- **Collaborating Across Sectors**: Foster collaboration among technology developers, industry stakeholders, policymakers, and researchers to drive collective efforts toward sustainable virtual ecosystems.
- **Promoting Awareness**: Involve users and stakeholders in sustainability projects by spreading knowledge and promoting ethical behavior in virtual environments.
- **Advocating for Policy Support**: Encourage the adoption of green technologies by advocating for strict laws and regulations that promote sustainability goals.

References

Chen JM (2021) Carbon neutrality: toward a sustainable future. Innovation 2:100127
Chen R, Zhang R, Han H (2021) Where has carbon footprint research gone? Ecol Indic 120:106882

Chen K, Yang M, Zhou X, Liu Z, Li P, Tang J, Peng C (2022) Recent advances in carbon footprint studies of urban ecosystems: overview, application, and future challenges. Environ Rev 30:342–356

Deshmukh MKG, Sameeroddin M, Abdul D, Abdul Sattar M (2023) Renewable energy in the 21st century: a review. Mater Today Proc 80:1756–1759

Hertwich EG (2021) Increased carbon footprint of materials production driven by rise in investments. Nat Geosci 14:151–155

Hu S, Yin J (2021) Global research on carbon footprint: a scientometric review. Environ Impact Assess Rev 89:106571

Hu M, Chen S, Wang Y, Xia B, Wang S, Huang G (2021) Identifying the key sectors for regional energy, water and carbon footprints from production-, consumption- and network-based perspectives. Sci Total Environ 764:142821

Jiang F, He W, Ju W, Wang H, Wu M, Wang J, Feng S, Zhang L, Chen JM (2022) The status of carbon neutrality of the world's top 5 CO_2 emitters as seen by carbon satellites. Fundam Res 2:357–366

Kabir M, Habiba U, Iqbal MZ, Shafiq M, Farooqi ZR, Shah A, Khan W (2023) Impacts of anthropogenic activities and climate change resulting from increasing concentration of carbon dioxide on environment in 21st century; a critical review. IOP Conf Ser Earth Environ Sci 1194:012010

Lobus NV, Knyazeva MA, Popova AF, Kulikovskiy MS (2023) Carbon footprint reduction and climate change mitigation: a review of the approaches, technologies, and implementation challenges. C 9:120

Nayak N, Mehrotra R, Mehrotra S (2022) Carbon biosequestration strategies: a review. Carbon Capture Sci Technol 4:100065

Obaideen K, Olabi AG, Al Swailmeen Y, Shehata N, Abdelkareem MA, Alami AH, Rodriguez C, Sayed ET (2023) Solar energy: applications, trends analysis, bibliometric analysis and research contribution to sustainable development goals (SDGs). Sustainability 15:1418

Olabi AG, Abdelkareem MA (2022) Renewable energy and climate change. Renew Sustain Energy Rev 158:112111

Pata UK, Kumar A (2021) The influence of hydropower and coal consumption on greenhouse gas emissions: a comparison between China and India. Water 13:1387

Probst B, Touboul S, Glachant M, Dechezleprêtre A (2021) Global trends in the invention and diffusion of climate change mitigation technologies. Nat Energy 6:1077–1086

Rabbi MF, Popp J, Máté D, Kovács S (2022) Energy security and energy transition to achieve carbon neutrality. Energies 15:8126

Rather RA, Wani AW, Mumtaz S, Padder SA, Khan AH, Almohana AI, Almojil SF, Alam SS, Baba TR (2022) Bioenergy: a foundation to environmental sustainability in a changing global climate scenario. J King Saud Univ Sci 34:101734

Razmjoo A, Gakenia Kaigutha L, Vaziri Rad MA, Marzband M, Davarpanah A, Denai M (2021) A Technical analysis investigating energy sustainability utilizing reliable renewable energy sources to reduce CO_2 emissions in a high potential area. Renew Energy 164:46–57

Steubing B, de Koning A, Merciai S, Tukker A (2022) How do carbon footprints from LCA and EEIOA databases compare? A comparison of ecoinvent and EXIOBASE. J Ind Ecol 26:1406–1422

Sun Y, Wang J, Wang X, Wei X (2023) Achieving energy justice and common prosperity through green energy resources. Resour Policy 81:103427

Wang F, Harindintwali JD, Yuan Z, Wang M, Wang F, Li S, Yin Z, Huang L, Fu Y, Li L et al (2021) Technologies and perspectives for achieving carbon neutrality. Innovation 2:100180

Wang D, Huangfu Y, Dong Z, Dong Y (2022) Research hotspots and evolution trends of carbon neutrality—visual analysis of bibliometrics based on citespace. Sustainability 14:1078

Wilberforce T, Olabi AG, Sayed ET, Elsaid K, Abdelkareem MA (2021) Progress in carbon capture technologies. Sci Total Environ 761:143203

Wu X, Li C, Guo J, Wu X, Meng J, Chen G (2021) Extended carbon footprint and emission transfer of world regions: with both primary and intermediate inputs into account. Sci Total Environ 775:145578

Xu P, Li J, Qian J, Wang B, Liu J, Xu R, Chen P, Zhou W (2023) Recent advances in CO_2 fixation by microalgae and its potential contribution to carbon neutrality. Chemosphere 319:137987

Yang S, Yang D, Shi W, Deng C, Chen C, Feng S (2022) Global evaluation of carbon neutrality and peak carbon dioxide emissions: current challenges and future outlook. Environ Sci Pollut Res 30:81725–81744

Yu B, Zhao Q, Wei Y-M (2021) Review of carbon leakage under regionally differentiated climate policies. Sci Total Environ 782:146765

Yuan X, Sheng X, Chen L, Tang Y, Li Y, Jia Y, Qu D, Wang Q, Ma Q, Zuo J (2022) Carbon footprint and embodied carbon transfer at the provincial level of the Yellow River Basin. Sci Total Environ 803:149993

Zhang L, Ling J, Lin M (2023) Carbon neutrality: a comprehensive bibliometric analysis. Environ Sci Pollut Res 30:45498–45514

Zou C, Xiong B, Xue H, Zheng D, Ge Z, Wang Y, Jiang L, Pan S, Wu S (2021) The role of new energy in carbon neutral. Pet Explor Dev 48:480–491

Exploring the Metaverse: Transforming Marketing Strategies in Virtual Environments

Ankita Pathak, Janvi Basawatia, and Maidul Islam

Abstract The metaverse, originally introduced in Neal Stephenson's 1992 novel "Snow Crash," set in a dystopian Los Angeles, has transformed into a worldwide phenomenon that combines virtual and real-life events. The marketing potential of the metaverse and its impact on consumer behavior is now gaining momentum. This chapter presents a multidisciplinary perspective on the Metaverse and outlines its potential impact on consumer behavior. The Metaverse is characterized by five important elements: digital media, spatiality, immersion, sharing, and real-time operation. This progress has garnered significant investment from businesses such as Google, Microsoft, Nvidia, and Meta. Early platforms such as Second Life have given way to more advanced virtual and augmented reality experiences. Walmart, Adidas, and McDonald's are pioneering this new frontier. The metaverse provides unparalleled prospects for immersive consumer interactions, which will revolutionize buying, socializing, and marketing techniques. This book chapter studies the metaverse's technology developments, market possibilities, and implications for future marketing and consumer behavior studies.

A. Pathak (✉)
GLA University, Institute of Business Management, Mathura, India
e-mail: ankitapthk76@gmail.com

J. Basawatia
IDBI Bank Ltd, Dhrangadhra, Gujarat, India
e-mail: janvibasawatia@gmail.com

M. Islam
Department of International Business (Keimyung Adams College), Keimyung University, Daegu, South Korea
e-mail: maidul@kmu.ac.kr

W. Leal Filho et al. (eds.), *Metaverse and Sustainability: Business Resilience Towards Sustainable Development Goals*, World Sustainability Series,
https://doi.org/10.1007/978-3-031-89545-6_14

1 Introduction

The metaverse's in marketing has sparked significant interest of companies due to its immense capacity to transform marketing strategy and consumer behaviour. The metaverse is a virtual world that exists beyond the physical world. It combines users imaginations with reality by merging social networking with realistic virtual environments. This unique idea of metaverse has resulted in significant expansion in the business. The market forecasts a remarkable growth from USD 22.79 billion in 2021 to USD 996.42 billion by 2030. The figures of growth are demonstrating an exceptional compound annual growth rate (CAGR) of 39.8% (Kumar 2023) Walmart, Adidas, and McDonald's have entered the metaverse, expanding beyond physical stores and internet purchasing, which has led to new and exciting frontier (Koohang et al. 2023a, b).

In a short period of time, the metaverse has sparked widespread interest. The phrase "metaverse" has quickly entered the lexicon of professionals, researchers, and consumers. It has elicited a wide spectrum of feelings, from excitement, interest, and optimism to uncertainty, scepticism, and even dread in the mind of researcher. Although there is no consensus on the exact concept of "the metaverse," it has definitely drawn substantial investment from corporations (Dwivedi et al. 2022). Prominent technological firms, Google, Microsoft, Nvidia, and, most notably, Meta, have made significant investments in building the hardware, software, and infrastructure necessary to support the metaverse (Huynh-The et al. 2023).

Industry analysts predict that the metaverse industry will soon be valued at $800 billion in the near future, perhaps contributing more than $1 trillion to global GDP (Hadi et al. 2024). Both corporations and academia have expressed interest in the metaverse, as it reflects consumers' growing participation in digital worlds like virtual products, services, and social interactions (Hennig-Thurau et al. 2023). The COVID-19 pandemic's social distancing strategies hastened the transition of consumer engagement in virtual worlds from a marginal curiosity to a major one. The success of online gaming platforms like Fortnite and Roblox has popularized a culture where users use avatars for more than just gaming, as well as for social interactions and financial transactions within virtual environments (Lee et al. 2021).

As the virtual world expands and progresses, it will change the way customers buy, socialise, and locate items. Recent study focuses on the technological components of the metaverse (Khan et al. 2022) and social interactions (Ghali et al. 2024). Studies have been carried out on the metaverse's impact on marketing practitioners. The chapter provides a introduction of the metaverse and presents a framework outlining its core components, applicability, possibilities, problems, application, and future research directions.

This chapter aims to explore the transformative potential of the metaverse within the marketing domain by delving into its core components, technological advancements, and market opportunities. It seeks to provide a detailed analysis of how the metaverse is changing the way consumers engage with brands, shifting traditional marketing strategies toward more immersive and interactive experiences. The chapter will also examine the challenges and strategic implications that businesses must navigate to effectively leverage the metaverse as a marketing platform.

The significance of the metaverse in today's marketing landscape cannot be overstated. With the rise of digital consumerism, companies are increasingly looking toward virtual environments as the next frontier for engaging consumers. The metaverse allows brands to go beyond traditional online experiences, offering immersive, real-time interactions that foster deeper connections with their audience. By integrating elements of social networking, gaming, and digital economies, the metaverse provides businesses with unique opportunities to create personalized, immersive brand experiences. This chapter will highlight how companies are capitalizing on these opportunities to shape the future of marketing.

In light of these advancements, this chapter posits that the metaverse represents a pivotal shift in marketing, where immersive technologies, real-time interactions, and virtual economies will redefine consumer behavior and brand strategies. Understanding how to navigate this evolving digital ecosystem will be essential for marketers aiming to stay competitive in an increasingly virtual world.

2 Literature Review

2.1 The Abstraction of Metaverse

The metaverse concept, first introduced in the fictional book Snow Crash in 1992, has developed into a significant worldwide occurrence of interest to both advertisers and scholars. (Pieters 2022). This virtual environment offers immersive experiences that are designed to closely match those seen in the real world. Virtual platforms that arose early on, Second Life and SIMS, are seen as forerunners of the metaverse (Pieters 2022).

These platforms introduced the concept of a virtual parallel existence, allowing users to partake in actions that closely match those in the realm of reality (Nleya and Velempini 2024). These early metaverse systems, developed during the web 2.0 era, were primarily intended to provide immersive virtual experiences, with a focus on gaming. Metaverse platforms have now included augmented reality, allowing users to engage in interactions based on real-world experiences, thanks to advancements in web 3.0 technology (Lee et al. 2021). Originally, these platforms offered users the choice of choosing avatars, vocations, and social connections, allowing them to engage in real-world activities within virtual worlds.

However, the metaverse has made substantial progress because to advancements in artificial intelligence (AI), deep learning, extended reality technologies, and content-related services (Ooi et al. 2023). These encompass immersive and interactive forms of media that are particularly tailored for the metaverse. In addition, it integrates payment systems such as digital currencies and cryptocurrencies, hence enhancing its immersive characteristics (Gadekallu et al. 2023).

(Bibri and Allam 2022) defines the evolved metaverse as a linked and long-lasting network of virtual settings in which people may interact with other creatures and objects in real time via their avatars.

Before fully interacting with the metaverse as a parallel world, it is vital to consider whether it is merely another transient event, similar to Second Life (Rostami 2022). Initial platforms were designed to offer an artificial experience. However, they were viewed as distinct from real-life interactions. The earlier platforms were built for specific interest groups, but modern metaverse platforms allow for real-life interactions and are decentralised, with the goal of preserving civility among individuals of all ages (Li et al. 2023). Furthermore, the metaverse enables avatars to explore a vast array of virtual environments, each offering diverse features.

The metaverse is a realm that goes beyond reality by mixing physical and digital dimensions, and it uses technologies like virtual reality (VR) and augmented reality (AR) (Park and Kim 2022). These technologies enable users to engage in immersive interactions with virtual worlds, objects, and persons by providing multimodal experiences. Augmented Reality (AR) integrates digital data into actual physical situations, whereas Virtual Reality (VR) creates artificial worlds. Previous research by Kim et al. (2023) has highlighted the incorporation of augmented reality (AR) and virtual reality (VR) in customer engagements. According to the literature, (Han et al. 2022) consumers have positive opinions about the metaverse and are interested in trying out virtual reality (VR) and augmented reality (AR) experiences. Furthermore, (Rane 2023) present thorough information regarding the customer experience journey in the metaverse and how it may influence attitudes and behaviour.

Academic research on the metaverse spans a variety of disciplines, including information technology (Dincelli and Yayla 2022), marketing (Wider et al. 2023), education (Alfaisal et al. 2024) tourism and hospitality (Özdemir Uçgun and Şahin 2023) and psychology (Barrera and Shah 2023). Although research on the metaverse in marketing is still in its nascent stages, it is swiftly growing across numerous sectors, including advertising, luxury goods, retail, sales, and brand development.

The metaverse offers a one-of-a-kind platform for comprehending user behavior and consumer culture, as well as novel measures and levels of intrusion. As companies venture into the metaverse, effective marketing approaches will be crucial to achieving success. The metaverse expands marketers' ability to communicate with existing and prospective customers, providing them with engaging and immersive experiences. Marketing within the metaverse is currently in its trial phase and encounters substantial obstacles. These limitations encompass technological and infrastructure issues regarding the accessibility and affordability of hardware and software, socio-cultural challenges, and strategic barriers (Table 1).

2.2 *Metaverse and Its Five Key—Elements*

The metaverse signifies an expansive and interlinked system of 3D virtual realms and settings that are generated in real-time.

Table 1 Summary on consumer behaviour with the metaverse

S. No.	Consumer behavior	Description	Themes
1	Awareness	Consumers who are tech-savvy, particularly those engaged in gaming and virtual reality, are eager to learn more about the metaverse	Inspired to discover the metaverse, acknowledge metaverse as a developing trend, technologically savvy individuals, engaged in games and virtual reality
2	Information search	Consumers discover the metaverse by watching videos, engaging with influencers, and interacting with others. They also take part in events and demonstrations to deepen their understanding	In their quest for knowledge about the metaverse, individuals watch YouTube and other resourceful videos, follow influencers and trendsetters, and interact with friends and fellow gamers
3	Evaluation of alternatives	Consumers encounter challenges in accessing the metaverse and look for alternatives like desktop versions, borrowing, or buying more affordable or second-hand equipment. They also gather with friends to borrow devices	Recognizing problems with accessibility and equipment, examining the desktop version, gathering with friends to borrow gear, and acquiring low-cost or pre-owned accessories
4	Engagement	Consumers participate in the metaverse through different means, such as spending extensive time, brand engagement, and skepticism about NFTs. Alternatives they explore include gaming and innovation, showcasing their readiness for the evolving virtual landscape	Engagement levels and time distribution, aspirations for involvement, mixed feelings and brand participation, and examining alternatives along with preparedness
5	Post-engagement evaluation	Consumers engage with the metaverse by sharing thoughts, creating content, exploring opportunities for longer stays, and hosting events. They offer tutorial sessions for trial access	Contemplating their time spent in the metaverse, organizing a schedule for metaverse events, developing tutorial sessions to offer trial access to the metaverse, and producing content to share their experiences

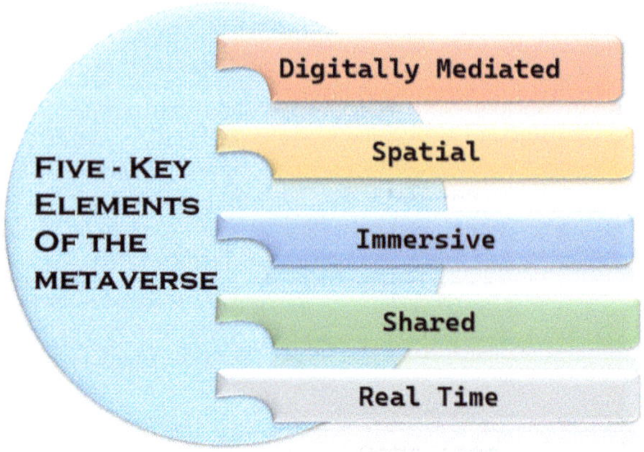

Fig. 1 Elements of metaverse

It allows an endless number of users to experience these virtual spaces simultaneously with each user having their own feeling of presence. This network ensures the uninterrupted flow of data, and elements such as identification, history, entitlements, objects, conversations, and payments. The concept of metaverse shows essential characteristic like digitally mediated, spatial, immersive, shared, and real-time. Figure 1 shows the characteristic of metaverse. The combination of these elements provide unique feature to metaverse.

2.2.1 The Metaverse is Digitally Mediated

Consumer experiences within the metaverse are enabled by digital technology, which acts as means for customers interact with other users and the actual content (Hennig-Thurau et al. 2023). The digital technologies include both hardware and software, which vary from each other significantly. Metaversal worlds are frequently created by using high-fidelity 3D computer graphics (Ball 2022). Furthermore, metaverse experiences frequently utilize extended reality (XR) technologies such as virtual reality (VR) and augmented reality (AR) to create and showcase digital representations of places, items, and individuals (Kumar et al. 2024).

Technological Drivers of the Metaverse

Several cutting-edge technologies are essential for building the metaverse, each playing a distinct role in enabling immersive, interactive, and secure digital environments. These technologies not only drive the technical aspects of the metaverse but also open new avenues for marketing innovation.

Blockchain

Blockchain technology ensures decentralized and secure transactions in the metaverse, enabling trustless operations without the need for intermediaries. This is particularly relevant in the context of non-fungible tokens (NFTs), which allow for the ownership and trading of digital assets, such as virtual real estate, fashion, or branded collectibles. From a marketing perspective, blockchain allows brands to create exclusive digital items (e.g., limited edition NFT sneakers by Nike) or reward loyal customers with verifiable digital assets. Additionally, blockchain-driven smart contracts facilitate seamless in-world transactions, further enhancing the customer experience by simplifying secure and transparent purchases.

Artificial Intelligence (AI)

AI is integral to creating responsive, personalized experiences in the metaverse. It powers virtual assistants, personalized content, and automated interactions, helping marketers better understand consumer behavior through data analysis. AI can generate intelligent virtual environments that adapt in real-time to user preferences. For example, AI-driven avatars can serve as brand representatives, engaging customers in interactive conversations to provide personalized product recommendations. In marketing, AI's role in analyzing vast amounts of user data allows companies to create more targeted campaigns, enhancing customer engagement and retention within the metaverse.

Virtual Reality (VR) and Augmented Reality (AR)

VR and AR technologies are foundational to the immersive experience of the metaverse. VR allows users to explore fully immersive digital environments, while AR blends virtual elements into the physical world, offering a unique hybrid experience. Brands are leveraging these technologies to create immersive shopping experiences, such as virtual fitting rooms where customers can try on products in a virtual environment or AR-powered product demos that showcase how items might fit into real-world spaces. These technologies enable consumers to interact with products in ways that are not possible in traditional digital marketing, significantly boosting brand engagement and enhancing consumer decision-making by providing lifelike experiences.

Virtual Reality (VR) completely immerses people in an artificial and digitally controlled environment, by isolating them from the real world (Wohlgenannt et al. 2020). AR, in contrast, adds digital material into the user's physical surrounding, and allow both digital and real-world components to be visible at the same time (Park and Kim 2022). Consumers interact with these experiences using a range of devices, such as desktops, laptops, and mobile screens, as well as immersive head-mounted displays like virtual reality (VR) headsets and augmented reality (AR) glasses (Xi et al. 2024). Digital technology acts as both the source and medium for consumer experiences. The Metaverse's digitally mediated nature allows for experiences that are not constrained by the limitations of physical reality. Users often use avatars to represent their actual look, but they can also create inventive personalities, such as avatars with unicorn horns or animal heads.

They can then interact with digital items that have inventive properties, such as wormholes that transfer them to different realities. Although digital mediation allows for infinite creative expression, it also raises the issue of dematerialization (Burrows 2022). The issue of dematerialization is common in digital environment. Thus, in the

Metaverse, buyers may buy virtual goods like clothing, real estate and pets (Bhasin et al. 2024). However, it is vital to emphasise that these goods exist only as digital reproductions of their physical counterparts.

Together, blockchain, AI, VR, and AR fuel marketing innovation by facilitating enhanced customer experiences, secure digital ownership, and dynamic interactions. By integrating these technologies, marketers can offer hyper-personalized experiences, drive consumer engagement through exclusive digital products, and create interactive environments that blur the line between virtual and physical realities. As the metaverse continues to evolve, these technologies will be at the forefront of revolutionizing marketing strategies, allowing brands to create deeper, more meaningful connections with consumers in virtual worlds.

2.2.2 The Metaverse is Spatial

The Metaverse is the blend of digitally enabled environment. Traditional social media platforms primarily involve the sharing of content through personal and institutional profiles, pages, and feeds (Sritong et al. 2024). In contrast, the Metaverse provides users with an immersive experience in three-dimensional "virtual worlds" similar to those found in video games (Wang et al. 2023). Spatiality in the metaverse refers to the concept of physical space being replicated or mimicked digitally. This allows users to experience and navigate virtual environments in ways that simulate real-world spatial interactions, such as proximity, distance, and movement. In marketing, spatiality is crucial as it influences consumer behavior by enabling virtual store layouts, 3D product exploration, and creating an immersive brand experience. For example, brands can design virtual showrooms that simulate the feeling of walking through a physical store, enhancing consumer engagement and interaction with the product. This spatiality allows marketers to recreate or invent physical spaces where customers can interact with their brand in ways that wouldn't be possible in traditional 2D advertising. Spatial marketing in the metaverse can enhance customer experience by enabling immersive product demonstrations, virtual tours, or interactive shopping environments, all of which can deeply affect consumer purchase decisions and brand perception.

Metaversal environments share resemblances with physical places. Interactions in virtual environments are constrained to persons and items that are in close proximity to the user at a particular point. In the Metaverse, there is limitation of one's direct perception towards virtual environment, through visual and auditory stimuli (Petit et al. 2022). User faces the similar constraints Contrary to the limitations imposed by physical reality, users are not confined by their real-world location, but rather by their virtual place inside the digital era at the given point of time (Kye et al. 2021). Behaviour inside the Metaverse is influenced by geographical and spatial factors, such as distance and closeness. The interaction of one's are determined by their virtual positioning within the Metaversal world. The Metaverse showcases spatial principles with its many transit choices. Some platforms allow users to travel virtual landscapes by walking. Even individuals also have the option to own automobiles and utilise public transportation, such as aeroplanes, to explore and move around different regions. The brain shows similar reaction to both virtual and physical stimuli (Saffari et al. 2023). It is anticipated that environmental signals would influence more consumer behaviour in the Metaverse as compare to their impact in the real world. According to (Allam et al. 2022) Within the Metaverse, individuals have the ability to digitally navigate and explore innovative homes and urban settings. The builder and planners can also analyse the socioeconomic consequences of groundbreaking architectural concepts.

2.2.3 The Metaverse is Immersive

Within the Metaverse, users are typically portrayed as being integrated into the internet rather than merely use it (Ball 2022). Therefore, a key feature of the Metaverse is its capacity to fully engross people in an interactive encounter. Although different researcher has defined immersive in different ways but (Ahn et al. 2024) stated immersion as an intrinsic quality of a technology, apart from how users perceive or respond to it. Immersion is achieved by a technology when it successfully delivers a realistic experience where user are that effectively places inside the simulated world (Shen et al. 2021). The first crucial step in creating a feeling of presence in a digital context is to create a "believable environment." This is achieved by actively involving the user inside that environment (Julian et al. 2023). In most of the cases, a greater level of immersion is associated with a higher probability that individuals would fee the presence in the virtual world (Han et al. 2022). Various visual characteristics enhance the sense of being fully engaged in a virtual environment. These include the sharpness of the images, the ability to perceive depth through stereoscopy, the extent of the display's panoramic view, and the perspective from which the user sees the environment. These factors were identified by Cummings and Bailenson (2015), Slater and Wilbur (1997). Several firms are now investigating olfactory devices that allow users to perceive aromas within the Metaverse (Panagiotakopoulos et al. 2022). Although these goods are niche still is consider in incorporation with visual and audio inputs. Researchers suggest that engaging several sensory modalities in the user increases the probability of genuine sense of immersion in the environment (Chen et al. 2024).

2.2.4 The Metaverse is Shared

The community character of the Metaverse is an important component, as users' experiences are modified by the presence and activity in the same digital space. This influence is evident when users engage as spectators, similar to public consumption in physical contexts. It is much more obvious on websites that allow all users to cooperate and develop Metaversal settings and experiences with one another.

The Metaverse is frequently used as a digital community in which users interact, trade, create friendships, and establish local communities through their avatars.

However, avatars in the Metaverse create social feelings equivalent to those experienced in in-person encounters. People may not experience these social sentiments as people live through avatars, there is no actual presence there. Physical co-presence is often used to establish genuine social ties (Voinea et al. 2022). Initial research suggests that, in contrast to in-person engagement, computer-mediated communication diminishes perceptions of social presence (Teng et al. 2023; Dulabh et al. 2023). Individuals that have less social presence are less likely to engage in parasocial behaviours via computer-mediated encounters, where there is less collaboration and more unfiltered emotional emotions. The research has identified crucial avatar qualities that improve social presence (Aseeri and Interrante 2021). One key aspect leading to increased social presence is an avatar's behavioural realism, which indicates knowledge of the user's avatar presence and activities.

Users in the Metaverse observe the avatars of others. The devices used to join the virtual world contain eye and face trackers that transfer a user's eye movements and facial emotions into their avatars.

2.2.5 The Metaverse is Real-Time

The distinguishing feature of the Metaverse is its operation in real time. This temporal component includes both temporal synchronisation and temporal persistence. Real-time operation in the metaverse refers to the ability of digital environments to operate without delay, where changes, interactions, and events happen instantaneously for all users. This means that the actions of one user are immediately visible and interactive for others, creating a shared, synchronous experience. In marketing, this real-time interaction allows for immediate consumer responses, live events, and real-time customer service, which can be crucial for campaigns that rely on instant feedback or audience engagement, such as product launches or flash sales. Real-time operation enhances the marketing potential of the metaverse by enabling live brand events, virtual meet-ups, and dynamic advertisements that can respond to user behavior instantly. This immediacy can increase customer engagement and create more impactful, personalized experiences. For instance, real-time product releases or interactive, live marketing campaigns can foster excitement and community interaction, strengthening the connection between the brand and the consumer.

The temporal synchrony focuses on the passage of time inside the Metaverse, whereas temporal persistence refers to the continuous flow of time in general (Turchet 2023). Users in the Metaverse interact in digitally mediated surroundings with others, and temporal synchronisation assures that they share experiences. In other words, experiences are temporally structured consisting of distinct occurrences with clear beginnings and ends (Tversky et al. 2008). Temporal synchronisation in the Metaverse ensures that all users experience events in the same order, length, and timing. Temporal synchronisation is essential for strengthening the sense of social presence (Cui et al. 2010). Studies have shown that when viewers engage with live-streamed content, they have a higher sense of social presence rather than pre-recorded material (Vincent and Frewen 2023) The ephemeral character of events typically boosts one's sense of absorption and the anticipate that material will vanish frequently will stimulates higher cognitive involvement (Lupu and Rokka 2022).

2.3 Application of Metaverse in Marketing

2.3.1 Metaverse and Digital Marketing:

Digitization and digital strategy advancements have recently offered marketers with strategic advantages in product and service promotion and advertising. The metaverse expands the scope of digital marketing (Nagarajan et al. 2023). Using 3D virtual environments, marketers may create complicated campaigns such as treasure hunts and team-building events to improve brand familiarity. RPG Enterprises, Epic Games, Nike, and Tinder are exploring the potential of leveraging the metaverse to attract new customers, retain current ones, and cultivate a customer-centric brand in their digital marketing strategies. According to Eggenschwiler et al. (2024). While the practical implications of the metaverse on digital marketing tactics and their success are yet not completely explored.

The availability of detailed data across several behavioural characteristics offers great potential. It will help to improve consumer interaction and give businesses with useful insights for running efficient digital marketing efforts. Facebook, Instagram, and LinkedIn are now exploring methods to include metaverse principles in order to improve the consumer experience (CX) (Gursoy et al. 2023). While the full integration of the metaverse with digital marketing presents a challenging endeavour for marketers, employing effective management strategies can result in favorable outcomes regarding customer experience (CX), engagement, and repeat usage. This has been observed on platforms such as Fortnite, Second Life, and Roblox (Jungherr and Schlarb 2022).

In terms of branding, displaying both static and dynamic brand logos on metaverse avatars and digital surroundings would increase brand exposure to the target demographic. (Bilgihan et al. 2024). Organizations can improve their understanding of client preferences by collecting data from metaverse participants and their reactions in various scenarios, thus boosting customer management. Businesses can utilize techniques such as social network analysis (SNA), natural language processing (NLP), and deep learning to analyze qualitative data from metaverse platforms. The metaverse allows them to develop more relevant branding, customer, and stakeholder management strategies for digital targeting (Buhalis et al. 2023). The data obtained from the metaverse can be utilised to create personalised digital adverts, such as banner ads, non-fungible tokens (NFTs), and marketing campaigns based on 3D avatars. Segmenting clients based on their behavior in the metaverse may have significant ramifications for a company's digital investments (Hackl et al. 2022). The investment of a company in the digital space could be significantly impacted by the segmentation of consumers based on their behaviour in the metaverse (Yoo et al. 2023).

2.3.2 Consumer Consciousness in the Metaverse

As per Sağkaya Güngör et al. (2022) consumers are increasingly adopting digital objects in order to keep up with technological developments. Business interest in the metaverse emphasises its marketing possibilities. The metaverse provides a more engaging experience compared to traditional and digital platforms (Gadekallu et al. 2023). Marketers can leverage the metaverse for a variety of purposes, including shopping, branding, advertising, and other forms of digital value creation. In general, the metaverse is an enhanced platform in the customer's buying process from a broad point of view (Bratu and Sabău 2022). Although the metaverse is expected to see significant expansion, there are still many undiscovered areas that hinder the understanding of consumer behaviour in this virtual world. Psychologically, it is uncertain to determine how consumers consciously perceive the metaverse environment. In contrast to the tangible universe, the metaverse offers a multitude of conscious distractions. Consumers' awareness can fluctuate or stay concentrated inside many expanded realities, including avatars, geometrics, items, brands, goods, tools, surroundings, and the visual haptics found in the metaverse (Colombo 2021). The qualities present in the metaverse provide appealing prospects for marketing (Petit et al. 2022). Nevertheless, the wide array of choices available in the metaverse might lead to conflict or subliminal attention for customers (Oliveira et al. 2022). This might potentially impact the entire consumer purchasing process without any concrete results (Weniger et al. 2023). For example, when a customer enters a metaverse, they come across several sensory touchpoints that might potentially inspire interest in marketing activities. At the same time, the multitude of items in the metaverse can potentially distract one's awareness (Guan 2022). Subsequent future research studies might investigate that marketers should control the quantity of sensory stimuli in the metaverse to ensure a deliberate customer experience (Huang and Chung 2024).

The metaverse can be linked to three levels of stimulation: seeing, hearing, and touch. The stimuli are sent through the interconnected tools inside the metaverse ecosystem; therefore it is important to strengthen consumers' awareness of technology. The presence of social interaction and the use of third-party avatars inside the metaverse significantly impact an individual's state of awareness and perception Wang et al. (2022). Consumers' decision-making might be impacted by the dominant awareness. Likewise, alterations in deliberate behaviours might generate cognitive disorder that could influence consumer decision-making Halkiopoulos et al. (2022). The experience can be altered in accordance with the conscious cycle, which can impact the overall consumer journey process, in decision-making. According to Hoyer et al. (2020) Marketers will have enormous challenges in guiding customers on a right conscious trip through the metaverse. It will influence consumer decision-making and improve the consumer experience.

2.3.3 Metaverse for Branding

The metaverse adds a 3D layer to the current 2D internet environment, which has gained significant attention in recent years. It provides a new opportunity for companies to succeed their brand in meeting the competition. According to Gupta et al. (2024), brands are cooperative entities that hold importance for multiple stakeholders. They usually originate from an organisation and are encountered by customers who either purchase them (Jones and Sasser 1995). It develop a strong emotional connections with them. Distributors play a role in facilitating the movement of brands from companies to customers, and other entities (Achrol and Kotler 1999). The creation of a brand, audience involvement, and experience are evolving processes that change over time due to numerous interactions with one another. The interaction have impacted by the interaction environment (Carleton and Collard 2020).

The metaverse can serve as the primary platforms for brand development. It depends on the nature of the brand (Catherine et al. 2024). The transition from conventional to electronic to metaverse has been documented over a period of time for specific businesses, particularly in retail brands (Allam et al. 2022). Nevertheless, the metaverse, is a unified platform for the development, distribution, and interaction of many brands (Jungherr and Schlarb 2022). It is not widely adopted but are found in virtual commerce, online virtual reality games. Typically, the metaverse serves as one of several worlds where brands coexist simultaneously.

The metaverse offers intangible software components known as haptics, which play a crucial role in consumer interactions and their engagement with brands and customers. Although the core principles of brand strategic thinking remain consistent across various environments and contexts (Amason and Ward 2020). The implementation of specific tactical choices is anticipated to be influenced by the specific 3D features. In reality, 3D metaverse environments allow companies to create brand content for consumers, who are often depicted as avatars. Avatars can engage in interactions with one another regarding brands (Elsharnouby et al. 2022).

Brands leverage the metaverse to engage with consumers, enhance awareness, and deliver pertinent information without emphasizing direct sales, while consumers focus on their own needs. Yadav et al. (2023) found that particular 3D software components improve communication quality and interactivity between parties.

Brand research and the metaverse bring unique traits and problems. Although a significant portion of the existing research concentrates on intangible offerings that are largely accessed and consumed through technology (Gómez and Vargas 2012). It is important to recognise and take into account that the metaverse is just one of several contexts in which most brands exist. The rapid evolution of metaverse-related immersive technologies, leads to frequent updates of equipment and 3D message types and content (Far et al. 2023). The updation require significant investments of time and financial resources. Researchers should exercise caution while exploring into this area of study since the context is always changing, which means that any discoveries that are tied to the context will likely be temporary (May and Perry 2022). The Further problem in expanding the metaverse is the coordination of resources across different contexts and the creation of cohesive interactions (Seidel et al. 2022).

It is important to additionally take into account the significance of the metaverse for certain specific audiences. Although the technology exists and has the potential to become more widely available and of better quality (Tofail et al. 2018). Consumers will necessarily be interested in metaverse-based brand experiences (De Wulf et al. 2001).

Researchers and practitioners should recognize that the metaverse is still an evolving environment, even though some individuals are heralding it as the next big breakthrough. Researchers and practitioners should explore the key elements of brand creation, management, experience, and engagement. It influence brand-related tactics rather focusing on the context features and tactics.

3 Opportunities

The numerous features and technologies of the metaverse, would enables companies to provide offers that would be unachievable in the actual world. Marketers have the ability to be highly imaginative and inventive in the virtual world. This allows them to offer products that are distinct and transcend the limits of reality. The virtual essence of the metaverse allows brands to engage with a wide variety of consumers at a more profound level of immersion. Ads within the metaverse can be extremely interactive. The content and features of these advertisements have the potential to transcend reality, by providing users with an ethereal experience. Consumers are able to effectively utilise products as a result of the high level of immersion and interactivity. It will assist brands in establishing more robust purchase patterns. Marketers are also afforded numerous opportunities to advertise their products in the metaverse. Since the inception, brands have implemented digital billboards to improve their visibility in the metaverse. This trend is expected to persist. Brands like Adidas virtual garments and Lamborghini artworks have also introduced NFT collections, to monetize their metaverse presence and generate extra revenue streams.

Marketers also need to pay close attention to brand communities in the metaverse. Brand loyalty is essential for sustained success over the long term. The metaverse allows brand community members to interact more effectively than through online blogs or chat rooms. Furthermore, events and competitions can be organised in the same way that they are in real life. It gives marketers new ways to connect with their customers. The metaverse's strategic tools also make it easier to analyse and evaluate the effectiveness of brand campaigns. It offers insights for future improvements. The metaverse is mostly used by Generation Alpha, Generation Z, and Generation Y. Therefore, the metaverse is extremely efficient for marketers who aim to reach these specific groups of users. The metaverse provides customers a unique options and experiences to interact with the brand. The metaverse allows customers to connect and speak with brands and other consumers. The metaverse is exceptionally interactive and possesses immersive characteristics that allow customers to view shopping as an exciting adventure, offering a hyper-personalized experience from the comfort of their own homes (Jim et al. 2023). Consumers will purchase a new range of virtual items in the metaverse, such as avatar clothing that match their real-world wardrobe (Breiter and Siegfried 2022). Numerous improvements are occurring in the metaverse to improve customers' virtual experiences. Furthermore, the metaverse will attract consumer segments that are influenced by the diffusion of innovation theory. As the virtual world increase, companies engage with their marketing tactics which is beneficial both brands and customers.

Challenges

In addition to the profusion of new marketing opportunities, the metaverse presents a variety of new challenges for markets. As the metaverse continues to develop, brands and platform companies must tackle a variety of technical and infrastructural challenges. The development of powerful hardware and software has a tremendous impact on the metaverse. The lack of cost and accessibility for VR or AR headsets and other devices could hinder companies. Metaverse influence user engagement and experience. The presence of ambiguity in consumer interactions and conduct inside the metaverse gives rise to socio-cultural concerns. Moreover, brands may face strategic challenges, such as choice, engage in a certain metaverse.

The decision-making process is more complex by the presence of several metaverses, each having distinct characteristics with different levels of immersion and interface quality. According to (Koohang et al. 2023a, b), marketers should initially assess the consumer viewpoint and ascertain whether customers have personal resources such as technology, talents, skills, and a wide range of interests before considering the metaverse platform. Brands have a hurdle since consumers have varying personal talents in different socio-cultural environments (Banu et al. 2021).

User preferences and behaviors are expected to impact the media reliance, applications, and satisfactions of brands as they shift to a new platform. Brand assets, services, and content are key facilitators of the metaverse. Marketers need to comprehend consumers' preferences and behaviors regarding content consumption in the metaverse (Rane 2023). Companies have to engage customers in a range of brand activities, such as virtual events and competitions.

The effectiveness of brands in the metaverse will be influenced by their brand communication and experiential marketing strategies. Gucci has previously engaged with the metaverse. Gucci Garden was a unique event held on the Roblox metaverse in 2021, which allow people to explore a virtual garden and experience a range of digital items. Specifically designed for the metaverse, brands like Nikeland, Samsung 837x, and Wendy's Food Fight game on Roblox have hosted similar events and introduced showrooms or games.

Marketers also need to pay attention to certain issues related to communication and branding. In the metaverse, branding has to match the brand mindset and the platform, even though brands have a name and a character in the real world.

Another significant challenge for marketers is the establishment of brand recognition. It will be challenging for brands to operate in both the physical realm and the virtual 3D universe. Within the metaverse, traditional media techniques cannot provide the same results. It require marketers to strategically design and adapt their marketing mixes to better connect with both domains. Marketers will have a difficulty in implementing a complete and all-encompassing strategy. As the metaverse progresses, it is crucial for companies to uphold ethical standards and adhere to regulatory laws about behaviour. Ultimately, the metaverse's business environment, interoperability, and scalability are all characterised by a lack of certainty. The advancements in the metaverse require marketers and organisations to exercise prudence and be vigilant.

Implication

The metaverse offers a blend of consumer-to-consumer social media interactions and business-to-consumer pleasurable virtual experiences. The metaverse has the ability to engage customers in real-time interactions through integrating virtual surroundings and simplifying access via particular technology. (Zhao et al. 2022). Nevertheless, virtual environments have been present in a variety of forms for decades. For instance, Second Life was launched in 2003 as a 3D social platform that offered the advantages of both virtual activities and social networking. Previous study has showed that virtual world experiential promotion (VWEP) provides greater degrees of atmospherics, social experience, and adaptive advertising in contrast to retail and e-commerce environments. According to Hoch and Deighton (1989) Consumers and marketers can experience beneficial as well as adverse outcomes as a result of this expanded exposure.

Marketers can improve consumer experiences by implementing augmented reality for consumer journey designs and improving narrative transportation (Sung et al. 2022). On the other hand, marketers need to address the possible negative impacts of the metaverse, including information overload, fear of missing out, data privacy and security concerns, as well as trust issues in technology-driven buyer–seller transactions. According Royo-Vela et al. (2021) Marketers are required to establish value co-creation processes in order to manage and support innovation within their functional units. This may be accomplished by applying a design thinking approach that takes into consideration the multidimensional vulnerabilities of consumer.

References

Achrol RS, Kotler P (1999) Marketing in the network economy. J Mark 63(4_suppl1):146–163

Ahn S, Jin BE, Seo H (2024) Why do people interact and buy in the metaverse? Self-expansion perspectives and the impact of hedonic adaptation. J Bus Res 175:114557

Alfaisal R, Hashim H, Azizan UH (2024) Metaverse system adoption in education: a systematic literature review. J Comput Educ 11(1):259–303

Allam Z, Sharifi A, Bibri SE, Jones DS, Krogstie J (2022) The metaverse as a virtual form of smart cities: opportunities and challenges for environmental, economic, and social sustainability in urban futures. Smart Cities 5(3):771–801

Amason A, Ward A (2020) Strategic management: from theory to practice. Routledge

Aseeri S, Interrante V (2021) The influence of avatar representation on interpersonal communication in virtual social environments. IEEE Trans Visual Comput Graphics 27(5):2608–2617

Ball M (2022) The metaverse: and how it will revolutionize everything. Liveright Publishing

Banu J, Baral R, Agarwal UA, Rastogi M (2021) Role of socio-cultural factors in shaping entrepreneurial decision and behavior: an Indian perspective. In: Women's entrepreneurship and culture. Edward Elgar Publishing, pp 140–172
Barrera KG, Shah D (2023) Marketing in the metaverse: conceptual understanding, framework, and research agenda. J Bus Res 155:113420
Bhasin C, Dilawari H, Yuvaa PK, Yadav SSK (2024) The metaverse: an exploration from the consumer consumption and economic viability perspectives. In: New trends in marketing and consumer science. IGI Global, pp 404–418
Bibri SE, Allam Z (2022) The metaverse as a virtual form of data-driven smart cities: the ethics of the hyper-connectivity, datafication, algorithmization, and platformization of urban society. Comput Urban Sci 2(1):22
Bilgihan A, Leong AMW, Okumus F, Bai J (2024) Proposing a metaverse engagement model for brand development. J Retail Consum Serv 78:103781
Bratu S, Sabău RI (2022) Digital commerce in the immersive metaverse environment: cognitive analytics management, real-time purchasing data, and seamless connected shopping experiences. Linguist Philos Investig 21:170–186
Breiter D, Siegfried P (2022) The metaverse: exploring consumer's expectations, their attitudes, and it's meaning to the fashion industry. eISSN 2683-5665
Buhalis D, Leung D, Lin M (2023) Metaverse as a disruptive technology revolutionising tourism management and marketing. Tour Manage 97:104724
Burrows G (2022) Your life in the metaverse: everything you need to know about the virtual internet of tomorrow. Gideon Burrows
Carleton WC, Collard M (2020) Recent major themes and research areas in the study of human-environment interaction in prehistory. Environ Archaeol 25(1):114–130
Catherine S, Kiruthiga V, Suresh NV, Gabriel R (2024) Effective brand building in metaverse platform: consumer-based brand equity in a virtual world (CBBE). In: Omnichannel approach to co-creating customer experiences through metaverse platforms. IGI Global, pp 39–48
Chen C, Zhang KZ, Chu Z, Lee M (2024) Augmented reality in the metaverse market: the role of multimodal sensory interaction. Internet Res 34(1):9–38
Colombo, D. (2021). unveiling the metaverse: paving the way for technological providers.
Cui N, Wang T, Xu S (2010) The influence of social presence on consumers' perceptions of the interactivity of web sites. J Interact Advert 11(1):36–49
Cummings JJ, Bailenson JN (2015) How immersive is enough? A meta-analysis of the effect of immersive technology on user presence. Media Psychol 19(2):272–309. https://doi.org/10.1080/15213269.2015.1015740
De Wulf K, Odekerken-Schröder G, Iacobucci D (2001) Investments in consumer relationships: a cross-country and cross-industry exploration. J Mark 65(4):33–50
Dincelli E, Yayla A (2022) Immersive virtual reality in the age of the metaverse: a hybrid-narrative review based on the technology affordance perspective. J Strateg Inf Syst 31(2):101717
Dulabh M, Vazquez D, Ryding D, Casson A, Caratù M (2023) The impact of virtual interactivity on shopper behaviour. In: The garment economy: understanding history, developing business models, and leveraging digital technologies. Springer International Publishing, Cham, pp 443–474
Dwivedi YK, Hughes L, Baabdullah AM, Ribeiro-Navarrete S, Giannakis M, Al-Debei MM, Wamba SF (2022) Metaverse beyond the hype: multidisciplinary perspectives on emerging challenges, opportunities, and agenda for research, practice and policy. Int J Inf Manag 66:102542
Eggenschwiler M, Linzmajer M, Roggeveen AL, Rudolph T (2024) Retailing in the metaverse: a framework of managerial considerations for success. J Retail Consum Serv 79:103791
Elsharnouby MH, Jayawardhena C, Liu H, Elbedweihy AM (2022) Strengthening consumer–brand relationships through avatars. J Res Interact Mark 17(4):581–601
Far SB, Rad AI, Bamakan SMH, Asaar MR (2023) Toward Metaverse of everything: opportunities, challenges, and future directions of the next generation of visual/virtual communications. J Netw Comput Appl 217:103675

Gadekallu TR, Wang W, Yenduri G, Ranaweera P, Pham QV, da Costa DB, Liyanage M (2023) Blockchain for the metaverse: a review. Futur Gener Comput Syst 143:401–419
Ghali Z, Rather RA, Khan I (2024) Investigating metaverse marketing-enabled consumers' social presence, attachment, engagement and (re) visit intentions. J Retail Consum Serv 77:103671
Gómez J, Vargas P (2012) Intangible resources and technology adoption in manufacturing firms. Res Policy 41(9):1607–1619
Guan J (2022) Extending the metaverse: exploring generative objects with extended reality environments and adaptive context awareness (Doctoral dissertation, Thesis. OCAD University)
Gupta R, Nair K, Mishra M, Ibrahim B, Bhardwaj S (2024) Adoption and impacts of generative artificial intelligence: theoretical underpinnings and research agenda. Int J Inf Manag Data Insights 4(1):100232
Gursoy D, Lu L, Nunkoo R, Deng D (2023) Metaverse in services marketing: an overview and future research directions. Serv Ind J 43(15–16):1140–1172
Hackl C, Lueth D, Di Bartolo T (2022) Navigating the metaverse: a guide to limitless possibilities in a Web 3.0 world. Wiley
Hadi R, Melumad S, Park ES (2024) The metaverse: a new digital frontier for consumer behavior. J Consum Psychol 34(1):142–166
Halkiopoulos C, Antonopoulou H, Gkintoni E, Aroutzidis A (2022) Neuromarketing as an indicator of cognitive consumer behavior in decision-making process of tourism destination—an overview. In: Transcending borders in tourism through innovation and cultural heritage: 8th international conference, IACuDiT, Hydra, Greece, 2021. Springer International Publishing, Cham, pp 679–697
Han DID, Bergs Y, Moorhouse N (2022) Virtual reality consumer experience escapes: preparing for the metaverse. Virtual Reality 26(4):1443–1458
Hennig-Thurau T, Aliman DN, Herting AM, Cziehso GP, Linder M, Kübler RV (2023) Social interactions in the metaverse: Framework, initial evidence, and research roadmap. J Acad Mark Sci 51(4):889–913
Hoch SJ, Deighton J (1989) Managing what consumers learn from experience. J Mark 53(2):1–20
Hoyer WD, Kroschke M, Schmitt B, Kraume K, Shankar V (2020) Transforming the customer experience through new technologies. J Interact Mark 51(1):57–71
Huang TL, Chung HF (2024) Impact of delightful somatosensory augmented reality experience on online consumer stickiness intention. J Res Interact Mark 18(1):6–30
Huynh-The T, Pham QV, Pham XQ, Nguyen TT, Han Z, Kim DS (2023) Artificial intelligence for the metaverse: a survey. Eng Appl Artif Intell 117:105581
Jim JR, Hosain MT, Mridha MF, Kabir MM, Shin J (2023) Towards trustworthy metaverse: advancements and challenges. IEEE Access
Jones TO, Sasser WE (1995) Why satisfied customers defect. Harv Bus Rev 73(6):88
Julian HLC, Chung T, Wang Y (2023) Adoption of metaverse in South East Asia: Vietnam, Indonesia, Malaysia. In: Strategies and opportunities for technology in the metaverse world. IGI Global, pp 196–234
Jungherr A, Schlarb DB (2022) The extended reach of game engine companies: how companies like epic games and Unity technologies provide platforms for extended reality applications and the metaverse. Social Media+ Soc 8(2):20563051221107641
Khan SW, Raza SH, Zaman U (2022) Remodeling digital marketplace through metaverse: a multi-path model of consumer neuroticism, parasocial relationships, social media influencers credibility, and ppenness to Metaverse experience. Pak J Commer Soc Sci (PJCSS) 16(3):337–365
Kim JH, Kim M, Park M, Yoo J (2023) Immersive interactive technologies and virtual shopping experiences: differences in consumer perceptions between augmented reality (AR) and virtual reality (VR). Telematics Inform 77:101936
Koohang A, Nord JH, Ooi KB, Tan GWH, Al-Emran M, Aw ECX, Wong LW (2023) Shaping the metaverse into reality: a holistic multidisciplinary understanding of opportunities, challenges, and avenues for future investigation. J Comput Inf Syst 63(3):735–765

Koohang A, Nord J, Ooi K, Tan G, Al-Emran M, Aw E, Wong L (2023) Shaping the metaverse into reality: multidisciplinary perspectives on opportunities, challenges, and future research. J Comput Inf Syst 63(3):735–765

Kumar S (2023) Sustainable agriculture development in india: emerging issues, challenges and opportunities. Asian J Adv Agric Res 23(3):64–74

Kumar J, Arora M, Erkol Bayram G (eds) (2024) Exploring the use of metaverse in business and education. IGI Global

Kye B, Han N, Kim E, Park Y, Jo S (2021) Educational applications of metaverse: possibilities and limitations. J Educ Eval Health Prof 18

Lee LH, Braud T, Zhou P, Wang L, Xu D, Lin Z, Hui P (2021) All one needs to know about metaverse: a complete survey on technological singularity, virtual ecosystem, and research agenda. arXiv preprint arXiv:2110.05352

Li J, Cai S, Yang Q, Huang H (2023) How to enrich metaverse? Blockchains, AI, and Digital Twin. In: From blockchain to web3 & metaverse. Springer Nature Singapore, Singapore, pp 27–61

Lupu I, Rokka J (2022) "Feeling in control": optimal busyness and the temporality of organizational controls. Organ Sci 33(4):1396–1422

May T, Perry B (2022) Social research: issues, methods and process. McGraw-Hill Education (UK)

Nagarajan G, Moorthy VS, Mohamed AK, Mohideen AS, Ishaq MM, Lakshmi MR (2023) The role of the metaverse in digital marketing. JETT 14(5):51–59

Nleya S, Velempini M (2024) Industrial metaverse: a comprehensive review, environmental impact and challenges

Oliveira PM, Guerreiro J, Rita P (2022) Neuroscience research in consumer behavior: a review and future research agenda. Int J Consum Stud 46(5):2041–2067

Ooi KB, Tan GWH, Al-Emran M, Al-Sharafi MA, Arpaci I, Zaidan AA, Iranmanesh M (2023) The metaverse in engineering management: overview, opportunities, challenges, and future research agenda. IEEE Trans Eng Manag

Özdemir Uçgun G, Şahin SZ (2023) How does Metaverse affect the tourism industry? Current practices and future forecasts. Curr Issues Tour 1–15

Panagiotakopoulos D, Marentakis G, Metzitakos R, Deliyannis I, Dedes F (2022) Digital scent technology: toward the internet of senses and the metaverse. It Professional 24(3):52–59

Park SM, Kim YG (2022) A metaverse: taxonomy, components, applications, and open challenges. IEEE Access 10:4209–4251

Petit O, Velasco C, Wang QJ, Spence C (2022) Consumer consciousness in multisensory extended reality. Front Psychol 13:851753

Pieters D (2022) Metaverse, new hype or second life fatigue. In: A context analysis of the rise of the metaverse juxtaposed to second life

Rane N (2023) Metaverse marketing strategies: enhancing customer experience and analysing consumer behaviour through leading-edge metaverse technologies, platforms, and models. Available at SSRN

Rostami S (2022) Experimental investigation of advanced metaverse realms of experience (Doctoral dissertation, Institut National de la Recherche Scientifique (Canada))

Royo-Vela M, Velasquez Serrano M (2021) Value co-creation process and measurement in 4.0 SMEs: an exploratory research in a B2B marketing innovation context. Adm Sci 11(1):20

Saffari F, Kakaria S, Bigné E, Bruni LE, Zarei S, Ramsøy TZ (2023) Motivation in the metaverse: a dual-process approach to consumer choices in a virtual reality supermarket. Front Neurosci 17:1062980

Sağkaya Güngör A, Ozansoy Çadırcı T (2022) Understanding digital consumer: A review, synthesis, and future research agenda. Int J Consum Stud 46(5):1829–1858

Seidel S, Berente N, Nickerson J, Yepes G (2022) Designing the metaverse

Shen B, Tan W, Guo J, Zhao L, Qin P (2021) How to promote user purchase in metaverse? A systematic literature review on consumer behavior research and virtual commerce application design. Appl Sci 11(23):11087

Slater M, Wilbur S (1997) A framework for immersive virtual environments (FIVE): speculations on the role of presence in virtual environments. Presence Teleoperators Virtual Environ 6(6):603–616. https://doi.org/10.1162/PRES.1997.6.6.603
Sritong C, Sawangproh W, Teangsompong T (2024) Unveiling the adoption of metaverse technology in Bangkok metropolitan areas: a UTAUT2 perspective with social media marketing and consumer engagement. PLoS ONE 19(6):e0304496
Sung E, Han DID, Choi YK (2022) Augmented reality advertising via a mobile app. Psychol Mark 39(3):543–558
Teng CI, Dennis AR, Dennis AS (2023) Avatar-mediated communication and social identification. J Manag Inf Syst 40(4):1171–1201
Tofail SA, Koumoulos EP, Bandyopadhyay A, Bose S, O'Donoghue L, Charitidis C (2018) Additive manufacturing: scientific and technological challenges, market uptake and opportunities. Mater Today 21(1):22–37
Turchet L (2023) Musical metaverse: vision, opportunities, and challenges. Pers Ubiquit Comput 27(5):1811–1827
Tversky B, Zacks JM, Hard BM (2008) The structure of experience. In: Understanding events: from perception to action, pp 436–464
Vincent A, Frewen P (2023) Being where, with whom, and when it happens: spatial, interpersonal, and temporal presence while viewing live streaming of collegiate sports in virtual reality. Front Virtual Rity 4:1167051
Voinea GD, Gîrbacia F, Postelnicu CC, Duguleana M, Antonya C, Soica A, Stănescu RC (2022) Study of social presence while interacting in metaverse with an augmented avatar during autonomous driving. Appl Sci 12(22):11804
Wang Y, Su Z, Zhang N, Xing R, Liu D, Luan TH, Shen X (2022) A survey on metaverse: fundamentals, security, and privacy. IEEE Commun Surv Tutor 25(1):319–352
Wang H, Ning H, Lin Y, Wang W, Dhelim S, Farha F, Daneshmand M (2023) A survey on the metaverse: the state-of-the-art, technologies, applications, and challenges. IEEE Internet Things J 10(16):14671–14688
Weniger A, Del Rosario P, Backes JG, Traverso M (2023) Consumer behavior and sustainability in the construction industry—relevance of sustainability-related criteria in purchasing decision. Buildings 13(3):638
Wider W, Jiang L, Lin J, Fauzi MA, Li J, Chan CK (2023) Metaverse chronicles: a bibliometric analysis of its evolving landscape. Int J Hum Comput Interact 1–14
Wohlgenannt I, Simons A, Stieglitz S (2020) Virtual reality. Bus Inf Syst Eng 62:455–461
Xi N, Chen J, Gama F, Korkeila H, Hamari J (2024) Acceptance of the metaverse: a laboratory experiment on augmented and virtual reality shopping. Internet Res 34(7):82–117
Yadav J, Yadav R, Sahore N, Mendiratta A (2023) Digital social engagements and knowledge sharing among sports fans: role of interaction, identification, and interface. Technol Forecast Soc Chang 195:122792
Yoo K, Welden R, Hewett K, Haenlein M (2023) The merchants of meta: a research agenda to understand the future of retailing in the metaverse. J Retail 99(2):173–192
Zhao Y, Jiang J, Chen Y, Liu R, Yang Y, Xue X, Chen S (2022) Metaverse: perspectives from graphics, interactions and visualization. Vis Inform 6(1):56–67

Creating Resilient and Sustainable Businesses: New Ideas for Reducing Digital Carbon Footprints

Pooja Kaushik and Kuldeep Singh Kaswan

Abstract The rapid expansion of digital technologies, while driving innovation and connectivity, has also escalated the global carbon footprint. This chapter explores creative solutions aimed at mitigating the environmental impact of digital activities, emphasizing sustainable practices within the tech industry. We delve into the multifaceted dimensions of digital carbon emissions, identifying key contributors such as data centers, streaming services and electronic waste. Innovative strategies including the adoption of energy-efficient algorithms, the utilization of renewable energy sources and the implementation of circular economy principles are some of the practical approaches to reduce carbon footprint. Furthermore, the chapter underscores the importance of fostering a culture of sustainability through education, policy advocacy and collaborative efforts among stakeholders. Case studies of successful initiatives demonstrate the potential for significant environmental benefits when creativity and technology intersect. By providing a comprehensive overview of the current landscape and future possibilities, this chapter aims to inspire stakeholders to prioritize and invest in sustainable digital practices, thereby contributing to a greener and more resilient digital future.

Keywords Carbon footprint · Sustainability · Digital · Carbon emission · Creative solution

1 Introduction

The digitalization of commercial organizations around the globe has called for a cocktail both opportunities and threats specifically on environmental responsibility. The data centers or utility servers, cloud services, dissemination of connected devices

P. Kaushik (✉)
Galgotias College of Engineering and Technology, Greater Noida, India
e-mail: poks3009@gmail.com

K. S. Kaswan
Galgotias University, Greater Noida, India

W. Leal Filho et al. (eds.), *Metaverse and Sustainability: Business Resilience Towards Sustainable Development Goals*, World Sustainability Series,
https://doi.org/10.1007/978-3-031-89545-6_15

in smart economy has lead to a massive consumption of energy and possess high level of carbon emissions which are called as the 'Digital Carbon Footprint'. This footprint measures the overall emission of green house gases that accrue from the various digital undertakings such as data storage, manipulation and transfer (Masanet et al. 2020). While the Internet and digital technologies have become critical to business functioning across the globe, there is discussion of the growing environmental effects of companies' digital activities. Cutting down ones digital carbon footprints is now setting pace as frameworks that organizations would embrace in trying to build the much needed resistance and sustainability (Shehabi et al. 2016). There are more reasons to address digital carbon emissions as a matter of the highest priority now, because data centers consume a lot of energy for their work, data according to the International Energy Agency (IEA 2021) on the use of electricity by data centers in 2020 indicates that it was approximately one percent of the total electric power worldwide. Continued with this trend, it is expected to rise sharply and become the next major threat to sustainable development of the entire world (Belkhir and Elmeligi 2018). Furthermore, since data traffic is projected to increase its growth rate steadily; initiatives to minimize the carbon impact of ICT innovations are crucial for compliance with the regulatory, market, and corporate sustainability standards (IRENA 2020). This is especially important in the present context where awareness of climate change and the call for companies to submit themselves to the cause of reducing Green house gas emissions (Google Sustainability 2021). In order to overcome these challenges, different techniques are being developed by the firms. Such are energy efficiency within digital processes, green IT, use of renewable power, and effective digital ecosystem (Murugesan 2008; Capra et al. 2019). Moreover, integration of such modern technologies as artificial intelligence (AI) and machine learning (ML) in predictive maintenance and energy consumption can provide the further opportunities to decrease digital carbon footprint (Evans et al. 2021). In addition, the organizations' culture that practices sustainability throughout the employees are also important in terms of sustainability (Loorbach et al. 2020). These strategies remain very relevant not just in cutting emissions but in building the capacity of businesses to be ready for the shocks posed by environmental or even regulatory forces.

Growing Concern about Digital Carbon Footprint

The advancement of technology and the massive spread of the internet has made it possible to talk about a new problem, namely the digital carbon footprint. This term is used to describe the effect of the internet or any human activity that uses technology, especially, the greenhouse gases which are emitted due to the use of energy by computers, data centers and networks. Internet traffic globally is predicted to account for two percent of the total global energy consumption, a proportion that is projected to rise due to the increased usage of cloud computing, video streaming and artificial intelligence applications (Andrae 2020). The backbone of our digital lives- the large data centers have emerged as some of the largest emitters of carbon. While attempts to boost the energy efficiency have had some impact, the need for data storage as well as processing is growing even more rapidly. A single data center can utilize as much power as a medium sized city (Masanet et al. 2020). Moreover, with the advances in the smart device industry and IoT, this problem will be further aggravated because more devices result in more data produced and stored. Further, ceaseless scrolling and binging on TV shows or movies or being constantly online through social media prints a good amount of carbon on the climate. For example, one hour of streaming video excludes the Netflix sends up to 55 g CO_2 which is equivalent to emitting 300 m in automobile (Shift Project 2019). So, solving this problem we need the introduction of new sources of energy, the improvement of energy consumption in data centers, and increasing people's awareness of sustainable ICT. Since there is a fast-growing use of digital technologies, the carbon footprint more than triples, which demands some invention in terms of decreasing the digital carbon footprint. The utilization of digital innovations and services continues expanding, so does the energy consumption, primarily served from fossil sources that spurge greenhouse gases. The so far practice of optimizing energy consumption rates in data centers and equipment also remains a critical yet insufficient fundamental remedy. System innovation is required mainly in relation to digital activities, individuals' digital behaviour, use of energy from renewable sources and creation of sustainable digital structures. On the same note, creativity can also be applied in new strategies of doing business through coming up with new business models, recycling of symbols of technology, for instance, mobile phones and tablets among others; or in integration of new technologies like artificial intelligence in the management of energy consumption. It is essential to step into these creative solutions as the first way to start achieving substantive and sustainable changes in reducing the negative influence of the digital world on the environment.

Objectives of the Study

- Review the current status of digital carbon footprint to understand the major areas that are having high emission and energy intensity.
- To explore innovative ideas concerning the reduction of overall digital carbon footprint comprising the possibilities of applying green technologies as well as practicing energy-efficient solutions and utilizing environmentally sustainable digital platforms.
- How reduction in digital carbon footprint in organisations can minimize operational vulnerabilities and conform to the guidelines that have been set in relation to emissions, all these because of improving their business reputations.
- Mobilise people towards the change in sustainable measures by stressing on the adjusted economical and environmental impacts of low digital carbon footprint.

2 Literature Review

The idea of developing or strengthening the longevity of businesses has become topical in recent years because of the issues with environmental sensitivity, especially computational impact. The digitization of organisations has brought many advantages but comes with the disadvantage of new carbon sources namely data centres, cloud, and devices. It has focuses on the emissions resulting from digital processes like data storage, communicating and using data, popularly known as digital carbon footprint. This footprint must be minimized in a bid to achieve sustainable solutions, and thus many organizations are seeking to implement novel approaches to minimize these effects. This literature review distils current literature on how digital carbon footprints can be minimised and reduced and its focus shall be energy conservation, green IT approaches, use of renewable energy, optimisation of this digital footprint and inculcation of sustainable practices within organisations. Sustainability is hence a critical approach used to curb the digital carbon footprint. Most of the digital activities that involve data processing and storage require energy, and this energy is usually sourced, and as we all know, energy production especially from thermal sources is a major cause of carbon emissions. The data derived from the IEA suggests that the data centers consumed around one per cent of the total electricity consumption at a global level in 2020. Emissions can be reduced when energy efficiency is improved by enhanced hardware, more effective cooling, and optimized software. For instance, replacing servers and storage devices with newer and more energy efficient ones can achieve energy reduction of up to 40% (Shehabi et al. 2016). Also, energy consumption reduction during data processing and transmission can be achieved through the use of efficient algorithms. In their study titled 'A deep reinforcement learning-based dynamic model-free control for energy saving in data centres', (Xu et al. 2016) explained that organizations' computing systems should incorporate machine learning algorithms with energy-conserving objectives that do not affect performance, hence achieving optimised computing. Some of the notable practices in green computing include the creation, implementation and operation of digital systems with minimal impact to the environment. Murugesan (2008) has described green computing as a concept where computing, comprising hardware, software, and pointed subsystems are fabricated, deployed and discarded without adverse effects on the physical environment. This approach involves avoidance of use of dangerous materials, conserving energy and ensuring that the product waste products or any other factory by-product is reusable or biodegradable. Recent research focuses on green software engineering which is mentioned to be the software engineering that aims at producing software which is less resource hungry so that it will not consume a lot of power. Code optimization can decrease energy consumption by as much as 20% which in turn decreases emissions related to software use (Capra et al. 2019). In addition, virtualization and cloud computing have been considered as the two major practices that define GC because they help to optimize the use of physical resources in computing systems by packing several applications on a single server which in turn reduces the cost of hardware and energy consumption (Belkhir and Elmeligi 2018). Another strategic area that has been emerged as an important factor for those organizations which want to achieve least digital carbon footprint is the shift to renewable energy. Previously, different components of data centers and other structures that facilitate digital technology were based on the use of fossil fuels, which are closely connected to carbon emissions. However, with the utilization of renewable energy like the solar, wind and hydroelectric power the emission of the same can greatly be minimized. For instance, in its sustainability report, Google and Alphabet's disclosed that all their data centers and other operations around the world used only renewable energy since 2017 (Google Sustainability 2021). Such a shift is enabled by advancements in renewable energy technologies and reduction of costs that make these technologies available to various companies (IRENA 2020). Besides, companies may buy RECs or engage in PPAs with green power suppliers as a way of reducing their carbon footprint. According to Masanet et al. (2020), that could bring down the carbon emission intensity of the global data center industry by between 15 and 25%. Artificial Intelligence (AI) and Machine Learning (ML) applied sciences have become favourable techniques to improve energy consumption and decreases carbon emission rates. On similar grounds, AI can assist in detecting and foreseeing trends in the consumption of energy within a business hence facilitating their enhancement of their energy management. In a study by Strubell et al. (2019), it was explained how AI algorithms can help in cutting down energy consumption in data centers through server utilization, cooling system among other infrastructure components. Also, another AI application for predicting maintenance can detect when there is a problem with digital networks and decide how to fix it before its energy consumption rises. One of the subfields within the ML framework is the reinforcement learning that may be instrumental in making data centers more efficient by dynamically tuning parameters to conserve power while maximizing the amount of raw computing performed (Evans et al. 2021).

Reducing Digital Carbon Footprint—A way towards Sustainability

Lowering digital carbon footprints is automatically connected to sustainability because it implies minimizing the negative effects of digital processes that form the basis of today's companies. Data are considered active digital items that occupy space, require power and consume energy in storage, processing, and transmission and are equally responsible for the augmentation of Greenhouse Gases emissions. This comes in line with the United Nations' Sustainable Development Goals of, specifically goal 7—affordable and clean energy and goal 13—Climate action, which call for efficiency and sustainable utilization of energy and reduced greenhouse emissions (United Nations 2015). Consequently, sustainable digital business practices are synergistic with global goals since they minimise the environmental footprint of companies while at the same time improving operational reliability in the long run.

Moreover, the sustainability aspect in digital processes improves a company's image as well as responds to the consumer and stakeholder expectations felt in the modern world (Loorbach et al. 2020). Organisations that aim to adopt lower levels of digital emissions, therefore stand to save more on costs as they seek to reduce energy usage and cut down on operating costs, besides preparing for new laws regulating emission levels in future (Masanet et al. 2020). Finally, positive sustainable digital practices promote creativity as well as new technology advancement and usage including green software, AI energy management, and renewable energy that is crucial for future business success and markets differentiation (Evans et al. 2021). Therefore, while concerns for shrinking digital carbon footprints are a responsibility for the environment; they are also about ensuring sustainable business models or growth that will be resilient to any shift in climate consciousness around the globe.

3 Scope of the Study

The chapter focuses towards discovering about other sustainable elements related to IT operations such as the energy efficient data centers, use of renewable energy to power the digital hardware, and decreasing energy utilization with efficient software and hardware. This paper will look into how the decrease of the e-carbon footprint can actively help the business become more sustainable. This entails looking at issues such as; the place of digital sustainability in the processes of managing risks and compliance with the provision of laws on the protection of the environment and business continuity during disruption events search for new technologies aiming to assess new technologies like the AI-based energy management system, blockchain for carbon recording and cloud solutions with less energy demand. It will also look at new organisational strategies that include sustainability and resilience of digital business models. The research is going to provide a global outlook to the topic, but at that same time, it will look at the regional factors such as regulation, the structure of energy market and so on. This will assist in determining regional specific approaches of minimising digital carbon imprint. This research will evaluate the applicability of the concept called digital carbon footprint reduction in the long-term perspective with respect to both the contribution to achieving the sustainability objectives, like carbon neutrality and organizational preparedness for future environmental and socioeconomic perturbations.

4 Sources of Digital Carbon Emissions

There is high awareness of the fact that the steep advancement of digital technologies has caused a sharp rise in carbon footprint. Most of these emissions are attributed to data centres, internet use, cloud computing, and blockchain applications.

Data Centers

A key fact that remains hidden from the general public is that data centers, which are the heart of the modern world, are huge polluters. They contain massive stockpiles of servers that work round-the-clock to store, process and transmit information. These servers need huge amounts of electricity to function and to cool them down as well. According to the IEA report on Global Energy Review, 2020, data centers accounts for approximately one percent of total electricity and generates over half of the world's information technology CO_2 emissions. It contributes 3% to the world's carbon emissions. And the use of the energy by these data centers is set to grow progressively with the growing need for digital services, forecasted to be twice as much by 2030.

Internet Usage

One of the ways by which our digital exposure contributes to carbon emissions therefore is through internet usage. The physical devices reaching to the global human audience through the internet, like cables, routers, and the other network equipment, do consume energy. Kurtz also incorporates the endpoint devices such as the mobiles, laptops, and tablets that use power on accessing the internet. An article in the journal Patterns in 2023 therefore suggested that the global internet footprint could generate up to 3 account for 7% of the total Green House Gas Emission by the year, 2025. Due to the energy intensity of internet services and the sharp growth of the volume of data transmitted worldwide, the adverse effects are deepened.

Cloud Services

Thus, although the cloud computing offers and provides companies with scalable and flexible IT resources it is also a source of carbon emissions. Cloud service providers administer large complex infrastructures for service delivery as storage, computing services, networking etc., on service basis. While, the inclusion of cloud services are normally more efficient in terms of energy usage when compared to conventional IT systems, their large-scale implementations, still generate a lot of energy. Over the course of the recent years, the increased use of cloud solutions has created additional pressure on data centers and has made them more emissions-heavy. Greenpeace in its report estimated that cloud computing by itself could be using as much as 20% of the global electricity by 2030.

Blockchain Technologies

Another large source of digital carbon emissions is blockchain—the underlying technology of cryptocurrencies such as Bitcoin. Minerals used in the block chain including bitcoins entail solving of complex mathematical problems that need a lot of electrical energy. According to the Cambridge Bitcoin Electricity Consumption Index, Bitcoin mining alone as of July 2020 consumes more energy as some countries like Argentina. This energy consumption can be easily converted to the amount of Carbon emissions, most especially when fossil energy is being used.

5 Impact of Digital Activities on Global Carbon Emissions

Internet use, data centers, cloud computing and blockchain technologies have topped the list of activities that are contributing to global carbon emissions. At the time of writing in 2023, digital sector is expected to contribute up to 3% globally as a sector they are equally responsible for 7–4% of global greenhouse gas emission as the aviation industry. This percentage is also believed to grow in the future as the process of digitalization does not slow down.

Data Centers

Facilities called data centers, where the servers and other essential equipment for digital activities are located, consume about one percent of the electric power and account for about 0. Today they are responsible for 3% of the total global carbon dioxide emissions. The further growth of the demand for data—fuelled by streaming, cloud, and artificial intelligence services—is expected to boost this effect even more.

Internet Usage

Owing to numerous activities that happen via the internet, and the use of various network devices, carbon emissions are clearly evident. It has been reported by Patterns that global internet usage which is at a continuous rise might be attributed to as much as 3. Green Touch to bring down the greenhouse gas emissions to 7% by the year 2025, in response to the rise of high-definition video streaming, online gaming and other data demanding services.

Blockchain Technologies

Cryptocurrency mining, more specifically, is another significant producer of digital carbons; that is, blocks. The CBECI revealed that only Bitcoin's mining process consumes more electricity per annum than some nations, and it raises greenhouse gas emissions mainly in areas using fossil energy (Fig. 1).

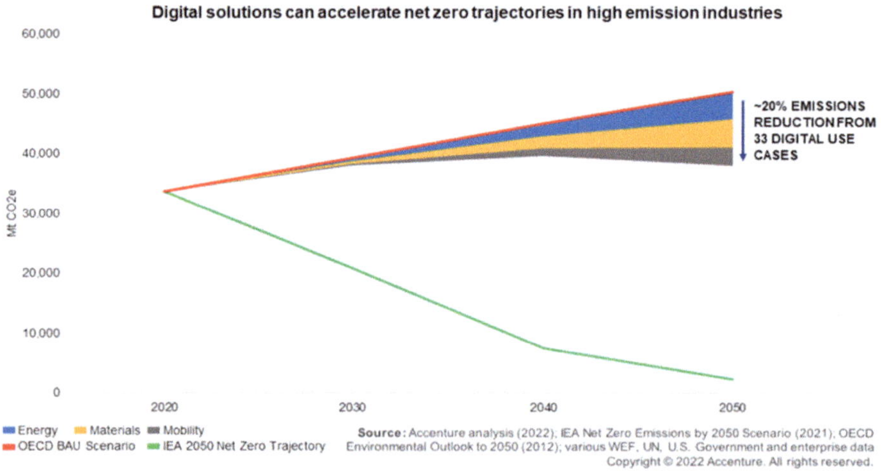

Fig. 1 Contribution of digital technology in carbon emission reduction when it comes to reaching net zero. *Source* World Economic Forum

6 Challenges in Reducing Digital Carbon Footprints

Reducing digital carbon intensity, an important subcategory of the overall fight against climate change, is a problem fraught with a multitude of issues at the technical, behavioural, economic, and policy levels.

Technical Barriers: The scaling of digital assets such as data centers, cloud solutions, and blockchain with associated requirements for energy and thus the emission of greenhouse gases. There is always a growth in the usage of internet services, computer services, and digital services in general, and the growth usually outperforms the overall energy efficiency achieved. As for now, it slowly becomes obvious that many data centers consume non-renewable energy sources and hence, have a great share in carbon emission (Aslan et al. 2018). Also, due to the nature of the complex, multiple and frequently concealed logistics systems of various digital services deep carbon footprints remain measurement and tracking challenges (Johnson 2021).

Behavioural Barriers: Consumer behaviour is another important component of the digital carbon footprint. Experiments showed a positive dependency of energy consumption from the number of video streaming, social networks and other online services. Nevertheless, the detectable rates of obtrusion as well as pollution are not typically realized by the users of the digital environment. Even where people are aware of the potential negative impact that comes with the use of specifically digital devices, convenience will most of the times offset the concerns for the environment thus making it rather difficult to observe any voluntary reduction on the usage of such devices (Hintemann and Hinterholzer 2019). Behavioral inertia and a belief in the irrelevance of contribution to the problem itself deepen it.

Economic Barriers: There are barriers to the cost of transition towards greener digital infrastructure. Large organization might be in a position to mobilize capital to implement renewable energy and improved technologies, while small organizations are unable to do that (GeSI 2020). However, financial motivations to stand for minimal e-impact costs are still low and Green Finance remains limited and slow to catalyze environment-friendly behaviours (World Economic Forum 2020). Economic competition creates pressure to achieve higher performance and higher profit, therefore, businesses do not want to spend money on more expensive but more environmental friendly technologies.

Policy Barriers: The nature is still rather weak when it comes to policies related to digital carbon footprints. Currently, there is no harmonized legislation or benchmarks for accounting for, and communicating the carbon footprint of digital services (Johnson 2021). Also, the international regulation is required since the digital infrastructure means may incorporate different countries and thus, the singular policy actions are not efficient. These are compounded by challenges that the government has to balance the drive to support digital innovation against risks on the environment resulting in regulation standards that may be inadequate in a bid to support the development of the digital innovations (GeSI 2020) (Fig. 2).

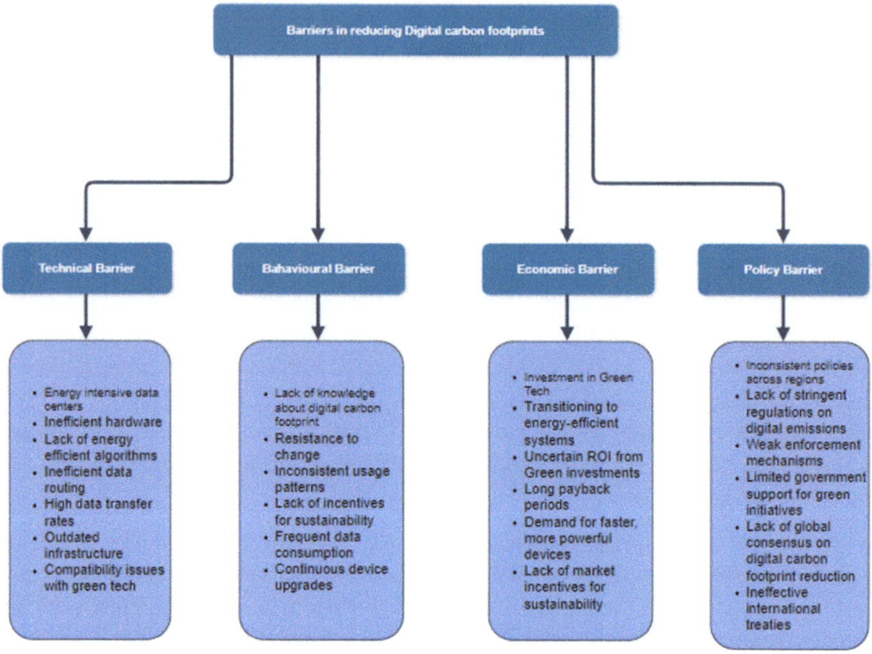

Fig. 2 Challenges in reducing digital carbon footprints. *Source* Author's Own

7 Creative Solutions for Reducing Digital Carbon Footprints

Carbon emissions in cyberspace need combating to meet the world's climate targets. It is only possible if the strategies adopted for such a purpose involve technology, policy, regulation, behavioural and cultural change.

Technological Innovations

Technology is here viewed as having striking potential to decrease digital emissions. Data center is one area, where energy efficiency is focused, and efficient data centers are being developed. Measures likes liquid cooling, dynamic voltage scaling and artificial intelligence optimizations, can decrease the energy density of data centers considerably (Greenpeace 2020). Currently, there is the plan and initiatives being taken by Google and Microsoft, amongst others, that are creating next-generation data centres that are powered by renewable energy, measures that significantly reduce the emission levels linked to business operations (Jones 2021). Decentralized cloud computing also offers a rather unique perspective. In other words, splitting computing load into several locations, closer to the sources and demands energy utilization can be minimized. A good example that is getting popular is edge computing where no huge data is required to be sent over long distances thus the energy needed to transport the data is eliminated (ITU 2019). Furthermore, it is crucial to optimise the software of a smart home. If developers make the software as simple as possible, then it is easy to reduce the energy that is used by digital services. Some of the strategies that may help to reduce the energy consumption but have no negative impact on the quality of the content delivered to users include use of code minification, good algorithm utilization and adaptive streaming (Hintemann and Hinterholzer 2019) (Fig. 3).

Policy and Regulatory Measures

Policy formulation and regulation are crucial since they are the only way through which systematic changes can be instigated. Carbon pricing is effective in triggering the minimization of digital carbon footprints. Assigning a value to CO_2 emissions gives the independent authority a tool to prod companies to use clean technologies and to develop environmentally friendly processes. However, for it to have the desired impact, carbon pricing has to be undertaken at the international level to avoid situation where the firms shift to areas where the rules are less stringent (GeSI 2020).

Another essential element is that regulations of energy efficiency concerning digital structures are also crucial. Some of the measures that governments may take to encourage optimal use of energy by these sectors include: The government can require efficiency standards to any data center and telecommunications network so that energy shall be used optimally throughout. For instance, the Energy Efficiency Directive in the European Union has provisions meant for the improvement of energy efficiency of digital services (European Commission 2020).

Another integral element of the regulation is connected with transparency and reporting. There should be calls for organisations to declare the carbon emissions of the digital products they offer, so that people buying those services can decide for themselves and so that investors can make informed decisions too. This can be further complemented by penalties for How companies that fail to observe the reduction

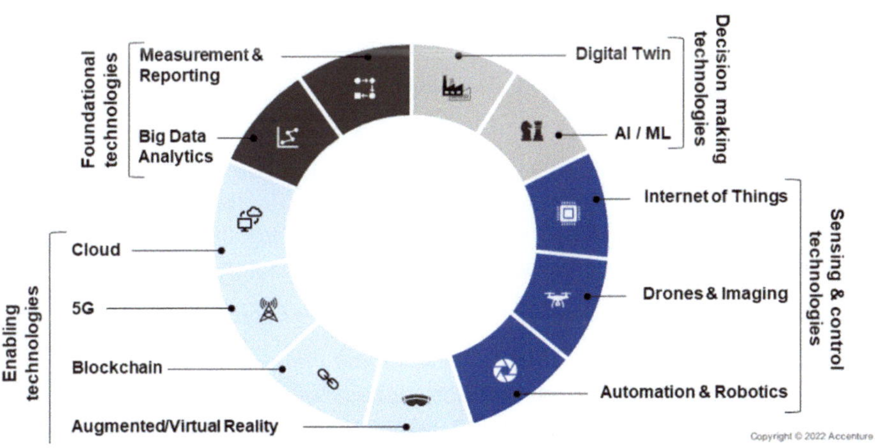

Fig. 3 Digital technologies that could drive decarbonisation in the energy, materials and mobility sectors. *Source* World Economic Forum

targets when it comes to their digital carbon footprint or bonuses for organizations that manage to achieve impressive goals, for instance, through a tax exemption or the right to participate in a public recognition program (Johnson 2021).

Behavioural and Cultural Changes

Behaviour and culture change is the best of the mentioned solutions on the effects of electronic activities on natural environment. Awareness programmes and advocacy can assist the user to appreciate the amount of carbon being accrued through regular usage practices as in streaming of videos, or the use of cloud storage and computing services among others (Aslan et al. 2018).

Another method that might be utilised is the encouragement of digital minimalism. In some cases, users should be encouraged to practice reductionism when it comes to their use of the available technology; do not store unnecessary data, do not stream at QHD when lower quality is enough, etc., this cuts down on consumption drastically. Organisations can assist these initiatives by architecting services that enable the users switch to more sustainable digital practices (Jones 2021). In addition, education for sustainability has to become embedded within the organisations' culture. Particularly, it is technology companies that can influence the pace of others' changes and can be an example themselves. In other words, through both sustainability of operations and sustainability in products and services, they can convince both the employees and consumers of its green digital initiatives. These range from making sustainable choices the norm and standard product configurations of server centres that are powered by renewable energy, for example, or cloud services that are carbon neutral at source (GeSI 2020). Digital sobriety whose goal was pioneered by the Shift Project is a way of using digital technologies more consciously, paying primary attention to the distinct activities that could be most crucial (The Shift Project 2019). The change in culture may be brought by people's behaviour and/or by change in the organisational culture that supports sustainable use of digital technologies.

8 Case Study and Example

Google can also be regarded as a rather good example of an organization that was successful in its efforts to minimize its digital carbon footprint. The technology company has deployed a broad policy aimed at reducing the imprint of digital processes on the natural environment and has achieved a lot in that direction.

Measures taken by Google to fight against Excessive Digital Carbon Emissions

Background: Google is one of the biggest digital companies on the planet with massive data centers that support its search engine, cloud, and more. Being a global company, Google understood the environmental implication of the company and has had to institute measures to cutting down on the amount of emissions.

Strategies Implemented:

Renewable Energy Investments: Google has been one of the dominant buyers of renewable energy. This year, Google meet 100% of the total electricity consumption it made around the world by buying renewable energy. This is because for each kilowatt-hour consumed by Google's data centres and offices, it has offset energy use with renewable energy like wind or solar power (Google 2021).

Energy-Efficient Data Centers: Currently, Google has created its data centers to be one of the best concerning energy efficiency. Some of developments that Google has incorporated include the use of machine learning algorithms in the cooling systems of these data centers as well as in-house servers that use less power than commercial ones In this regard, Google has halved the energy usage per floor area of its data centers compared to the industry norm (Evans 2018).

Carbon–Neutral Cloud: For its part, in October the Californian multinational announced that its operations in the cloud fell to zero emissions in the same year. The company was able to do this through the purchase of what was effectively carbon offsets for its cloud computing business and by further expanding its use of renewable energy. With this, Google Cloud customers can deploy their applications in the "greenest" way thus contributing to minimization of the total carbon emissions in the digital economy sector (Google Cloud 2020).

Circular Economy Practices: Another aspect that defines Google's business model is circular economy: The company invests huge efforts in the longevity of its servers. Servers and other equipment are recycled in the data centers and therefore causes minimal demand for new materials to be manufactured that in turn emits minimal carbon (Cook and Jardim 2021).

Commitment to Carbon-Free Energy: Beyond mere carbon neutrality, another more specific target that Google has put forward relates to resource of its data centers as well as offices to carbon-free energy all through the remaining part of each day of 2030. This goal is rather Rothschild, as it implies the use of different types of renewable energy sources as well as the effective work of energy storage systems, providing Google with clean energy at all scales of its functioning (Pichai 2020).

Outcomes: These efforts have seen Google attain carbon neutrality in its operation in all the offices around the globe while it is also working towards the reduction of future greenhouse gas emission. It has also placed the company at the head of sustainable practices for other organisations as a way of emulating the firm.

One of the ongoing projects which showed sustainable digital solutions is Ocean Protocol. Consisting of three components, this distributed marketplace is designed to provide a set of solutions for building a fairer and longer-lasting digital economy by providing a way to securely exchange data while preserving the identity of the transacting parties. Keeping this in mind, the goal of the project is to make data more available to people while at the same time making as less of a negative impact on the environment as possible using blockchain.

Example of a Project Focusing on Sustainable Digital Solutions

The Powers of the Ocean Protocol is PoA or rather Proof-of-Authority, a consensus mechanism which is less resource-demanding than PoW, the kind used by Bitcoin, for example. Being tied to a lower energy consumption of data transactions, Ocean Protocol brings a perspective of a lower digital carbon footprint (Ocean Protocol 2021).

In addition, Ocean Protocol encourages people and companies to share data responsibly, or rather, allows the masses to share data only on specified conditions, thus making data usage by companies to be socially responsible. It not only paves a way to incite innovation but also promotes sustainabity concepts as there will be less requirement to store same data and process it multiple times (Hartman et al. 2020). The project is especially important for industries such as healthcare, where the issues of data protection and green computing are especially important. It is quite inspiring to envisage Ocean Protocol's model as a look into the future of sustainable and responsible digital innovation.

9 Future Directions and Targets

Emerging Technologies

Novel technologies such as quantum computing, blockchain with environment friendly standards, artificial intelligence indeed have promise in bringing down the digital CO2 emissions. Therefore, improving general efficiency and the use of energy, as well as fostering the development of more environmentally friendly digital systems, these kinds of technologies can be used to significantly reduce the negative impact of the digital economy on the environment.

Quantum Computing

Quantum computing is a step up from most commonly used devices seen in today's world as it holds better capability in solving complex computing issues as compared to classical computers. Hailed for its capacity to perform intense computations with minimal energy, one of the environmental impacts ascribed to the use of quantum computing is efficiencies' creation. Quantum computers can perform calculations that would take a classical computing system many years and in a much more efficient way (Preskill 2018). This might lead to systems having significantly smaller carbon footprints than currently data centers that are among the giants in the circuit regarding consumption of electrical energy. Moreover, this type of computing could enhance renewable energy systems, as it could enhance the energy distribution and storage, which would contribute to the shift to a low carbon energy grid (Bova et al. 2021).

Blockchain with Green Protocols

Cryptographic enthusiasts have pointed towards the fact that despite being breakthrough in terms of technology, blockchain is comparably energy-greedy especially in setups that employ PoW system of consensus. But as we have seen, blockchain can also support and promote sustainability in adopting mechanisms such as PoS or PoA, which are much more energy-efficient. These protocols help in decreasing carbon emissions in the blockchain systems since they possibly remove the mining processes that require enormous amounts of energy (de Vries 2020).

Furthermore, in carbon trading markets, blockchain can help increase transparency and accelerate the processes associated with carbon credits making it easier to monitor carbon credits while helping the global efforts to reduce emissions. For instance, through the use of blockchain, production of decentralized energy nets can be accomplished which facilitates the direct trade of renewable energy from consumer to producer (Zheng et al. 2020) elaborated the potential use case of blockchain such as decentralized renewable energy grid. Successful implementation of environmentally friendly blockchain protocols can be achieved hence promoting for this technology's adaption in environmental sustainability.

Artificial Intelligence (AI)

AI has unlimited uses in improving efficiency and lowering energy use in many fields. Another area where AI can be implemented in the digital economy, to optimize the consumption of power, is through taking effective control on data center cooling, workloads to be accomplished, or via predictive maintenance necessary for avoiding inefficient utilization of energy (Xu et al. 2020). Machine learning is useful in the analysis of large sets of data able to point out resource optimization and result to low carbon footprint in digital procedures.

Smart grids and systems of renewable energy are also another aspect where AI has a critical role to play. This can be used in forecasting energy demand and supply changes, in improving the incorporation of renewable electricity sources and minimizing loss in energy transmission (Goh et al. 2021). Besides, supply chain management can also benefit from the use of AI to automate the distribution network by decreasing transportation pollution and waste while trying to contribute to the reduction of CO_2 emissions.

Synergistic Impact

In combination, such technologies possess synergistic characteristics through which they can enhance the work of each other in the process of achieving sustainability. For instance, the application of quantum computing improves the AI algorithms and results in better improvements of energy use. Likewise, AI can make improvements in the blockchain networks and, in this way, it can minimize the environmental cost (Andoni et al. 2019).

Future targets in IEA's Net Zero Scenario up to 2050

See Fig. 4.

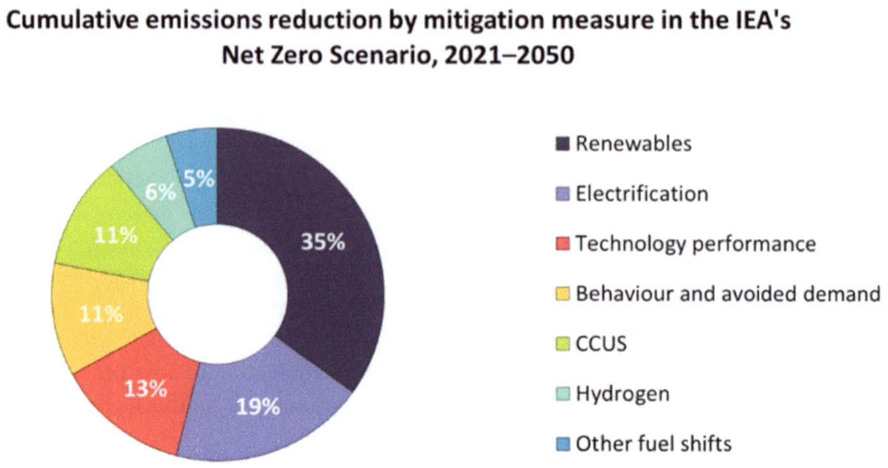

Fig. 4 Cumulative emissions reduction target in the IEA's Net Zero Scenario, 2021–2050. *Source* International Energy Agency

10 Need of Global Collaboration

The international cooperation is highly pertinent to the policy-making, technologies and knowledge at present and for the future, for coping with numerous global issues such as climate change, cybersecurity, and sustainable development. These challenges apply across the globe hence one requires to engage in multilateralism in order to come up with a round solution.

Policy-Making

Increased international cooperation within the policy-making domain is the best way to make the problems that require collective actions to be addressed in a coherent manner. When it comes down to changes in climate for instance, there is only one solution, which is when everyone has to act as one since one country's pollution affects the entire world. International cooperation is provided by such instruments as the Paris Agreement which unites countries aimed at achieving the minimum level of greenhouse gas emissions and exchanging their best practices and strategies in order to fulfil the international climate objectives (UNFCCC 2015). In the absence of such collaboration, there will be a lack of synergy in ways of fighting climate change and the world as a whole will be more exposed to the change than is desirable. Moreover, international cooperation helps not only coherence but also fairness of policies taken because they correspond to the needs and capacities of developed and developing countries (Grubb 2014).

Technological Development

Similar to the view above, cooperation in technological development cannot be underscored. Global issues, some of which are urgent, like energy, touch issues that demand technologies that cannot be developed or deployed by a single country. Multilateral cooperation in the sphere of R&D can enhance the rate of technology development, including renewable energy sources, carbon capture, and storage, as well as energy-efficient structures. For example, projects such as the CERN and the International Space Station (ISS) are depicting how huge technological endeavours that are concerted can advance mankind universally (Levy 2017). Also, coordination enables the sharing of technologies since best practice is promoted leading to technologies that can work in multiple parts of the world hence enhancing the overall global value of all the technologies developed.

Knowledge Sharing

One of key forms of cooperation is knowledge sharing because different countries can benefit from other's success and experience. This is why in today's world, one needs to train in fields such as: Public health, Education, the Environment, and a whole list of other sciences, which are useful today. For instance, in the current COVID-19 pandemic, the sharing of knowledge on the virus, the development of vaccines, and measures to undertake were from international platforms (WHO 2020). Likewise, communicating information on sustainable activities and technology can assist nations to achieve their goals for green results. Global institutions and conclaves thus provide a very important support to this extent, to ensure that knowledge effectively circulates, and that all countries, especially the developing ones, are able to access knowledge in the many fields where they can learn from others around the world (Berkman 2017).

11 Findings and Discussion

The paper stresses that development of strong and sustainable enterprises calls for unique strategies of minimizing one's digital footprints. If the venturing firms embrace these new approaches, they stand to benefit from an increased firm resilience, global sustainability and long-term organisational success within the emerging green conscious market. The research finds that data storage and usage of cloud and other digital devices account for a large portion of companies' carbon impact. Some of the trends include the ability to show data centers that are green and made from renewable energy, cloud services optimized to show efficiency as far as energy is concerned and software optimization to take a shorter span of time and less energy. The research conducted provides evidence that retrofitting may greatly benefit the process of attaining other kinds of sustainability. If a business manages to reduce its digital consumption then it is operating in closer alignment to the UN's Sustainable Development Goals, specifically those that are aligned to climate (SDG 13) and responsible production and consumption (SDG 12). This shows that digital carbon reduction is one way through which companies can participate actively in the fight against Climate Change and promote Corporate Sustainability. A few recommendations for any business that is looking forward to minimizing their digital carbon footprint are proposed in the final discussion. They include; Investing in Renewable sources of energy, Supply chain partnerships with technology companies to ensure better energy consumption management, Promoting best practices and standards and policies on digital sustainability across the industry. In the same regard, the study emphasizes the need to monitor the progress of Digital Carbon Reduction and reporting this regularly.

12 Practical and Social Implications

The chapter "Creating Resilient and Sustainable Businesses: New Ideas for Reducing Digital Carbon Footprints" is a practically and socially study that can go a long way to help extend discussions of sustainability beyond the dominance of climate change concerns to embrace digital sustainability as a key area of focus.

Practical Implications: Chapter focuses on shifting to efficient energy consumption technology, for instance, the reliable data centers and the software optimization that is likely to cause low TEC in the digital economy. They do not only decrease organisational expenses but they also help to decrease the overall detrimental effect on the environment of digital services. Furthermore, the integration of green blockchain protocols along with coming up with AI-powered energy management systems can be of great opportunities for improving the efficiency and sustainability of infrastructures. In implementing such solutions, organizations can be able to meet set regulatory measures, foster sustainable development as envisaged globally, as well as open up a competitive edge certified by the growing conscience of businesses and consumers on environmental sustainability.

Social Implications: On a societal level, the chapter points to a need of creating awareness involving citizens on the harms done to their environment due to digital activities and therefore encourage a change in behaviour. The promotion of digital minimalism and raising awareness among the users might foster enhanced environmental discussions and therefore, promote green culture in the technology industry. This can promote a sustainable information technology society through encouraging users to use lesser energy as they conduct their business online. Also, the chapter looks at the aspect of corporate responsibility and points out that firms should model green responsibility as a way of managing the consequences and therefore ensuring that people conclude to adopt environmentally friendly practices in the digital world.

13 Conclusion

The need to understand and minimize digital carbon footprints has become important in today's digital world in order to support sustainable business. Considering the fact that digital activities significantly affect the world's carbon footprint, it is crucial for companies to develop concepts that improve the energy intensity of IT solutions, implement green IT initiatives, switch to renewables and optimize digital assets. Indeed, the application of newer technologies including the AI and Machine learning has vast potential for decreasing energy use and managing digital resources more proactively. In addition, focus on the development of organizational cultural approaches that encourage people to adopt sustainable behaviour by changing their habits and their life styles contributes also to the decrease in the emissions and to the fulfilment of the goals of sustainability in the world. Ignoring their digital carbon footprints not only still costs businesses money and leaves them vulnerable to an ever-increasing tide of regulation and consumer demand for sustainability but it also offers businesses the solution to increasing efficiency across their industries. These efforts underpin the long-term goal of ensuring that business activities across the world become compliant with the UN Sustainable Development Goals more specifically, with the climate change and clean energy initiatives. Therefore, being able to practice reducing the digital carbon footprint is another factor of sustainable business as it helps the businesses become more environmentally friendly and make them ready for an increasingly environmentally conscious market.

References

Andoni M, Robu V, Flynn D, Abram S, Geach D, Jenkins D, Peacock A (2019) Blockchain technology in the energy sector: a systematic review of challenges and opportunities. Renew Sustain Energy Rev 100:143–174

Andrae ASG (2020) New perspectives on internet electricity use in 2030. Engineering 6(3):287–291

Aslan J, Mayers K, Koomey JG, France C (2018) Electricity intensity of internet data transmission: untangling the estimates. J Ind Ecol 22(4):785–798

Belkhir L, Elmeligi A (2018) Assessing ICT global emissions footprint: trends to 2040 & recommendations. J Clean Prod 177:448–463. https://doi.org/10.1016/j.jclepro.2017.12.239

Berkman PA (2017) International environmental governance: perspectives on knowledge sharing and diplomacy. Springer

Bova F, Goldfarb A, Melko R (2021) Commercial applications of quantum computing. Nature 595(7867):359–367

Cambridge Centre for Alternative Finance (2021) Cambridge Bitcoin electricity consumption index (CBECI)

Capra E, Morisio M, Sprugnoli R (2019) Assessing the energy efficiency of green software: results from some case studies. J Syst Softw 147:66–77. https://doi.org/10.1016/j.jss.2018.10.027

Cook G, Jardim E, in Greenpeace report (2019) Clicking clean Virginia

Cook G, Jardim E (2021) Greenpeace's clicking clean report: who is winning the race to build a green internet?. Greenpeace

Corinne C (2023) The internet is eating the planet. https://doi.org/10.58704/dmnx-1r61

de Vries A (2020) Bitcoin's energy consumption is underestimated: a market dynamics approach. Energy Res Soc Sci 70:101721

European Commission (2020) Directive 2012/27/EU on energy efficiency

Evans S et al (2021) Energy-efficient computing: the role of reinforcement learning. Energy Rep 7:874–882. https://doi.org/10.1016/j.egyr.2021.08.017

Evans P (2018) How Google's data centers are reducing their carbon footprint. TechCrunch

Google Cloud (2020) Google cloud is now carbon-neutral. Retrieved from https://cloud.google.com/blog/products/identity-security/google-cloud-is-now-carbon-neutral
GeSI (2020) Digital with purpose: delivering a SMARTer 2030. Global Enabling Sustainability Initiative
Goh J, Sim S, Lim G (2021) AI in the energy sector: challenges and opportunities for sustainability. J Clean Prod 297:126697
Google (2021) Google sustainability: 100% renewable energy. Retrieved from https://sustainability.google/projects/announcement/
Google Sustainability (2021) Powering our operations with 100% renewable energy. Retrieved from https://sustainability.google/
Greenpeace (2020) Clicking clean: who is winning the race to build a green internet?
Grubb M (2014) Planetary economics: energy, climate change, and the three domains of sustainable development. Routledge
Hartman C, van der Laan H, Kang T (2020) Sustainable digital economies: the role of ocean protocol. J Blockchain Res 5(2):104–116
Hintemann R, Hinterholzer S (2019) Energy consumption of data centers worldwide. In: Proceedings of the 2nd international conference on information and communication technologies for sustainability
IEA (2020) Data centres and data transmission networks. International Energy Agency
International Energy Agency (IEA) (2021) Data centres and data transmission networks. Retrieved from https://www.iea.org/
IRENA (2020) Renewable power generation costs in 2019. International renewable energy agency. Retrieved from https://www.irena.org/
ITU (2019) Measuring digital development: facts and figures
Johnson B (2021) Measuring the Digital Carbon Footprint: Challenges and Opportunities. Int J Environ Stud 78(5):673–692
Jones N (2021) How to stop data centres from gobbling up the world's electricity. Nature 561:163–166
Levy J (2017) The cooperative edge: the role of international scientific collaboration in addressing global challenges. Sci Publ Policy 44(2):179–188
Loorbach D et al (2020) Navigating sustainability transitions. Environ Innov Soc Trans 36:174–183. https://doi.org/10.1016/j.eist.2020.01.001
Masanet E et al (2020) Recalibrating global data center energy-use estimates. Science 367(6481):984–986. https://doi.org/10.1126/science.aba3758
Murugesan S (2008) Harnessing green IT: principles and practices. IEEE IT Prof 10(1):24–33. https://doi.org/10.1109/MITP.2008.10
Ocean Protocol (2021) Ocean protocol overview: empowering data ownership & privacy. Retrieved from https://oceanprotocol.com
Pichai S (2020) Our third decade of climate action: realizing a carbon-free future. Google Blog. Retrieved from https://blog.google/inside-google/company-announcements/our-third-decade-climate-action-realizing-carbon-free-future
Preskill J (2018) Quantum computing in the NISQ era and beyond. Quantum 2:79
Shehabi A, Smith SJ, Sartor DA et al (2016) United States data center energy usage report. Lawrence Berkeley Nat Lab. https://doi.org/10.2172/1372902
Shi W, Cao J, Zhang Q, Li Y, Xu L (2016) Edge computing: vision and challenges. IEEE Internet Things J 3(5):637–646. https://doi.org/10.1109/JIOT.2016.2579198
Shift Project (2019) Climate crisis: the unsustainable use of online video. The shift project
Strubell E, Ganesh A, McCallum A (2019) Energy and policy considerations for deep learning in NLP. In: Proceedings of the 57th annual meeting of the association for computational linguistics, 3645–3650. https://doi.org/10.18653/v1/P19-
The Shift Project (2019) Towards digital sobriety: a path to sustainable ICT
UNFCCC (2015) Paris agreement. United Nations Framework Convention on Climate Change

United Nations (2015) Sustainable development goals. https://www.un.org/sustainabledevelopment/sustainable-development-goals/

WHO (2020) Managing epidemics: key facts about major deadly diseases. World Health Organization

World Economic Forum (2020) Harnessing the fourth industrial revolution for the earth

World Economic Forum (2022) https://www.weforum.org/agenda/2022/05/how-digital-solutions-can-reduce-global-emissions/

Xu H, Chen C, Zhang C (2016) A deep reinforcement learning-based dynamic model-free control for energy saving in data centres. In: Proceedings of the 2016 IEEE international conference on smart grid communications (SmartGridComm), pp 1–6. https://doi.org/10.1109/SmartGridComm.2016.7778756

Xu Y, Ding W, Zhao Z, Zhang X (2020) Energy-efficient data centers via AI: challenges, solutions, and future trends. IEEE Access 8:70172–70191

Zheng Z, Xie S, Dai H, Chen X, Wang H (2020) An overview of blockchain technology: Architecture, consensus, and future trends. In: IEEE international conference on big data, pp 557–564

Leveraging Virtual Collaboration: Metaverse Routes to Achieve Sustainable Development Goals

Pooja Kaushik, Tanya Chouhan, and Avinash Chopra

Abstract This chapter examined the crucial role of virtual collaboration in accomplishing the sustainable development goals (SDGs). As the globe becomes more interconnected, digital platforms provide creative answers to challenging global issues. The first section of the chapter looks at the theoretical foundations of virtual collaboration, emphasizing how it may promote equitable participation and cut across geographic borders. Virtual collaboration technologies have been successfully implemented in a number of industries including environmental management, healthcare and education. The research highlights the advantages of virtual cooperation such as increased information exchange, wider stakeholder participation and improved resource efficiency. The chapter also discusses the drawbacks and restrictions of remote cooperation, including cybersecurity and digital divide problems. The chapter highlights the ways to get over these challenges, focusing on the need of having a strong digital infrastructure and fair access to technology. This chapter presents a thorough overview of how virtual collaboration may be used to propel sustainable development projects by combining insights from various viewpoints. It ends with suggestions on how scholars, practitioners and politicians might use virtual cooperation to create a more just and sustainable future.

Keywords Virtual collaboration · Metaverse · Sustainable development · Social network theory

---

P. Kaushik · T. Chouhan (✉) · A. Chopra
Galgotias College of Engineering and Technology, Greater Noida, India
e-mail: tanyachouhan511@gmail.com

W. Leal Filho et al. (eds.), *Metaverse and Sustainability: Business Resilience Towards Sustainable Development Goals*, World Sustainability Series,
https://doi.org/10.1007/978-3-031-89545-6_16

1 Introduction

1.1 Overview About Virtual Collaboration

Virtual collaboration can be defined as the process in which people or teams are connected in an organization and they share common objectives through using telecommunications and computer technologies. This mode of collaboration is also characterized by geographical distribution where employees or teams can exchange work-related data, information, and knowledge in real time or/and on demand through the use of technology enhanced tools. Technological enhancement has been noted as one of the main drivers for the changes in virtual collaboration. During the first level of adoption the major means of remote communication were email and low end teleconferencing. However, these methods were also somewhat less effective when it came to interaction with, and involvement of, the intended audience. The coming of the internet and other related technologies made virtual collaboration more functional as opposed to the previous static and less efficient form. In implementing schedules, Skype and later the Zoom application were introduced to offer the visual perception and improved the awareness of the parties involved in the meetings (Davidson 2019). The platforms enabled the provision of real-time 'face-to-face' communication, which is vital in the creation of trust and collaboration especially in a virtual setting. Since then, there have been improvements in features of cloud computing which has been relevant in the pioneers of virtual collaboration. One example is Google Workspace and Microsoft Teams where document sharing, editing, and planning of projects can be done from any part of the world at any time (Gratton 2020). These tools have made evolution from mere communication to collaboration possible where entities are able to work on complicated projects regardless the geographical location of the members. Other ecosystems that have facilitated this evolution are other social media and business applications such as Slack, Trello among others that are specifically designed to allow for team and project communications. These platforms are both real-time and asynchronous encouraging participation by the anyone even though the participants cannot be present at the same time (Leonardi 2018).

The COVID-19 outbreak played the key role to accelerate and transform the techniques of virtual collaboration. As organisations shifted to working from home, they quickly adopted better and higher forms of Virtual Collaboration Tools hence quick enhancement and inclusion in working environments (Spataro 2021).

1.2 Contribution of Virtual Collaboration Towards Sustainable Development

Virtual collaboration or the procedure of performing collaborative work by means of information technology irrespective of the geographical distance emphasizes great developmental potential in the sphere of sustainable development. Therefore, it is established that through the use of technology, virtual collaboration can be used to work towards multiple UN sustainable development goals.

SDG 4: Quality Education

Virtual collaboration makes it easier to acquire good education through distance learning and promotes cross-section educational environments. Zoom and Microsoft Teams, Google Classroom enables a student as well as an educator to collaborate across the continents hence enabling students to learn and educators to teach irrespective of the geographical location of any region (Means et al. 2014). For example, the COVID-19 crisis revealed the significance of distance collaboration during education for maintaining continuity on the international level (UNESCO 2020).

SDG 8: Decent Work and Economic Growth

The COVID-19 breakout also forced organizations to adopt remote work culture, indicating how virtual means can support the objective of decent work and economic development. Thus, fostering flexibility of working circumstances with ICT tools contributes to the creation of employment for persons in distant and weakly populated regions, thus decreasing polarization in income distribution (ILO 2020). Also, virtual collaboration contributes to a higher efficiency and idea generation since individuals from different groups can work together.

SDG 9: Deals with Industry, Innovation, and Infrastructure

Remote cooperation has become the essence of the contemporary digitalized market and economic environment, as it contributes to innovation and infrastructure creation. Through adoption of cloud solutions and applied use of digital media, the companies are in a better position to cut expenses, enhance efficiency and encourage disruption culture (McKinsey and Company 2020). For instance, the communication app such as Slack and Asana helps the project teams to organize their work resulting in better business results.

SDG 13: Climate Action

In fact, virtual collaboration may be one of the most straightforward benefits as it holds the capability to decrease climate change. It is estimable that virtual meetings and remote work drastically cut back carbon emissions tied to transportation (Global e-Sustainability Initiative 2015a, b). Carbon Trust (2014) established that replacing as few as one-fifth of business travel with the conventional video conference could lead to emission cuts.

SDG 17: Partnership for the Goals

Building of effective partnerships for the achievement of SDGs can be highly benefited from virtual collaboration. Resources, knowledge, and the latest practices and trends can be shared easily and quickly over the global digital networks among the stakeholders (UN 2015). For instance, intergovernmental organizations and non-governmental organizations exploit virtual collaboration tools as a means of planning, reporting, and executing efficient and sustainable development projects.

1.3 History of Virtual Collaboration Tools and Their Evolution

Virtual collaboration platforms have been around for a few decades and were first inspired by early computer networks and uses of the internet. In 1960s, the time-shared systems got introduced which enabled several users to interact with a central computer at one time which paved for cooperation (Ceruzzi 2003). Establishment of ARPANET in the late 1960s under funding by the U.S. Department of Defense was also a milestone since it is the ancestor of modern world internet and contributed to remote communication and distribution of data (Leiner et al. 1997). Simple and straightforward examples are the development of new means of communication such as Electronic Mail or 'Email' that could be used to transfer messages electronically when it was developed in the 1970s and 1980s. An early technology related to communication was email, which had been invented by Ray Tomlinson in 1971 and later became essential in the business world (Hauben and Hauben 1997). At the same time, there was the appearance of BBS wherein individuals can disseminate as well as transmit data; these initiated simple forms of social networking as well as collaboration (Rheingold 1993).

There of course where some major development in the 1990s with the coming of commercialization of Internet and the invention of World Wide Web. This period presented better tools that enabled the provision of services online. Thus, the first site was launched by Tim Berners-Lee in 1991 and by mid 90s website based email services like Hotmail were introduced more facilitating the communication phase (Gillies and Cailliau 2000). Also, presence of graphical browsers, in particular with the advent of ICQ instant messaging in 1996, provided for text based real time communication that was crucial for cooperative work (O'Neill 2010).

This early 2000 was a major turning point with the introduction of the Web 2. 0 technologies that centred on information creation by people, ease of use and compatibility. Later, with the help of the newest technologies, and changing our model for collaborative content creation, sites such as Wikipedia, which was created in 2001, appeared (Tapscott and Williams 2006). As a result of increased internet usage, sites like LinkedIn started in 2003 and Facebook in 2004 meant for professional and personal networking contributed in coordinating the increased collaboration (Boyd and Ellison 2007).

At the same time, the business application of project management and collaboration tools emerged. A tool that was launched before the widespread of the currently popular tools is Basecamp, which was founded in the year 2004, and offered Project management and collaboration interface for groups and organizations, paving the way for the development of other tools (Fried and Heinemeier Hansson 2010). In 2008, google came up with Google Docs which was a web-application for sharing documents and editing them concurrently by many users; this revolutionized how groups of people coped with documents and even spreadsheets (Griffith 2008).

The trend during the first decade of this year was a massive increase in the collaboration tools mainly due to ever-growing popularity of the Cloud computing

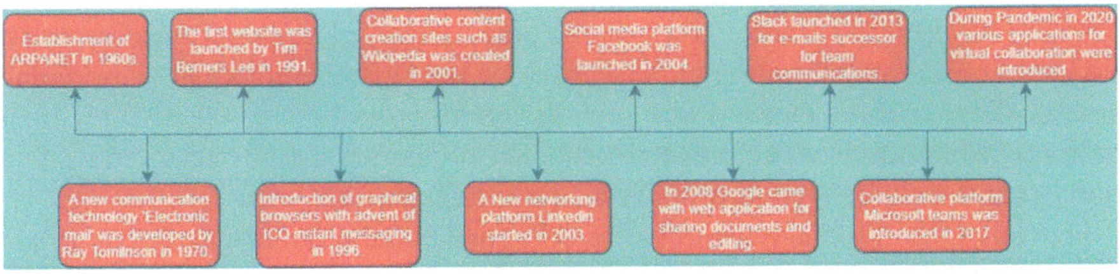

Fig. 1 History of virtual collaboration (Author's own)

and Mobile technologies. Slack launched in 2013 and has become an e-mails' successor for team communication; This service-connected multiple services permitting synchronous work (Butterfield 2017). Likewise, with Microsoft Teams, introduced in 2017, tied Office 365 offering a communication tool, collaboration platform, and file sharing and Project management program (Microsoft 2017).

The pandemic, in particular COVID-19, that took place in 2020 greatly boosted the application and development of virtual collaboration means due to increasing reliance on remote work. Applications like Zoom, Microsoft Teams, and Google Meet garnered sudden attention, which actually make such tools an important part of sustaining the business and letting employees work from home (Forbes 2020) (Fig. 1).

1.4 Current Trends in Virtual Collaboration

Virtual collaboration existing now a days is marked by high popularity and constant changes in technology, which has been shifted even more with the COVID-19 pandemic. The pandemic brought the idea of work-from-home and necessitated many organizations to shift their work process online which led to the popular culture of virtual meeting including zoom, teams, meet among others. They have condensed into themselves essential roles of continuity of business and key to the communication and collaboration among team irrespective of the distance that may exist between them.

The forced transition to remote work showed that in most cases the availability of stable and efficient digital communication platforms plays crucial role. Consequently, most of the platforms experienced their users' numbers skyrocketing and the system's optimization to improve such indices. For example, Zoom users connected daily in meetings rose from 10 million in December 2019 to more than 300 million in April 2020 (Zoom 2020). Tools such as breakout sessions, virtual backgrounds, and changes in security levels were quickly incorporated to satisfy various users' requirements.

The pandemic also gave an insight into to how work can be done online and different pros and cons associated with virtual collaboration. Advantages: schedule and time saving from no or less travelling, wider pool of talent not bounded by location. Nevertheless, some issues like 'Zoom fatigue', threats of cyber security, and issues concerning the unity of the team came up. Organizations had to overcome these challenges through measures such as learning the best practices for virtual meetings, which engaged employees and better protecting the business and its data from cyber threats, as well as promote a firm virtual culture.

In detail, COVID-19 has greatly affected virtual collaboration in the present state making it more advanced due to its effects. Despite the significant focus on the impact of remote work on the relationships between employees and managers, the manifestations of strengths and weaknesses of online collaboration have probably contributed to its role as a fixture in the future of work.

1.5 Importance of Virtual Collaboration for Sustainable Development

They are crucial for the sustainable development objectives since they encourage effective and accessible methods of working that require the use of technology. Here's why it's essential:

Reduction of Environmental Impact

Lower Carbon Footprint: Instead of physically coming together for meetings, time and resources are used through virtual means and this contributes to cutting down; traveling which is a major factor for increased carbon footprint. Through provision of e-meetings, e-conferences and e-workshops, organizations can significantly reduce on the greenhouse gases emitted in place of air and road transport, thus contribute to the realization of the SDG 13 on Climate Action.

Efficient Resource Use: Due to the use of the internet and related technologies, resource sharing and information sharing are done electronically and thus eliminating the use of paper, which in turn saves resources.

Global Inclusivity and Access

Bridging Geographical Gaps: Online collaborative applications enable persons and/or institutions from different locations to achieve a common task. Consequently, this makes it possible for anyone, including those in the rural areas or parts of the globe where access to technological devices is quit limited, to contribute towards the achievement of the various set goals and objectives. In this sense, the promotion of inclusiveness assists in the achievement of SDG 10 which is Reduced Inequality.

Diverse Perspectives: Since virtual teamwork encourages the exchange of ideas and solutions across the borders they are a valuable source of diverse perspectives needed to tackle the global issues in the sphere of sustainable development.

Cost-Effective Solutions

Reduced Operational Costs: Virtual modes of collaboration remove expensive structures, physical distribution, and communication expenses thus make it cheaper for the organizations to engage with sustainable development projects.

Scalability: Influence of stakeholders can be obtained with relative ease, online environments allow for the inclusion of large quantities of participants in the processes of decision making without costing much.

Enhanced Collaboration and Innovation

Real-Time Communication: Other applications like video conferencing, instant messaging, and collaborative document editing improve the speed of timely communication and decision making enhance the efficiency of implementation of sustainable undertakings.

Innovation Through Collaboration: Virtual environments for the sharing of knowledge and practice promoting interdisciplinary and cross-sectoral solutions to sustainable development issues.

Resilience and Continuity

Adaptability in Crisis: The global outbreak of the COVID-19 pandemic underlined the significance of remote collaboration in preserving the operations of businesses and projects in emergencies. This is because of the necessity of constant working on sustainability goals irrespective of the occurring global disruptions.

Support for Remote and Hybrid Work: Virtual collaboration serve remote and hybrid modes of working which are gradually becoming obligatory to implement as sustainable business environment strategies. These models help in achieving a better work-life balance and hence professionals, being more productive and satisfied, with work correlate with SDG 8: Decent Work and Economic Growth.

Fostering Global Partnerships

Collaboration Across Sectors: ICTs enable stakeholders in the global community to share information and work together resulting in partnerships which are essential in attainment of the goals (SDG 17) partnerships for sustainable development. Accelerated Knowledge Sharing: These tools help in quickly reaching out to society, new policies, and various other practices that would foster the development of sustainable development in a short time.

2 Theoretical Framework

2.1 Theories Related to Virtual Collaboration and Sustainable Development:

Virtual collaboration and sustainable development are concepts that are related in several ways because there are several main theories to address virtual collaboratives, which focus on the processes that people, organizations, and communities engage in, in order to attain sustainable development objectives. Some of the related theories include; Social Network Theory, Collaborative Theory, Actor-Network Theory, and Theory of Collective Action.

Social Network Theory

Berners-Lee's collaboration of Social Network Theory (SNT) is fundamental to virtual collaboration. This premise assumes that nodes (people or organizations) and links (connections) create networks, impact people's behaviour, information transfer, and resources (Wasserman and Faust 1994). Regarding virtual collaboration, SNT assists in evaluating how the information technology supports building and sustaining the networks along with the dissemination of these networks geographically. Web 2.0 applications, for instance, social network sites and discussion boards for the creation of densely connected but dynamic networks that facilitate knowledge creation, generation, sharing and solution finding on sustainable development (Wellman et al. 2001).

Here, we witness SNT's focus on weak ties as a main principle in establishing a base for sustainable development: these are relatively loose social connections between people, who can be sources of new ideas and information for others within a particular network (Granovetter 1973). Hence, these weak ties are important in virtual environments where stakeholders, including; INGOs, governments, and businesses, convene to solve global problems like climate change and poverty. In this regard, the theory stresses that virtual networks can effectively mobilize resources and coordinate actions in order for the SDGs to be attained.

Finally, in the context of the present section that addressed virtual collaboration and sustainable development, network theory helps argue deeply that decentralised and interconnected systems support collective action toward sustainable development goals. Social media tools that facilitate the sharing of information and the creation of actions and initiatives cancelling geographical differences give the diverse actors freedom of expression and allow cooperation and innovation. Castells also refers to these networks in his network society theory stating they improve the flow of information improving communication and empowering groups that have been sidelined while supporting other sectors to work in harmony. This integration can help to drive the advancement of the United Nations Sustainable Development, especially due to the coordination on climate activities around the world (Castells 2010).

Collaborative Theory

Collaborative Theory deals with the activities and effects of the common undertaking by more than one party. Combined with the perspective of virtual collaboration, theory highlights the importance of social capital, including such factors as trust, communication, and organizational goals, in promoting successful team work (Gray 1989). The components mentioned above are complemented by virtual collaboration tools, which are the technology environment that facilitates communication of the stakeholders in spite of distance.

Out of all the theories identified, Collaborative Theory finds application in sustainable development most suits the objectives of the SDGs because of its significance in partnerships. Cohesion, according to the theory, improves the quality of decisions since different people from different perspectives work together (Ansell and Gash 2008). In enhancing this, the virtual platforms support real time communication, project management as well as sharing of resources which are very vital in dealing with multi-faceted sustainability issues.

Collaborative governance model promotes sustainability for decision making among different stakeholders in government, private sectors and civil societies via technology. Because virtual collaboration can allow an income of as many stakeholders as necessary, it stands for inclusiveness and openness. This model strengthens overall stewardship for the achievement of sustainable development goals or the SDGs including climate action SDG 13. When decision-making is an asynchronous process that is supported by technology, there are continual objectives of open communication and gatherings of information that are more suitable to environmental conditions. Particularly important in sustaining a long-term cooperation are such values as trust, openness and accountability (Ansell and Gash 2008).

Actor-Network Theory

Increased attention to technology is provided by Actor-Network Theory (ANT) that represents an approach that postulates actors both human and non-human (Latour 2005). According to ANT, digital tools are not only mundane objects, which are utilised during the collaboration for the sustainable agenda, but rather actants, which not only take part in the processes but affect them.

For instance, data analytics and AI in virtual platforms can support decision-making processes, shape the activities of projects, and even change the type of power relations in networks. ANT focuses on the relations between actors and asserts the need to recognize 'the structure of engagement' as a way to direct and enhance virtual collaboration for sustainable development.

Theory of Collective Action

According to Mancur Olson's Theory of Collective Action, which deals with a scenario where people or entities unite in a cause and work towards achieving it, the focus is directed towards the action, benefitting all the members, at large. This is where different collaborative technologies help to address the difficulty of the collective action, including coordination dilemmas and the presence of freeloaders

(Olson 1965). In sustainable development it is used when coming up with projects like climate action whereby everyone needs to pull in the same direction in order to arrive at the set goals. Communication technologies support the improvement of the openness, responsibility, and inclusion since they allow all the interested parties to contribute to the achievement of the SDGs.

3 Interdisciplinary Approach

This chapter's topic—virtual collaboration for sustainable development is cross-disciplinary in nature as it involves the use of technology, environmental science and social sciences to solve global issues. Primarily, this symbiosis of fields is essential to integrate effective and efficient innovations into practice.

Technological Support

It is critical of virtual collaboration given that it offers the requisite means and ways through which folks affiliated to different organizations can interact and work on a project based on a common set of tenets from different geographical locations. Examples are in the usage of advanced computing such as cloud computing, artificial intelligence, and organizational collaborative platforms such as Microsoft Teams and Slack (Hoffman 2020). These technologies also improve the potentiality, extendibility, accessibility of the actual communication in virtual space to engage different stake holders on sustainability processes.

Environmental Science Support

Environmental science is important for this cross-disciplinary approach as it provides important information about the environmental problems that virtual collaboration seeks to solve. Knowledge on the state of the environment which includes matters such as climate change, loss of biological diversity and depletion of natural resources is useful in formulating sustainable strategies (Rockström et al. 2009). That way, virtual collaboration enables environmental scientists in engaging policymakers, businesses, and communities in different parts of the world to embrace a systems approach in designing the potential solutions to tackle the impacts of the environment. For example, virtual collaboration makes it possible to limit the extent of travel by the global teams hence achieving climate actions objectives (Global e-Sustainability Initiative 2015a, b) (Fig. 2).

Social Science Support

Business is better understood with an integration of the social sciences, as source of context for culture, organization, and behaviour of people in virtual collaboration. Scientific concepts that elaborate how virtual teams are formed, how they work, and how they get things accomplished include social network theory and collaborative governance (Borgatti and Foster 2003). These elements hold up the principles of multiculturalism to be a perfect guideline for the tailored virtual cooperation and

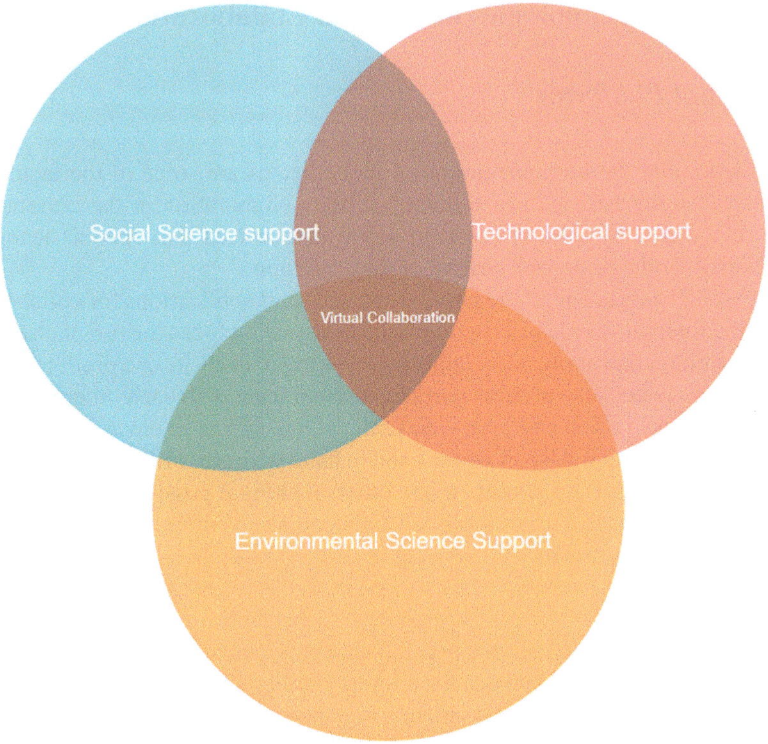

Fig. 2 Interdisciplinary approach (Author's own)

interaction. In addition, social sciences contribute towards the consideration of stakeholders and participation which are central to 'development' in order to ensure that 'sustainability' is socially acceptable across the society (Reed 2008).

Technology, environmental science as well as social sciences when implemented through virtual collaboration successively represent the various aspects of sustainable development. It enables the formulation of solutions unique and original in concept and practice, technically, environmentally and socially sustainable and implementable. When based on the qualities of both academic disciplines, the virtual collaboration is apt to meet the compound and intertwined issues of sustainability in the conditions of the continually evolving world.

4 Tools and Technologies for Virtual Collaboration

4.1 Digital Platforms

Remote work and the implementation of global teams are some of the factors that have increased the necessity of utilizing virtual collaboration in the current work setting. It is eased by different platforms and instruments that connect teams and allow them to collaborate irrespective of their location.

Microsoft Teams is one of the best known and used applications for collaboration through the Internet. These offers communication that includes the use of chat, video calls, file sharing, and application inclusion, making it an all-in-one product for team collaboration. Teams also permits users to communicate with each other using short or long messages via chat and even video call, which creates an environment of face-to-face communication. Also, it works hand in hand with other Microsoft Office suits like the Word, Excel, PowerPoint among others making it possible to work together on the same documents. For example, several people can edit the same text at a time, which contributes to the smooth working in a team (Microsoft 2023).

Another tool is Slack which has simple design and integration options. Slack divides the team communication into channels, so that within any channel the communication is quite specific and to the point. It also provides features for direct communication like messages and video calls; interoperability with many other applications such as Google Drive, Trello, and GitHub. Such integration is beneficial as it allows the members of a given team to have all the tools integrated within the interface of Slack and thus increase productivity (Slack 2023).

The application is now popularly referred to as the only application for video conferencing especially since the year 2020 when the COVID-19 pandemic started. Zoom can provide the video meetings, breakout rooms are perfect for discussion of the small groups, and webinar functionality therefore Zoom can be used for small teamwork and large webinars. The simplicity of operation and high performance has attributed to it being widely used as a tool for Virtual collaboration (Zoom 2023).

In the sphere of project management, there are such instruments as Trello or Asana. For instance, Trello works on the card system that contains tasks; the cards can be dragged to different project statuses. This picture-based method of working lets teams see and comprehend the state of affairs and keep things on the same page. Thus, Asana has a more elaborate set of tools for project management enabling teams to plan projects, define tasks, set timeframes, and track the progress. All these tools can work with other collaboration apps, such as Slack and Microsoft Teams for effective work automation (Trello 2023; Asana 2023).

Google Workspace is another strong tool of the virtual collaboration managing system. These are Gmail, Google Drive, Google Docs, Sheets, and Slides all of which that allow for collaboration in real time. Of these tools, Google Meet that is part of this suite is a video conferencing application that is very compatible with Google Calendar whereby, making appointments and joining the meetings is very simple. The fact that collaborative editing is possible on documents, spreadsheets, and presentations with the team members even when they are geographically located in different areas makes Google Workspace suitable for use in virtual teams (Google Workspace 2023).

Finally, Miro is a digital whiteboard software that can be used by teams to create Top Design visuals. It is effective in real-time writing, drawing, and diagramming which helps especially in mind mapping and ideation sessions. Thus, built-in connectivity to instruments like Slack, Microsoft Teams, and Zoom guarantees that visible collaboration contains a multifaceted workflow (Miro 2023).

All these platforms and tools have their own features that allow the approach to virtual collaboration from the angle of communication and project management, sharing of documents, and visual collaboration, respectively. Thus, with help of these tools, teams can manage the difficulties of remote work environment and coordinate the work effectively across distances.

4.2 Innovations for Sustainable Projects

In sustainability projects, there is need to have the involvement of different people in different specialties and offices. Thus, effective virtual collaboration in these elaborate undertakings has been supplemented by novel tools and technologies that address requirements particular to sustainability-focused ventures.

EcoChain is one such innovation which is a sustainability management tool aimed at assisting business organizations and other institutions on how they can rate and manage their ecological footprints. EcoChain facilitates collaboration because the users can input, share, and coordinate efforts towards achieving the requisite low carbon profile. Another feature that might be beneficial for sustainability projects is the platform's LCA tool, which integrates a system to review the environmental profile of products and services throughout their lifecycle (EcoChain 2023).

Sustain Lab another great tool that uses AI to enable sustainability reporting and data integration. Sustain Lab's collaborative tool works as a platform to assimilate the data from different sources that could be used by the teams to better understand the trends and review the achievements in sustainability. The insights derived through the use of AI tools in project teams supported by Sustain Lab help the teams in understanding where improvements can be made and what decisions should be taken in order to meet some or all of the sustainability goals set out for the project as discussed in more detail in Sustain Lab, 2023.

As mentioned, WeLOOP is a common working platform covering the aspect of product sustainability and eco-design. This tool enhances virtual working in the sense that involves the creation of a common ground where teams conduct environmental evaluation and eco-design related tasks. Through WeLOOP, teams are able to collaborate and design sustainability analysis on a product, develop potentials layouts of the life cycle of a product and even design their product to achieve the best environmental benefits. It further adds the applicability of the platform in other big sustainability projects due to its interoperability with other tools and datasets (WeLOOP 2023).

Discussing the opportunities, it's crucial to notice that Hive innovates and offers an interesting perspective of using project management in communication and collaboration for sustainability projects along with sustainability measurements. Hive combines all the features of task, file, and team collaboration and messaging, as well as a sustainability tracking system for the teams. It also guarantees that the concept of sustainability remains at the forefront throughout the project work while encouraging the organizations to embrace change (Hive 2023).

5 Case Study

Technological solutions that foster a virtual collaboration have extended their potential to support of the sustainable development goals (SDGs), especially if it comes to the ultimate global impacts such as climate change and inequality. A good example is the Global Alliance for Sustainable Development Goals (GASDG) which is an international consortium of associations, governments, and academic institutions that has used the virtual platforms to advance and achieve sustainable development goals across the globe.

GASDG at present performs its function through a network of virtual working groups that deal with various SDGs. These groups use video conference to meet frequently, work on documents and use project management tools to advance projects. For example, the group has mobilized for clean energy (SDG 7): this has spurred research and development of pilot projects in renewable energy sources in parts of the world with little-to-no access to such technologies to date.

A great example of virtual work in GASDG is the "Virtual Renewable Energy Conference" that took place in 2021. The event was fully virtual and attracted well over 5,000 attendees from across the world, comprising mainly of policy makers, researchers, and business managers. Due to its virtual format, lots of people engaged during this time without the detrimental effects produced by international travel, thereby addressing SDG 13 (Climate Action). Several countries began new policies and partnership after the conference in order to increase the implementation of policies on Renewable Energy.

The achievement of GASDG has proven that the conventional geographical and economic limitations hinder participation and strategic progress towards the common development of sustainable development goals. Given that many global activities are being shifted to the online platform in the wake of the COVID-19 pandemic, the case of GASDG demonstrates how effective collaboration for the achievement of the SDGs may be efficiency enhanced by innovation.

6 Strategies for Effective Virtual Collaboration

Many establishments are taking advantage of the possibilities presented by new technologies to increase their effectiveness and productivity. One of the latest methods to be employed is "virtual team collaboration." These are teams composed of individuals who do not work together in the same place or at the same time. They can be found all over the world and span across multiple time zones. The ability to work together on projects without being co-located or present at the same time has been made possible due to advances in computer and software-mediated communication. Remote teams can coordinate their efforts more effectively when working on shared projects and learning objectives when using virtual collaboration. To increase employee engagement and retention, team members can easily communicate with each other via instant messaging, voice, or video when separated by great distances. Companies can efficiently manage each individual's learning experience according to their specific needs using the NovoEd learning platform. Through team assignments, the NovoEd online learning platform provides an effective way to learn about virtual collaboration, enabling the creation of immersive online learning events. The platform facilitates teamwork by offering a private workspace for the team as well as features for file sharing, synchronous chat, and meeting scheduling.

7 Impact of Virtual Collaboration on Sustainable Development Goals

Several transformations have been realized in the world today due to the various developments in industries as well as technology. Today competition turns into volatility, unpredictability, certainly, and uncertainty, and these are the driving forces of the so-called digitalization. About the link between sustainability and DT at the corporate level, Gomez and Gonzalez provided the information (Han et al. 2022). In the research study the authors employed a systematic review of 89 published studies. Strict content filtering procedures were used with regards to the contents. Thus, the research found that it is feasible to develop a research perspective of the key argument of DT as antecedent and enabler of sustainability. According to the report, there were things that companies needed to do to be able to weather the digital storm and they included; (i) strengthening digital competency and (ii) ensuring that there was digital optimality in economic, (iii) environmental, and (iv) social aspects. The more ICT is used in industrial production, the more crucial is the preventive risk assessment and political awareness of the risks and opportunities of ICT for environmental sustainability. Kunkel and Matthes conducted in 2020 an evaluation of the policies in the industrial and digital sectors of three East Asian and Pacific Island countries and four Sub-Saharan African countries specifically in terms of the predictive ICT impacts for businesses for environmental sustainability (Lian et al. 2022). This research demonstrated that the policies are full of a broad spectrum of rather generic and quite often expressed expectations and hallow the use of information and communication technologies. As analysed by Bican and Brem (2020) [Bican 2020] digitalization notably supports the United Nations Sustainable Development Goals. It can be stated that future economic and environmental implications cannot be effectively managed without transitioning existing businesses. In an effort to derive at a common understanding and definition that would form the premise for interrelatedness in a conceptual framework the study also reviewed literature.

The present research started with a literature review of the field of management and economics to define the concept of digitalization and seven related terms, namely digital, business model, digital business model, digital technology, digital innovation, digital transformation, and entrepreneurship digital.

DT can offer numerous enhancements in operational excellence, enhance customer satisfaction, and foster innovation and substantial, long-term sustainability advantages for organisations (Zhang et al. 2021). DT can be useful for improving the companies and decreasing their effects on the environment. With the help of digital tools, it becomes possible to rationalise the organizational processes, as well as automate many types of work and/or minimise the requirement for tangible assets. This lowers energy usage, lowers carbon output and cuts down on waste production. In developing organizations, DT creates the organizational culture that fosters innovation and agility. DT unites members of an organization and teams up with other organizations in order to share knowledge. Through the usage of integration of digital work platforms and communication solutions, business can minimize on the number of times that employees are required to travel which in effect decreases the transportation related emissions as noted in (Fang et al. 2023). DT helps organizations to be sustainable in their ways of relating to the stakeholders and the customers. Leveraging computerized platforms, organizations are afforded the opportunity of sharing messages of sustainability initiatives, marketing and or advocating for greener products and services and increasing the public's green consumer awareness (Guinan et al. 2019). That will guarantee more customer base loyalty, brand creation and therefore a better a sustainable business climate.

DT seek to create transformations of and through emerging technologies including use of artificial intelligence, machine learning, cloud computing, data analytics, and the IoT to transform operations and produce greater consumer value. The strength of this technique, therefore, kicks from its ability to gather a large volume of data that can then be analysed in order to get useful information. Organizations can use big data for enhancing their operations, understanding the customers' needs and discovering opportunities for business development based on information analysis. Also, DT results in increased ability by organizations to adapt and respond effectively to market demands. Use of digital age tools and automation enables organizations to enhance their operational effectiveness, cut expenses, enhance the pace of new product delivery and be more responsive to the buyers (Dutta et al. 2020).

Thus, through digital, such as social media and email, preferred channels, and individualized communications and multichannel, companies are able to reach out to customers to meet their needs and offer tailored solutions, thus creating a uniqueness in the marketplace. Competitive advantage in DT is evident when firm's technology investments are in harmony with strategic goals, where there is constant embracing of digital innovations, and the culture of embracing change and being customer centric (Khanom 2023). It is with these considerations that sustainability in digital options can be useful to score various goals by the organizations. DT initiatives can pave the way for energy efficiency and resource optimized (Judith et al. 2022). Implementing the e-government approaches and applying the digital technologies, businesses can decrease the paperwork and eliminate the paper; Businesses can optimize the energy consumption; The businesses will also embrace digital teamwork to minimize tele work-related carbon emission; Thus, DT will promote sustainable work environment and minimize travelling related carbon emissions (Kranz et al. 2022). DT implementation of organisational sustainability goals allows organisations to assume production processes that are friendly to the environment, effectively use resources and have reduced negative impacts to the environment therefore becoming sustainable.

8 Implications

Practical implications: In the framework of the Sustainable Development Goals (SDGs) of the UN, virtual cooperation has emerged as a crucial tactic for attaining company resilience. Businesses can sustain operations, encourage innovation, and guarantee continuity even amid disruptive situations like pandemics or natural catastrophes by utilising technology. Therefore, virtual cooperation gives firms the means to adapt and prosper in unpredictable situations by improving supply chain management, supporting innovation, lowering environmental impact, increasing operational continuity, and encouraging talent mobility. By fostering talent mobility and lowering operational risks, remote work has been shown to increase organisational resilience and employee happiness (George et al. 2021). Businesses that have a robust virtual collaboration infrastructure are more adaptable and better able to respond to market changes with innovation (Ghauri et al. 2022).

Theoretical and Social implications: It contributes to the theoretical understanding of integrating sustainability into digital collaboration frameworks, refining models of remote work dynamics with a focus on eco-efficiency and virtual team management. The potential of virtual collaboration to promote sustainable development by promoting remote work as a means of reducing carbon footprint, increasing access to global opportunities, and supporting broader societal change towards more sustainable practices.

9 Future Scope of Research

Future research on virtual collaboration for sustainable development could explore several promising areas. A key focus could be on enhancing virtual collaboration tools through AI and machine learning, optimizing communication, productivity, and environmental impact. It is also important to examine the long-term effects of reduced travel and resource use on global sustainability goals, including comparing virtual and traditional collaboration methods. Furthermore, examining the social effects of virtual collaboration on organizational culture, employee well-being, and team dynamics can provide insights into how to improve remote work environments. Interdisciplinary studies could integrate sustainability with technology and organizational behaviour, exploring how virtual collaboration can be used effectively across different sectors, especially in developing countries. Finally, research on policy frameworks could identify how regulations can support and improve sustainable virtual collaboration practices, while addressing potential barriers and promoting widespread adoption.

10 Conclusion

The chapter "Virtual Collaboration for Sustainable Development" highlighted the important role of various ICTs in enhancing the SDGs. Through the use of virtual collaboration, it is possible to combat geographical isolation, thereby enhancing a better way of addressing global issues by organizations and even individuals. Due to the fact that the subject involves technology, environmental sciences, and social sciences in general, one cannot underestimate the challenges and opportunities of virtual collaboration to make the world a better place. That is why virtual collaboration remains essential as the world grapples with its worst environmental and social problems. Additionally, it encourages emerging practices to address complex issues and equalizes everyone's chances of contributing to reducing carbon emissions as a result of physical meetings. However, it can be concluded that effective virtual collaboration depends on a good incorporation of cultural, organizational, and technological factors. Thus, virtual collaboration is not just a method, but a new approach and way to achieve sustainable development in all areas of human activity. Therefore, through this approach, new guarantors of safer systems and societies are developed to create a more sustainable future.

References

Ansell C, Gash A (2008) Collaborative governance in theory and practice. J Publ Adm Res Theory 18(4):543–571

Asana (2023) Asana: work on big ideas, without the busywork

Bican PM, Brem A (2020) Digital business model, digital transformation, digital entrepreneurship: Is there a sustainable "digital"? Sustainability 12:5239
Borgatti SP, Foster PC (2003) The network paradigm in organizational research: a review and typology. J Manag 29(6):991–1013
Boyd DM, Ellison NB (2007) Social network sites: definition, history, and scholarship. J Comput-Mediat Commun 13(1):210–230
Butterfield S (2017) Slack: the future of work. Harvard Business Review
Castells M (2010) The rise of the network society. Wiley-Blackwell
Ceruzzi PE (2003) A history of modern computing. MIT Press
Davidson R (2019) The evolution of video conferencing technology. J Commun Technol 15(2):34–47
Dutta G, Kumar R, Sindhwani R, Singh RK (2020) Digital transformation priorities of India's discrete manufacturing SMEs–a conceptual study in perspective of Industry 4.0. Compet Rev Int Bus J 30:289–314
EcoChain (2023) EcoChain: sustainability management platform
Fang M, Liu F, Xiao S, Park K (2023) Hedging the bet on digital transformation in strategic supply chain management: a theoretical integration and an empirical test. Int J Phys Distrib Logist Manag 53:512–531
Forbes (2020) Zoom's growth soars as remote work revolutionizes communication. Forbes
Fried J, Heinemeier Hansson D (2010) Rework. Crown Publishing Group
George et al (2021) Digital sustainability and entrepreneurship: how digital innovations are helping tackle climate change and sustainable. Development. https://doi.org/10.1177/104225871989 9425
Ghauri et al (2022). The role of multinational enterprises in achieving sustainable development goals. https://doi.org/10.46697/001c.31077
Gillies J, Cailliau R (2000) How the web was born: the story of the World Wide Web. Oxford University Press
Global e-Sustainability Initiative (2015) #SMARTer2030: ICT solutions for 21st century challenges. Retrieved from https://gesi.org
Global e-Sustainability Initiative (2015) SMARTer 2030: ICT solutions for 21st century challenges
Granovetter M (1973) The strength of weak ties. Am J Sociol 78(6):1360–1380
Gratton L (2020) The new long life: a framework for flourishing in a changing world. Harper Business
Gray B (1989) Collaborating: finding common ground for multiparty problems. Jossey-Bass
Griffith E (2008) Google docs: how cloud computing is changing the world. PC Magazine
Guinan PJ, Parise S, Langowitz N (2019) Creating an innovative digital project team: levers to enable digital transformation. Bus Horiz 62:717–727
Han Z, Yang Y, Wang W, Zhou L, Gadekallu TR, Alazab M, Gope P, Su C (2022) RSSI map-based trajectory design for UGV against malicious radio source: a reinforcement learning approach. IEEE Trans Intell Transp Syst 24:4641–4650
Hauben M, Hauben R (1997) Netizens: on the history and impact of Usenet and the Internet. Wiley-IEEE Computer Society Press
Hive (2023) Hive: project management & sustainability tracking
Hoffman DL (2020) Remote work and digital transformation: a new era of collaboration. Harvard Business Review
https://asana.com/
https://carboncare.org/
https://ecochain.com/
https://hive.com/
https://sustainlab.co/
https://www.weloop.org/
International Labour Organization (ILO) (2020) Teleworking during the COVID-19 pandemic and beyond: a practical guide, ilo.org

Judith AM, Priya SB, Mahendran RK, Gadekallu TR, Ambati LS (2022) Two-phase classification: ANN and A-SVM classifiers on motor imagery BCI. Asian J Control 25:3318–3329
Khanom MT (2023) Business strategies in the age of digital transformation. J Bus 8:28–35
Kranz J, Zeiss R, Beck R, Gholami R, Sarker S, Watson RT, Whitley EA (2022) Practicing what we preach? Reflections on more sustainable and responsible IS research and teaching practices. Commun Assoc Inf Syst 51:42
Latour B (2005) Reassembling the social: an introduction to actor-network-theory. Oxford University Press
Leiner BM et al (1997) The past and future history of the Internet. Commun ACM 40(2):102–108
Leonardi PM (2018) Social media and the changing information environment: a study of virtual teams. J Manage Stud 55(7):1201–1223
Lian Z, Zeng Q, Wang W, Gadekallu TR, Su C (2022) Blockchain-based two-stage federated learning with non-IID data in IoMT system. IEEE Trans Comput Soc Syst 10:1701–1710
McKinsey & Company (2020) The future of work after COVID-19
Means B, Bakia M, Murphy R (2014) Learning online: what research tells us about whether. Routledge, When and How
Microsoft (2017) Introducing Microsoft Teams. Retrieved from https://www.microsoft.com/en-us/microsoft-365/microsoft-teams/group-chat-software
Microsoft (2023) Microsoft Teams. https://www.microsoft.com/en-us/microsoft-teams/group-chat-software
Miro (2023) Miro: the visual collaboration platform. https://miro.com/
Moghrabi IA, Bhat SA, Szczuko P, AlKhaled RA, Dar MA. Digital transformation and its influence on sustainable manufacturing and business
O'Neill N (2010) The history of ICQ, the original instant messenger. SocialTimes
Olson M (1965) The logic of collective action: public goods and the theory of groups. Harvard University Press
Reed MS (2008) Stakeholder participation for environmental management: a literature review. Biol Cons 141(10):2417–2431
Rheingold H (1993) The virtual community: homesteading on the electronic frontier. Addison-Wesley
Rockström J, Steffen W, Noone K et al (2009) A safe operating space for humanity. Nature 461(7263):472–475
Slack (2023) Slack features. https://slack.com/features
Spataro J (2021) The next great disruption is hybrid work—are we ready? Microsoft Work Trend Index
Tapscott D, Williams AD (2006) Wikinomics: how mass collaboration changes everything. Portfolio
Trello (2023) Trello: visual project management. https://trello.com/
Carbon Trust (2014) The carbon trust: home. Retrieved from carbontrust.com
UNESCO (2020) Education: from disruption to recovery, unesco.org
United Nations (UN) (2015) Transforming our world: the 2030 agenda for sustainable development
Wasserman S, Faust K (1994) Social network analysis: methods and applications. Cambridge University Press
Wellman B, Quan-Haase A, Boase J, Chen W, Hampton K, de Diaz II, Miyata K (2001) The Internet in everyday life: an introduction. Am Behav Sci 45(3):436–459
WeLOOP (2023) WeLOOP: eco-design & environmental assessments
Google Workspace (2023) Google Workspace: collaboration and productivity apps for business. https://workspace.google.com/
Zhang J, Long J, von Schaewen AME (2021) How does digital transformation improve organizational resilience? Findings from PLS-SEM and fsQCA. Sustainability 13:11487
Zoom (2020) A message to our users. https://blog.zoom.us/a-message-to-our-users
Zoom (2023) Zoom video communications. https://zoom.us/

Digital Twin Technology: A Step Towards the Future of Intelligent Electric Vehicles

Parul Asati, Sandeep Raghuwanshi, Arif Hasan, Prashant Raj Singh, Mohammad Suhail, and Neetu Mishra

Abstract The aim of this chapter is to discuss how the integration of technologies of sector 4.0, including the concept of Digital Twin and smart material, may help the EV sector improve the sales figure of EV in the Indian market. Digital twins are believed to be a disruptive technology that has the potential to change the face of many markets. The various possibilities of application of digital twin are in aviation, medicine, smart factories, urban planning, etc. With the growth and rapid rise of electric vehicles, the demand for newer and better technology is growing high. Various factors come in the way of widespread use of electric vehicles in India; these include, among others, a general lack of easy access to charging stations and an intricate internal wiring system that makes any necessary repairs and general maintenance an absolute pain and a very expensive one at that. This research has focused on a promising future of EV repair and maintenance in the marketplace of India, which is focused on simplification of the procedure. Various challenges regarding EV repair and maintenance are included in this paper. When these obstacles are met using different kinds of smart materials and Industry 4.0 technology, EVs would make an efficient and seamless transition into the Indian market. It also discusses the future potential of the study in the industry and huge research obstacles faced by the study. The results of this study enabled us to identify various intelligent materials that the electric vehicle industry could employ in their vehicles. These materials will save exorbitant costs from dent and scratch repairs whenever an accident occurs, thus making the overall maintenance of EVs far cheaper. Also, applying Digital Twin to electric vehicle servicing and repair will make the process much more effective by reducing service time, thereby reducing associated expenses at a very significant rate. To reach the point we wanted with this

P. Asati (✉) · S. Raghuwanshi · A. Hasan · N. Mishra
Amity University, Gwalior, Madhya Pradesh, India
e-mail: nitinagarwal714.na@gmail.com

P. R. Singh
Avan and Associates, Aligarh, India

M. Suhail
Samarkand State University, Boulevard, Uzbekistan, Samarkand 140104

W. Leal Filho et al. (eds.), *Metaverse and Sustainability: Business Resilience Towards Sustainable Development Goals*, World Sustainability Series,
https://doi.org/10.1007/978-3-031-89545-6_17

study, we went into great explanation regarding how DT and smart materials operate in an electric vehicle market.

Keywords Digital twin · Electric vehicles · Sustainability · Industry 4.0

1 Introduction

Digital Twin: A technological version that utilizes real-time data to simulate with automated learning, reasoning, and emulation of a real-world item was popularised in the year 1991 by the Mirror Worlds of David Gelernter commercially available since 2002. NASA'S John Vickers used the term "digital twin" in the year 2010 (Piascik et al. 2012). The basic theory of using digital twin is to do work easier. It will, therefore, be right in saying that NASA started the idea of digital twin technology in the 1960s over their space exploratory missions when for every rocket that went into space, a clone was provided with the same accuracy in its physical counterpart used to predict and analyze all those issues that may arise during the time of the mission.

The digital twin provides insight into probable output of many researches and the design of products, hence increasing efficiency. They can also help organizations in product improvements prior to production; they can also monitor production systems after the production for maximum efficiency during manufacturing processes, as noted by John et al. (2025). When a product has reached its end of life and needs final processing, the digital twin can assist producers in making decisions about the products' disposal (Zhuang et al. 2018). They will help in determining the materials necessary for the production of products.

2 Digital Twin

The use of digital systems facilitates industrial transformation and achieve intelligent manufacturing through the factors of Industry 4.0. Intelligent Smarter/advanced manufacturing strategy aims to address industrial manufacturing demands aimed at being socializing, service-oriented, and environmentally friendly (Asati et al. 2024a, b, c). To do this, it mixes cutting-edge technology and develops innovative production lines. Recognising the integration of information and physical production is crucial for reaching such aims. These days, a lot of businesses have set up their first intelligent manufacturing lines. Due to poorly actual knowledge and digital data are merged technically (Venkatesan et al. 2019).

Regarding Industry 4.0, the **benefits of digital twin** is seen as the perfect concept. With the help of robot's combination, data exploration, and the Internet of Things, a new integrated environment called a "digital twin" is being created. Using everyday life designs, sensor updates, and multiscale aspects is known as a "digital twin" modelling method. The Manufacturing Twin, which is used while an item is being

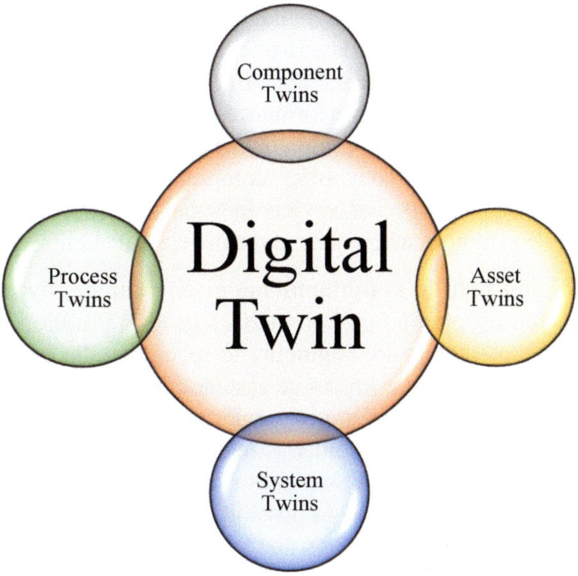

Fig. 1 Smart production and manufacturing of electric vehicles (Vandana and Panigrahi 2021)

manufactured, the Product Twin, which is used once it is ready for use, and the Service Twin, which is used while the product is not being used, work together to form the digital twin, which is simply the digital version of the real systems (Bhatti et al. 2021). **Industry 4.0** uses the digitization of things that exist to build automation at the R&D, manufacturing, and storage levels. The development stage combines information and technology for manufacturing to implement an effective digital twin. Digital twins have already had a substantial impact on product development, cutting development times by 20–50% overall, as well as the costs related to pre-production prototypes, problem solving, and the likelihood of commercial success (Bednarz et al. 2024) (Fig. 1).

There are four major categories of digital twin technologies, each having its characteristic and benefit. Four of them include component, asset, system, and process parts. Let's look closer at each of them.

A component twin is a digital twin that precisely represents the motors, sensors, switches, and valves. Component twins provide detailed information on the behavior of a component over time and real-time about the performance based on its state. In this manner, it will be easy for the companies to monitor the performance and health of these components and make changes if needed.

The Asset twin is the digital version of actual assets, such as structures, machinery, and cars. It will grant real-time data on operating status, performance metrics, and other environmental factors of an asset. In this way, business operational efficiency can be enhanced while downtime will be reduced.

The system twin is a digital representation of the overall systems or processes. It allows the business to keep a track, assess, and find faults in systems to make improvements in their performances. An organization can easily manage the workflow and can enhance overall performance with the support of a system twin.

Digital copies of complete customer journeys or corporate processes are called **process twins**. They offer comprehensive data on how clients engage with a company's goods and services in actual time (Perno et al. 2022). This aids businesses in determining where they can enhance the client experience.

Digital twin in manufacturing gives estimates by suggesting that the market will increase from $6.5 billion in 2021 to $125.9 billion in 2040, it is clear that a wide range of sectors are beginning to recognize the benefits of digital twins (Alptekin et al. 2021). Here are some instances of how businesses are now utilizing digital twins. Digital twins are particularly relevant in the manufacturing sector, which is a massive business. Ford was really among the first businesses to use a digital twin approach. The business makes sure that resources are spent wisely and that items are made as efficiently as possible by using digital twins to monitor and maintain its manufacturing operations on reality fact.

Digital twins are also being used in electric automobiles to enhance product design. Businesses have used them, for instance, to model how **various manufacturing techniques** could affect how long their products last. Digital twins have been used by General Electric (GE) for a number of years. The business has enhanced the performance and design of many different goods, such as engines and gas turbines, by using digital twins (Naseri et al. 2023). To test fresh concepts and tactics before putting them into practice in the real world, GE has developed a "virtual engine" using digital twins. This expedites the development process and lowers the possibility of a failed product. The business has saved clients 1.6 billion dollars by using digital twins' real-time monitoring abilities. By taking an early, end-to-end perspective of the product, they have also been able to cut waste from the product by as much as 75% and total expenses by up to 30% (Huang et al. 2016).

The role of enabling technologies for the development of electric vehicles in the automotive sector, with the advancement of IoT and network innovations, offline digital models can now be converted to digital twins (DT), which offers several advantages like smart tracking system, ability to predict and reschedule upcoming maintenance events, faults in location, their endurance, and their remaining usefulness (Shihab et al. 2024). DTs have been recognized as facilitators for additional enhancement of the dependability and efficiency of electric vehicles (EVs). Additionally, the Internet of Things, cloud computing, and artificial intelligence, with virtual reality techniques are included as five contributing advances in technology (Qi et al. 2021).

Energy storage system for electric cars, digital twin can extend battery life by combining data from all technical domains into a single model. It can improve manufacturing and production processes, detect power faults during operation, and accurately model equipment fault status using large data samples. Digital twin recycling can solve battery material scarcity issues. A digital twin can enhance energy system life cycle management, recognize client needs, and provide an effective operation method, ultimately saving operating expenses (Shoukat et al. 2022). It can also address the issue of battery material scarcity through digital twin recycling.

3 Smart Material

Although the idea of smart materials is particularly fresh, their practical application is expanding because of how widely they are being applied in virtually every modern industry, from the automobile to the healthcare, security, and industrial sectors. When exposed to external encouragement, smart materials have the capacity to instantly transform their configuration or structure. Their development across almost every sector functioning in the current environment can be made possible by technology (Schwartz 2008). A significant idea of Smart material advancement in Industry 4.0 with Digital Twin seeks to establish a real-time connection among an actual property to its virtual counterpart, enabling the exchange of information in real-time. The smart materials help to improve the effectiveness, lifespan, and performance of their commodity. India is the world's most population nation (Abdullah and Al-Alwan 2019). There are many more cars on Indian roadways due to the country's dense population. The majority of Indians also have unethical and bad driving behaviours. Due to these two reasons, cars in India are more likely to sustain unintentional dents and scratches.

Types of Smart Materials

Piezoelectric Materials

While pressure is applied to certain materials, or vice versa, electricity can be produced. They may be utilised in the development of intelligent sensors and device that detect and control situation like braking force, engine vibration, and tyre pressure (Suhaib Kamran et al. 2022).

Chromoactive Materials

Smart materials known as chromoactive materials react to light, humidity, hardship, and various inputs by changing colour. The electric vehicle sector can employ them for adaptive stealth and interactive displays (Qader et al. 2019).

Magnetostrictive Materials

Magnetostrictive is a thing of magnetic field that causes them to change their shape and dimensions during the process. Materials that are magneto strictive include conducting polymers, nanomaterials, polymers that are electroactive, thermoelectric materials, magnetorheological fluids, and optical fibres which helps in shaping alloys (Bucchi 2014).

Graphene Materials

Graphene is a material made of pure carbon that is one atom thick and arranged in a hexagonal lattice. Graphene is the thinnest, lightest, and strongest material known

(Aissa et al. 2015). It is also transparent, flexible, and an excellent conductor of electricity and heat, which makes them perfect for lowering vehicle weight and increasing fuel economy.

Photoactive Materials

Light-absorbing substances known as photoactive materials have the ability to transform light into biological or electrical power. An electric current is produced by the expulsion of electrons from a photoactive material by light. Battery cells that create rechargeable photo assistance can be produced using photoactive materials while utilised in electric vehicles (Qazi et al. 2022). In addition to having excellent charging and moving capabilities, these kinds of batteries can have a high energy density.

Limitation of Widespread Adoption of Digital Twin Technology

First Monetary Commitment

Digital twin implementation can come at a considerable cost, including for hardware, software, and training. Small firms or initiatives with tight budgets may find this difficult. Team members who are not familiar with digital technologies may find it difficult to keep up with the high learning curve of digital twin technology (Tzachor et al. 2022).

Lack of Standardized Strategies

Standardised approaches to using digital twin technology are lacking. Data security issues may arise from data exchanged between digital models and real-world systems. To secure the data, businesses must invest in strong security procedures (Augustine 2020).

Connecting Infrastructure

The complex method of integrating digital twins with current systems may result in disparities in information and difficulties with interoperability (Mishra et al. 2024a, b, c, d). It's possible that the regulations and rules already established for the deployment of digital twins are inadequate to handle their use in intricate systems.

Exemplary Management

Implementing strong model governance is essential to avoiding weaknesses in safeguarding information. Complex object replication calls for a lot of data and computing power (Mishra et al. 2023).

Barriers to Digital Twin Technology

Pricing

Since digital twin technology requires specialised equipment, software, and knowledge, it can be pricey. The pricing is a main factor of EV batteries and dependent on the availability and demand for raw materials (Bhatti et al. 2021).

Safety and Confidentiality

Digital twins are an attractive target for security threats to data due to their complexity. In the event of an accident, EV batteries may catch fire. EV battery manufacture generates a lot of carbon dioxide and consumes heavy metals. Used electric vehicle batteries can be challenging to recycle and may contaminate water supplies.

Teamwork

Throughout the battery value chain, it may be challenging to connect people due to shortage of connectivity. Some individuals might have to wait in queue to charge since there aren't enough charging stations (Mishra et al. 2024a, b, c, d). Apartment dwellers might not have easy access to a charging station. So, teamwork and proper knowledge sharing team base is required to help and support customers.

Batteries Issues

Under certain scenarios, batteries used in electric vehicles may malfunction very quickly, and reusing and recycling batteries can present difficulties. The lifespan of EV batteries fluctuates. Making batteries more affordable, dependable, and safe can increase the practicality of EVs (Roy et al. 2023). Lack of electric vehicles charging station is required as compared to petrol pumps, which is the main concerns nowadays. Charging time: Depending on the kind of charger, an EV battery can take two to six hours to completely charge. Range anxiety may arise from an EV's constrained driving range (Sharma et al. 2022).

Digital Twin Technology Assistance

Battery Management

Digital twins can aid in battery management by providing real-time information on the performance of batteries. Predictive upkeep is made possible via both of these factors, which aid in the identification of breakdown mechanisms (Bello et al. 2024).

Supply Line Control

Challenges in the supply chain, such as fleet management, route efficiency, and packaging performance, can be addressed by digital twins (Bello et al. 2024). Automakers can produce intricate simulations of powertrain components with the use of digital twins.

Workflow Improvement and Training

Since there is intense rivalry in the market, electric vehicles technology is developing swiftly and efficiently to better meet the needs of consumers. The engineering adjustments may call for modifications to layout requirements these significant specification modifications may affect the reliability, security, durability, and life-cycle efficacy of the electric vehicle. In addition to providing information on the technical characteristics of electric vehicles, training will include maintenance, diagnostics, and repair. To obtain practical understanding of the system and vehicle, manufacturers should also offer hands-on training.

Technological Infrastructure

AI-based Advanced Driver Assistance Systems (ADAS) can enhance EV safety by identifying possible risks and implementing preventative measures. Technology is improving thanks to power driving and smooth communication. Our manufacturing, and maintenance businesses now operate entirely differently thanks to Industry 4.0 (Mihai et al. 2022; Mishra et al. 2024a, b, c, d). The incorporation of robots and cognition in these processes has risen with the advancement of industrial revolutions. The convenience and efficiency of many industrial processes have grown thanks to these new technologies, which have also lowered expenses and performance requirements.

Future Trends of Market

In 2023, there were 14 million electric automobiles registered worldwide, bringing the total number of automobiles to 40 million. It is anticipated that the market for electric vehicles (EVs) will continue to expand by 2030. EV sales would represent a sizable portion of all automotive sales.

Sales: By the end of 2024, 17 million electric vehicles are expected to have additionally be sold, maintaining the upward trend in sales. This would represent an increase of almost 20% every year (Alcaraz and Lopez 2022).

Funding: The purchase of electric vehicles and the infrastructure necessary for EV charging will probably increase. By 2030, it is projected that the worldwide demand for infrastructure to charge electric vehicles will reach $229.1 billion (Minerva et al. 2020).

Sustainable Benefits: It is anticipated that electric vehicles would eventually replace conventional gasoline automobiles due to their lower overall cost of ownership and benefits (Mishra et al. 2024a, b, c, d).

Autonomous Driving: EVs are an excellent match for autonomous driving since they have fewer moving parts than cars with internal combustion engines.

Prices for Batteries: It is projected that as time goes on, battery prices will decline, bringing down the price of EVs.

Government Policies: It is projected that the EV business would grow in the presence of favourable government policies (Ali et al. 2023).

Public Curiosity: Growing environmental curiosity is expected to drive growth in the electric vehicle (EV) sector. The digital twin market in India is growing quickly. It is propelled by things like increased industry-wide digitisation, improved technology adoption, and the need for efficient asset management. In the upcoming years, the sector is expected to develop at a considerable compound annual growth rate (CAGR) (Bhatti et al. 2021).

Transformation of Industry

Batteries: Toyota says that it has made substantial progress in this domain. These technologies, which are more potent and smaller than conventional batteries, allow electric automobiles to go 1200 km on just one charge although they only take 10 min to finish the charge.

Infrastructure for Charging

The purpose is to speed out the charging procedure in order to eliminate anxiety over range. EV charging is more effective, inexpensive, and ecological with smart charging stations. Infrastructure for charging is necessary to enable vehicles to go long distances without compromising. In order to accelerate the usage of EVs, smart grids make it easier to integrate renewable energy.

Battery Innovation

Improvements in the battery sector have improved power density, recharging velocity, and overall efficiency. Batteries made from solid-state materials are considered a game-changer in the evolution of electric vehicles, since they provide faster charging times and greater range (Das et al. 2024).

Automatic Driving

Electric cars usually incorporate modern equipment like radar detectors, and sensors, which are defining the field for autonomous driving (Gao et al. 2023).

Conductive Charging Technologies

Through metal-to-metal contact, power is transferred straight from the grid to the EV using this method. It enables charging both inside and outside of the ship (Naseri et al. 2023).

Energy Recovery Systems

The primary objective of energy management solutions in EV charging systems is to maximise the efficient utilisation of readily accessible energy resources.

4 Building Electric Vehicles with Digital Twin

The Fourth Industrial Revolution has highly impacted the product and the way they are manufactured. In the emerging fast development, new ways of working are utilized in the field of digital twin version of physical items. Time and price are two main factors of the strategy. For this reason, a digital twin model is developed to test and represent the behavior of an electric vehicle. Thus, this has led to the thought of converting data into a virtual to physical electric vehicle model. The era of Industry 4.0 has impregnated the element of chaos related to the automobile sector. New and several innovations have been inducted with the manufacturing processes as well as within the products. The rise in burning of fossil fuel along with greenhouse gas emissions is also more (Asati et al. 2024a, b, c). On the other hand, electric vehicles are getting a rapid success among people. As per Zhang and Zhang (2020), the automotive industry is being driven into a massive transition into sustainable and greener alternatives. The process of creating an electric vehicle, designing a product with digital twins, is going to be put to use before the product cycle begins.

Objectives and strategies of data collection in the traditional method for electric vehicles creation. The operation of EVs is to be carried out in a race or training ring, this is an inefficient technique since the cost, time, and effort required are very high. A possible remedy to these problems is the testing of the efficiency of electric vehicles using the digital twin approach by a digital twin model. This solution will effectively reduce time to improve on plans or implant the technology before competition. EVs have been participating in competitions for years now and, nowadays, most features and services are being realized through smart technology. Conventional vehicles, when not electric, carry an internal combustion engine, which contributes further to the world's fossil fuel consumption and greenhouse gas emissions. Electric vehicle has been advanced to handle this issue of green fuel (Ali et al. 2023; Asati et al. 2024a, b, c).

The electric mobility sector, which has been placed or situated now with the help of the adoption of The Internet of Things by many industrial domains to make this revolution intelligent in performance and improvement. Thanks to the sensors underlying the networks of the Internet of Things, vehicles are going to be more in the nature of smart objects (Agarwal et al. 2022; Shankar et al. 2024; Spencer et al. 2004). Therefore, the idea of electric vehicle (EV) management software is going to change quite dramatically because of the volume of data that these sensors can gather. The detectors would be monitoring every area of the vehicle, its internal parts, and motions and changes taking place while monitoring engines' charging (Drosu et al. 2019). The data generated by these sensors is required to be collected, synchronized, analysed, and processed for the betterment of the EV support quality and to maintain the EV control system for making decisions. In the year 2019, DT revolutionized the top 10 strategic innovations across the world involving the Technology improvement, which facilitated innovative technology with the likes of quantum computing, LED panels, and Driver Less Cars (Lim et al. 2020). Embracing the DT concept might facilitate the development of the electric industry and those of business activities. Digital data might improve the intelligence of an electric vehicle system towards the engine optimization. The results of the assessment can be used to enable the product and the process to perform better in the actual environment.

"Digital Twin and Eco-Friendly System" will support the transport industry. DT technology employs intelligent technology support powerful strengths to control effective emergency plans, traffic awareness, and safety informative systems. Moreover, it can provide a new route and transport solutions such as smart handle with productivity and security and enable easy jams control (Liu et al. 2021). Many innovations have been triggered by the explosive expansion of smart control systems. The effect and the information delved are by the growth of the main causes of the smart electric vehicle transportation sector and this expansion is produced through it. An electric vehicle is charged by connecting it to a charging cable attached to one's home or public area. With automation taking over manual work, automatic charging systems are compulsory because manual charging would just be ineffective. Driverless cars with electric cars have become a cliché. With regard to these statements, now more than ever, an electric vehicle now requires a battery management system to ensure the longevity, safety, and dependability of the batteries (Gandoman et al. 2019; Ali et al. 2024).

On the other hand, the digital twin is among the newest forms of transportation services, which include data exchange, vehicle security, and traffic information reporting in smart transportation (Bhatti et al. 2021). With the rapid level of advancement in the electric mobility sector, we should use as well as connect DT technology with electric vehicles to keep up the pace in cities that are smart, whether they are autonomous or not. This will also create a multitude of managing and enhancing opportunities for mobile transportation systems locally and nationally (Chiarini 2021).

Innovations in digital twin and internet of things in electric and smart transportation are being facilitated with real, tested analytics and selection through data interpretation. This incorporation of DT technology handles challenges, enabling connectivity and intelligent access to physical equipment, thus transforming the electric transportation revolution (Botín-Sanabria et al. 2022).

User-friendly applications in the process, including digital twins and autonomous electric vehicles, also known as driverless vehicles or fully automatic driving vehicles that can self-drive without any human intervention. Also, since electric propulsion is decidedly simpler to handle, they are electric. Advanced car technology will calculate the direction to steer and drive safely based on sensors. The navigation heeds and inputs past information on the car, which regulates the movement or decision process. The intelligent electric vehicles market is expanding and holds the potential benefits (Elvas and Ferreira 2021).

Prospective of this technology for solutions in ecological and sustainable development problems—It should include a solution to the depletion of fossil fuels, increasing air pollution, and the issues with global warming and changes in the climate. EVs are alternatives to such solutions and show some promise for reducing these problems. Increasing EV uptake is expected to further promote sustainable solutions to transport (Hull 2008).

The "Role of digital twin in industry 4.0" states the industries manufacturing smart batteries of fourth-generation electric vehicles. This has to do with the understanding of the ability to use the digital twin in battery production and developing charging stations within industry 4.0 (Ashwin 2024; Zema et al. 2022). The sector 4.0 automotive is manufacturing electric vehicles by a distinct system that has an ability of turning gasoline into energy. This may include a battery, solar panels, and fuel cells (Athanasopoulou et al. 2023; Yang et al. 2020).The industry has undergone a serious change with the coming of the technology of additive manufacturing, which means creating things layer by layer in a variety of forms, sizes, and materials. Development in 3D printing has further created the manufacturing of a wide variety of products from industrial level to commercial to the even personal level (Franke et al. 2019). These range from printing little stuff to big printing facilities to even rockets.

5 Conclusion

The idea of digital twin technology and smart materials is now accessible and can be used in the Indian electric vehicle automation market. We coupled the idea of the digital twin with the application of smart materials in EV bodies to identify several obstacles to the EV industry's adoption in India. This research primarily offers policies for utilizing Smart Materials and the Digital Twin idea to facilitate EV assistance, adjustment, and conservation. As technology is updated more and more societies are fully dependent on Autonomous Vehicles. The development of autonomous EVs will heavily rely on digital twins. To guarantee the security and dependability of autonomous systems, they will model and examine a vast array of driving situations. Vehicle electrification is a global trend, and in the near future, it will be the pivot point of conveyance. AI, big data, online applications, IoT, and blockchain form part of the development which is emerging on which the research in Digital Twin and battery technologies is dependent. Coming to digital twin technology, this area is being utilized for upgrading the functionalities in electric vehicles. Battery and smart energy storage systems are only in their infancy now, a new period of investigation. With the huge development of new electric vehicles and growth in sustainable energy, there's a need for advanced technology at all levels, enabling intelligent system connections and better transportation. Digital transformation occurs in the asset-intensive industry; it has been called for by advanced insight into assets, equipment, facilities, and processes. They constitute a part of that reform in that the continuous upgrading of skills and capability means that digital twins will always create insights necessary in the development of better products and processes.

References

Abdullah YS, Al-Alwan HA (2019) Smart material systems and adaptiveness in architecture. Ain Shams Eng J 10(3):623–638

Agarwal P, Mittal M, Ahmed J, Idrees SM (eds) (2022) Smart technologies for energy and environmental sustainability. Springer International Publishing

Aissa B, Memon NK, Ali A, Khraisheh MK (2015) Recent progress in the growth and applications of graphene as a smart material: a review. Front Mater 2:58

Alcaraz C, Lopez J (2022) Digital twin: a comprehensive survey of security threats. IEEE Commun Surv Tutorials 24(3):1475–1503

Ali WA, Fanti MP, Roccotelli M, Ranieri L (2023) A review of digital twin technology for electric and autonomous vehicles. Appl Sci 13(10):5871

Ali ZM, Calasan M, Gandoman FH, Jurado F, Aleem SHA (2024) Review of batteries reliability in electric vehicle and E-mobility applications. Ain Shams Eng J 15(2):102442

Alptekin B, Tunaboylu B, Zaim S, Perlo P (2021) Smart manufacturing of electric vehicles. In: Digital conversion on the way to industry 4.0: selected papers from ISPR2020, September 24–26, 2020 Online-Turkey. Springer International Publishing, pp 767–773

Asati P, Raghuwanshi S, Hasan A, Zeffer A (2024b) The role of electric vehicles in the agriculture industry using IoT: turning electricity into food. Digit Agric Ecosyst Revolutionary Adv Agric 369–379

Asati P, Raghuwanshi S, Vashisht A, Muley D, Khare VK (2024) Exploring the green energy: factors influencing battery electric vehicles adoption in Madhya Pradesh Region. Int Rev Manag Mark 14(5):151–157

Asati P, Raghuwanshi S, Vashisht A, Vyas R, Singh PR (2024) Factors influencing health and environmental intentions to adopt electric vehicles from the perspective of business school students. Int J Relig 5(1):743–750

Ashwin KV (2024) Industry 4.0: intelligent routing and scheduling algorithm for EV charging based on EV battery management system. In: Industry 4.0 key technological advances and

design principles in engineering, education, business, and social applications. CRC Press, pp 307–328
Athanasopoulou L, Bikas H, Papacharalampopoulos A, Stavropoulos P, Chryssolouris G (2023) An industry 4.0 approach to electric vehicles. Int J Comput Integr Manuf 36(2):334–348
Augustine P (2020) The industry use cases for the digital twin idea. In: Advances in computers, vol 117, No 1. Elsevier, pp 79–105
Bednarz T, Baier A, Paprocka I (2024) A framework for communicating and building a digital twin model of the electric car. Appl Sci 14(5):1776
Bello AO, Olanrewaju OI, Gbenga PO, Khan AA, Isa RB (2024) Exploring the barriers to digital twin adoption in the Nigerian construction industry: a structural equation modelling approach. J Eng Des Technol
Bhatti G, Mohan H, Singh RR (2021) Towards the future of smart electric vehicles: digital twin technology. Renew Sustain Energy Rev 141:110801
Botín-Sanabria DM, Mihaita AS, Peimbert-García RE, Ramírez-Moreno MA, Ramírez-Mendoza RA, Lozoya-Santos JDJ (2022) Digital twin technology challenges and applications: a comprehensive review. Remote Sens 14(6):1335
Bucchi F (2014) Smart materials and vehicle efficiency. Design and experimentation of new devices. Available at: https://core.ac.uk/pdf/80276398.pdf
Chiarini A (2021) Industry 4.0 technologies in the manufacturing sector: are we sure they are all relevant for environmental performance?. Bus Strategy Environ 30(7):3194–3207
Das O, Zafar MH, Sanfilippo F, Rudra S, Kolhe ML (2024) Advancements in digital twin technology and machine learning for energy systems: A comprehensive review of applications in smart grids, renewable energy, and electric vehicle optimisation. Energy Convers Manag X 100715
Drosu A, Suciu G, Scheianu A, Petre I (2019) An analysis of hybrid/electric vehicle monitoring systems and parameters. In: 2019 electric vehicles international conference (EV). IEEE, pp 1–5
Elvas LB, Ferreira JC (2021) Intelligent transportation systems for electric vehicles. Energies 14(17):5550
Franke T, Bühler F, Cocron P, Neumann I, Krems JF (2019) Enhancing sustainability of electric vehicles: a field study approach to understanding user acceptance and behaviour. In: Advances in traffic psychology. CRC Press, pp 295–306
Gandoman FH, Jaguemont J, Goutam S, Gopalakrishnan R, Firouz Y, Kalogiannis T, Omar N, Van Mierlo J (2019) Concept of reliability and safety assessment of lithium-ion batteries in electric vehicles: basics, progress, and challenges. Appl Energy 251:113343
Gao X, Simeone A, Zhang J (2023) Smart decision-making for design adaptation of electric vehicles using big sales data. Procedia Cirp 119:710–715
Huang CJ, Hu KW, Chen HM, Liao HH, Tsai HW, Chien SY (2016) An intelligent energy management mechanism for electric vehicles. Appl Artif Intell 30(2):125–152
Hull A (2008) Policy integration: what will it take to achieve more sustainable transport solutions in cities. Transp Policy 15(2):94–103
Kamran SS, Haleem A, Bahl S, Javaid M, Nandan D, Verma AS (2022) Role of smart materials and digital twin (DT) for the adoption of electric vehicles in India. Mater Today Proceed 52:2295–2304
Lim KYH, Zheng P, Chen CH (2020) A state-of-the-art survey of Digital Twin: techniques, engineering product lifecycle management and business innovation perspectives. J Intell Manuf 31(6):1313–1337
Liu M, Fang S, Dong H, Xu C (2021) Review of digital twin about concepts, technologies, and industrial applications. J Manuf Syst 58:346–361
Mihai S, Yaqoob M, Hung DV, Davis W, Towakel P, Raza M, Karamanoglu M, Barn B, Shetve D, Prasad RV, Venkataraman H (2022) Digital twins: a survey on enabling technologies, challenges, trends and future prospects. IEEE Commun Surv Tutorials 24(4):2255–2291
Minerva R, Lee GM, Crespi N (2020) Digital twin in the IoT context: a survey on technical features, scenarios, and architectural models. Proc IEEE 108(10):1785–1824

Mishra NK, Chakraborty A, Singh S, Ranjan P (2023) Efficiency analysis of engineering colleges in India: decomposition into parallel sub-processes systems. Socioecon Plann Sci 89:101708
Mishra NK, Pande Sharma P, Chaudhary SK (2024b) Redefining agile supply chain practices in the disruptive era: a case study identifying vital dimensions and factors. J Glob Oper Strateg Sour
Mishra NK, Raj A, Jeyaraj A, Gupta R (2024) Antecedents and outcomes of blockchain technology adoption: meta-analysis. J Comput Inf Syst 64(3):342–359
Mishra NK, Sahoo S, Agarwal S, Sharma PP, Ilahi F (2024d) Impact of institutional pressures and security on blockchain technology adoption and organization performance: an empirical study. J Technol Transf 1–26
Mishra N, Vashisht A, Raghuwanshi S (2024a) Transforming Indian agriculture: unleashing the potential of digital agriculture using efficiency analysis. Digital agricultural ecosystem: revolutionary advancements in agriculture, pp 123–143
Naseri F, Gil S, Barbu C, Çetkin E, Yarimca G, Jensen AC, Larsen PG, Gomes C (2023) Digital twin of electric vehicle battery systems: comprehensive review of the use cases, requirements, and platforms. Renew Sustain Energy Rev 179:113280
Perno M, Hvam L, Haug A (2022) Implementation of digital twins in the process industry: a systematic literature review of enablers and barriers. Comput Ind 134:103558
Piascik B, Vickers J, Lowry D, Scotti S, Stewart J, Calomino A (2012) Materials, structures, mechanical systems, and manufacturing roadmap. NASA TA 12–2
Qader IN, Mediha KÖK, Dagdelen F, Aydoğdu Y (2019) A review of smart materials: researches and applications. El-Cezeri 6(3):755–788
Qazi AM, Mahmood SH, Haleem A, Bahl S, Javaid M, Gopal K (2022) The impact of smart materials, digital twins (DTs) and Internet of things (IoT) in an industry 4.0 integrated automation industry. Mater Today Proceed 62:18–25
Qi Q, Tao F, Hu T, Anwer N, Liu A, Wei Y, Wang L, Nee AY (2021) Enabling technologies and tools for digital twin. J Manuf Syst 58:3–21
Roy R, Bhakta D, Roy S (2023) Applications and types of smart materials. In: Modeling, characterization, and processing of smart materials. IGI Global, pp 266–296
Schwartz M (2008) Smart materials. CRC press, taylorfrancis.com. https://doi.org/10.1201/9781420043730, p 554
Shankar U, Kaul V, Gupta V, Mishra NK (2024) Workplace ostracism and counterproductive work behaviour: parallel mediation of felt accountability and emotional exhaustion. J Manag Develop
Sharma A, Kosasih E, Zhang J, Brintrup A, Calinescu A (2022) Digital twins: state of the art theory and practice, challenges, and open research questions. J Ind Inf Integr 30:100383
Shoukat MU, Yan L, Du C, Raza MUM, Adeel M, Khan T (2022) Application of digital twin in smart battery electric vehicle: industry 4.0. In: 2022 International conference on IT and industrial technologies (ICIT). IEEE, pp 1–7
Spencer BF Jr, Ruiz-Sandoval ME, Kurata N (2004) Smart sensing technology: opportunities and challenges. Struct Control Health Monit 11(4):349–368
Tzachor A, Sabri S, Richards CE, Rajabifard A, Acuto M (2022) Potential and limitations of digital twins to achieve the sustainable development goals. Nat Sustain 5(10):822–829
Vandana GA, Panigrahi BK (2021) Multi-dimensional digital twin of energy storage system for electric vehicles: a brief review. Energy Storage 3(6):e242
Venkatesan S, Manickavasagam K, Tengenkai N, Vijayalakshmi N (2019) Health monitoring and prognosis of electric vehicle motor using intelligent-digital twin. IET Electr Power Appl 13(9):1328–1335
Yang C, Zha M, Wang W, Liu K, Xiang C (2020) Efficient energy management strategy for hybrid electric vehicles/plug-in hybrid electric vehicles: review and recent advances under intelligent transportation system. IET Intel Transp Syst 14(7):702–711
Zema T, Sulich A, Grzesiak S (2022) Charging stations and electromobility development: a cross-country comparative analysis. Energies 16(1):32
Zhang X, Zhang Z (2020) How do smart villages become a way to achieve sustainable development in rural areas? Smart village planning and practices in China. Sustainability 12(24):10510

Zhuang C, Liu J, Xiong H (2018) Digital twin-based smart production management and control framework for the complex product assembly shop-floor. Int J Adv Manuf Technol 96:1149–1163

John HB, Agrawal M, Nema P, Hasan A, Rakhra IK, Singh A (2025) Green growth: a bibliometric analysis of digital innovation and Sustainable Development Goals (SDGs). J Innov Entrep 14(1):15

Shihab KK, Mir FA, Hasan A, Agarwal M, Dixit A (2024) Pitch competitions–catalyzing innovation through alternative funding. In Alternative Finance (pp. 198–211). Routledge.

The Role of Digital Twins in Toward Sustainable Urbanization: The Case of Smart Cities

Garima Sharma, Sandeep Raghuwanshi, Mohammad Suhail, Trapati Narwaria, and Arif Hasan

Abstract Innovation in "smart cities" is accelerating at a rate never seen before in modern history. From a development, operations, and maintenance standpoint, smart cities are undergoing rapid change of all kinds. To achieve this advanced level of urbanization and fully utilize digital infrastructure for rapid urban growth, today's Smart Cities are striving. When it comes to making their cities better places to live and providing inhabitants with more comfortable and joyful living options, Smart Cities are utilizing a variety of tools and technologies to become smarter and more connected with their stakeholders through technology. The relentless expansion of urban areas has brought with it a host of new social, economic, and environmental problems, all of which must be tackled in concert if humanity is to achieve its goal of a more sustainable global community. It is crucial to provide planners with the necessary tools to tackle these issues, with sustainability being one of the primary goals. Incorporating digital twins with modeling and simulation capabilities can greatly enhance the use of these technologies. Integrating these models and technology improves data-driven environmental planning processes, which in turn offers new ways to tackle difficult environmental problems. However, to get the intended results, one must carefully navigate and mitigate the many complicated obstacles that this undertaking presents. For sustainable urban development, this study provides an all-inclusive reference that will inspire future studies, initiatives, and policymaking. This book chapter goes into detail on a number of these important technologies that are having a beneficial impact on Smart Cities. The manner in which these technologies are influencing the growth and functioning of Smart Cities is also emphasized.

Keywords Smart cities mission · Digital twins · Sustainable development · Technologies · Sustainable development goals

---

G. Sharma (✉) · S. Raghuwanshi · T. Narwaria · A. Hasan
Amity Business School, Amity University, Gwalior, Madhya Pradesh, India
e-mail: garimaonesharma@gmail.com

M. Suhail
Samarkand State University, Samarkand, Uzbekistan

W. Leal Filho et al. (eds.), *Metaverse and Sustainability: Business Resilience Towards Sustainable Development Goals*, World Sustainability Series,
https://doi.org/10.1007/978-3-031-89545-6_18

1 Introduction

"A smart city is one that has digital technology embedded across all city functions," is the simple but accurate description offered by the Smart Cities Council (Javed et al. 2022). A growing number of "Smart Cities" are springing up all over the world, and the ones that already exist are undergoing fast change. also, India isn't lagging behind. The Smart City Mission is an initiative by the Indian government that aims to transform 100 existing cities into "Smart Cities" within the next few years (Ministry of Urban Development Government of India Smart Cities Mission Statement and Guidelines Government of India Ministry of Urban Development 2015). In addition, a number of other developing nations around the world are constructing or transforming their current urban areas into Smart Cities. We should attempt to grasp the concept of a "Smart City" and how these communities are improving the quality of life for their inhabitants before delving further into the topic (De Guimarães et al. 2020), While the idea of "smart cities" has grown in popularity over the past decade or two, meticulously planned and engineered urban areas have been there for much longer. Smart cities are defined by their use of cutting-edge information and communication technologies in every stage of their creation, from planning to development to management and operation (Sánchez-Corcuera et al. 2019). In the modern era of industry 4.0, cities have improved their intelligence by utilizing a wide range of cutting-edge technologies, tools, and information technology infrastructures (Mhlanga 2021). Thus, the idea of a Smart City might vary from nation to country and even from city to city depending on the infrastructure, tools, and technology used, as well as the culture and openness to change of the residents (Fig. 1).

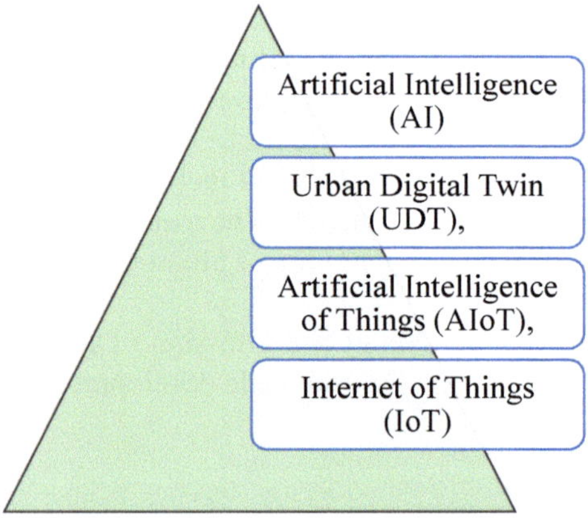

Fig. 1 Technologies in sustainable urban development

2 Smart and Sustainable Cities

Smart cities and sustainability both are associated with the same goal of society where they both deal with the use of data-driven solutions and cutting-edge tech that rediscover the new society full of new technology and comfortless however with this the cities improve the efficiency, quality of life, and environmental sustainability. New sustainable smart cities are springing up in reaction to the criticism leveled at older smart city programs (Ahvenniemi et al. 2017; Evans et al. 2019). These communities aim to achieve the Sustainable Development Goals (SDGs) by utilizing data-driven technology (Sharifi et al. 2024). Smart city enablers were thoroughly examined (Ahad et al. 2020) critical issues spanning technical, social, and environmental dimensions and offered solutions for making smart cities sustainable. In their study, del Mar Martínez-Bravo and Labella-Fernández (2024) sought to clarify "sustainable smart cities" and offer a more comprehensive interpretation. A sustainable smart city, according to this model, must exhibit sustainability-coherence, moderate expansion, increased awareness, and a mix of bottom-up and top-down efforts. The proliferation of sustainable smart city projects around the world makes it critical to assess their relative merits. That is why it's crucial to have evaluation instruments that can supply key performance indicators for all the parties involved in city planning (Leal Filho 2022).

A comprehensive strategy for sustainability, with a focus on technological and environmental aspects, is required for the transition from smart cities to sustainable smart cities. The current literature offers helpful suggestions for how to create smarter, more eco-conscious cities in the future (Blasi et al. 2022).

By utilizing (AI) and (IoT) solutions for renewable energy, energy conservation, water resource management, waste management, sustainable transportation, pollution control, c, and intelligent infrastructure, sustainable smart cities showcase the integration of technological advancements and environmental strategies (Bibri et al. 2024b). A plethora of research has investigated how (UDT), in addition to (AI) and (AIoT), may contribute to sustainable smart city design and management that is both innovative and environmentally friendly (Efthymiou and Egleton 2023). Through the integration of state-of-the-art technology, sustainable smart cities have the potential to transform into adaptive entities that can handle the increasing demands of urbanization, environmental changes, and citizen wants. Sustainable smart city ideas and practices offer a basic road map for navigating the complicated terrain of environmental sustainability (Lopez and Castro 2021). Incorporating urban digital twin (UDT) with artificial intelligence (AI) and the artificial intelligence of Things (AIoT) in a coordinated fashion allows for the creation of data-driven planning systems that are both technologically advanced and ecologically resilient (Baran et al. 2022).

3 Digital Twin for Urban Areas

One innovative technological idea that is changing the face of sustainable urban development is the urban digital twin (Samarakkody et al. 2023). Achieving the environmental goals of sustainable smart cities (Weil et al. 2023) is a top priority, and urban digital twin aims to be a potent instrument for improving urban planning, management, and governance processes and practices to contribute to the Sustainable Development Goals (SDGs) (Caprari et al. 2022). urban digital twin allows city planners and stakeholders to monitor, model, simulate, and visualize complex urban systems. It then uses AIe models and techniques to analyze the data, which is updated continuously with real-time data from IoT sensors and devices. With urban digital twin, stakeholders can understand behavioral patterns, test scenarios, predict environmental changes, optimize urban infrastructure design, and enhance services (Gkontzis et al. 2024).

The integration of AI computational and analytical capabilities into IoT devices throughout urban infrastructures is the multi-faceted and critical role of artificial intelligence of Things in urban digital twin (Hannan et al. 2021). As a result, internet of things systems and apps can gain intelligence and decision-making capabilities, and they can acquire, analyze, and anticipate data in real-time (Gade 2021). A smarter, more resilient, and sustainable urban ecosystem that enhances the quality of life for inhabitants while minimizing environmental damage can be achieved through this integration, which is crucial in building a responsive and dynamic framework. Using examples from sustainable smart cities throughout the world, urban digital twin has proven it can deliver a comprehensive and user-friendly platform for a wide range of data-driven planning and design tasks, such as (Dembski et al. 2020).

- Modeling and simulation: urban digital twin allows for dynamic and accurate three-dimensional (3D) city simulations using physical, social, economic, historical, and real-time data.
- This paves the way for planners to model and evaluate possible solutions and scenarios to determine their efficacy.
- Data-informed decision-making: Extracting valuable insights from integrated data promotes data-driven decision-making. Sustainable city planning can benefit from planners' analysis of patterns and trends (Bokhari and Myeong 2022).
- Collaborative planning: it encourages inclusive and participatory planning by bringing together several stakeholders in a virtual setting, such as community members, architects, engineers, city planners, and people.
- Allocating resources efficiently: It helps us understand how to make our cities' infrastructure, energy, and transportation systems more sustainable by revealing how we might use these things more effectively.
- Urban areas can be better prepared for climate change, natural catastrophes, and emergency responses through resilience planning, which also allows for simulations of these future events (Deng et al. 2021).

- Management and monitoring: It is possible to manage urban services proactively with the help of real-time urban digital twin monitoring. City infrastructure can run well because planners can quickly spot problems and fix them.

Urban Design Studio (UDS) serves as a digital proving ground for city planning initiatives and policies. To facilitate more efficient, adaptable, data-driven, and cooperative methods of sustainable urban development, it makes use of cutting-edge technology to build a detailed and evolving model of a city (Ferré-Bigorra et al. 2022).

3.1 Urban Planning

The complex and multi-disciplinary process of urban planning includes the design, regulation, and management of land use and resources in towns and their underlying infrastructure. Improving people's quality of life, combating the negative effects of urbanization and environmental degradation, and helping to achieve the SDGs are its main intentions (Ahmed et al. 2022). It includes a wide range of conceptual domains that are sometimes categorized into several types, such as environmental planning, transportation planning, land use planning, urban planning, regional planning, and sustainable planning (Thite 2011). Many different facets of environmentally responsible city planning are covered by these academic fields. Nevertheless, the complex issues that come from cities' ever-changing character are what urban planners constantly face (Bibri et al. 2024a). Urban planning is marked by the complexity of multidimensional and systemic aspects (Li et al. 2017).

Urban systems are complicated, and this complexity has implications for urban planning, which have been investigated by Marshall (2012). Emphasizing the ever-changing character of cities, they brought attention to the move away from conventional top-down planning methods and toward a bottom-up approach. Planning strategies that embrace and exploit complexity to build functional urban complexity are advocated for in these works, along with a more nuanced understanding of urban complexity. Expanding on these findings, Batty and Marshall (2012) investigated how data-driven technology may be used to design smart cities that are environmentally friendly.

By viewing urban planning as a general computational problem, Bettencourt (2014) explored the potential benefits of big data in this field. In general, the author showed that basic engineering and applied science ideas can be used to create new, more efficient solutions to urban problems by coordinating new data sources with municipal policy. The research concluded that big data plays a critical role in easing the flow of information, coordinating various processes, encouraging teamwork in communication, and bolstering the learning and sharing activities that take place among varied groups and individuals that act as data agents (Truby 2020). But according to Yigitcanlar and Cugurullo (2020) a realistic way to think about AI being used in sustainable planning is to find strategic areas where it can be used in conjunction with common planning tasks: (a) community visioning, (b) making plans, (c) standards, policies, and incentives, (d) development work, and (e) public investments.

To tackle the complicated and ever-changing problems that cities face today, a data-driven approach to planning is essential in the field of urban development Vinuesa et al. (2020). Inefficiencies, ambiguities, and a dearth of up-to-the-minute information plague traditional approaches to city planning. A game-changing approach is to incorporate data-driven techniques into planning through the use of digital models and cutting-edge technology (Botequilha-Leitão and Díaz-Varela 2020).

3.2 Data in City Planning

Modern urban planning is characterized by a dynamic, data-driven approach that goes beyond conventional wisdom and is driven by creative tactics and cutting-edge methods. Decisions in data-driven urban planning are based on thorough examination, interpretation, and use of data in all its forms. Urban digital twin (UDT) is a data-driven urban planning system that aims to comprehend urban dynamics, find patterns, and forecast future trends (Shekarappa et al. 2023). Utilizing data-driven technology, this method assists urban planners in formulating better strategies to solve sustainability and resilience issues, improve public services, and upgrade urban infrastructure (Azzari et al. 2018) One approach that prioritizes social, environmental, and economic sustainability is smart city planning, which Mersal (2016) expanded upon from the dominant theory of urban development. The emphasis on technology integration, active user interaction, and the investigation of future potential in this approach reflects the developing nature of cities and constitutes a break from previous planning paradigms (dos Santos 2016). Integral tools in urban planning include real-time data, advanced data analytics models, AI algorithms, and spatial analysis. To create sustainable smart cities that are sensitive to environmental concerns and meet the requirements of their residents, data-driven urban planning has developed into an iterative process that uses cutting-edge models and techniques from AI and IoT (Alreshidi 2019). Urban planning techniques are made more efficient and effective with the help of AI-driven decision support systems. By linking infrastructure and physical things to the internet, the IoT enables real-time data collection and monitoring, which contributes to data-driven urban (AlMetwally et al. 2020) Internet of Things (IoT) sensors strategically positioned around the city can record data on a range of environmental factors.

To optimize urban networks and improve the timeliness of information accessible to city planners, AI-enabled decision-making procedures use this real-time data (Toan and Nhu 2020). The internet of things adds to data-driven urban planning by facilitating adaptive methods, smarter decision-making and automation, real-time monitoring, efficient use of resources, sophisticated analytics, predictive modeling, and adaptable techniques. With its integration of AI, IoT, and AIoT and UDT enhances data-driven planning through the ability to visualize urban systems and dynamics, conduct scenario predictions, and facilitate simulations (Barns 2018). By combining the UDT with AI, IoT, and AIoT, conventional urban planning is revolutionized into a data-driven, all-encompassing strategy.

3.3 Environmental Planning

Sustainable urban development is a top priority for environmental planners, who must solve pressing issues with resource management, pollution prevention, and ecosystem preservation (Nativi et al. 2021). Using knowledge of the two-way street that is human–environment interaction, environmental planners facilitate communication between human actions and natural processes (Shu-Yang et al. 2004). Integral to this framework are methods, plans, and strategies for studying, controlling, and bettering ecosystems and natural resources. The goals of this comprehensive strategy are to promote long-term growth, lessen negative impacts on the environment, and boost community welfare (Bhatti et al. 2021). An important part of sustainable urban development is environmental and ecological planning, which ensures that natural resources are preserved, ecosystems are protected, and urban growth occurs in harmony with the environment (Mersal 2016). Different historical and comparative approaches to this planning process were investigated by Nativi et al. (2021), who highlighted their unique viewpoints on the relationship between human activities and the environment. However related to this research, environmental planning seeks to build eco-friendly, resource-efficient, and climate-resilient urban areas (van Wynsberghe, 2021). This is in line with the goals of smarter eco-cities, which include enhancing public transit, controlling pollution, conserving water, reducing waste, protecting biodiversity, and optimizing energy use (Bibri et al. 2024b).

Environmental planners have seen a sea change in the last several years due to the proliferation of cutting-edge innovation, which has provided new resources for addressing long-standing problems and paving the way for smart cities that can continue to thrive in the future (Sellak et al. 2017). Applying cutting-edge technologies such as machine learning and deep learning models, expert systems, decision support systems, GIS, remote sensing, big data analytics, and planning support systems is essential (Nishant et al. 2020). For instance, ML and DSS have been useful in species conservation efforts (Salcedo-Sanz et al. 2016), and research on biodiversity and ecosystem services has used ML to predict competition and population dynamics, with an emphasis on ecosystem services. With these technical developments, planners can quickly and efficiently collect, analyze, and understand massive volumes of environmental data and somehow society should be thankful for the technology because the one they have an abundance of resources but now the new society has been changed into to new 4.0 Era. There is a potent synergy between cutting-edge technology and environmental planning that is rapidly approaching convergence (Sharps et al. 2017).

3.4 Approaches to Environmental Sustainability Informed by Data

As Sustainablity the plans are aimed at preserving and protecting the natural environment in urban areas are which is connected with the environmental sustainability strategies in sustainable smart cities.Renewable energy adoption, energy efficiency upgrades (in buildings, transportation, and infrastructure), programs to reduce and recycle waste, policies to preserve and expand green spaces, steps to conserve water, green building standards, transportation policies, and plans to mitigate and adapt to climate change are all examples of such strategies (He et al. 2022). Smart sustainable cities can improve their environmental performance with data-driven techniques and AI/AIoT (Herath and Mittal 2022). To reduce their influence on the environment and increase sustainability in smart cities and beyond, they depend substantially on data, AI algorithms, and AIoT tools to learn about the present, foretell the future, and make educated decisions (Wang et al. 2021). With the ability to track, analyze, and optimize resource consumption in real-time, AIoT can improve environmental sustainability initiatives within the framework of sustainable smart cities.

Adopting data-driven solutions allows planners to acquire a better understanding of environmental concerns, identify improvement areas, and create focused interventions by utilizing data from many sources. Sustainability smart cities can have their environmental performance tracked and evaluated using data-driven environmental sustainability policies that are built into UDT platforms (Saadah 2021). Sustainable smart cities can integrate data-driven urban planning, UDT, and AIoT to build healthier, more resilient urban environments and develop more effective and responsive approaches to complex challenges (Ahn and Chen 2020).

4 Challenges

Focussing on the real-world implications, this section delves into the complex range of challenges and opportunities presented by sustainable smart cities as they relate to (AI), (IoT), (UDT), data-driven urban planning, and environmental sustainability. The overarching goal is to educate readers on the mechanisms of technological convergence and how different models, domains, and technologies can work together for smart urban development that is both sustainable and efficient.

4.1 Data Security and Privacy Issues

The security and privacy of the infrastructures and systems must be prioritized in sustainable smart cities due to the prevalence of interconnected devices and data integration. Defending against cyberattacks and maintaining the secrecy of citizens' data requires the implementation of robust privacy and security measures. Embedding (UDT) frameworks in sustainable smart cities is a critical but challenging step in urban development, and the increasing use of (AI) and (AIoT) technologies. Implementation of urban digital twin (UDT) (Nishant et al. 2020) or AI/AIoT-driven (UDT) in other domains is imminent, and with it, worries over the privacy and security of citizens' data, which have grown in importance as these technologies become ingrained in the urban fabric. As per the study is examined the architecture, security, and privacy of the Internet of Things, which includes both the Internet of Things (IoT) and artificial intelligence of things (Truby 2020). Given the inherent weaknesses and decentralized structure of IoDT, the authors stated that security and privacy concerns are impeding its deployment (Liengpunsakul 2021). This, in turn, affects AIoDT. Innovative defense methods within the framework of IoDT were also showcased. The significance of the privacy and security concerns raised in IoDT, and its consequences reach into the wider field of (UDT), which is fundamental to data-driven ecological planning in sustainable smart cities. This study reinforces the groundwork for (UDT) implementation by locating and fixing security holes and privacy breaches in IoDT (and thus AIoDT). The success of data-driven environmental planning projects is enhanced when UDT systems are secure and privatized, where they are vital for monitoring and managing a variety of environmental indicators. In this way, the study by (Rieder et al. 2023) helps to promote environmental planning practices by strengthening the digital infrastructure that is necessary for sustainable urban growth.

Citizens may be at risk from the massive amounts of data acquired by sensors, devices, and linked systems since personal information is becoming more vulnerable to abuse and illegal access. It has become an urgent task for city planners, policy-makers, and tech developers to find a balance between using data for informed urban planning decisions through (UDT) and protecting individuals' privacy rights (Kassens-Noor and Hintze 2020). It is practically impossible to implement privacy-by-design principles due to the complexity of UDT and the volume of accessible data, which compound data privacy concerns. Nobody can say for sure that their integration will keep users' identities secret. Finding a way to improve urban environments' functionality, efficiency, sustainability, and liveability while also addressing individuals' legitimate concerns about the privacy and security of their personal information is the actual issue. To put urban digital twin into action, one must integrate diverse datasets from many sources to build accurate virtual representations of the physical and functional features of urban environments (Voda and Radu 2019). The privacy and security situation is made more complicated by the fact that these data are diverse, including topics such as energy consumption habits, travel behaviour, mobility mode, and trash management, among many others.

The crucial relevance of preserving personal data and ensuring ethical and responsible usage of AI and IoT technologies has been highlighted by the considerable separate discussion of security and privacy risks linked with artificial intelligence (AI) and artificial intelligence of things (AIoT). To ensure the future of sustainable smart cities is shaped by artificial intelligence (AI) and artificial intelligence of things (AIoT) technologies, urban planners, tech developers, and policymakers must work together and create strong frameworks that include privacy-enhancing mechanisms, strict data protection measures, and transparent governance structures. This will help build confidence among citizens and make the most of these transformative technologies. Ethical design, informed consent, data minimization, openness, accountability, and transparency are some of the suggestions. We are calling on policymakers to prioritize security and privacy, work with stakeholders, and create flexible legislation. Nevertheless, the study highlights the need for regulatory compliance and governance frameworks as well as the difficulties in achieving it (Dash and Sharma 2022). This includes the legal and licensing complications that arise from urban digital twin (UDT) with partially accessible data (Klar et al. 2023). Ethical and regulatory hurdles exist in urban digital twin (UDT) when it comes to collecting and distributing sensitive information while ensuring data privacy and security.

4.2 Social and Ethical Factors

Modelling, simulation, and data-driven decision-making capabilities undergo a revolutionary leap with the integration of artificial intelligence (AI), artificial intelligence of things (AIoT), urban digital twin (UDT). Nevertheless, beyond concerns about privacy and security, this unprecedented power raises important social and ethical questions that must be thoroughly investigated.

There are far-reaching consequences for decision-making procedures that fail to take ethical and social factors into account. These factors include prejudice, lack of transparency, and accountability. The complex relationship between artificial intelligence (AI), artificial intelligence of things (AIoT), urban digital twin (UDT) calls for a thoughtful ethical and social approach that takes into account the many ways in which these technologies could affect sustainable smart city management, planning, governance, and services. When it comes to monitoring, modeling, and simulating urban environments and processes within urban digital twin frameworks, one of the biggest challenges is making sure AI algorithms are fair (Ford and Wolf 2020). When it comes to sustainable smart cities, algorithmic bias could make things even worse by contributing to uneven representation and results (Ullah et al. 2020). In the urban digital twin setting, the concept of fairness—which has received a lot of attention in the wider AI area and artificial intelligence of things (AIoT) domain becomes vital. Difficulty in sharing data (Goh and Vinuesa 2021) and complicated urban digital twin models are two barriers to attaining transparency. Studies have shown that when there isn't enough information available, people are less likely to participate in UDT.

Understanding decision-making processes and identifying and mitigating potential areas of bias requires the responsible and transparent use of artificial intelligence and artificial intelligence of things in the development and implementation of urban digital twin. This is crucial for urban planners and policymakers. Research on data fairness and transparency is crucial to increase public acceptance and citizen participation in urban digital twin (UDT) projects in sustainable smart cities (Ramirez Lopez and Grijalba Castro 2020), even if many researchers argue that urban digital twin (UDT) designs often disregard social and ethical concerns artificial intelligence, the Internet of Things, and artificial intelligence of things. Furthermore, to hold individuals or entities accountable for the outcomes of decisions and activities driven by artificial intelligence (AI) and the artificial intelligence of things (AIoT) within the urban digital twin (UDT) framework, systems need to be put in place to track the origins of AI algorithmic decisions. Officials and citizens in the city may be resistant to urban digital twin (UDT) implementation due to concerns such as legal uncertainties, unclear accountability roles, and competing stakeholder goals.

It is critical to address these concerns to promote trust and acceptance of urban digital twin (UDT) for sustainable smart city adoption. The incorporation of urban digital twin (UDT) into urban planning processes necessitates that ethical and social factors be given due consideration while developing and implementing artificial intelligence (AI) and artificial intelligence of things (AIoT) technologies, rather than being considered as an afterthought. Responsible deployment of artificial intelligence (AI) and urban digital twin (UDT) driven by artificial intelligence of things (AIoT) requires strong social and ethical frameworks, which can only be achieved through extensive collaboration between technologists, urban planners, ethicists, and legislators. Specifically, planners should be aware of the implications of AI to use it wisely and prevent problems, like widening social gaps, according to Corrado et al. (2022) Responsible artificial intelligence adoption, improved planning processes, and a focus on human-centered approaches are all goals that the authors highlight in their call to action for planners. Furthermore, they stress the importance of planners actively shaping AI-based planning tools to promote inclusive and equitable community outcomes and call for ethical AI practices. For sustainable smart cities to make the most of the opportunities presented by combining urban digital twin (UDT) with artificial intelligence (AI) and artificial intelligence of things (AIoT) it is essential to tackle social and ethical concerns in a way that is fair, equal, transparent, and accountable. In sustainable urban development, this preventative measure makes sure that tech progress is in line with social and ethical concerns, encouraging innovation in tech while also looking out for society's best interests.

4.3 Threats to the Environment

Sustainable urban development stands to benefit greatly from the new technical era that has dawned thanks to the fast development and convergence of artificial intelligence, the internet of things (IoT), and urban data technology urban digital twin. We must carefully consider the environmental hazards and costs of these technologies, notwithstanding their potential for improved connectivity, intelligence in urban planning, data-driven insights, and efficiency (Mersal 2016). Recognizing the possible ecological effects of integrating (AI), (AIoT) and (UDT) into smart cities is of utmost importance as these communities seek new solutions to achieve environmental sustainability goals (Rathore et al. 2018). The ecological problems caused by these technologies' broad use, as well as possible solutions to these problems.

Thorough deliberation is required in light of the environmental concerns raised by the integration of (AI), (AIoT), and (UDT). There needs to be a coordinated effort to reduce these hazards to balance the advantages of technological advancement with the preservation of environmental quality (Naseer et al. 2023). As they face these difficulties, urban development stakeholders must remain committed to reducing the environmental impact of technological advancements (Bhatt et al. 2020). Striking a balance between technological landscapes in cities and environmental health calls for well-thought-out plans and recommendations.

4.4 Engagement with the Community and Inclusivity

Several factors beyond technology considerations must be considered to ensure that UDT driven by artificial intelligence and the Internet of Things contributes to inclusive urban development and meets the demands of various communities. Solid community engagement initiatives are an essential part of a holistic approach. Sustainable smart city initiatives, especially those looking to execute urban digital twin, run the danger of unintentionally contributing to or exacerbating preexisting socioeconomic inequities and social injustices unless they are well-planned and deliberate. Prioritizing socioeconomic factors is the goal of the people-centric urban digital twin paradigm, this perspective argues that urban digital twin should actively work to improve the lives of all citizens. There are some present barriers to inclusion, such as the lack of ethnic variety in data (Publication Dipak & Gade 2021) and the developers' bias in selecting information transmission methods. There is a lack of clarity regarding the advantages of urban digital twin among citizens, and researchers have pointed out a distinct lack of effective approaches for engaging individuals and communities (Mylonas et al. 2021). A crucial part of solving this problem in sustainable smart cities is making sure that the community is involved in urban digital twin (UDT) projects from the very beginning. An important part of this is making sure that people from all walks of life, especially those from under-represented groups, get a chance to voice their opinions (Magliacani 2023). By including locals in the decision-making processes around urban planning and development, community-based participatory techniques can ensure that citizens' voices are heard and valued (Publication Dipak & Gade 2021).

Sharing relevant information and facilitating opportunities for constructive discussion and teamwork are the cornerstones of effective community participation. It calls for a change to co-creation, where community members actively participate in making urban digital twin UDT's objectives, features, and implementation plans. So that people of all income levels and levels of technology knowledge may reap the benefits of urban digital twin, politicians, and urban planners must think about how easy it is for people to use digital technologies. Using urban-scale digital twin as a tool for simulating urban settings and constructing scenarios for policy concerns, Kumar et al. (2020) investigated how the virtual public sphere affects urban experiences in an increasingly data-driven society.

The author delves into the transition from a technical to a socio-technical view of smart cities, arguing that while urban digital twin aims to increase public input into urban planning decisions, it relies on too few variables and processes, which raises concerns. The study sheds light on the difficulties of abstracting variables in digital twin at the urban scale and offers helpful insights into the conflict between realism and idealism in this process. However, urban digital twin (UDT) and other tech-driven solutions aren't universally well-received or adopted due to differences in the socio-cultural environment of various groups, calling for customized approaches.

The key to successful community engagement is getting the word out and making it easy for people to talk to each other and work together. To achieve this, there must be a change towards co-creation, in which community members actively participate in defining the objectives, features, and methods for deploying urban digital twin. To make sure that all communities and individuals, regardless of their socioeconomic situation or level of technology literacy, can reap the benefits of urban digital twin, urban planners, and lawmakers should also think about how easy it is for people to use digital technologies. With an eye on urban-scale digital twin, a tool for modelling urban settings and creating scenarios for policy issues, Kumar et al. (2020) investigated how the virtual public realm affects urban experiences in a new data-driven society.

The author delves into the transition from a technological to a socio-technical view of smart cities, arguing that while urban digital twin aims to increase public input into urban planning decisions, it relies on too few variables and procedures, which raises issues (Rana et al. 2019). The study sheds light on the difficulties of abstracting variables in digital twin at the urban scale and offers helpful insights into the conflict between realism and idealism in this process. Urban digital twin (UDT) and other tech-driven solutions aren't universally well-received; therefore, specific strategies for engaging stakeholders and communicating with them are required to make society better for tomorrow.

5 Limitations

There are several limitations that must be acknowledged, although this extensive systematic research greatly aids in comprehending the convergence of artificial intelligence and the internet of things, urban planning, and environmental sustainability. The analysis and synthesis of the selected research may have overlooked new dimensions and implications due to the continuously growing nature of (AI), (AIoT) and (UDT) technologies.

Also, there is a risk of oversimplification are found in individual studies due to the abstract nature of synthesizing research results. Any thorough evaluation and synthesis of current research is inherently fraught with the possibility of biases in the review process and the interpretation of results, as acknowledged in the study. This study does recognize these further limitations, but it also highlights the importance of continuing research to account for new information, improve methods, and deal with biases. Future studies can expand upon these recognized limitations for the researchers where they have to focus on the innovation and updated technology of the society and also explore how the Indian society and Indian cities are converted into the new world that is 4.0 and try to stand India as a most powerful country above all country.

6 Recommendations for Future Research

There is a wealth of uncharted territory for future studies in the ever-changing field of sustainable smart cities, which this extensive systematic review delves into as it investigates the complex relationship between (AI), (AIoT), urban digital twin (UDT), environmental sustainability and data driven urban planning. To advance sustainable smart cities, this chapter lays out a plan by identifying key gaps and the implications of those gaps. Strategic initiatives can be put in place to address these gaps and advance sustainable smart urban development. The recommendations can be used as a starting point for further research into the intersection of data-driven technology, urban planning, and environmental sustainability, which holds great transformative potential.

This extensive list of research priorities is derived from a thorough examination of the problems that arise from the complex interaction of AI, AIoT, UDT, environmental sustainability and data driven urban planning. In addition to addressing current knowledge gaps, the suggested research directions aim to create a future where cities are more welcoming, secure, and ecologically sustainable for everybody. Researchers, urban planners, technologists, and policy-makers must work together across fields to explore these uncharted frontiers and build sustainable smart cities that are egalitarian and resilient.

To further improve the thoroughness and reliability of systematic reviews in this field, future studies can tackle the identified constraints. To begin with, researchers can embrace more adaptable approaches to account for the ever-changing nature of (AI), (AIoT) and (UDT) technologies. This will enable them to include new trends and real-time updates. Furthermore, going beyond the specified timeframe can include new information and make sure the literature review is up-to-date. Applying stringent methods to research selection and data analysis can also help reduce bias that may have crept into the review process. To improve the interpretation of results and reduce the possibility of subjective evaluations, it is beneficial to embrace multidisciplinary teamwork and include other points of view. New approaches to combining diverse data sets and extracting subtle insights from individual investigations can be investigated in subsequent studies. To better understand sustainable smart cities, future studies should focus on these areas; doing so will allow for more educated decision-making and more resilient approaches to urban development.

7 Conclusion

A deep and complex comprehension of the consequences of integrating (AI), (AIoT) and (UDT) into data-driven environmental planning processes to promote sustainable urban development practices is crucial, given the rising prevalence of these technologies in urban settings. In the ever-changing world of sustainable smart cities, the need to accept and utilize data-driven solutions that are both integrated and comprehensive has never been greater in tackling the intricacies of urban planning and environmental sustainability. In light of this, it is essential to conduct in-depth analyses and syntheses of different research threads with the main goal of producing coherent new insights and viewpoints.

An important gap in our understanding was filled by this thorough systematic review. This entailed delving into the complex ecology of sustainable smart cities and revealing the deep interplay of data-driven urban planning, environmental sustainability, AI, AIoT and UDT. These models and technologies work together to improve data-driven environmental planning processes, which in turn provide new ways to tackle difficult environmental problems. However, there are complexities and obstacles involved with this activity that must be properly managed and addressed if desired outcomes are to be achieved.

This study has made significant contributions to both theoretical and applied understanding. The study provides a thorough understanding of the interdependent relationship and collaborative interaction among (AI), (AIoT), (UDT), sustainability, and other related fields by integrating different viewpoints, addressing contradictions, maximizing synergies, analyzing obstacles, recognizing limitations, finding research gaps, and offering recommendations for future research. As a result of these findings, researchers, practitioners, and politicians now have a road map to follow in their pursuit of resiliently planned, technologically sophisticated, ecologically conscious urban spaces.

References

Ahad MA, Paiva S, Tripathi G, Feroz N (2020) Enabling technologies and sustainable smart cities. Sustain Cities Soc 61:102301. https://doi.org/10.1016/j.scs.2020.102301

Ahmed I, Noguchi M, O'brien D, Tucker C, Vahanvati M (2022) Resilience and sustainability in architecture and urban planning policies, practices, strategies and visions. www.mdpi.com/journal/architecture

Ahn MJ, Chen YC (2020) Artificial intelligence in government: potentials, challenges, and the future. ACM international conference proceeding series. https://doi.org/10.1145/3396956.3398260

Ahvenniemi H, Huovila A, Pinto-Seppä I, Airaksinen M (2017) What are the differences between sustainable and smart cities? Cities 60:234–245. https://doi.org/10.1016/j.cities.2016.09.009

AlMetwally SAH, Hassan MK, Mourad MH (2020) Real time Internet of Things (IoT) based water quality management system. Procedia CIRP 91:478–485. https://doi.org/10.1016/j.procir.2020.03.107

Alreshidi E (2019) Smart sustainable agriculture (SSA) solution underpinned by Internet of Things (IoT) and artificial intelligence (AI). Int J Adv Comput Sci Appl 10(5). https://doi.org/10.14569/IJACSA.2019.0100513

Azzari M, Garau C, Nesi P, Paolucci M, Zamperlin P (2018) Smart city governance strategies to better move towards a smart urbanism, pp 639–653. https://doi.org/10.1007/978-3-319-95168-3_43

Baran M, Kłos M, Chodorek M, Marchlewska-Patyk K (2022) The resilient smart city model–proposal for polish cities. Energies 15(5). https://doi.org/10.3390/en15051818

Barns S (2018) Smart cities and urban data platforms: designing interfaces for smart governance. City Cult Soc 12:5–12. https://doi.org/10.1016/j.ccs.2017.09.006

Batty M, Marshall S (2012) The origins of complexity theory in cities and planning. In: Complexity theories of cities have come of age. Springer, Berlin, pp 21–45. https://doi.org/10.1007/978-3-642-24544-2_3

Bettencourt LMA (2014) The uses of big data in cities. Big Data 2(1):12–22. https://doi.org/10.1089/big.2013.0042

Bhatt JG, Jani OK, Bhatt CB (2020) Automation based smart environment resource management in smart building of smart city. In: Advances in 21st century human settlements. Springer, pp 93–107. https://doi.org/10.1007/978-981-13-6822-6_3

Bhatti G, Mohan H, Raja Singh R (2021) Towards the future of smart electric vehicles: digital twin technology. Renew Sustain Energy Rev 141:110801. https://doi.org/10.1016/j.rser.2021.110801

Bibri SE, Huang J, Jagatheesaperumal SK, Krogstie J (2024) The synergistic interplay of artificial intelligence and digital twin in environmentally planning sustainable smart cities: a comprehensive systematic review. In: Environmental science and ecotechnology, vol 20. Editorial Board, Research of Environmental Sciences. https://doi.org/10.1016/j.ese.2024.100433

Bibri SE, Krogstie J, Kaboli A, Alahi A (2024) Smarter eco-cities and their leading-edge artificial intelligence of things solutions for environmental sustainability: a comprehensive systematic review. Environ Sci Ecotechnol 19:100330. https://doi.org/10.1016/j.ese.2023.100330

Blasi S, Ganzaroli A, De Noni I (2022) Smartening sustainable development in cities: strengthening the theoretical linkage between smart cities and SDGs. Sustain Cities Soc 80:103793. https://doi.org/10.1016/j.scs.2022.103793

Bokhari SAA, Myeong S (2022) Use of artificial intelligence in smart cities for smart decision-making: a social innovation perspective. Sustainability (Switzerland) 14(2). https://doi.org/10.3390/su14020620

Botequilha-Leitão A, Díaz-Varela ER (2020) Performance based planning of complex urban social-ecological systems: the quest for sustainability through the promotion of resilience. Sustain Cities Soc 56:102089. https://doi.org/10.1016/j.scs.2020.102089

Caprari G, Castelli G, Montuori M, Camardelli M, Malvezzi R (2022) Digital twin for urban planning in the green deal era: a state of the art and future perspectives. Sustainability 14(10):6263. https://doi.org/10.3390/su14106263

Corrado CR, DeLong SM, Holt EG, Hua EY, Tolk A (2022) Combining green metrics and digital twins for sustainability planning and governance of smart buildings and cities. Sustainability 14(20):12988. https://doi.org/10.3390/su142012988

Dash B, Sharma P (2022) Role of artificial intelligence in smart cities for information gathering and dissemination (a review). Acad J Res Sci Publish 4(39):58–75. https://doi.org/10.52132/Ajrsp.e.2022.39.4

De Guimarães JCF, Severo EA, Felix Júnior LA, Da Costa WPLB, Salmoria FT (2020) Governance and quality of life in smart cities: towards sustainable development goals. J Clean Prod. https://doi.org/10.1016/j.jclepro.2019.119926

del Mar Martínez-Bravo M, Labella-Fernández A (2024) Sustainable smart cities, pp 125–140. https://doi.org/10.4018/979-8-3693-3567-3.ch006

Dembski F, Wössner U, Letzgus M, Ruddat M, Yamu C (2020) Urban digital twins for smart cities and citizens: the case study of Herrenberg, Germany. Sustainability 12(6):2307. https://doi.org/10.3390/su12062307

Deng T, Zhang K, Shen Z-J (2021) A systematic review of a digital twin city: a new pattern of urban governance toward smart cities. J Manag Sci Eng 6(2):125–134. https://doi.org/10.1016/j.jmse.2021.03.003

dos Santos MJPL (2016) Smart cities and urban areas—aquaponics as innovative urban agriculture. Urban Forestry Urban Greening 20:402–406. https://doi.org/10.1016/j.ufug.2016.10.004

Efthymiou I-P, Egleton TE (2023) Artificial intelligence for sustainable smart cities, pp 1–11. https://doi.org/10.4018/978-1-6684-6821-0.ch001

Evans J, Karvonen A, Luque-Ayala A, Martin C, McCormick K, Raven R, Palgan YV (2019) Smart and sustainable cities? Pipedreams, practicalities and possibilities. Local Environ 24(7):557–564. https://doi.org/10.1080/13549839.2019.1624701

Ferré-Bigorra J, Casals M, Gangolells M (2022) The adoption of urban digital twins. Cities 131:103905. https://doi.org/10.1016/j.cities.2022.103905

Ford DN, Wolf CM (2020) Smart cities with digital twin systems for disaster management. J Manag Eng 36(4). https://doi.org/10.1061/(ASCE)ME.1943-5479.0000779

Gade DS (2021) Disruptive technologies for efficient and sustainable smart cities. Int J Manag Technol Soc Sci 47–64. https://doi.org/10.47992/IJMTS.2581.6012.0152

Gkontzis AF, Kotsiantis S, Feretzakis G, Verykios VS (2024) Enhancing urban resilience: smart city data analyses, forecasts, and digital twin techniques at the neighborhood level. Future Internet 16(2):47. https://doi.org/10.3390/fi16020047

Goh HH, Vinuesa R (2021) Regulating artificial-intelligence applications to achieve the sustainable development goals. Discov Sustain 2(1). https://doi.org/10.1007/s43621-021-00064-5

Hannan MA, Al-Shetwi AQ, Ker PJ, Begum RA, Mansor M, Rahman SA, Dong ZY, Tiong SK, Mahlia TMI, Muttaqi KM (2021) Impact of renewable energy utilization and artificial intelligence in achieving sustainable development goals. Energy Rep 7:5359–5373. https://doi.org/10.1016/j.egyr.2021.08.172

He P, Almasifar N, Mehbodniya A, Javaheri D, Webber JL (2022) Towards green smart cities using Internet of Things and optimization algorithms: a systematic and bibliometric review. Sustain Comput Inf Syst 36:100822. https://doi.org/10.1016/J.SUSCOM.2022.100822

Herath HMKKMB, Mittal M (2022) Adoption of artificial intelligence in smart cities: a comprehensive review. Int J Inf Manag Data Insights 2(1). https://doi.org/10.1016/j.jjimei.2022.100076

Javed AR, Shahzad F, Rehman S, ur, Zikria, Y. Bin, Razzak, I., Jalil, Z., & Xu, G. (2022) Future smart cities: requirements, emerging technologies, applications, challenges, and future aspects. Cities 129:103794. https://doi.org/10.1016/j.cities.2022.103794

Kassens-Noor E, Hintze A (2020) Cities of the future? The potential impact of artificial intelligence. AI 1(2):192–197. https://doi.org/10.3390/ai1020012

Klar R, Fredriksson A, Angelakis V (2023) Digital twins for ports: derived from smart city and supply chain twinning experience. IEEE Access 11:71777–71799. https://doi.org/10.1109/ACCESS.2023.3295495

Kumar H, Singh MK, Gupta MP, Madaan J (2020) Moving towards smart cities: solutions that lead to the smart city transformation framework. Technol Forecast Soc Chang. https://doi.org/10.1016/j.techfore.2018.04.024

Leal Filho W (2022) Pursuing sustainable development across disciplines. In: World sustainability series. Springer Science and Business Media, Deutschland GmbH, pp 495–504. https://doi.org/10.1007/978-3-031-04764-0_27

Li F, Liu H, Huisingh D, Wang Y, Wang R (2017) Shifting to healthier cities with improved urban ecological infrastructure: from the perspectives of planning, implementation, governance and engineering. J Clean Prod 163:S1–S11. https://doi.org/10.1016/j.jclepro.2016.11.151

Liengpunsakul S (2021) Artificial intelligence and sustainable development in China. Chin Econ 54(4):235–248. https://doi.org/10.1080/10971475.2020.1857062

Lopez LJR, Castro AIG (2021) Sustainability and resilience in smart city planning: a review. In: Sustainability (Switzerland), vol 13, Issue 1. MDPI AG, pp 1–25. https://doi.org/10.3390/su13010181

Magliacani M (2023) How the sustainable development goals challenge public management. Action research on the cultural heritage of an Italian smart city. J Manag Governance 27(3):987–1015. https://doi.org/10.1007/s10997-022-09652-7

Marshall S (2012) Planning, design and the complexity of cities. In: Complexity theories of cities have come of age. Springer, Berlin, Heidelberg, pp 191–205. https://doi.org/10.1007/978-3-642-24544-2_11

Mersal A (2016) Sustainable urban futures: environmental planning for sustainable urban development. Procedia Environ Sci 34:49–61. https://doi.org/10.1016/j.proenv.2016.04.005

Mhlanga D (2021) Artificial intelligence in the industry 4.0, and its impact on poverty, innovation, infrastructure development, and the sustainable development goals: lessons from emerging economies? Sustainability (Switzerland) 13(11). https://doi.org/10.3390/su13115788

Ministry of Urban Development Government of India Smart Cities Mission Statement & Guidelines Government of India Ministry of Urban Development (2015)

Mylonas G, Kalogeras A, Kalogeras G, Anagnostopoulos C, Alexakos C, Munoz L (2021) Digital twins from smart manufacturing to smart cities: a survey. In: IEEE access, vol 9. Institute of Electrical and Electronics Engineers Inc, pp 143222–143249. https://doi.org/10.1109/ACCESS.2021.3120843

Naseer F, Khan MN, Altalbe A (2023) Telepresence robot with DRL assisted delay compensation in IoT-enabled sustainable healthcare environment. Sustainability 15(4):3585. https://doi.org/10.3390/su15043585

Nativi S, Mazzetti P, Craglia M (2021) Digital ecosystems for developing digital twins of the earth: the destination earth case. Remote Sens 13(11):2119. https://doi.org/10.3390/rs13112119

Nishant R, Kennedy M, Corbett J (2020) Artificial intelligence for sustainability: challenges, opportunities, and a research agenda. Int J Inf Manage 53:102104. https://doi.org/10.1016/j.ijinfomgt.2020.102104

Publication Dipak S, Gade SS (2021) Disruptive technologies for efficient and sustainable smart cities. Technology, and Social Sciences (IJMTS) A Refereed International Journal of Srinivas University, India. Gade, Dipak S, 6(2), 2581–6012. https://doi.org/10.5281/zenodo.5221206

Ramirez Lopez LJ, Grijalba Castro AI (2020) Sustainability and resilience in smart city planning: a review. Sustainability 13(1):181. https://doi.org/10.3390/su13010181

Rana NP, Luthra S, Mangla SK, Islam R, Roderick S, Dwivedi YK (2019) Barriers to the development of smart cities in Indian context. Inf Syst Front 21(3):503–525. https://doi.org/10.1007/s10796-018-9873-4

Rathore MM, Paul A, Ahmad A, Chilamkurti N, Hong W-H, Seo H (2018) Real-time secure communication for smart city in high-speed big data environment. Futur Gener Comput Syst 83:638–652. https://doi.org/10.1016/j.future.2017.08.006

Rieder E, Schmuck M, Tugui A (2023) A scientific perspective on using artificial intelligence in sustainable urban development. Big Data Cognitive Comput 7(1). https://doi.org/10.3390/bdcc7010003

Saadah M (2021) Artificial intelligence for smart governance; towards Jambi Smart City. IOP Conf Series Earth Environ Sci 717(1). https://doi.org/10.1088/1755-1315/717/1/012030

Salcedo-Sanz S, Cuadra L, Vermeij MJA (2016) A review of computational intelligence techniques in coral reef-related applications. Eco Inform 32:107–123. https://doi.org/10.1016/j.ecoinf.2016.01.008

Samarakkody A, Amaratunga D, Haigh R (2023) Technological innovations for enhancing disaster resilience in smart cities: a comprehensive urban scholar's analysis. Sustainability (Switzerland) 15(15). https://doi.org/10.3390/su151512036

Sánchez-Corcuera R, Nuñez-Marcos A, Sesma-Solance J, Bilbao-Jayo A, Mulero R, Zulaika U, Azkune G, Almeida A (2019) Smart cities survey: technologies, application domains and challenges for the cities of the future. Int J Distrib Sensor Netw 15(6). SAGE Publications Ltd. https://doi.org/10.1177/1550147719853984

Sellak H, Ouhbi B, Frikh B, Palomares I (2017) Towards next-generation energy planning decision-making: an expert-based framework for intelligent decision support. Renew Sustain Energy Rev 80:1544–1577. https://doi.org/10.1016/j.rser.2017.07.013

Sharifi A, Allam Z, Bibri SE, Khavarian-Garmsir AR (2024) Smart cities and sustainable development goals (SDGs): a systematic literature review of co-benefits and trade-offs. Cities 146:104659. https://doi.org/10.1016/j.cities.2023.104659

Sharps K, Masante D, Thomas A, Jackson B, Redhead J, May L, Prosser H, Cosby B, Emmett B, Jones L (2017) Comparing strengths and weaknesses of three ecosystem services modelling tools in a diverse UK river catchment. Sci Total Environ 584–585:118–130. https://doi.org/10.1016/j.scitotenv.2016.12.160

Shekarappa GS, Badi M, Raj S, Mahapatra S (2023) An overview of smart city planning—the future technology. Artificial intelligence and machine learning in smart city planning, pp 319–335.https://doi.org/10.1016/B978-0-323-99503-0.00016-8

Shu-Yang F, Freedman B, Cote R (2004) Principles and practice of ecological design. Environ Rev 12(2):97–112. https://doi.org/10.1139/a04-005

Thite M (2011) Smart cities: implications of urban planning for human resource development. Hum Resour Dev Int 14(5):623–631. https://doi.org/10.1080/13678868.2011.618349

Toan NQ, Nhu DT (2020) Smart urban governance in smart city. IOP Conf Series Mater Sci Eng 869(2). https://doi.org/10.1088/1757-899X/869/2/022021

Truby J (2020) Governing artificial intelligence to benefit the UN sustainable development goals. Sustain Dev 28(4):946–959. https://doi.org/10.1002/sd.2048

Ullah Z, Al-Turjman F, Mostarda L, Gagliardi R (2020) Applications of artificial intelligence and machine learning in smart cities. In: Computer communications, vol 154. Elsevier B.V., pp 313–323. https://doi.org/10.1016/j.comcom.2020.02.069

van Wynsberghe A (2021) Sustainable AI: AI for sustainability and the sustainability of AI. AI and Ethics 1(3):213–218. https://doi.org/10.1007/s43681-021-00043-6

Vinuesa R, Azizpour H, Leite I, Balaam M, Dignum V, Domisch S, Felländer A, Langhans SD, Tegmark M, Fuso Nerini F (2020) The role of artificial intelligence in achieving the sustainable development goals. Nat Commun 11(1). https://doi.org/10.1038/s41467-019-14108-y

Voda AI, Radu LD (2019) How can artificial intelligence respond to smart cities challenges? In: Smart cities: issues and challenges mapping political, social and economic risks and threats. Elsevier, pp 199–216. https://doi.org/10.1016/B978-0-12-816639-0.00012-0

Wang K, Zhao Y, Gangadhari RK, Li Z (2021) Analyzing the adoption challenges of the internet of things (Iot) and artificial intelligence (ai) for smart cities in china. Sustainability (Switzerland) 13(19). https://doi.org/10.3390/su131910983

Weil C, Bibri SE, Longchamp R, Golay F, Alahi A (2023) Urban digital twin challenges: a systematic review and perspectives for sustainable smart cities. Sustain Cities Soc 99:104862. https://doi.org/10.1016/j.scs.2023.104862

Yigitcanlar T, Cugurullo F (2020) The sustainability of artificial intelligence: an urbanistic viewpoint from the lens of smart and sustainable cities. Sustainability (Switzerland) 12(20):1–24. https://doi.org/10.3390/su12208548

Virtual Storytelling for Sustainable Development: Engaging Audiences with SDG Narratives

S. Mahalakshmi, H. Bharath, Sandeep Kautish, and Deepmala Singh

Abstract The Sustainable Development Goals (SDGs) can be effectively promoted using virtual storytelling, which allows viewers to interact meaningfully with important sustainability tales. This chapter examines the ways in which virtual storytelling strategies, such as interactive online narratives and immersive virtual reality experiences, might drive significant change toward the SDGs, inspire action, and increase awareness. Virtual storytelling, which turns difficult sustainability concerns into relatable tales, builds a personal bond with the audience and increases comprehension and emotional commitment in subjects like clean energy, climate change, and less inequality. The chapter looks at how storytelling shapes public attitudes and behaviors, presents examples of how virtual storytelling has successfully communicated SDG-related ideas, and talks about the technology advancements that make these experiences more widely available. The chapter also discusses potential ethical issues and problems, like eliminating misinformation and guaranteeing diversity, while employing virtual storytelling for SDG advocacy. This chapter concludes by arguing that virtual storytelling has a great deal of potential to create a more active, knowledgeable, and involved global community that is prepared to assist and take part in sustainable development initiatives. Virtual storytelling has the potential to be a game-changing tool in the pursuit of the SDGs by utilizing the combined power of narrative and technology.

S. Mahalakshmi (✉)
Faculty of Management Studies, CMS Business School, Jain (Deemed-to-be University), Bengaluru, India
e-mail: dr.mahalakshmibharath@gmail.com

H. Bharath
Jain (Deemed-to-be University), Bengaluru, India

S. Kautish
AIT CSE, Chandigarh University, Mohali, Punjab, India

D. Singh
Symbiosis Centre for Management Studies, Symbiosis International (Deemed University), Nagpur Campus, Pune, India

W. Leal Filho et al. (eds.), *Metaverse and Sustainability: Business Resilience Towards Sustainable Development Goals*, World Sustainability Series,
https://doi.org/10.1007/978-3-031-89545-6_19

Keywords Virtual storytelling · Sustainable development goals (SDGs) · Virtual reality (VR) · Augmented reality (AR) · Climate actions · Sustainability narratives

1 Introduction

The Role of Storytelling in Advocacy: Storytelling has enormous influence in advocacy, especially when dealing with complicated, global issues like sustainable development. People from various societies and eras have shared ideals, elicited empathy, and impacted one other's behaviors through storytelling. While numbers, research, and policies are important in promoting the Sustainable Development Goals (SDGs), storytelling bridges the gap between awareness and meaningful action by converting abstract concepts into relatable and urgent narratives (Palakshappa and Dodds 2020). Stories personalize large-scale concerns, helping audiences to empathize with the individuals and communities affected.

In sustainable development, storytelling is about more than just transmitting knowledge; it is also about connecting distant issues to the audience's own experiences, viewpoints, and ideals. People are more inclined to empathize and relate to human stories concerning the effects of climate change on coastal areas or the difficulties women confront in accessing safe drinking water (Rivero Moreno 2020). Empathy may be a powerful motivation, encouraging people to support legislation, volunteer, give, or adopt sustainable behaviors in their own lives. Storytelling humanizes SDG-related concerns by introducing faces and voices that bring these issues to life in ways that numbers and figures cannot.

Character-driven narratives, moments of personal victory, and realistic depictions of struggle are all examples of effective advocacy storytelling features. Storytelling inspires audiences to believe in and work towards solutions by highlighting hope, resilience, and the possibility of good change. Stories of successful forestry operations, community-led conservation projects, and renewable energy programs, demonstrate that improvement is attainable, instilling hope and urgency (Akter et al. 2017).

Virtual Storytelling Defined: Virtual storytelling uses digital technologies such as virtual reality (VR), augmented reality (AR), and 360-degree video to create immersive experiences that change the way audiences interact with tales. Unlike traditional storytelling, which is based on passive observation, virtual storytelling adds an interactive dimension by allowing individuals to "step into" stories and experience them firsthand. This sensory engagement strengthens emotional ties to the tale and increases audience immersion, making complicated themes like sustainable development more relevant and compelling (Vassilopoulos et al. 2023).

VR users wear headsets that take them to fully designed virtual environments where they may explore scenes in 3D, interact with components, and immerse themselves in the tale. For example, virtual reality can mimic strolling through a coral reef affected by climate change or experiencing a futuristic city powered by renewable energy (Tea et al. 2021). This type of immersion allows individuals to experience circumstances that are far removed from their daily lives, which improves understanding and empathy.

In contrast, augmented reality (AR) superimposes digital information on real-world environments, combining physical and virtual aspects. It's especially useful for contextual storytelling, in which users can scan real-world things to access stories or data on sustainable practices, or to visualize a landscape with anticipated climate change effects. AR's versatility makes it useful and interesting in a variety of settings, including museums, classrooms, and outdoor installations. 360-degree video, which is available on a variety of platforms, allows viewers to explore scenes by shifting their field of view, creating a semi-immersive experience that combines standard video and VR (Jayaswal and Parida 2023). This format is particularly beneficial for virtual storytelling on social media or digital platforms, which allows for a larger audience reach.

Relevance to SDGs: The 17 Sustainable Development Goals (SDGs) are a worldwide roadmap for tackling crucial challenges like poverty, inequality, climate change, and renewable energy, with the goal of making the world more sustainable and equitable by 2030 (Vaid 2024). These goals are inextricably linked, with accomplishment in one area, such as quality education, having a favorable impact on others, such as economic growth and reduced inequality. However, the intricacy and magnitude of these aims can make them appear remote or overwhelming to many people, necessitating the development of effective methods for communicating their significance. This is where virtual storytelling becomes an essential tool (Mason et al. 2022).

Virtual storytelling brings the SDGs to life by providing immersive experiences that enable audiences to connect with these challenges on a personal level. Virtual storytelling allows audiences to "experience" SDG-related landscapes and challenges through technologies such as virtual reality (VR), augmented reality (AR), and 360-degree video. For example, users might use VR to immerse themselves in a simulation of increasing sea levels, immediately observing how climate change threatens coastal areas, emphasizing the urgency of SDG 13 (Climate Action). Similarly, augmented reality (AR) may overlay digital data onto real-world landscapes, allowing audiences to visualize the impact of renewable energy solutions on SDG 7 (Affordable and Clean Energy) (Nhamo et al. 2020).

This sensory engagement improves knowledge and empathy, building a stronger connection to the goals and motivating actions that are more meaningful. Virtual storytelling also allows for storylines that demonstrate the interrelated nature of the SDGs, highlighting how achieving gender equality (SDG 5) Virtual storytelling raises awareness and inspires people to take action by making global issues concrete and personal, leading to a collective movement toward a more sustainable future. It transforms the concept into the experiential, inspiring audiences to envision themselves as active contributors in attaining the SDGs (Chakraborty and Sengupta 2023).

2 Virtual Storytelling Platforms and Technologies

Virtual Reality (VR): Virtual Reality (VR) is an effective technique for engaging audiences in the Sustainable Development Goals (SDGs) by immersing them directly in situations that emphasize crucial global concerns. Unlike traditional media, which frequently gives a passive experience, VR allows people to actively explore scenarios linked to the SDGs, encouraging a deeper awareness of complex difficulties and the importance of sustainable solutions. VR allows users to digitally "step into" events that they may never have had in person. For example, students can visit climate-change-affected coral reefs and observe coral bleaching and reductions in marine biodiversity. This interactive experience promotes emotional connections, making the importance of SDG 14 (Life below Water) feel immediate and personal (Gupta 2023).

Similarly, VR enables viewers to imagine future cities that promote sustainability and clean energy, in line with SDGs 11 (Sustainable Cities and Communities) and 7 (Affordable and Clean Energy). Walking around metropolitan landscapes with solar-powered infrastructure and efficient transit systems allows consumers to see the benefits of environmentally friendly development, prompting them to advocate for sustainable practices (Gupta 2023).

VR's capacity to break down geographical and abstract barriers is especially useful for difficult, remote concerns. Audiences are more inclined to empathize with impacted populations and comprehend the consequences of their actions after firsthand experience with these difficulties (Gupta 2023). Finally, VR becomes a bridge that connects people to the realities of the SDGs, inspiring them to participate in sustainable change through unforgettable and intimate experiences into the world of sustainable development.

Augmented Reality (AR): Augmented Reality (AR) is a revolutionary technology that improves our interactions with the world by superimposing digital information over real-world situations. This capacity is especially useful for communicating the Sustainable Development Goals (SDGs), as it gives real-time data visualizations and enhances storytelling about crucial challenges in specific locations. AR can provide engaging and interactive experiences that enable audiences to examine SDG themes in context, making the material more relevant and meaningful (Jayaswal and Parida 2023).

AR can also improve storytelling by enabling interactive narratives that are tied to specific locations. For example, as users travel through a city, they may come across AR experiences that depict the consequences of climate change on urban areas or promote renewable energy programs in real time. Users could get information on energy-efficient buildings or community gardens by pointing their cellphones toward specific landmarks, in line with SDGs 11 (Sustainable Cities and Communities) and 13 (Climate Action).

Furthermore, AR's interactivity increases involvement and engagement by allowing users to explore, ask questions, and go further into topics that are meaningful to them. This interactive method not only makes learning more pleasant, but it also empowers people to take action to promote sustainability (tom Dieck et al. 2024). AR plays an important role in raising awareness and motivating good change by linking people to their environment and the global concerns addressed in the SDGs.

360-degree Video and Interactive Platforms: 360-degree video and interactive platforms offer a novel way to engage audiences with critical issues related to the Sustainable Development Goals (SDGs) by allowing viewers to navigate immersive environments while hearing personal stories from people directly affected by issues such as hunger, gender inequality, and climate change. This style increases storytelling by allowing a multi-dimensional experience that mixes visual engagement with emotional narratives (Balaji et al. 2024).

Viewers can tour a variety of locations using 360-degree movies, such as a rural community dealing with food scarcity or an urban neighborhood dealing with the effects of climate change, while listening to personal experiences from persons experiencing these realities. This immersive method lets viewers to observe individuals' problems and resiliency in real time, promoting empathy and a better understanding of the nuances surrounding each topic (Balaji et al. 2024).

This experience is enhanced further by interactive platforms, which allow viewers to choose what to look and what narratives to explore, resulting in a more customized journey through the narratives. This degree of engagement enables users to consider the interconnection of the SDGs, demonstrating how concerns such as hunger (SDG 2) and gender equality (SDG 5) are frequently linked (García-Sánchez et al, 2024).

By putting viewers in the shoes of those affected, 360-degree video and interactive storytelling not only raise awareness but also motivate action. Audiences are more likely to feel responsible and motivated to contribute to solutions, making these technologies critical tools for promoting the SDGs and creating significant social change.

Benefits and Limitations: Virtual storytelling has great potential for effect, especially in terms of raising awareness and participation with the Sustainable Development Goals (SDGs). Its immersive aspect can elicit significant emotional responses from audiences, encouraging empathy and motivating action. Virtual storytelling, which uses technology such as VR, AR, and 360-degree video, can translate abstract global issues into relatable experiences, allowing people to engage with the narratives of those affected by topics such as poverty, climate change, and gender inequality (Chen et al. 2023).

However, there are certain restrictions to consider. Accessibility is a big challenge, as not all audiences have access to the appropriate gadgets or high-speed internet to have an optimal experience. This digital divide might stymie the effectiveness of virtual storytelling, especially among underrepresented groups (Chen et al. 2023). Furthermore, developing high-quality virtual content frequently necessitates major logistical needs, such as technological skills, production resources, and financing, which might limit project scale and reach.

A balanced strategy is required to promote greater inclusion. This entails making virtual storytelling experiences accessible to a wide range of audiences, including individuals with impairments and those living in low-income communities. Using several storytelling modes, such as conventional media and in-person events, in addition to digital experiences, can help reach a wider audience (Samaddar and Mondal 2023). Recognizing the strengths and constraints of virtual storytelling allows creators to develop experiences that not only maximize engagement and effect, but also promote diversity, ultimately promoting greater awareness and support for the SDGs across various cultures.

3 Designing Effective SDG Narratives for Virtual Storytelling

The Elements of a Compelling Narrative: A engaging narrative is vital for successful storytelling, especially when dealing with complicated subjects such as the Sustainable Development Goals (SDGs). Character-driven stories, relatable challenges, and optimistic resolutions that inspire hope and action are all important aspects of such narratives.

Character-driven stories focus on individuals, making them approachable and humanizing the topics at hand. Audiences can relate emotionally with the characters by hearing their own experiences with issues like as poverty, climate change, and gender injustice. This connection promotes empathy and comprehension, motivating customers to invest in the journey and its outcome (Rohmah et al. 2024).

Relatable conflicts are essential for engaging the audience. These confrontations highlight the characters' hardships and concerns, making the topics feel timely and relevant. A narrative, for example, may represent a community working to protect the environment or a young lady overcoming educational difficulties. Such conflicts strike a chord with viewers, expressing universal themes of resilience, perseverance, and hope for a better future. Finally, positive resolutions are critical for instilling optimism and driving action. While it is critical to recognize the gravity of the issues, show-casing solutions—whether through community efforts, breakthrough technologies, or personal triumphs—shows that change is achievable. This optimistic viewpoint empowers audiences, allowing them to perceive their role in pushing progress towards the SDGs (Bandola-Gill et al. 2022).

Storytellers may develop narratives that not only enlighten but also motivate action by weaving together character-driven storylines, relatable challenges, and positive resolves, eventually contributing to a more sustainable and fair future.

Framing SDG Goals: Storytelling can be an excellent way to demonstrate the linked nature of the Sustainable Development Goals (SDGs). Multi-layered narratives that show how specific goals influence broader systemic change help audiences grasp the intricate relationships between the SDGs, resulting in a more comprehensive understanding of sustainable development. Each SDG is part of a larger, interconnected system in which development in one area has a substantial impact on others. For example, access to clean water and sanitation (SDG 6) not only improves health

(SDG 3), but it also promotes gender equality (SDG 5) by reducing the amount of time women and girls spend gathering water. Storytelling that weaves together these parts demonstrates how efforts in one sector can have ripple effects that uplift several areas, providing a higher quality of life and societies that are more egalitarian (Rohmah et al. 2024).

Multi-layered tales, which follow persons or communities affected by various SDGs, can effectively highlight these connections. A tale, for example, could focus on a rural community's journey to increased water access, demonstrating how this transition leads to greater health, educational outcomes, and economic stability. By highlighting these cumulative advantages, audiences have a better understanding of how accomplishing one goal helps others and leads to systemic transformation. Such tales not only improve knowledge but also motivate action by stressing the communal impact of individual actions. This integrated framing allows audiences to perceive the SDGs as a coherent framework, enabling them to support complete solutions that meet several targets, thereby promoting a stronger commitment to a sustainable and inclusive future (Bandola-Gill et al. 2022).

Emotional Appeal and Empathy: Virtual storytelling is based on emotional appeal and empathy, which translate complex global challenges into personal, relevant experiences. Virtual storytelling transcends basic awareness by engaging audiences in the lives and problems of people affected by issues such as poverty, climate change, and injustice. It appeals to emotions, making these difficulties feel more immediate and personal, resulting in a stronger commitment to action.

Research findings suggest that emotionally compelling narratives raise awareness and motivate people to take effective action. VR experiences that imitate living in a drought-stricken environment, for example, allow viewers to see and feel the daily struggles of populations that lack consistent water availability. These experiences not only enlighten, but also elicit empathy, allowing audiences to comprehend the human importance of Sustainable Development Goals (SDGs) such as clean water (SDG 6) and climate action (SDG 13). According to research, participants in such VR simulations are more likely to later support water conservation measures or advocacy projects (Obura 2020).

Similarly, 360-degree movies or AR experiences that feature human tales from those fighting for gender equality (SDG 5) or education access (SDG 4) can help audiences connect deeply to these issues. Users get more invested in supporting solutions as they hear and witness actual individuals describe their struggles and accomplishments. By engaging emotionally, virtual storytelling not only promotes awareness but also creates an empathy bridge, making listeners feel like they are a part of the solution. This method has proved to lead to better engagement rates, pushing viewers to take concrete activities aligning with the SDGs (Obura 2020).

Behavioral Influence Techniques: Virtual storytelling's behavioral impact approaches employ persuasive narrative aspects to persuade viewers to adopt long-term behaviors. These strategies make sustainability relatable and feasible by including testimonials, simulations, and visions of prospective futures, motivating viewers to commit to actions that promote the Sustainable Development Goals

(SDGs). Testimonials are an effective method for sharing genuine voices and real-life experiences with audiences. Individuals who share personal experiences about how sustainability affects their lives—for example, farmers who embrace eco-friendly techniques or youth who advocate for climate justice—humanize the SDGs. Testimonials show audiences the tangible advantages of sustainable behaviors, which increases their likelihood of embracing comparable practices. This relevant approach fosters a sense of shared responsibility and demonstrates that everyone, regardless of background, can contribute to global goals (Helgeson et al. 2022).

Another effective strategy is simulations of prospective futures, which allow audiences to see outcomes based on various actions and choices. For example, virtual reality can show two opposing scenarios: one in which sustainable practices have changed a community into one with clean energy and thriving ecosystems, and another in which inaction leads to environmental damage. This representation of future possibilities not only highlights the stakes, but it also empowers audiences by demonstrating the positive influence that their actions can have (Fallah Shayan et al. 2022).

These emotional and cognitive engagement-based approaches transform passive awareness into active commitment. Audiences are more motivated to adopt sustainable choices after witnessing the experiences of others and seeing different futures, sustaining a collective drive toward achieving the SDGs. Virtual storytelling transforms intention into action through intentional narrative use, establishing a proactive sustainability culture (Helgeson et al. 2022).

4 Strategies for Implementing Virtual Storytelling in SDG Advocacy

Audience-Centric Design: An audience-centric design approach is critical for effective virtual storytelling in Sustainable Development Goal (SDG) advocacy because it ensures that storytelling methods are accessible to a wide range of audiences. Advocates may make their messages more accessible, inclusive, and impactful by adapting them to target audiences' preferences, technology access, and cultural contexts. The first stage is to understand the preferences of the audience. Interactive and immersive formats such as VR, AR, or 360-degree media can increase engagement and interest among younger, more tech-savvy viewers. Simpler forms, such as mobile-friendly AR or web-based interactive storytelling, are more feasible for audiences with limited technological access. Ensuring interoperability with commonly available devices, such as smartphones, allows larger involvement without the need for specialist gear (Dessart and Standaert 2023).

Cultural sensitivity is also essential in audience-centered storytelling. Different cultures have unique values, customs, and experiences that influence their opinions on global concerns. Narratives that reflect these many perspectives build authenticity and respect, making the SDGs feel more relevant and attainable. Stories about climate action (SDG 13) can, for example, incorporate local examples of resilience and adaptation, connecting global goals to familiar, meaningful circumstances. Furthermore, considering accessibility, such as providing subtitles, audio descriptions, or reduced language, ensures that all users are included. This inclusive design strategy extends the scope and effect of virtual storytelling by removing barriers to full interaction (Mondejar et al. 2021).

In the end, an audience-centric design approach to virtual storytelling enables advocates to build more personalized and captivating experiences. It fosters closer links to the SDGs and stimulates collective action toward a sustainable future by reflecting people's individual perspectives and experiences (Dessart and Standaert 2023).

Collaborative Partnerships: Collaborative partnerships are essential for promoting virtual storytelling for the Sustainable Development Goals (SDGs), as they combine experience, resources, and technology to generate powerful and unique content. Governments, non-governmental organizations (NGOs), and commercial technology corporations all contribute distinct qualities to these collaborations, resulting in high-quality, relevant tales that appeal with broad audiences and motivate real action.

Governments play an important role in ensuring that storytelling matches with policy goals and targets the most pressing concerns by granting access to data, resources, and local networks. For example, government engagement might promote virtual experiences that emphasize important issues such as water scarcity or climate change in specific locations, in line with national SDG priorities and activities. Governments can also assist scale these tales by including them into publicity initiatives or educational programs, so increasing their range and influence (Stott and Murphy 2020).

Private IT firms provide the technical skills required to make these narratives immersive and available. They allow immersive storytelling by creating cutting-edge VR, AR, and interactive platforms. Partnerships with technology companies also provide long-term funding strategies, promoting the creation of impactful content. Governments, non-governmental organizations (NGOs), and technology companies can collaborate to create virtual storytelling experiences that not only inform but also motivate individuals to take action toward attaining the SDGs, thereby establishing a collaborative framework for sustained change (Leal Filho et al. 2024).

Long-Term Engagement Tactics: Long-term engagement strategies are required to ensure that virtual storytelling experiences for the Sustainable Development Goals (SDGs) have a long-term impact and lead to action. These strategies convert one-time encounters into continuous commitments to sustainable change by encouraging users to return stories, share their experiences, and participate in SDG-related activities beyond their initial connection. One effective way is to create tales with numerous levels or updates that people may return to. For example, a VR experience focused on climate action (SDG 13) may grow to include new scenarios based on current events, scientific discoveries, or community actions, encouraging people to return on a frequent basis. This ongoing narrative strategy keeps the content relevant and entertaining while highlighting the importance of the concerns over time (Alshaketheep et al. 2024).

Encouraging people to share their experiences on social media or community forums expands the story's reach and promotes a sense of shared responsibility. Users can exchange thoughts, discuss actions taken, and even participate in SDG-related challenges, resulting in a community of engaged advocates. This collective sharing also generates social proof, encouraging others to join in and raising awareness. Engaging users in SDG-related activities following the encounter strengthens their commitment. Organizations can encourage users to perform basic actions, such as filing petitions, volunteering, or making sustainable choices in their daily lives, so linking virtual experiences to real-world effects. Linking to educational resources, local events, or existing campaigns can help users turn awareness into action. Long-term engagement methods transform virtual storytelling into a strong instrument for ongoing advocacy and progress toward the SDGs by encouraging revisits, sharing, and actionable follow-ups (Grainger-Brown and Malekpour 2019).

Monitoring and Evaluation: Monitoring and evaluation are critical for determining the performance of virtual storytelling efforts that promote the Sustainable Development Goals (SDGs). Organizations can measure the effectiveness of their storytelling initiatives and make data-driven adjustments by tracking important indicators such as engagement rates, knowledge retention, behavior changes, and advocacy adoption. Engagement rates, which include metrics such as view counts, duration of engagement, and return visits, are important indicators of audience interest and influence. High engagement indicates that the content is meaningful to viewers, whereas low engagement may highlight areas where the narrative method should be improved to better capture and hold attention. Tracking these stats helps producers determine which forms and topics attract the most interest (Saner et al. 2020).

Another key metric is knowledge retention, which is frequently tested via pre- and post-experience surveys or quizzes to assess users' comprehension of the issues presented. Effective storytelling should leave viewers with a better understanding of SDG-related issues, and analyzing retention rates helps determine how successfully the content teaches and informs.

Long-term influence can be better understood through behavioral changes and advocacy uptake. Metrics such as alterations in user attitudes, self-reported behavior changes (for example, adopting sustainable practices), and participation in follow-up actions such as advocacy campaigns or community projects show that the storytelling has sparked real-world action. Surveys, interviews, and internet activity tracking can provide useful information about these consequences. By routinely monitoring these measures, organizations may assess the reach, effectiveness, and long-term viability of virtual storytelling projects, ensuring that plans to generate greater awareness, participation, and action toward attaining the SDGs are constantly refined (Saner et al, 2020).

5 Ethical and Practical Considerations in Virtual Storytelling for SDGs

Ethics of Virtual Experiences: The ethics of virtual experiences are becoming increasingly important, especially as these technologies spread into domains like education, tourism, and social campaigning. Virtual experiences can elicit strong emotional responses by immersing users in locations or situations that would otherwise be unavailable, but this power also carries the potential of emotional manipulation. When VR experiences are intended to raise awareness or inspire empathy, the line between compelling narrative and inappropriate psychological influence may become blurred. Users may leave feeling pressured or emotionally misled, rather than guided to informed, balanced conclusions (Barbara et al. 2021).

Another tenet of ethics is accuracy. Misrepresentations can cause confusion, propagate harmful stereotypes, or transmit false information about the cultures, people, or historical events depicted. Virtual experiences must attempt to deliver information responsibly, with well-researched content that represents multiple points of view, especially when dealing with sensitive or difficult issues. Furthermore, vulnerable communities must be shown in a respectful manner. Using VR to highlight topics like poverty, conflict, and environmental degradation might help to raise awareness and mobilize resources. However, there is a narrow line between increasing awareness and exploiting those groups for dramatic effect. To ensure appropriate depiction, virtual experiences must minimize sensationalism, stress dignity, and, if possible, involve members of those communities in content creation (Hyrynsalmi et al. 2021).

Overall, virtual experience makers should adhere to ethical norms that emphasize transparency, truth, and inclusivity, striking a balance between the potential for empathy building and the obligation toward viewers and the communities that are depicted. This can help guarantee that virtual experiences are used to effect beneficial, informed change rather than manipulation (Barbara et al. 2021).

Data Privacy and Accessibility: Data security and accessibility are critical to developing inclusive technology environments, particularly as digital services become increasingly integrated into everyday life. Ensuring accessibility entails more than just delivering technology; it also entails addressing the profound disparities that exist for underserved or rural audiences, who may encounter challenges such as cost, internet connectivity, or digital literacy. Access discrepancies impede these groups from fully realizing the potential of technology, exacerbating socioeconomic divides (Peña-Acuña and Rubio-Alcalá 2024).

Providing secure access is a fundamental aspect of inclusion. Data privacy is especially important for underprivileged audiences, who are typically unaware of their data gathering rights and may be more vulnerable to exploitation. Ethical data practices include transparency in how data is collected, handled, and used, with a focus on individual rights and privacy. Companies and policymakers should prioritize simple, easy-to-understand privacy policies and provide tools that educate consumers on how to protect their information (Wakunuma et al. 2020).

Meanwhile, overcoming accessibility gaps necessitates infrastructure investment, such as low-cost internet in rural regions, as well as solutions developed with inclusivity in mind, such as offline capabilities, multi-language options, and simple interfaces. Collaboration with local organizations can help align technical solutions with the requirements of individual communities, promoting trust as well as utility. Technology providers can ensure that innovation benefits all groups equally by actively addressing both privacy and accessibility, resulting in a more inclusive digital landscape in which privacy rights are respected and technological benefits are available to everyone, regardless of location or socioeconomic status (Peña-Acuña and Rubio-Alcalá 2024).

Balancing Engagement with Sensitivity: Balancing participation and sensitivity is critical in storytelling, especially in an age where online platforms have expanded the reach and effect of shared experiences. Effective narrative can promote awareness and elicit empathy, yet sensationalism frequently overshadows the genuine message. When arguments are exaggerated for impact, they can desensitize viewers or cause emotional weariness, reducing their willingness to engage meaningfully in crucial causes (Kang 2018).

Avoiding sensationalism entails prioritizing accuracy and respect in the way tales are written and delivered. This method encourages storytellers to focus on fact-based information that respects the dignity of the individuals and communities involved, rather than depending on dramatic or unpleasant elements to keep viewers interested. Creators can develop informed empathy by grounding their stories in reality and nuance, allowing audiences to be touched emotionally while also being pushed to grasp and critically think on the concerns addressed.

Furthermore, responsible storytelling requires consideration of audience overload. Constant exposure to powerful or painful storylines can create emotional exhaustion, lowering engagement over time. To avoid this, stories should be structured in a way that emphasizes resilience, progress, or concrete steps in addition to problems. Instead of feeling overwhelmed or helpless, this balanced perspective enables listeners to respond positively.

Storytellers may create powerful narratives that connect with audiences on a deep level by combining engagement with sensitivity, fostering persistent attention and support for critical causes without abusing emotions. This respectful narrative technique can have a more profound and long-term influence on both viewers and the communities being represented (Fallah Shayan et al. 2022).

6 Conclusion

Virtual storytelling provides a revolutionary technique to engaging global audiences with the storylines of the Sustainable Development Goals. VR and AR technologies, which use immersive experiences, can assist bridge-understanding gaps in complex global concerns such as poverty, climate change, and social inequality. These technologies allow viewers to not only learn about the SDGs, but also experience them in a way that promotes empathy, awareness, and a stronger emotional connection. Unlike traditional techniques, virtual storytelling has the unique capacity to immerse participants in the places and scenarios impacted by these concerns, inspiring more proactive and persistent commitment to sustainability.

In order to become really powerful, virtual narrative producers must adhere to ethical criteria that balance participation with accuracy and respect for the communities represented. Sensationalism must be avoided; instead, storytelling should focus on depicting realistic narratives that highlight the resilience and agency of affected individuals and communities. Furthermore, maintaining data protection and accessibility is critical, especially for underprivileged audiences who could gain the most from exposure to SDG tales. Addressing these issues ensures that virtual experiences are not only enjoyable, but also inclusive and courteous.

As technology advances, collaboration among content creators, governments, and local communities will be critical in building authentic and relevant virtual stories. When driven by ethical storytelling principles, virtual experiences may be great educational and advocacy tools, inspiring people all around the world to take meaningful activities in support of the SDGs. Finally, virtual storytelling has the ability to connect global audiences by motivating a shared commitment to sustainable development and instilling a sense of communal responsibility for addressing these vital global issues. Through engaging, well-crafted tales, the vision of a sustainable, egalitarian future is transforms into a story that everyone can see and feel, converting awareness into action.

References

Akter S, Fu X, Bremermann L, Rosa M, Nattrodt V, Väätänen J, Teplov R, Khairullina I (2017) MNEs' contribution to sustainable energy and development: the case of 'light for all' program in Brazil. International Business and Management, pp 195–224. https://doi.org/10.1108/s1876-066x20170000033010

Alshaketheep K, Mansour A, Deek A, Zraqat O, Asfour B, Deeb A (2024) Innovative digital marketing for promoting SDG 2030 knowledge in Jordanian universities in the Middle East. Discov Sustain 5(1). https://doi.org/10.1007/s43621-024-00419-8

Balaji MS, Jiang Y, Dwivedi YK, Behl A, Jain K (2024) Virtual reality vs traditional previews: the role of self-construal in customer decision-making. Int J Contemp Hosp Manag. https://doi.org/10.1108/ijchm-10-2023-1578

Bandola-Gill J, Grek S, Tichenor M (2022) Scripting the SDGs: the role of narratives in governing by goals. In: Sustainable development goals series, pp 69–96. https://doi.org/10.1007/978-3-031-03938-6_4

Barbara J, Koenitz H, Bakk ÁK (2021) The ethics of virtual reality interactive digital narratives in cultural heritage. In: Lecture notes in computer science, pp 288–292. https://doi.org/10.1007/978-3-030-92300-6_27

Chakraborty A, Sengupta S (2023) Equitable pathways for a sustainable future: the case for mainstreaming gender across sustainable development goals (SDGs). In: Gender inequality and its

implications on education and health, pp 191–201. https://doi.org/10.1108/978-1-83753-180-620231017
Chen J, Xi N, Pohjonen V, Hamari J (2023) Paying attention in metaverse: an experiment on spatial attention allocation in extended reality shopping. Inf Technol People 36(8):255–283. https://doi.org/10.1108/itp-09-2021-0674
Dessart L, Standaert W (2023) Strategic storytelling in the age of sustainability. Bus Horiz 66(3):371–385. https://doi.org/10.1016/j.bushor.2023.01.005
Fallah Shayan N, Mohabbati-Kalejahi N, Alavi S, Zahed MA (2022) Sustainable development goals (SDGs) as a framework for corporate social responsibility (CSR). Sustainability 14(3):1222. https://doi.org/10.3390/su14031222
García-Sánchez I-M, Aibar-Guzmán B, Raimo N, Vitolla F (2024) Embracing tomorrow: global insights on corporate commitment to the 2030 agenda. Meas Bus Excell. https://doi.org/10.1108/mbe-04-2024-0044
Grainger-Brown J, Malekpour S (2019) Implementing the sustainable development goals: a review of strategic tools and frameworks available to organisations. Sustainability 11(5):1381. https://doi.org/10.3390/su11051381
Gupta VP (2023) Integration of new-age technologies in education system to achieve sustainable development goals (SDGs) in emerging economies. In: Fostering sustainable businesses in emerging economies, pp 259–280. https://doi.org/10.1108/978-1-80455-640-520231016
Helgeson J, Glynn P, Chabay I (2022) Narratives of sustainability in digital media: an observatory for digital narratives. Futures 142:103016. https://doi.org/10.1016/j.futures.2022.103016
Hyrynsalmi S, Kimppa KK, Smed J (2021) Ethics of interactive storytelling. In: International series on computer entertainment and media technology, pp 143–153. https://doi.org/10.1007/978-3-030-81538-7_10
Jayaswal P, Parida B (2023) Past, present and future of augmented reality marketing research: a bibliometric and thematic analysis approach. Eur J Mark 57(9):2237–2289. https://doi.org/10.1108/ejm-05-2022-0397
Kang S (2018) Communicating sustainable development in the digital age: the relationship between Citizens' storytelling and engagement intention. Sustain Dev 27(3):337–348. https://doi.org/10.1002/sd.1905
Leal Filho W, Dibbern T, Pimenta Dinis MA, Coggo Cristofoletti E, Mbah MF, Mishra A, Clarke A, Samuel N, Castillo Apraiz J, Rimi Abubakar I, Aina YA (2024) The added value of partnerships in implementing the UN sustainable development goals. J Clean Prod 438:140794. https://doi.org/10.1016/j.jclepro.2024.140794
Mason P, Augustyn M, Seakhoa-King A (2022) Tourism destination quality and the UN sustainable development goals: tourism agenda 2030. Tour Rev 78(2):443–460. https://doi.org/10.1108/tr-05-2022-0259
Mondejar ME, Avtar R, Diaz HL, Dubey RK, Esteban J, Gómez-Morales A, Hallam B, Mbungu NT, Okolo CC, Prasad KA, She Q, Garcia-Segura S (2021) Digitalization to achieve sustainable development goals: steps towards a smart green planet. Sci Total Environ 794:148539. https://doi.org/10.1016/j.scitotenv.2021.148539
Nhamo G, Nhemachena C, Nhamo S, Mjimba V, Savić I (2020) Energy in the context of the 2030 Agenda for sustainable development. In: SDG7—ensure access to affordable, reliable, sustainable and modern energy, pp 1–31. https://doi.org/10.1108/978-1-78973-799-820201003
Obura DO (2020) Getting to 2030—scaling effort to ambition through a narrative model of the sdgs. Mar Policy 117:103973. https://doi.org/10.1016/j.marpol.2020.103973
Palakshappa N, Dodds S (2020) Mobilising SDG 12: co-creating sustainability through brands. Mark Intell Plan 39(2):265–283. https://doi.org/10.1108/mip-08-2018-0360
Peña-Acuña B, Rubio-Alcalá FD (2024) Ethical approach to the use of immersive technologies. Advance about digitalisation of multilingual programs in the EHEA. Front Virt Real 5. https://doi.org/10.3389/frvir.2024.1357595

Rivero Moreno LD (2020) Sustainable city storytelling: cultural heritage as a resource for a Greener and Fairer urban development. J Cult Herit Manag Sustain Dev 10(4):399–412. https://doi.org/10.1108/jchmsd-05-2019-0043

Rohmah Z, Makrakis V, Kostoulas-Makrakis N, Hidayati L, Fitriyanto NA, Projosasmito SR, Auwibi BR, Prijambada ID (2024) Sustainable development goals through participatory video and digital storytelling. Sustain Econ 2(3):176. https://doi.org/10.62617/se.v2i3.176

Samaddar K, Mondal S (2023) Ar and VR-based travel: a responsible practice towards sustainable tourism. Int J Tour Cities 10(1):105–128. https://doi.org/10.1108/ijtc-05-2022-0135

Saner R, Yiu L, Nguyen M (2020) Monitoring the SDGs: digital and social technologies to ensure citizen participation, inclusiveness and transparency. Dev Policy Rev 38(4):483–500. https://doi.org/10.1111/dpr.12433

Stott L, Murphy DF (2020) An inclusive approach to partnerships for the SDGs: using a relationship lens to explore the potential for transformational collaboration. Sustainability 12(19):7905. https://doi.org/10.3390/su12197905

Tea S, Panuwatwanich K, Ruthankoon R, Kaewmoracharoen M (2021) Multiuser immersive virtual reality application for real-time remote collaboration to enhance design review process in the social distancing era. J Eng Technol 20(1):281–298. https://doi.org/10.1108/jedt-12-2020-0500

tom Dieck MC, Krey N, Cranmer E (2024) The value of augmented reality: exploring hedonic and utilitarian augmented reality experiences. Int Res. https://doi.org/10.1108/intr-11-2023-1048

Vaid J (2024) Sustainable tourism: a roadmap for the 2030 sustainable development agenda. In: Contemporary studies in economic and financial analysis, pp 21–26. https://doi.org/10.1108/s1569-37592024000113b002

Vassilopoulos A, Papadaki L, Koundouri P (2023) Can virtual reality nudge toward green investing? An experiment with small business entrepreneurs. Rev Behav Financ 16(2):322–340. https://doi.org/10.1108/rbf-01-2023-0005

Wakunuma K, Jiya T, Aliyu S (2020) Socio-ethical implications of using AI in accelerating SDG3 in least developed countries. J Responsib Technol 4:100006. https://doi.org/10.1016/j.jrt.2020.100006

Virtual Workspace: A Way to Decent Work? Aligning Practices with SDG8

Archana Yadav, Abhilasha Dixit, Arif Hasan, Umesh Gautam, and Deepali Soni

Abstract Virtual workspaces are transforming the global employment market by removing geographical barriers and enabling flexible work arrangements. These digital environments enable the establishment of new job opportunities in a variety of fields, such as digital content creation, virtual event coordination, community management, metaverse progress, virtual property, and virtual commodities and services. Furthermore, they encourage the gig economy by offering freelance and short-term project opportunities, as well as developing roles in technical assistance, training, and induction. The cheap operational costs associated with virtual workspaces encourage entrepreneurial ventures and new business models, extending their benefits to traditional industries such as education and telemedicine. Furthermore, virtual workspaces contribute to sustainable growth by reducing the need for physical infrastructure and promoting environmentally friendly employment. These settings promote collaboration, enhance productivity, and strengthen team relationships by including technology such as Slack, Zoom, Trello, Asana, Google Drive, and advanced systems such as Microsoft Mesh and Gather. Microsoft Mesh provides immersive mixed-reality experiences, whereas Gather enables spontaneous interactions and individualized virtual workplaces, both of which promote a cohesive and innovative work culture. Virtual workspaces eventually support economic growth and contribute to the achievement of the SDGs by leveraging a varied global workforce, stimulating innovation, and creating an inclusive and flexible labor market.

Keywords Virtual workspace · Decent work · SDGs · Equal employment opportunity

---

A. Yadav
GLA University, Mathura, Uttar Pradesh, India

A. Dixit (✉) · A. Hasan · U. Gautam
Amity University, Gwalior, Madhya Pradesh, India
e-mail: adixit@gwa.amity.edu

D. Soni
Jiwaji University, Gwalior, Madhya Pradesh, India

W. Leal Filho et al. (eds.), *Metaverse and Sustainability: Business Resilience Towards Sustainable Development Goals*, World Sustainability Series,
https://doi.org/10.1007/978-3-031-89545-6_20

1 Introduction

The concept of 'Decent work' has gain lot of attention in recent years, largely with the acceptance of the United Nations' Sustainable Development Goals. SDG 8.5 is set to achieve: "Full and productive employment and decent work for all women and men, including for young people and persons with disabilities, and equal pay for work of equal value" (Duffy et al. 2016). Decent work is the idea of having a job that provides one with an income sufficient not only to live but also to extend opportunities for advancement, security, dignity, and the development of human capacities in consonance with a reasonable quality of life. It is not only about pay but also means a productive job with possibilities for growth and development and social protection. This will involve a safe and healthy work environment that is free from hazards and harassment. Allows employees to form unions and negotiate for good working conditions. Ensures non-discrimination and fair opportunity to all. The International Labour Organization (ILO) further defined, "Decent work encapsulates four main pillars: employment creation, rights at work, social protection, and social dialogue." In this highly transforming era, virtual workplaces can be an excellent method to attain some of the benefits of decent work. Four pillars of decent work can easily be mapped with virtual work places (Hughes et al. 2024). Virtual employment enhances opportunities for persons with disabilities and open global talent reach. The influx of virtual workplaces has sharply altered the nature of work, spawning new challenges and opportunities in attaining decent work. Virtual workspaces represent this paradigm shift within the international labor market, eliminating boundaries that are usually geographical and leading to flexible forms of work (Andriessen and Vartiainen 2005). Such virtual environment has opened new spheres of employment in areas like digital content creation, organizing virtual events, community administration, metaverse development, virtual property management, providing virtual goods and services, and many more. Gig opportunities related to freelance and short-term projects are made available within the virtual workplaces, with further opportunities in the creation of technical support, training, and induction processes (Kuhn et al. 2021). This can allow people to achieve an appropriate work-life balance that is more deserving as they can do other types of responsibility aside from the job. The decent work principle of virtual workspaces also includes the demand for rights at work, which refers to the provision for remote workspace or metaverse. This also created work-life integration instead of work-life balance. However, decent work issues in the virtual workplaces need attention to correspond with SDG 8 (Majumder and Biswas 2023). It highlights developing overall well-being and meaningfulness of work for all population. Target of SDG 8 is to promote long-term, inclusive and sustainable economic growth, full and productive employment, and decent work for all. This wide aim has been broken down into ten targets. Therefore, more importance is given to the workplace by considering social acceptance in the study, within which target 8.5 under SDG 8 discusses full employment and dignified work with equal pay. The purpose of this research is to identify the profound impact of virtual workspaces over employment and its contribution in promoting sustainable development.

1.1 Objectives of the Chapter

1. To investigate the impact of virtual workspaces on decent work principles.
2. To identify how virtual workspaces used to fit with SDG 8 and generative equal employment opportunities.
3. To showcase the importance of virtual workspaces in developing inclusive work place.

2 Theoretical Background

The Psychology of Working Theory (PWT) (Duffy et al. 2016) is a comprehensive framework that makes it possible to understand experiences, especially among people struggling to secure decent work. For instance, individuals living in poverty or those being discriminated against by their societies. Duffy et al. (2016) developed the Psychology of Working Theory, which shows how work can meet basic needs to survive, connect with others, and have self-determination. The theory used social and economic factors to shape work experiences and outcomes. It also showed contextual and individual factors that affect access to decent work, such as discrimination, economic conditions, and social support. It explains how work would contribute to the welfare, and how interventions would facilitate the processes of accessibility to meaningful and decent work. The theory also depicts the psychological and economic mediators and moderators that influence the relationship between these contextual characteristics and the acquisition of decent job (Song and Lee 2023).

Another well accepted theory based on solid research is Inclusive Growth Theory (Lin 2004). This indicates that economic growth should benefit all segments of society, rather than just a few privileged individuals. Advocating for policies and practices that foster widespread economic involvement guarantees that the advantages of growth are distributed evenly. The foundation of the philosophy emphasizes job generation. It is based on the expansion of employment possibilities in all sectors. In this sense, inclusive growth theory means that all people, regardless of birth orientation, have equal access to resources and opportunities. Social protection It protects vulnerable people with safety nets. Human capital It raises the degree of education and skill. This reduces existing disparities while also promoting long-term economic growth (Ranis et al. 2000).

2.1 Virtual Workspaces: A Double-Edged Sword

Virtual workplaces offer a variety of benefits that can help you achieve your career goals (Murphy 2024). For example, they may be able to improve work-life balance and create flexible work arrangements to help employees manage with dual tasks. Furthermore, it provides possibilities for persons with impairments and from remote places, fostering accessibility and diversity in the workplace. Organizations will also benefit from lower overhead expenses associated with the accommodation of huge physical office spaces. Slack, Zoom, Trello, Asana, Google Drive, and advanced platforms such as Microsoft Mesh and Gather all improve virtual communication, productivity, and team relations (Stefanic 2024). Microsoft Mesh, known for its immersive mixed-reality experiences, and Gather, renowned for its capacity to promote spontaneous encounters and personalized virtual workspaces, both help to foster a cohesive and innovative workplace. Virtual workspaces boost economic growth and help achieve the Sustainable Development Goals (SDGs) by leveraging a diversified global workforce, encouraging innovation, and establishing an inclusive and adaptable labor market (Pereira et al. 2024). SDG 8 aspires to achieve long-term, comprehensive, and productive employment, and adequate labor for all (Asaju 2022). Virtual workspaces have the advanced technology to match these goals by changing work methods, giving new opportunities, and removing some traditional barriers to successful employment.

However, the inherent adaptability of virtual labor may allow employees to blur the lines between work and personal time, thus leading to longer working hours and burnout (Haque and Mu 2023). Furthermore, legal and regulatory concerns also arouse, while managing virtual employees from different countries. In some circumstances, there may be ambiguity or lack of clarity regarding labor standards and safeguards. To fully realize the promise of virtual workspaces in furthering SDG 8, it is critical to address the issues connected with virtual employment. These problems include ensuring social support for all workers, balancing work and personal life, and protecting workers' rights in a digital environment. Regulators, employers, and employees can work together to address these challenges and maximize the benefits of virtual workspaces while avoiding hazards (Chen 2023; Wang and Parker 2023). This will help to achieve the goals of sustainable development and creating decent jobs for everyone.

3 Impact of Virtual Workplaces on Decent Work Principles

Virtual workspaces have significantly increased job opportunities by removing geographical obstacles and allowing firms to tap into a global talent pool (Olawale et al. 2024). This feature makes it easier for people, particularly those who live in remote or disadvantaged locations, to have access to previously unavailable job opportunities. However, this dynamic heightens concerns about job security and exposes employment to the gig economy.

According to Estevez-Abe et al. (2001), social protection policies and initiatives aim to improve worker well-being and welfare. The trend to virtual workspaces raises further concerns about social protection. Remote employees may have difficulty obtaining older forms of social security, health insurance, and other benefits connected with physical work places (Charalampous et al. 2018). Furthermore, the blurring of the lines between work and personal life has the potential to increase stress and contribute to burnout. Companies must adjust by providing more flexible work arrangements and ensuring that remote employees have suitable support networks.

Additionally, it is critical to develop legal safeguards for employees' rights in the virtual workplace. Data privacy, digital surveillance, and cyber security concerns are growing in importance (Dunn Cavelty 2014). Employers must be legally required to follow legislation governing labor rights in the digital sphere, including the ability to disconnect and precautions against online harassment. This shows good social discourse in virtual workspaces has several problems. Face-to-face interaction is necessary for effective communication and collaboration. However, digital tools can enable real-time feedback and inclusive decision-making processes, which are beneficial for new engagement. The employer should cultivate an environment characterized by open and honest communication and a culture centered around trust and collaboration. Virtual workspaces are internet-based workplaces that are available on demand, at any time, and without limitations. They are used for work, projects, and tasks (Desouza 2005).

3.1 *Alignment with SDG 8*

Virtual workplaces can facilitate the attainment of objectives by promoting efficiency, fostering creativity, and cultivating emerging economic prospects (Cascio and Montealegre 2016). Virtual workplaces enhance productivity by streamlining procedures, resulting in reduced time and resource requirements (Chudoba et al. 2005). It has greatly reduced travel time, decreased operational expenses, and enhanced the flexibility of work schedules. It fosters economic progress by promoting the development of flexible and adaptable enterprises.

SDG 8 encompass principles such as inclusion and fair chances inside the workplace. Virtual workspaces can be utilized to generate employment opportunities for marginalized individuals, such as those with disabilities, caretakers, or individuals residing in remote areas (Gupta and Vegelin 2016). Technology has facilitated the creation of a workforce that is more extensive, varied, and inclusive for companies (Daher and Ziade 2023).

3.2 *Issues*

Although virtual workspaces aligned with SDG 8 have potential benefits, there are several hurdles to their realization. Additional supportive measures are ensuring proficiency in digital skills, providing access to appropriate technology, and closing the gap in digital accessibility. Policymakers and businesses should establish mutually agreed-upon plans to address the obstacles that hinder inclusive economic growth (Leal Filho et al. 2021). Virtual workspaces facilitate the creation of more inclusive workplaces by removing physical and geographical obstacles, allowing varied talent to participate more easily (Khindri and Tanwar 2022). This provision will foster a culture characterized by diversity and inclusivity, among other attributes. It enables the use of various hiring strategies, novel work methodologies, and inclusive digital communication tools.

Enhancing virtual offices have numerous benefits, such as enhanced accessibility for individuals with disabilities. In this manner, the design may create digital tools and platforms that cater to all the requirements of the employees, thereby providing the organization with the highest quality service. These accommodations encompass assistive technologies such as screen readers, closed captioning, and user interfaces that offer high contrast or large print options.

Promoting and facilitating a healthy equilibrium between work and personal life. Virtual workspace facilities offer significant flexibility in terms of working hours and location, which can contribute to a better work-life balance for individuals, particularly caregivers, parents, and those with personal responsibilities (Rathnaweera and Jayathilaka 2021). Job satisfaction and staff retention are positively correlated with flexible work arrangements.

Establishing Diverse and Welcoming Work Environments Establishing inclusive working environments in virtual firms requires deliberate and purposeful efforts. Employers ought to establish policies, practices, and communication tactics that promote and uphold workplace inclusion. The training program should encompass instruction on digital etiquette, the development of virtual team-building activities, and the fostering of an inclusive mindset among employees. In conclusion Adopting virtual workspaces has both advantages and potential difficulties in applying fair work principles in line with SDG 8 and achieving the goal of inclusive job opportunities (Ribeiro-Duthie 2019). Virtual workplaces will undoubtedly enhance productivity, inclusivity, and accessibility, necessitating the development of new frameworks for social protection, workers' rights, and social discussion. Addressing these difficulties would enable the realization of a more inclusive and sustainable future of work by harnessing the promise of digital technologies.

The research investigated covers a significant temporal range, from 2001 to 2024, and offers a comprehensive view of theme evolution. The compilation comprises contributions from 363 distinct sources, including journals and books, resulting in a total of 734 items. This vast assortment enables a thorough and all-encompassing thematic cartography. The dataset has a substantial yearly growth rate of 18.28%, suggesting a considerable rise in interest and expansion of research effort in this particular topic. The data maintains its currency and reflects current research trends, with an average document age of 3.87 years. The average citation count for each document is 13.1, highlighting their significance and influence in the academic community.

Table 1 contains a wide range of themes, as shown by the presence of 1155 Keywords Plus and 2167 author-provided keywords. The diversity of issues examined in the papers is a reflection of their extensive and complex character. The area of decent work has a significant research community, consisting of 1848 authors. It is worth mentioning that a total of 187 authors have created documents on their own, indicating a clear preference for joint research endeavors. These findings are

Table 1 .

Description	Results
Main information about data	
Timespan	2001:2024
Sources (Journals, Books, etc.)	363
Documents	734
Annual growth rate (%)	18.28
Document average age	3.87
Average citations per doc	13.1
References	0
Document contents	
Keywords plus (ID)	1155
Author's keywords (DE)	2167
Authors	
Authors	1848
Authors of single-authored docs	187
Authors collaboration	
Single-authored docs	197
Co-authors per doc	2.87
International co-authorships (%)	24.93
Document types	
Article	734

reinforced by the fact that, on average, each document has 2.87 co-authors and 24.93% of these co-authors are from international institutions, highlighting the global and collaborative aspect of the research (Fig. 1).

The bar chart showcases the preeminent contributors depending on the quantity of papers they have contributed to the dataset. Duffy R. D. is at the top of the leaderboard with 15 documents, demonstrating a noteworthy contribution to the field. Allan B. A. is closely followed by 10 documents, which demonstrate a prominent research presence. Autin K. L., Kim H. J., and Rossier J. are all tied for third place, with each of them having 8 documents. This indicates their significant contribution to the research landscape. Blustein D. L. has made a significant but slightly less influential contribution by providing 7 documents, which is not as impactful as the contributions made by the top contributors. Both Di Fabio A. and Piper N. have six documents that serve as evidence of their substantial contributions. Finally, the roster of top contributors is completed by Dos Santos N. R. and Masdonati J., both of whom have contributed five documents. This figure emphasizes the significant influence that these authors have on the thematic development of the field, with Duffy R. D. being the most productive, followed by a competitive group of other notable contributors.

The topics are categorized into four quadrants by the thematic analysis diagram based on their progression and significance (Fig. 2). "Human," "article," and "human

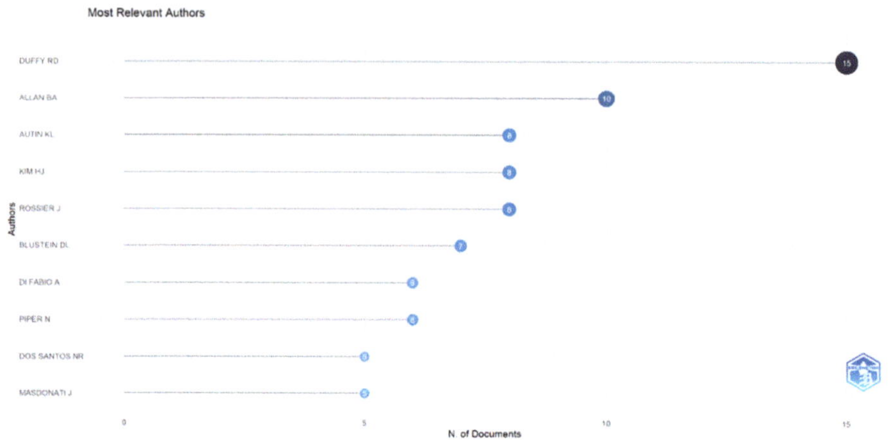

Fig. 1 .

rights" are prominent motifs in the thematic network, indicating their central significance and significant development. Conversely, subjects such as "employment," "working conditions," and "labor market" are regarded as fundamental issues, indicating that they are significant but necessitate additional investigation and development. The themes of "sustainable development goal," "sustainable development," and "sustainability" are not well-established and do not occupy a central position. This suggests that they are either older themes that are losing significance or emerging topics that are acquiring attention.

Visual representation of key words (Fig. 3) on sustainable development, employment, and economic growth are reflecting the frequency of their occurrence, in the existing literature. Prominent terms include "sustainable development," "employment," "sustainability," "economic growth," "working conditions," "labor market," "human rights," and "United Nations." This suggests strong attention to the issues of labor standards, economic policies, human rights, and sustainability within the domain of decent work. Other prominent occurrence shows "human," "governance approach," "social policy," and "education," which further illustrate the multi-dimensionality of these discussions, involving different aspects of socioeconomic activities, the process of decisions making, and the role of international organizations.

3.3 Opportunities of Virtual Workplaces Aligning with SDG 8

SDG 8 emphasise on advancing economic growth that is sustained, inclusive, and sustainable, while ensuring full and productive employment and decent work opportunities for everyone. This study explored the alignment of virtual workspaces with the objectives of SDG 8, drawing on evidence from various case studies and research findings.

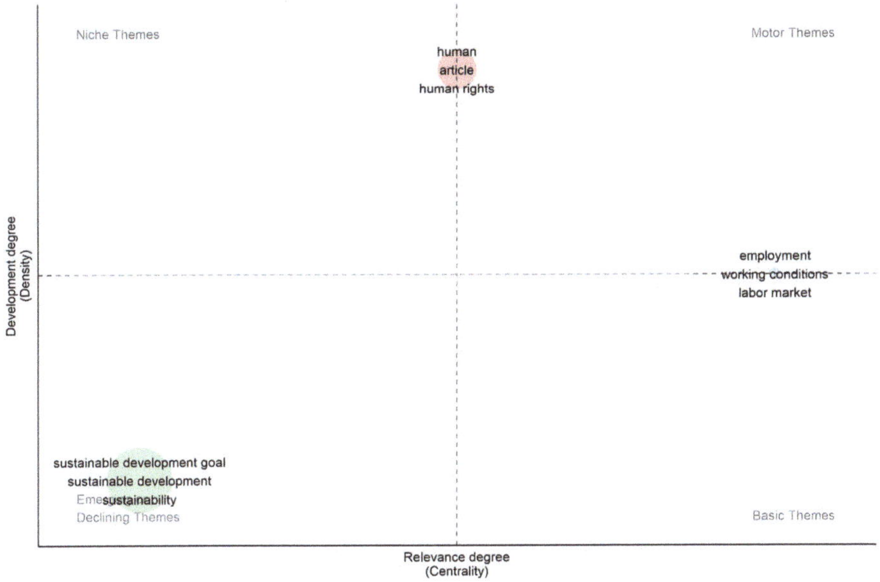

Fig. 2 .

Fig. 3 .

1. **Enhancing Economic Development by Boosting Productivity** Remote work settings enable individuals to operate in conditions that frequently optimise their productivity. A study conducted by Harvard Business School (Choudhury 2020) found that remote workers recorded more hours and demonstrated a 4.6% increase in productivity compared to those working in the office during the pandemic.

The increase in productivity is a significant factor in driving economic growth, aligning with the objectives outlined in SDG 8.

2. **Promoting Quality Employment by Emphasising Flexibility and Inclusiveness** A significant advantage of virtual workspaces holds relevance for underrepresented populations, including women with caregiving duties, who often encounter limitations due to conventional work hours and commuting demands. A survey conducted by Owl Labs in 2021 indicated that remote work resulted in a 22% decrease in turnover rates, primarily due to the increased autonomy it provided employees in managing their schedules (Gupta 2024; Russo et al. 2023). This plays a crucial role in achieving the objectives of SDG 8.5, which aims to guarantee full employment and decent work for everyone, encompassing women and individuals with disabilities. Dell Technologies's dedication to remote work has enabled the employment of individuals with disabilities who encounter physical obstacles in conventional office environments. Dell's provision of remote opportunities has broadened access to quality employment for individuals who might otherwise face exclusion, thereby directly contributing to the inclusivity objectives outlined in SDG 8.5.
3. **Advancing Comprehensive Employment Opportunities and Tackling Disparities** Virtual workspaces enable organisations to connect with skilled individuals from various locations around the world. This has the potential to enhance employment rates in developing nations where skilled individuals may lack access to nearby job prospects. Embracing remote work allows companies to engage employees from low-income nations, offering them higher-paying opportunities than what is usually accessible in their local economies. The emergence of remote freelancing platforms like Upwork and Fiverr has enabled individuals in countries such as the Philippines and India to participate in global labour markets, enhancing their earning opportunities. A report from 2020 by the World Bank emphasised the ways in which virtual work is broadening opportunities for individuals in the Global South. Individuals engaged in digital platforms in nations such as Kenya, Bangladesh, and Indonesia have gained access to employment opportunities that were once out of reach, thereby fostering local economic development and advancing the objectives of SDG 8 (Open knowledge repository 2020).
4. **Minimising Informal Employment and Encouraging Sustainable, Formal Job Opportunities** A significant number of freelance workers, particularly in developing economies, have traditionally functioned within the informal sector, lacking job security and benefits. Platforms like Toptal and Upwork offer a structured environment for workers to connect with global clients, allowing them to earn regulated wages and gain access to formal employment benefits, thereby contributing to the advancement of SDG 8, which focusses on the formalisation and growth of small enterprises. A report by Oxford University's Online Labour Index indicates that 70% of online gig work is performed in low- and middle-income countries, thereby creating more formal employment opportunities in areas characterised by high levels of informal work.

5. **Fostering Innovation and Advancing Technological Development** The rise of virtual workspaces has propelled the integration of digital tools and the pursuit of upskilling programs. Digital-first work environments foster the enhancement of new competencies associated with technology, communication, and project management. A report from McKinsey Global Institute in 2021 highlighted that companies focussing on remote work experienced a rise in employee engagement in professional development programs, which aided workers in gaining new skills that boost productivity and competitiveness (McKinsey Global Surveys 2021).
6. **Quality Employment and Opportunities for Under-represented Populations** Remote workspaces create pathways for marginalised groups, including women and individuals with disabilities, by dismantling physical barriers that have traditionally hindered their participation in the workforce. Women, facing significant challenges due to caregiving responsibilities, gain from the ability to harmonise their professional and personal lives. Individuals with disabilities experience improved access to employment opportunities through the elimination of physical commuting requirements and the necessity for workplace accommodations. GitLab: GitLab, operating as a fully remote organisation, has successfully enhanced workforce diversity through global recruitment and the provision of remote-first roles, which foster greater inclusion of women, individuals with disabilities, and minority groups. This fosters greater equity in opportunities, contributing to the advancement of SDG 8 which aim to reduce the percentage of young individuals who are not engaged in employment, education, or training.

3.4 *Challenges and Risks Associated with the Objectives of SDG 8*

Although virtual workspaces contribute to the progress of various SDG 8 targets, it is important to acknowledge the challenges that arise. It has been observed that unequal access to digital infrastructure can worsen inequality, especially in areas with inadequate internet connectivity. Individuals in rural or underserved regions might not experience the same advantages from remote work, potentially compromising the inclusivity objectives of SDG 8. The emergence of freelance platforms and gig economy roles presents opportunities for flexibility and access to international markets; however, it may also result in increased job insecurity (Chigbu and Nekhwevha 2023; Roy and Shrivastava 2020). A significant number of virtual workers function as independent contractors, lacking the social protections and benefits associated with conventional employment. This situation could impede advancements related to SDG 8 which focusses on safeguarding labour rights and fostering safe working conditions.

3.5 Managerial, Practical and Policy Implications

Virtual workspaces aligning with SDG 8 serves several implications such as providing enforcing labour standards through smart regulations and coalition of stakeholders for safe and secured work place. Remote work also enable integration of persons with disabilities and women into labour market and promote lifelong learning among employees and other related individuals. SDG 8 can be linked with improving policies for health and safety at work and strengthen workplace cooperation. It will also assist in setting requisites for maintaining decent work among organisations.

4 Conclusion

Virtual workspaces closely align with various objectives, providing innovative avenues for decent employment, inclusivity, and economic advancement. Eliminating geographical constraints, fostering adaptability, and supporting formal structures, virtual work settings contribute to the progress of sustainable development initiatives. Nonetheless, it is essential to tackle the inequalities in technology access and the threat of job instability to ensure that virtual work environments can genuinely fulfil their promise of fostering inclusive and sustainable economic growth for everyone.

The Decent work under Sustainable Development Goal 8 (SDG 8) has few encounters in its implementation and interpretation. SDG 8 aims to promote fair employment opportunities and sustainable economic growth. Yet, there are concerns regarding its ability to address systemic inequalities and protect the rights of laborers (Frey and Osborne 2017; Chigbu and Nekhwevha 2023). The objective of the aim is not explicitly defined, which allows for the incorporation of both human rights monitoring and a business-focused approach. This may compromise its usefulness as a human rights obligation (Frey and Osborne 2017). Despite some progress in addressing gender inequality, the presence of gender prejudice and income disparities continues to impede inclusive economic development and the availability of fair employment opportunities (Chigbu and Nekhwevha 2023). Additionally, the inherent inequalities within market economies, precarious employment situations, and exploitative labor practices, particularly within the informal sector, are not adequately addressed by SDG 8 (Chigbu and Nekhwevha 2023). The objective also demonstrates an insufficient integration of environmental sustainability, emphasizing the need for a more comprehensive approach to the promotion of sustainable development and satisfactory employment (Chigbu and Nekhwevha 2023). The significance of decent work in virtual workspaces has increased in tandem with the emergence of digital technologies. Workspace VR and similar virtual reality applications provide collaborative telework environments designed to achieve a harmonious combination of individual productivity and teamwork (Ortiz and Cruz-Neira 2023). The emergence of the digital gig economy brings about many prospects and difficulties for workers, especially in developing nations. To tackle concerns like employment stability and equitable remuneration, it is imperative to enforce decent work standards (Heeks 2017). Grid computing enables the creation of virtual workspaces that can be customized and deployed on different resources, allowing for increased flexibility and configurability. The advent of information and communication technology (ICT) has had a profound effect on the way work is organized, productivity is achieved, and employment prospects are created. ICT has the potential to improve global competitiveness, but it also gives rise to problems regarding decent work, which includes employment, social security, workers' rights, and social discourse. These advancements emphasize the necessity of adjusting labor procedures and regulations to the changing digital environment.

References

Andriessen JHE, Vartiainen M (2005) Mobile virtual work: a new paradigm? Springer Science & Business Media

Asaju K (2022) Achieving the sustainable development goals (SDGS) and the intricacies and dynamics of development administration. J Contemp Soc Issues 2(2):173. https://doi.org/10.19184/csi.v2i2.27890

Cascio WF, Montealegre R (2016) How technology is changing work and organizations. Annu Rev Organ Psych Organ Behav 3(1):349–375. https://doi.org/10.1146/annurev-orgpsych-041015-062352

Charalampous M, Grant CA, Tramontano C, Michailidis E (2018) Systematically reviewing remote e-workers' well-being at work: a multidimensional approach. Eur J Work Organ Psy 28(1):51–73. https://doi.org/10.1080/1359432x.2018.1541886

Chen Z (2023) Metaverse office: exploring future teleworking model. Kybernetes 53(6):2029–2045. https://doi.org/10.1108/k-10-2022-1432

Chigbu BI, Nekhwevha F (2023) Exploring the concepts of decent work through the lens of SDG 8: addressing challenges and inadequacies. Front Sociol 8:1266141

Choudhury P (2020) Our work-from-anywhere future. Harv Bus Rev. https://hbr.org/2020/11/our-work-from-anywhere-future

Chudoba KM, Wynn E, Lu M, Watson-Manheim MB (2005) How virtual are we? Measuring virtuality and understanding its impact in a global organization. Inf Syst J 15(4):279–306. https://doi.org/10.1111/j.1365-2575.2005.00200.x

Daher MM, Ziade F (2023) Technology, workforce, and the future of sustainable work. In: Navigating the intersection of business, sustainability and technology. Springer Nature Singapore, pp 119–136. https://doi.org/10.1007/978-981-99-8572-2_6

Desouza KC (2005) The internet in the workplace: how new technology is transforming work the internet in the workplace: how new technology is transforming work, by Wallace Patricia. Cambridge: Cambridge University Press, 2004. Acad Manag Rev 30(2):442–445. https://doi.org/10.5465/amr.2005.16387901

Duffy RD, Blustein DL, Diemer MA, Autin KL (2016) The psychology of working theory. J Couns Psychol 63(2):127–148. https://doi.org/10.1037/cou0000140

Dunn Cavelty M (2014) Breaking the cyber-security dilemma: aligning security needs and removing vulnerabilities. Sci Eng Ethics 20(3):701–715. https://doi.org/10.1007/s11948-014-9551-y

Estevez-Abe M, Iversen T, Soskice D (2001) Social protection and the formation of skills: a reinterpretation of the welfare state. In: Varieties of capitalism. Oxford University Press, Oxford, pp 145–183. https://doi.org/10.1093/0199247757.003.0004

Frey CB, Osborne MA (2017) The future of employment: how susceptible are jobs to computerisation? Technol Forecast Soc Chang 114:254–280

Gupta VP (2024) Leveraging digital transformation in marketing of home-stay businesses to promote tourism in the hospitality industry. In: Technology and Luxury hospitality. Routledge, pp 93–115
Gupta J, Vegelin C (2016) Sustainable development goals and inclusive development. Int Environ Agree Polit Law Econ 16(3):433–448. https://doi.org/10.1007/s10784-016-9323-z
Haque S, Mu S (2023) The impact of remote work on HR practices: navigating challenges, embracing opportunities. Eur J Hum Resourc Manag Stud 7(1). https://doi.org/10.46827/ejhrms.v7i1.1549
Heeks R (2017) Decent work and the digital gig economy: a developing country perspective on employment impacts and standards in online outsourcing, Crowdwork, etc. In: Development informatics working paper, vol 71
Hughes D, Warhurst C, Benger E, Ifans M (2024) Building better futures: decent work, inclusion and careers support services in the UK. In: Decent work, inclusion and sustainability. Routledge, pp 73–87. https://doi.org/10.4324/9781032662251-8
Khindri A, Tanwar S (2022) Challenges in the virtual workspace by the inception of industry 4.0: a conceptual frame work. In: Computational modelling in industry 4.0. Springer Nature Singapore, pp 23–34. https://doi.org/10.1007/978-981-16-7723-6_3
Kuhn KM, Meijerink J, Keegan A (2021) Human resource management and the GIG economy: challenges and opportunities at the intersection between organizational HR decision-makers and digital labor platforms. In: Research in personnel and human resources management. Emerald Publishing Limited, pp 1–46. https://doi.org/10.1108/s0742-730120210000039001
Leal Filho W, Azul AM, Brandli L, Lange Salvia A, Wall T (2021) Decent work. In: Encyclopedia of the UN sustainable development goals. Springer International Publishing, pp 188–188. https://doi.org/10.1007/978-3-319-95867-5_300034
Lin JY (2004) Development strategies for inclusive growth in developing Asia. Asian Dev Rev 21(02):1–27. https://doi.org/10.1142/s0116110504000053
Majumder S, Biswas D (2023) Decent work-life in business: essential tool for sustainable development. Taylor & Francis
McKinsey Global Surveys (2021) McKinsey global surveys, 2021: a year in review. https://www.mckinsey.com/~/media/mckinsey/featured%20insights/mckinsey%20global%20surveys/mckinsey-global-surveys-2021-a-year-in-review
Murphy L (2024) Wellbeing in the age of virtual teams and workplace automation—a systematic review and future research agenda. Int J Organ Anal. https://doi.org/10.1108/ijoa-08-2023-3938
Olawale O, Ajayi FA, Udeh CA, Odejide OA (2024) Remote work policies for it professionals: review of current practices and future trends. Int J Manag Entrepr Res 6(4):1236–1258. https://doi.org/10.51594/ijmer.v6i4.1056
Open Knowledge Repository (2020) https://openknowledge.worldbank.org/bitstream/handle/10986/34406/9781464816192
Ortiz JA, Cruz-Neira C (2023) Workspace VR: a social and collaborative telework virtual reality application. In: Companion publication of the 2023 conference on computer supported cooperative work and social computing, pp 381–383
Pereira J, Rodrigues RG, Veiga PM (2024) Entrepreneurship among social workers: implications for the sustainable development goals. Sustainability 16(3):996. https://doi.org/10.3390/su16030996
Ranis G, Stewart F, Ramirez A (2000) Economic growth and human development. World Dev 28(2):197–219. https://doi.org/10.1016/s0305-750x(99)00131-x
Rathnaweera D, Jayathilaka R (2021) In employees' favour or not?—The impact of virtual office platform on the work-life balances. PLoS ONE 16(11):e0260220. https://doi.org/10.1371/journal.pone.0260220
Ribeiro-Duthie AC (2019) SDG 8 decent work and economic growth. In: Science for sustainable societies. Springer, Singapore, pp 117–133. https://doi.org/10.1007/978-981-32-9927-6_9
Roy G, Shrivastava AK (2020) Future of GIG economy: opportunities and challenges. Imi Konnect 9(1):14–27

Russo D, Hanel PH, Altnickel S, van Berkel N (2023) Satisfaction and performance of software developers during enforced work from home in the COVID-19 pandemic. Empir Softw Eng 28(2):53

Song JY, Lee K-H (2023) Influence of economic and academic barriers on perception of future decent work: a moderated mediation model of work volition and social support. J Vocat Behav 141:103842. https://doi.org/10.1016/j.jvb.2023.103842

Stefanic D (2024) Maximize efficiency with virtual team collaboration. Hyperspacemv—the Metaverse for Business Platform. https://hyperspace.mv/virtual-team-collaboration

Wang B, Parker SK (2023) Embracing the digital workplace: a SMART work design approach to supporting virtual work. In Handbook of virtual work. Edward Elgar Publishing, pp 403–424. https://doi.org/10.4337/9781802200508.00032

Business and Marketing in the Metaverse: Exploring SDG Integration and Consumer Engagement

Md. Faiz Ahmad, Mohd Danish Kirmani, Md. Asadul Haque, and Mohammad Riyazuddin

Abstract The advent of Metaverse offers tremendous avenues for businesses specially marketers, metamorphosing engagement with consumers and implementation of corporate strategies. The present chapter tries to delve into the exploration of sustainability development goals (SDGs) and practices of marketing in the Metaverse environment. It also unearths as to how business can bank upon the trending technologies viz augmented reality (AR), virtual reality (VR), blockchain (BC) and artificial intelligence (AI) aiding to change the perception and behaviour of consumers and provide superior customer engagement and experiences. It is evident that metaverse's impact is profound on customer engagement and experiences whether its healthcare, medicine, entertainment, retail by proposing customization, personalization, unique and innovative techniques that breaks the jinx between virtual and real world. The chapter provides insight to various stakeholder such as academia, industry and practitioner about the advancement of Metaverse and hinderances towards successful implementation of SDGs into marketing practices and business.

Keywords Metaverse · SDG marketing · Virtual reality · Blockchain technology

1 Introduction

The concept "Metaverse" increasingly gaining prominence not only in technological domain but also getting traction in the discourse of business. It represents a conjoint virtually shared arena. It is a promising digital forefront in today's world resulting as

Md. F. Ahmad (✉) · Md. A. Haque
Paari School of Business, SRM University-AP, Amravati, Andhra Pradesh, India
e-mail: mdfaizahmad61@gmail.com

M. D. Kirmani
School of Management, IILM University, Greater Noida, Uttar Pradesh, India

M. Riyazuddin
NIIT, Shanghai, China

W. Leal Filho et al. (eds.), *Metaverse and Sustainability: Business Resilience Towards Sustainable Development Goals*, World Sustainability Series,
https://doi.org/10.1007/978-3-031-89545-6_21

a synergy among various technological advancements such as artificial intelligence (AI), augmented reality (AR), virtual reality (VR) and blockchain (BC). It paves the way by providing an immersive experience and engaging environment through which users will be able to perform numerous tasks from socializing with others and playing games to doing shopping on various platforms and conducting virtual events.

The purpose of the present chapter are as follows.

1. To explore the synergy between business and marketing in the realm of the Metaverse.
2. To analyse as to how SDGs (Sustainable Development Goals) can be integrated into Metaverse strategies and
3. To examine consumer engagement in the Metaverse.

The objectives of the present chapter help in achieving different United Nations (UN) Sustainability Development Goals (SDGs). These objectives resonate with SDG 8 i.e., promoting innovation, come up with new business models and sustainable economic growth within digital economies; SDG 9 i.e., exploring digital transformation, virtual platforms and innovative marketing approaches within the metaverse; SDG 12 i.e., encouraging stakeholders such as customers and businesses to engage in sustainable and responsible virtual consumption in the metaverse, and SDG 17 i.e., fostering collaborations among consumers, businesses and institutions to promote sustainable engagement in virtual sphere.

The emergence of Metaverse could be traced way back in the year 2003 with the introduction of virtual worlds "Second Life", which makes it possible for the users to create their own avatars, engage with fellow users and construct assets virtually. Over the past two decades, advancements in the technology have significantly empowered virtual worlds, resulting in development of an immersive and sophisticated atmosphere. The inception of VR and AR technology has fuelled the progress of Metaverse which offers end users an immersive and engaging experiences. environments.

The concept "Metaverse" was coined in the year 1992 by Neal Stephenson in science fiction novel 'Snow Crash'. The word metaverse is combination of two words viz 'meta which means beyond' and 'verse which means universe'. The term started gaining prominence in industry as well as in academia. Metaverse is explored through various scholars who belongs to different school of thoughts (Dwivedi et al. 2022, 2023; Koohang et al. 2023). As per Buhalis et al. (2023), metaverse is still in its nascent conceptual stage and there is no clear execution of the concept. Lundmark (2022) in the financial times article explicates "By 2030, every physical device that can be digitally connected will be. Eventually, every action in the digital world will have an effect on the physical world, and vice versa. So, the metaverse isn't dependent on a virtual reality headset. Rather, it's the coming together of complementary technologies, including cloud and edge (near the data source) computing, artificial intelligence, blockchain, the internet of things, virtual reality, augmented reality and digital twins." The author has captured the concept of metaverse in business and marketing and presented in Table 1.

Table 1 Definitions of the concept metaverse

Contributor(s)	Definition	Perspective
Buhalis et al. (2023)	"The convergence of physical and digital universes, where users can seamlessly traverse between them for working, education and training, health, exploring interests and socialising with others"	Tourism and marketing
Barrera and Shah (2023)	"A technology-mediated network of scalable and potentially interoperable extended reality environments merging the physical and virtual realities to provide experiences characterized by their level of immersiveness, environmental fidelity, and sociability"	Marketing
Buhalis et al. (2023)	"Metaverse as a confluence of virtual and physical universes, radically different from past technologies in three dimensions viz (a) Functionality (b) Interoperability and (c) Immersiveness"	Tourism and marketing
Buhalis and Karatay (2022)	"A seamless convergence of digital and physical universes that use ambient intelligence to enhance physical spaces, products and services"	Tourism
Moy and Gadgil (2022)	"A seamless convergence of our physical and digital lives, creating a unified, virtual community where we can work, play, relax, transact and socialize"	Business
Park and Kim (2022)	"Metaverse is a compound word of transcendence meta and universe and refers to a three-dimensional virtual world where avatars engage in political, economic, social, and cultural activities"	–
Ball (2021)	"An expansive network of persistent, real-time rendered 3D worlds that support the continuity of identity and objects"	Business
Collins (2021)	"A massive virtual world where millions of people—or their avatars—will interact in real time"	Business
Caulfield (2021)	"A shared virtual 3D world, or worlds, that are interactive, immersive, and collaborative"	Gamification
Herrman and Browning (2021)	"A fully realized digital world that exists beyond the analog one in which we live"	Business
Damar (2021)	"3D virtual shared world where all activities can be carried out with the help of augmented and virtual reality services" describing the metaverse as "the layer between you and reality"	–
Kim (2021)	"An interoperated persistent network of shared virtual environments where people can interact synchronously through their avatars with other agents and objects"	Advertising
Sparks (2021)	"A shared online space that incorporates three-dimensional graphics, either on a screen or in virtual reality"	–

1.1 Key Technologies Driving Metaverse

The driving force behind the success of metaverse environment are various advanced technologies which foster an immersive and experiential experiences for the users. The major technologies enhancing the capabilities of metaverse are discussed in the below paragraphs.

1. Augmented Reality (AR): It edges over the digital information onto the physical world. It enhances the user's perception while performing different activities. In the realm of Metaverse, augmented reality (AR) helps to create combined reality engagement and experience in which physical and virtual aspects parallelly go hand in hand to provide interactive environment.
2. Virtual Reality (VR): It engulf users completely in digital space specially through the usage of VR headsets. In the realm of Metaverse, VR has the capabilities to simulate the real world as well as illusory surroundings, which allow users to foresee and engage within this fictional environment and feels that they are into the physical reality.
3. Blockchain (BC): It facilitates decentralized virtual economies by offering translucent and secure platform to create, own and transfer of digital possessions. In the realm of Metaverse, BC allows to create virtual currencies such as Bitcoins, NFTs (non-fungible tokens), and decentralized market.
4. Artificial Intelligence (AI): It empowers the interactivity aspect of Metaverse for example through chatbots and NPCs (non-player characters) and virtual chatbots. It facilitates the various activities through different mechanism such as mechanical AI which helps to provide solutions for repeated task, thinking AI to aid in decision-making and feeling AI to provide personalization of user experience.

The above-mentioned technologies are regularly developing due to which Metaverse also transforming in day and day out through expansion and enhancement. The maturity of these technologies will bring easy accessibility, more immersiveness, and engulfed into our day-to-day life, offering new avenues for various stakeholders such as consumers, marketers and business enterprises.

1.2 Metaverse Importance in Today's Business

The inception of Metaverse have changed the way in terms of carrying out business operations, customer engagement and creation of value proposition. As per the report by Citi (2022), metaverse market economy could grow between $8 trillion to $13 trillion by 2030 (refer Table 2).

The following factors aid to enhance the progression of Metaverse market.

- Expanding User Base
- Digital Economies
- Virtual Real Estate

Table 2 Metaverse market by 2030 (in $ trillions)

Metaverse TAM, 2030 ($ trillions)		Digital economy as % of GDP			
		15%	20%	25%	30%
Metaverse as % of digital	10%	1.9	2.6	3.2	3.8
	20%	3.8	5.1	6.4	7.7
	30%	5.8	7.7	9.6	11.5
	40%	7.7	10.2	12.8	15.3
	50%	9.6	12.8	16.0	19.2

Source Citi (2022)
TAM total addressable market

- Advertising and Branding.

2 Metaverse and the Sustainable Development Goals (SDGs)

United Nations (UN; comprising of 193 members) endorsed a visionary plan in 2015 aimed at creating a brighter future for all. This plan, known as "Agenda 2030," aims to eradicate extreme poverty, combat inequality, and safeguard the environment. The 'Agenda 2030' focused on 17 goals which are known SDGs. These objectives outline a shared vision for a better world, applicable to all nations and communities.

2.1 Importance of SDG Integration in Business

The SDGs gave a clear direction to the governments and also to the business organizations and significantly helped them to strategize and focus on environmental conservation in their business actions. They are now having an opportunity to tactically communicate, and leverage connect with their target customers. They are now compelled to work for social change and contribute towards the SDGs. Beyond responding to society's demands for increased transparency and accountability, integrating purpose with profit can create a distinctive competitive edge that aligns with the expectations of discerning consumers, investors, and employees. Companies are driven by five key factors (UNDP 2024) to embed social impact and SDG alignment into their core business strategies:

1. Generate new revenue by exploring opportunities for market uniqueness, expansion, and growth, including innovating to tap into emerging markets that are either undeveloped or in their early stages.
2. Improve firm attractiveness, improving recruitment and retention efforts.

3. Boost supply chain resilience by improving sustainability and operational efficiency within supply chains.
4. Stay "ahead of the curve" by ensuring compliance with regulations and effectively managing risks to secure a license to operate.

Achieving the SDGs is projected to unlock at least USD 12 trillion in business opportunities. According to World Investment Report (2014), there is a compelling business rationale for the private sector to invest in SDG implementation. For example, the Business and Sustainable Development Commission estimates that achieving the SDGs could create around USD 12 trillion in market opportunities within four key economic systems: food and agriculture, cities, energy and materials, and health and well-being (UNDP 2024). These sectors, which constitute approximately 60 per cent of the real economy, are essential for realization of the SDGs. The overall economic impact and business potential of implementing the SDGs could be two to three times larger if the benefits extend across the entire economy and are supported by significantly higher labour and resource productivity.

2.2 Use of Virtual Technologies to Achieve SDGs

Virtual technologies such as metaverse can significantly accelerate efforts towards achieving SDGs. The interactive and immersive nature of the metaverse opens up a realm of possibilities for addressing complex challenges across various sectors, from poverty and hunger to climate change and inequality. Table 3 delves into specific examples of how virtual technologies can offer innovative solutions to achieve each of the 17 SDGs.

2.3 Sustainable Challenges in Metaverse and Other Virtual Technologies

The metaverse and other virtual technologies face criticism for their adverse effects on society and the environment. Research has highlighted various concerns related to technology use, including privacy and security risks such as cyber targeting, information manipulation, and societal impacts like heightened expectations, overreliance on technology, and technological addiction (Zolfagharian and Yazdanparast 2017; Jabeen et al. 2023). The sustainability challenges in the metaverse have been discussed in Table 4.

Table 3 Virtual technologies and SDGs

Specific SDG	Virtual solution to achieve SDG
Goal 1	Foster a deeper understanding of poverty mitigation through immersive virtual experiences, and informed decision-making
Goal 2	Leverage VR technologies to modernise agriculture and enhance food production efficiency
Goal 3	Construct virtual environments to enable therapeutic interventions
Goal 4	Use of virtual reality in imparting information and education
Goal 5	Train women in the use of VR technology
Goal 6	Employ VR tools to impart effective water resource management techniques
Goal 7	Provide workforce development programs on renewable energy through VR technology
Goal 8	Expand economic scale related to virtual reality
Goal 9	Develop infrastructure and acquire technical interoperability through virtual reality
Goal 10	Foster inclusivity and combat biases in virtual reality spaces by ensuring a diverse representation of avatars
Goal 11	Deliver public services and facilitate civic engagement through virtual platforms
Goal 12	Implement VR-based educational programs to minimize food waste at every stage of food chain
Goal 13	Employ VR tools to create eco-tourism experiences and promote responsible consumption
Goal 14	Use of VR tools to create training, awareness and capacity-building programs
Goal 15	Leverage VR experiences to encourage sustainable human behaviors
Goal 16	Improved training, rehabilitation and understanding of criminal justice practices through virtual reality environments
Goal 17	Promote virtual reality collaboration and knowledge exchange

Source Kunitake et al. (2022)

Table 4 Sustainability challenges in metaverse

S. No.	Impact	Description
1	Privacy Issues	Metaverse technology poses several risks, including cyberbullying, data misuse, sexual harassment, misinformation, child abuse, financial crimes, and digital terrorism
2	Environment Issues	The metaverse contributes to environmental degradation through carbon and greenhouse gas emissions, as well as the generation of massive electronic waste
3	Health Issues	Prolonged metaverse use can lead to various health issues, including vision impairment, disrupted sleep patterns, obesity, blurred consciousness, anxiety, stress, and social isolation

Source Rosenberg (2023), Kshetri et al. (2023), Dwivedi et al. (2023) and Oh et al. (2023)

3 Consumer Engagement in the Metaverse

Consumer engagement in the metaverse can be referred to as the positive cognitive, social, and behavioural activities a consumer demonstrates concerning a brand during or related to interactions with that brand (Hollebeek et al. 2020). These dimensions have been presented in Table 5.

The cognitive dimension represents the value consumers perceive from their experiences in the Metaverse (Pamucar et al. 2023; Queiroz et al. 2023). The social dimension refers to the brands' and consumers' co-create engagement. The behavioural dimension refers to positive interactions between consumer and brand.

Simulated environments like the Metaverse offer consumers a way to interact with their preferred brands in the virtual world (Arya et al. 2024). According to Arya et al. (2024), users can experience brand association through the building of brand communities offering quality products in the metaverse environment. Brand association refers to the connection of consumers with a brand's virtual assets, often created through NFTs and their associated intangible benefits and usefulness. A strong brand association with the customers ensures higher acceptance of products/services by the customers.

3.1 Tools and Techniques Used by Brands for Consumer Engagement in Metaverse

According to a McKinsey report (2022), the following strategies have been employed by the brands for higher consumer engagement in the metaverse:

- Building Immersive Brand Experiences and Virtual Worlds:

This includes creating virtual spaces where users can interact with the brand and its products. For example, Nike's Nikeland on Roblox allows users to play games, socialize, and purchase virtual Nike products. Similarly, the brands can also develop virtual showrooms or stores to showcase products in a 3D environment and allow

Table 5 Dimensions of consumer engagement in the metaverse

Dimension	Description	Key concepts
Cognitive	The value consumers derive from their metaverse experiences	Experiential value, perception
Social	Co-created engagement between brands and consumers	Social exchange theory, interaction
Behavioral	Positive interactions between consumer and brand	Augmentation, immersiveness and co-development

Source Arya et al. (2024)

users to try virtual items before purchasing. In the same vein, virtual events like concerts, fashion shows, or product launches, offering unique and memorable brand experiences can be hosted in the metaverse.

- Leveraging Virtual Goods and NFTs:

Brands can sell virtual goods and collectables like clothing, accessories, or art, either as standalone items or tied to physical products. Also, they can leverage NFTs (non-fungible tokens) to allow brands to tap into the growing market for digital collectables. NFTs can also be used to offer exclusive access or benefits to holders, building loyalty and community.

- Partnering with Influencers and Creators:

Brands can collaborate with virtual influencers, having a strong following within the metaverse, to leverage their reach and engagement to promote products. They can also do partnerships with creators to design and develop virtual experiences. The events and activities can be sponsored on Metaverse to gain visibility and traction.

- Gamification and Entertainment:

Games can be integrated into the virtual world. The mini-games provide an interactive experience to the users and thereby, promote products and entertain users.

- Utilizing Data and Analytics:

The metaverse environment generates a large amount of users' data that can leveraged to design personalize experiences and tailor marketing efforts. The tracking of sales and engagement metrics can be used to measure the effectiveness of marketing campaigns and hence optimize strategies. Also, artificial intelligence (AI) and machine learning can be employed to analyse users' data and identify the challenges and new opportunities and as a result, design product development and marketing strategies.

4 Conclusions

The present chapter delve into the importance of metaverse in business specially in marketing. Metaverse found to be transformative technology which offers enormous opportunities as well as challenges for the corporations. New avenues for Metaverse are emerging as technology advances. AR, VR, BC and AI technologies offer unique and innovative ways for the users to engage in more immersive experiences. Since the Metaverse is increasingly integrated into sustainable development goals (SDGs) and marketing operations, academia, policymakers and practitioner can play a vital role in optimizing its impact while adhering to social and ethical norms. Business houses must ensure that Metaverse implementation must considered with actual sustainability objectives instead of just aiming towards profit making. It is critical to assess the enduring impact of Metaverse. The impact assessment practices need to be carried out by various stake holders for the long-term sustainability, marketing and fruitful growth of the business.

References

Arya V, Sambyal R, Sharma A, Dwivedi YK (2024) Brands are calling your AVATAR in metaverse—a study to explore XR-based gamification marketing activities & consumer-based brand equity in virtual world. J Consum Behav 23(2):556–585. https://doi.org/10.1002/cb.2214

Ball M (2021) Framework for the metaverse. MatthewBall.vc. https://www.matthewball.vc/all/forwardtothemetaverseprimer

Barrera KG, Shah D (2023) Marketing in the Metaverse: conceptual understanding, framework, and research agenda. J Bus Res 155:113420

Buhalis D, Karatay N (2022) Mixed reality for generation Z in cultural heritage tourism towards metaverse. In: Stienmetz JL, Ferrer-Rosell B, Massimo D (eds) Information and communication technologies in tourism 2022. Springer, Cham, pp 16–27

Buhalis D, Leung D, Lin M (2023) Metaverse as a disruptive technology revolutionising tourism management and marketing. Tour Manage 97:104724

Caulfield B (2021) What is the metaverse? The Official NVIDIA Blog. https://blogs.nvidia.com/blog/2021/08/10/what-is-the-metaverse/

Citi (2022) Metaverse and money: decrypting the future. https://ir.citi.com/gps/x5%2BFQJT3BoHXVu9MsqVRoMdiws3RhL4yhF6Fr8us8oHaOe1W9smOy1%2B8aaAgT3SPuQVtwC5B2%2Fc%3D

Collins B (2021) The metaverse: how to build a massive virtual world. Forbes Mag. https://www.forbes.com/sites/barrycollins/2021/09/25/the-metaverse-how-tobuild-a-massive-virtual-world/?sh=60735bad6d1c

Damar M (2021) Metaverse shape of your life for future: a bibliometric snapshot. J Metaverse 1(1):1–8

Dwivedi Y, Hughes L, Baabdullah A, Ribeiro-Navarrete S, Giannakis M, Al-Debei M, Dennehy D, Metri B, Buhalis D, Cheung C, Conboy K, Doyle R, Goyal DP, Gustafsson A, Jebabli I, Kim Y-G, Kim J, Koos S, Kreps D, Wamba F et al (2022) Metaverse beyond the hype: multidisciplinary perspectives on emerging challenges, opportunities, and agenda for research, practice and policy. Int J Inf Manag 66. Article 102542

Dwivedi Y, Hughes L, Wang Y, Alalwan AA, Ahn S, Balakrishnan J, Barta S, Belk R, Buhalis D, Dutot V, Felix R, Filieri R, Flavi'an C, Gustafsson A, Hinsch C, Hollensen S, Jain V, Kim J, Krishen A, Wirtz J et al. (2023) How metaverse will change the future of marketing: implications for research and practice. Psychol Mark. https://doi.org/10.1002/mar.21767

Herrman J, Browning K (2021) Are we in the metaverse yet? The New York Times. https://www.nytimes.com/2021/07/10/style/metaverse-virtual-worlds.html?.%3Fmc=aud_dev&ad-keywords=auddevgate&gclid=CjwKCAjwzaSLBhBJEiwAJSRoksQC2o3eU9nurJhxNMp1dNXKX6QYwHqa-Fyot3sCemL1fkhEWekNDehoC0ZAQAvD_BwE&gclsrc=aw.ds

Hollebeek LD, Clark MK, Andreassen TW, Sigurdsson V, Smith D (2020) Virtual reality through the customer journey: framework and propositions. J Retail Consum Serv 55:102056. https://doi.org/10.1016/j.jretconser.2020.102056

Jabeen F, Tandon A, Sithipolvanichgul J, Srivastava S, Dhir A (2023) Social media-induced fear of missing out (FoMO) and social media fatigue: the role of narcissism, comparison and disclosure. J Bus Res 159:113693. https://doi.org/10.1016/j.jbusres.2023.113693
Kim J (2021) Advertising in the metaverse: research agenda. J Interact Advert 21(3):141–144
Koohang A, Nord J, Ooi K, Tan G, Al-Emran M, Aw E, Baabdullah A, Buhalis D, Cham T, Dennis C, Dutot V, Dwivedi Y, Hughes L, Mogaji E, Pandey N, Phau I, Raman R, Sharma A, Sigala M, Ueno A, Wong L (2023) Shaping the metaverse into reality: multidisciplinary perspectives on opportunities, challenges, and future research. J Comput Inf Syst 63 (in print). https://www.academia.edu/94457087/
Kshetri N, Dwivedi YK, Davenport TH, Panteli N (2023) Generative artificial intelligence in marketing: applications, opportunities, challenges, and research agenda. Int J Inf Manag 102716. https://doi.org/10.1016/j.ijinfomgt.2023.102716
Kunitake Y, Rights V, Field VR (2022) The potential of virtual reality for the SDGs: infrastructure development through content and cultural policies. Available at: https://sdgs.un.org/sites/default/files/2023-05/A9%20-%20Kunitake%20-%20The%20Potential%20of%20Virtual%20Reality%20for%20the%20SDGs%20Infrastructure%20Development.pdf. Accessed 18 Aug 2024
Lundmark P (2022) The real future of the metaverse is not for consumers. Retrieved from https://www.ft.com/content/af0c9de8-d36e-485b-9db5-5ee1e57716cb
McKinsey Report (2022) Marketing in the metaverse: an opportunity for innovation and experimentation. McKinsey Quarterly. Available at: https://www.mckinsey.com/~/media/mckinsey/business%20functions/marketing%20and%20sales/our%20insights/marketing%20in%20the%20metaverse%20an%20opportunity%20for%20innovation%20and%20experimentation/marketing-in-the-metaverse-an-opportunity-for-innovation-and-experimentation-vf.pdf?shouldIndex=false. Accessed 18 Aug 2024
Moy C, Gadgil A (2022) Opportunities in the metaverse: how businesses can explore the metaverse and navigate the hype vs reality. JPMorgan
Oh HJ, Kim J, Chang JJ, Park N, Lee S (2023) Social benefits of living in the metaverse: the relationships among social presence, supportive interaction, social self-efficacy, and feelings of loneliness. Comput Hum Behav 139:107498. https://doi.org/10.1016/j.chb.2022.107498
Pamucar D, Deveci M, Gokasar I, Delen D, Köppen M, Pedrycz W (2023) Evaluation of metaverse integration alternatives of sharing economy in transportation using fuzzy Schweizer-Sklar based ordinal priority approach. Decis Support Syst 171:113944. https://doi.org/10.1016/j.dss.2023.113944
Park SM, Kim YG (2022) A metaverse: taxonomy, components, applications, and open challenges. IEEE Access 10:4209–4251
Queiroz MM, Wamba SF, Pereira SCF, Jabbour CJC (2023) The metaverse as a breakthrough for operations and supply chain management: implications and call for action. Int J Oper Prod Manag 43(10):1539–1553. https://doi.org/10.1108/IJOPM-01-2023-0006
Rosenberg L (2023) Even thought it's virtual, the metaverse does actually impact the environment. World Economic Forum, 16 Feb 2023. Available at: https://www.weforum.org/agenda/2022/02/how-metaverse-actually-impacts-the-environment/. Access 10 Mar 2024
Sparks M (2021) What is a metaverse? New Sci 251(3348):18
UNDP (2024) Sustainable development goals: business and the SDGs. Available at: Business and the SDGs | United Nations Development Programme (undp.org). Accessed 18 Aug 2024
World Investment Report (2014) Investing in the SDGs: an action plan for promoting private sector contributions. Available at: https://unctad.org/system/files/official-document/wir2014ch4_en.pdf. Accessed 18 Aug 2024
Zolfagharian M, Yazdanparast A (2017) The dark side of consumer life in the age of virtual and mobile technology. J Mark Manag 33(15–16):1304–1335. https://doi.org/10.1080/0267257X.2017.1369143

Building Resilient Virtual Markets: Connecting Ethical and Legal Consumerism with Sustainable Development Goals in the Metaverse

Kusum Pal and Poonam Gulati

Abstract The metaverse, defined as a collective, virtual, shared space, is becoming a significant part of the global digital economy. Evaluating Sustainable Development Goals (SDGs), marketing techniques, consumer perceptions of virtual reality, and the impact of the metaverse on customer experiences across various industries is crucial for discussing critical performance measures for sustainability and the future role of public sector organizations specifically in developing countries like India. In reference to Responsible Consumption and Production (SDG 12), Peace, Justice, and Strong Institutions (SDG 16), and marketing techniques, consumer perceptions of virtual reality, as well as the impact of the metaverse on customer experiences across various industries are crucial for discussing critical performance measures for the sustainable growth of the world. Exploring the intersection of consumer rights, ethical considerations, and the sustainable development of cyber consumers vis-à-vis virtual market vendors this study examines the applicability of existing consumer protection laws on the global front as well as in India, to have a comparative understanding of developed and developing nations. Identifying the challenges in current legal frameworks and proposing innovative solutions, the study provides policy recommendations and strategies for businesses to adopt ethical and sustainable practices, contributing significantly to the discourse on consumer protection in virtual markets.

Keywords Metaverse · Sustainable development goals (SDGs) · Ethical practices · Consumer protection · Virtual markets

---

K. Pal · P. Gulati (✉)
Amity University Punjab, Mohali, India
e-mail: pgulati@pb.amity.edu
URL: https://www.linkedin.com/in/poonam-gulati-46b26a219; https://vidwan.inflibnet.ac.in/profile/354360

K. Pal
e-mail: kpal@pb.amity.edu
URL: https://vidwan.inflibnet.ac.in/profile/250311

W. Leal Filho et al. (eds.), *Metaverse and Sustainability: Business Resilience Towards Sustainable Development Goals*, World Sustainability Series,
https://doi.org/10.1007/978-3-031-89545-6_22

1 Introduction

Virtual markets through e-commerce have revolutionized the way people engage in commercial activities, enabling them to buy and sell goods, transfer funds, and access services online via the Internet. This process has seamlessly connected sellers and consumers on a global scale, contributing significantly to the rapid growth of the digital market. While this commercial movement in the global digital economy is commendable, it also has a deceptive side where consumer protection becomes a critical concern that impacts sustainable development. Irrespective of the fact that cyber-consumer laws are designed to address these concerns, focusing on safeguarding consumers in the online marketplace, these legislations still suffer from great ambiguity regarding securing the cyber customers' interests appropriately (Kumar and Singh 2021).

Today when we are talking about the integrated growth of the world as a single unit, it becomes pertinent to imbibe a few sustainable development goals (SDGs) in our policies, 'Responsible Consumption and Production',[1] and 'Peace, Justice, and Strong Institutions' (Sen and Das 2021).[2] Reference to these SDGs is essential since they are direct correlates of cyber-consumer behaviours.[3] In the present era, virtual markets have become a household term, and consumers prefer making deals with purchasing goods and services through such platforms (Blasio 2008). The convenience of shopping from anywhere and at any time has driven the tremendous growth of e-commerce. However, this feature has also opened doors for online vendors to engage in many malpractices that extend beyond commercialism's moral and ethical boundaries (Shaik and Poojasree 2021).

If we are determined to achieve the objectives of 'Peace and Prosperity for People and The Planet…', the main aim behind achieving sustainable development goals to enjoy social and economic prosperity, we must work on attaining the protection ensuring ethical legal and sustainable development of consumer rights in the metaverse market (https://www.undp.org/sustainable-development-goals, n.d.).

The present research is an outreach to various publications, notable of which are essential to be discussed. Snellman and Vihtkari (2003) contribute to the literature on customer complaining behavior by demonstrating that, contrary to common assumptions, the type of service encounter (traditional vs. technology-based) does not significantly influence the overall complaining rates. However, the study highlights that technology-based services are more efficient in responding to complaints and that customers' perception of responsibility plays a crucial role in their decision to complain. These findings offer valuable insights for service industry professionals aiming to improve customer experience through better-designed service recovery systems.

[1] Sustainable Development Goal No. 12.

[2] Sustainable Development Goal No. 16.

[3] The term 'cyber-consumer behaviour' is a conceptual analysis that explores the nature of commercial transactions between online consumers and vendors.

Rust and Chung (2006) provide a comprehensive overview of the evolving field of marketing models for service and customer relationships. Their work highlights the shift toward more dynamic and personalized service interactions, driven by advances in technology and a deeper understanding of customer behavior. As the service sector continues to grow and digital interactions become more prominent, future research will need to focus on the delicate balance between customization, privacy, productivity, and financial outcomes. The increasing importance of CLV and customer equity underscores the long-term financial impact of effective service management and relationship models, making this area of study crucial for both academia and practice.

Blasio (2008) provides a comprehensive analysis of the urban–rural differences in internet usage, e-commerce, and e-banking in Italy. The study highlights how the digital divide limits rural populations' access to the benefits of ICTs, particularly in terms of economic participation through e-commerce and financial inclusion via e-banking. While urban residents are more active in online spaces, rural residents face challenges due to limited infrastructure and digital literacy. The study underscores the need for targeted policies to improve rural access to ICTs and suggests that doing so could significantly enhance rural economic opportunities and close the gap in digital participation.

Thakur et al. (2023) present a comprehensive overview of the opportunities and challenges posed by the Metaverse as a virtual form of smart cities. While the Metaverse holds significant potential to transform urban living through enhanced governance, resource management, and economic opportunities, it also raises important concerns related to privacy, inclusivity, job displacement, and environmental sustainability. The paper emphasizes the need for a collaborative approach among policymakers, businesses, and communities to ensure that the benefits of the Metaverse are equitably distributed and that its development contributes to a sustainable and inclusive urban future.

2 Methods

Three major concerns affect the sustainable goals of 'Responsible Consumption and Production' and 'Peace, Justice, and Strong Institutions', in virtual markets operating within the metaverse. These three considerations are (a) Threats to the sustainable growth of virtual markets (b) Ethical issues within the virtual markets and (c) Legal concerns required to be attended to for sustainable growth of metaverse markets. To find solutions that address these issues, it becomes essential to delve prior into these three considerations, which will be subsequently analysed following the approach of doctrinal research.

2.1 Understanding the Concept of Consumers in Metaverse Market

The Metaverse, described as a persistent and immersive digital world, is increasingly recognized as a cutting-edge platform for achieving the objective of catering to the needs of consumers right at their doorsteps. The Metaverse can support smart city initiatives by reducing the need for physical travel, which could decrease energy consumption and carbon emissions. Virtual environments enable better resource management, modeling, and testing of urban infrastructure through digital twin technology, allowing for more informed decision-making and innovation in urban planning. Additionally, the Metaverse could provide new avenues for entertainment, education, and commerce, helping to reshape urban economies (Thakur et al. 2023).

The Metaverse presents numerous advantages, particularly in areas like urban resource management, where it can optimize operations and reduce testing costs through digital twin technology (Thakur et al. 2023). It also has the potential to lower energy consumption and carbon emissions by minimizing the need for physical travel, while simultaneously enhancing social interactions and inclusivity. The tourism sector stands to benefit from phygital experiences, the preservation of heritage sites, and increased financial resources for businesses (Patel and Joshi 2022). Furthermore, the Metaverse offers promising opportunities for improving urban planning, education, healthcare, entertainment, and overall quality of life. However, there are significant drawbacks. Economic disparities may not be alleviated and could even deepen, exacerbating urban inequality. Inclusivity may be hampered by the high cost of hardware, limiting access for marginalized groups and reinforcing social segregation. Additionally, ethical concerns, such as privacy and security, pose challenges to widespread adoption. Accessibility issues for individuals with disabilities, such as the blind and deaf, further complicate interaction within the Metaverse. Finally, the virtual realm cannot fully replicate physical experiences, and serious concerns such as surveillance, marginalization, and dystopian risks require careful governance and public participation to ensure democratic development and control over the Metaverse's evolution (Pellegrino et al. 2023).

The scope of such a digital world has grown so vast that it captures not only the wide sectors of health, education, and financial institutional services but also provides the day-to-day necessities to the consumers. Conventional consumers are rapidly preferring cyber consumerism to save time and enjoy the cost advantages offered by a variety of virtual markets. However, the flexibility provided to the cyber vendors has prompted concerns about the sustainability of their burgeoning virtual market[4] regarding the protection of the rights of the consumers they cater to (Nogoev et al. 2011).

The vendors in the virtual market promote their goods and services with the assistance of websites, social media, mobile applications, or even search engines. Such a market relies more upon digital technology rather than using conventional

[4] The concept of promoting the brands, products or even services through digital market can be termed as virtual market.

marketing methods. In the present era, such markets have become the household term where consumers prefer making deals for their purchases of goods and services through such platforms (Blasio 2008).

There are many factors responsible for the growth of such markets. It has been observed that such markets offer benefits to the consumer like being cost-effective, time-saving, wider choices, and providing personalisation. However, the picture is just not all rosy. There are many ethical, sustainable, and legal concerns that have emerged with the growth of the metaverse. Consumers in such kind of virtual market are constantly threatened by privacy breaches, financial loss, and deception. Until the interests of such consumers are safeguarded, it might pose a constant hazard to responsible consumption and production, as well as peaceful growth, **affecting** the components of sustainable growth and development (Blasio 2008).

2.2 Sustainable and Ethical Threats Hovering Over the Virtual Markets

Online consumers frequently encounter various ethical and legal challenges when engaging in e-commerce transactions which can immensely impact the sustainable development goals set for better future growth of the world (Snellman and Vihtkari 2003).

The foremost challenge faced is in regard to 'Responsible Consumption and Production'[5] which arises in cases of deceptive product descriptions, selling counterfeit products, failing to deliver goods, failure to address issues related to damaged goods (which leaves the consumer dissatisfied), unexpected shipping costs and price changes during the checkout process, improper response to monetary refunds, etc (Chen and Lee 2020; Sharma and Gupta 2020).

Along with the threat to privacy, consumers also become victims of various fraudulent practices adopted by vendors, such as selling counterfeit products or failing to deliver goods, or failing to address issues related to damaged goods, leaving consumers dissatisfied. Additional issues include unexpected shipping costs, price changes during checkout, and improper monetary refunds.

However, it is not only consumers who face threats in the virtual market; vendors also become prey in the loop, which touches on the sustainable development goals of 'Decent Work and Economic Growth'[6] and 'Industry, Innovation, and Infrastructure'.[7] Issues such as shopping cart abandonment, insignificant customer loyalty, and competing against established retailers and manufacturers are commonly witnessed. Often, consumers initiate a purchase but fail to complete the transaction, leading to potential losses for online vendors (Chou and Chou 2000). The lack of loyalty among online consumers can lead to unstable market dynamics. Furthermore, online

---

[5] Sustainable Development Goal No. 12.

[6] Sustainable Development Goal No. 8.

[7] Sustainable Development Goal No. 9.

vendors frequently struggle to compete with established retailers and manufacturers, particularly in terms of pricing and product availability, which results in regular instability in their economic growth.

Another major challenge faced in e-commerce is related to the sustainable development goal of 'Peace, Justice, and Strong Institutions'.[8] Many consumers lack the necessary digital skills and financial literacy to navigate the complexities of e-commerce effectively. This leads to challenges in resolving disputes with online vendors, especially in cross-border transactions. Determining the applicable law and jurisdiction for cross-border e-commerce transactions can be challenging, leading to legal uncertainties (Blasio 2008).

Inclusivity is another major challenge. Access to the Metaverse requires expensive hardware, such as VR headsets, which may limit participation for economically disadvantaged populations. This could exacerbate existing social inequalities in urban settings, where the digital divide is already a concern (Thakur et al. 2023).

2.3 Understanding Legal Concerns Required for Sustainable Growth of Virtual Markets

In the rapidly evolving digital marketplace, particularly within the expansive realms of the metaverse, digital marketers face a myriad of legal challenges. These challenges, if not properly managed, can result in significant reputational damage, financial losses, and legal liabilities. The growing dependence on digital platforms for marketing and consumer engagement has introduced new risks, such as unauthorized data breaches, deceptive advertising practices, copyright infringement, and defamatory content. This complex landscape necessitates a proactive approach to legal risk management to safeguard brand reputation and ensure compliance with relevant laws and regulations (Rust and Chung 2006). Legal Issues in The Virtual Markets are depicted in Fig. 1.

(a) **Data Privacy and Security Threats**

The collection, storage, and utilization of consumer data have become integral to digital marketing strategies. However, this reliance on data introduces significant privacy and security threats, especially as marketing activities extend into the metaverse. Unauthorized access, disclosure, or misuse of sensitive consumer information can lead to severe legal consequences, including reputational damage, financial losses, and legal liabilities. Data breaches, often resulting from cyber-attacks like phishing, ransomware, and malware, are particularly concerning. The consequences of such breaches are far-reaching, potentially resulting in regulatory fines, class-action lawsuits, and a loss of customer trust (Rust and Chung 2006).

[8] Sustainable Development Goal No. 16.

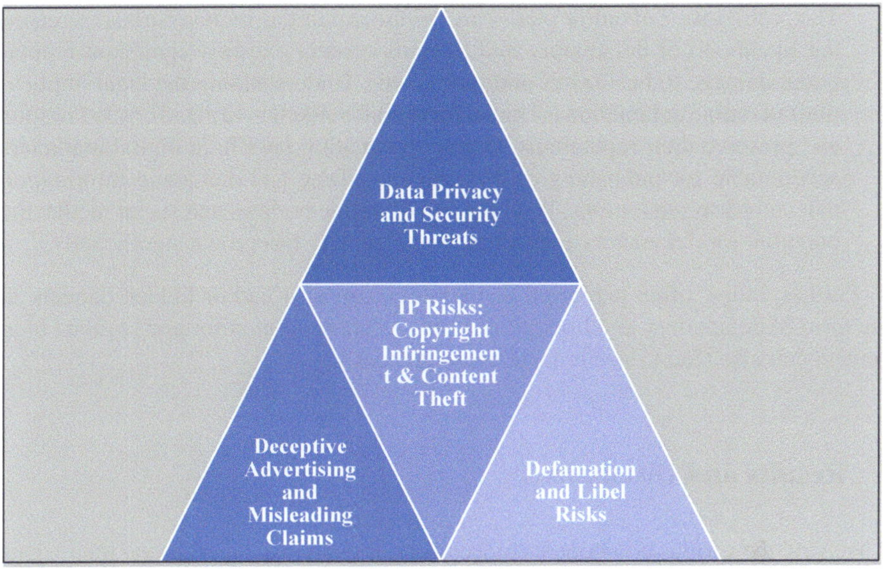

Fig. 1 Legal issues in the virtual markets. *Source* Authors' own tabulation

(b) **Deceptive Advertising and Misleading Claims**

Deceptive advertising and misleading claims are critical legal risks in the digital market. These practices can lead to regulatory enforcement actions, civil lawsuits, and severe damage to brand reputation. Digital marketers must ensure that their advertising campaigns are transparent, accurate, and compliant with relevant laws and regulations. Misleading claims or false promises can attract serious legal consequences, including fines, legal action, and long-term reputational damage.

(c) **Intellectual Property Risks: Copyright Infringement and Content Theft**

In the digital marketing arena, including the emerging metaverse, navigating intellectual property rights is essential to avoid legal risks such as copyright infringement and content theft. Unauthorized use of copyrighted images, videos, or other creative works in marketing campaigns can have serious repercussions. One critical aspect of this is image licensing, where marketers must obtain the necessary permissions or licenses to use copyrighted materials. Failure to do so can result in legal action and financial penalties. Content scraping, or the unauthorized extraction and use of content from other websites, is another significant risk. This practice not only violates copyright laws but also undermines the intellectual property rights of the original creators.

(d) **Defamation and Libel Risks**

> The proliferation of online platforms and the expanding metaverse has increased the likelihood of defamatory and libelous content causing significant reputational damage to businesses and individuals. Understanding the legal implications of online defamation is crucial for digital marketers to avoid costly lawsuits and preserve their reputations. Online defamation laws hold digital marketers accountable for publishing or disseminating false and damaging information that can harm reputations. With the rise of online reviews and social media, the potential for defamatory content to spread rapidly has grown significantly.

Further, blogs, often perceived as harmless, can also harbor hidden dangers in the form of defamatory and libelous content. A failure to monitor and regulate blog content can also lead to severe legal consequences.

3 Results and Discussions

Analysing the various components of a consumer in a virtual market and their rights to sustainable growth, it is well observed that numerous advantages offered by metaverse markets are contributing to their endless expansion. However, the parallel threats and hazards that metaverse markets are giving birth to, need considerable attention.

The objective of ensuring consumers' sustainable growth in such a market can be achieved through formulating and securing a cyber-consumer redressal system[9] and cyber-consumer protection[10] laws. By enhancing consumer protection and redress mechanisms, the global digital market, including emerging spaces like the metaverse, can offer a safer and more trustworthy environment for online transactions (Panchal et al. 2021).

On the international as well as national forefront efforts have been made to reconcile the conflicting interests of the metaverse vendors and cyber consumers on the path of sustainable growth, which are commendable and need due attention. These strides adopted from global and Indian perspectives address various issues affecting the sustainable development goals specifically related to 'Responsible Consumption and Production', and 'Peace, Justice, and Strong Institutions' (Rao and Bhatia 2021).

---

[9] It refers to the legal remedies available to consumers who are aggrieved by issues such as non-delivery, incorrect delivery of goods, or fraudulent behaviour by merchants. Redress is a critical concern, but it is part of the broader concept of cyber-consumer protection.

[10] Cyber-consumer protection encompasses the regulation of the bargaining phase, contract terms, and access to justice issues. Effective consumer protection can reduce the likelihood of disputes, making them less costly and easier to resolve.

3.1 *Cyber-Consumer Protection and Redressal in a Global Perspective*

(a) **General Civil Laws**: The global landscape for consumer protection in the digital market, particularly in the context of e-commerce and the emerging metaverse, is shaped by varying legal frameworks across countries. Some nations, like El Salvador, Mexico, and Peru, rely on general civil law to regulate e-commerce activities. In contrast, countries such as Chile, Colombia, France, the Russian Federation, and the United States have established specific provisions under their consumer protection laws to address the unique challenges of e-commerce.

(b) **The Organization for Economic Cooperation and Development (OECD)**: OECD has played a crucial role in establishing core principles for the effective functioning of e-commerce and the protection of online consumers. The OECD's recommendations emphasize key areas such as privacy, security risks, payment protection, and the safety of goods. A significant aspect of these guidelines includes the revised OECD recommendation concerning the protection of privacy and the cross-border flow of personal data. These guidelines incorporate essential principles like collection limitation, data quality, purpose specification, security safeguards, individual participation, accountability, implementation, and interoperability.

(c) **World Trade Organisation (WTO)**: Further, the WTO members have identified several key conditions that need attention to achieve the goals of a robust global e-commerce framework like 'Trade Liberalization of E-Commerce-Related Sectors' to enhance consumer convenience and interests (Mehta and Kapoor 2022); promoting specific 'Principles for Domestic Regulations'[11] which includes transparency in publishing regulatory measures, non-discriminatory treatment, avoidance of unnecessary trade restrictions, adherence to globally harmonized rules and standards, and recognition of the equivalence of foreign regulatory measures; and 'Cooperation with Relevant International Organizations,'[12] by harmonizing existing international rules and ensuring that these rules are reflected in domestic laws and regulations.

(d) **Cyber-Consumer Redressal**: Regarding Cyber-Consumer Redressal in a Global Perspective, E-commerce and cyber-consumer protection present unique challenges, particularly in resolving disputes that arise in the global digital market. International jurisdictional issues are addressed through common legal frameworks, such as the Rome Convention of 1980 on the applicable law in

[11] The principles governing domestic regulations for e-commerce are determined by frameworks like the General Agreement on Tariffs and Trade (GATT), the General Agreement on Trade in Services (GATS), the Agreement on Trade-Related Aspects of Intellectual Property Rights (TRIPS), and the OECD guidelines.

[12] International forums like the World Intellectual Property Organization (WIPO), the United Nations Commission on International Trade Law (UNCITRAL), the International Organization for Standardization (ISO), the Hague Conference on Private International Law, and the OECD are actively involved in shaping international rules related to e-commerce.

Fig. 2 Requirements for achieving a strong global e-commerce framework. *Source* Authors own tabulation

contractual matters and the Brussels Convention, now known as the 44/2001 European rules on international jurisdictions. These frameworks include provisions on conflicts of laws, often allowing consumers to choose the jurisdiction of their residence. Courts play a critical role in applying these frameworks to specific disputes arising from e-commerce activities (Fig. 2).

3.2 Cyber-Consumer Protection and Redressal Under Indian Regime

Cyber-consumer protection and redressal under the Indian regime are vital aspects of the legal framework designed to protect consumers from the challenges and threats posed by the digital marketplace. With the rapid growth of e-commerce and digital transactions in India, the need for robust consumer protection mechanisms has become increasingly important (Fig. 3).

An overview of how cyber-consumer protection and redressal are addressed under the Indian legal regime is as follows:

(a) **Consumer Protection Act, 2019**: The Consumer Protection Act, 2019, is the cornerstone of consumer protection in India. It was enacted to replace the

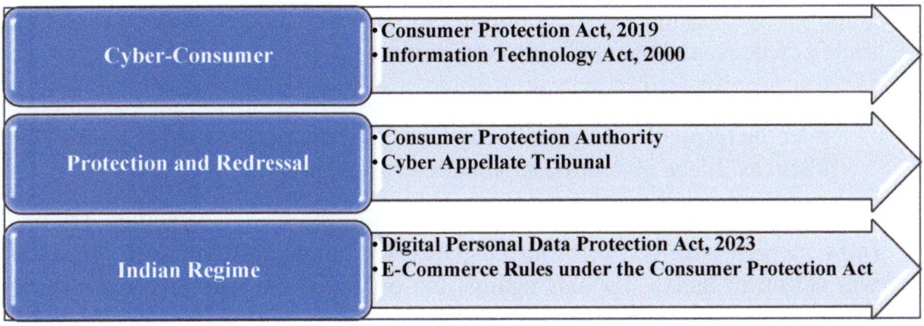

Fig. 3 Cyber-consumer protection and redressal in the Indian legal regime. *Source* Authors own tabulation

Consumer Protection Act of 1986 and introduced several provisions to address the challenges of the digital age, including e-commerce and online transactions (Rai and Mehra 2021).

a. **E-Commerce Regulations**: The Act includes specific provisions for e-commerce entities, defining their obligations toward consumers. It mandates the disclosure of information related to products, services, refund policies, and other important aspects that impact consumer rights (Press Note No. 3 2020).
b. **Unfair Trade Practices**: The Act defines unfair trade practices and ensures that consumers have remedies against deceptive practices by online sellers.
c. **Consumer Grievance Redressal**: The Act establishes a three-tier consumer dispute redressal mechanism at the district, state, and national levels. It allows consumers to file complaints electronically and provides a simplified process for filing complaints.

(b) **Information Technology Act, 2000**: The Information Technology Act, of 2000, is the primary legislation that governs cyber activities in India, including issues related to cybercrimes and electronic commerce.

a. **Legal Recognition of Electronic Transactions**: The IT Act provides legal recognition to electronic transactions and electronic contracts, ensuring that digital agreements are enforceable.
b. **Data Protection**: Though the IT Act primarily focuses on cybercrimes, it also addresses data protection to some extent. Sections 43A and 72A of the Act deal with compensation for unauthorized access to personal data and breach of confidentiality.
c. **Penalties for Cybercrimes**: The Act prescribes penalties and punishments for various cybercrimes, such as hacking, phishing, and unauthorized access to computer resources, which could affect consumers.

(c) **Consumer Protection Authority (CPA)**: The Consumer Protection Act, 2019, establishes the Central Consumer Protection Authority (CCPA), which is empowered to regulate matters relating to violations of consumer rights, unfair trade practices, and misleading advertisements.

a. **Powers of CCPA**: The CCPA has the authority to investigate complaints, order the recall of unsafe goods, and order the discontinuation of unfair trade practices. It can also impose penalties on violators.

(d) **Cyber Appellate Tribunal**: Under the IT Act, the Cyber Appellate Tribunal (now merged with the Telecom Disputes Settlement and Appellate Tribunal) was set up to handle appeals against the orders of the Adjudicating Officers concerning contraventions under the Act.

(e) **Digital Personal Data Protection Act, 2023**: India has been moving towards a comprehensive data protection regime with the introduction of the Digital Personal Data Protection Act, 2023. This Act is expected to enhance consumer protection by imposing stringent obligations on entities handling personal data and providing consumers with greater control over their data.

(f) **E-Commerce Rules under the Consumer Protection Act**: In 2020, the Indian government notified the Consumer Protection (E-Commerce) Rules, which mandate specific duties for e-commerce entities, such as transparency in advertising, grievance redressal mechanisms, and preventing unfair trade practices.

4 Key Findings

There is a constant threat looming over cyber consumers that their personal data is at threat and can be mishandled. To avoid the same, stringent data protection strategies and incident response plans must be implemented ensuring that they are prepared to address any potential breaches promptly and effectively.

Strict guidelines on deceptive advertising such as the Federal Trade Commission (FTC) guidelines on deceptive advertising should be adopted in every national policy ensuring that all advertising content is truthful, non-misleading, and substantiated. By maintaining transparency and honesty in their marketing practices, businesses can avoid legal pitfalls and preserve a positive brand image.

To avoid the risk of Intellectual Property Rights infringement, marketers must ensure that all content used in their campaigns is legally sourced and properly attributed. Implementing thorough due diligence processes and effective content tracking systems can help minimize the risk of copyright infringement and maintain ethical marketing practices.

The threat of frivolous blogs and their misleading content is common in a virtual market. To mitigate these risks, digital marketers must verify the accuracy of the information before publishing or sharing it and develop strategies to address negative or false online reviews and mitigate the impact of malicious online activity. This proactive approach can help identify and address potential legal issues before they escalate, ensuring that marketing content remains high-quality, engaging, and legally compliant.

The effectiveness of e-commerce depends on enacting relevant legislation and ensuring the enforceability of redress options for online consumers who have been deceived.

National reports indicate that significant difficulties arise in determining the localization of incriminated acts and assessing whether these acts pertain to a specific market. These challenges are not confined to national borders, and the absence of uniformity in legal approaches can raise doubts about the effectiveness of traditional tools in cyber consumer protection. The rise of unfair commercial practices on the Internet necessitates specific reactions from states, including the development of specialized e-commerce rules within national legal systems. To address jurisdictional challenges in e-commerce disputes, member states should rely on international treaties or reciprocal enforcement agreements. These agreements typically focus on the connections that a foreign entity has with a sovereign state's interests, particularly in matters of authentication.

5 Conclusion and Suggestions

The journey toward sustainability within the Metaverse is dynamic, presenting both challenges and opportunities. By addressing concerns, fostering inclusivity, and leveraging virtual experiences for environmental advocacy, the Metaverse has the potential to align with global sustainability goals. As businesses continue to integrate the Metaverse into their operations, the path to a greener future becomes increasingly tangible, offering a bridge between the digital market and the real world.

The efficiency of e-commerce is pivotal to the sustainability of economic growth, enhancing public welfare, and fostering social cohesion in the digital age. It is increasingly recognized as a critical factor in the multilateral trading system. In addition to providing a more liberal approach to both consumers and vendors, the rapid and expansive growth of e-commerce is expected to help developing nations achieve vital economic goals, such as poverty reduction and improved education. However, since nothing comes absolute in nature, the expansion of virtual markets is not without its limitations. Thus, to build consumer confidence and promote fair-trade practices in e-commerce, the following factors are essential to be considered:

(a) **Comprehensive Legal and Institutional Framework**: It is crucial to establish proper legislation and an institutional framework that specifically regulates e-commerce activities. This framework must also include enforceable laws and provide consumers with easy access to dispute-resolution mechanisms specifically when the transnational virtual markets are approached. For instance, in India, though we have the 'Consumer Protection Act', legislation protecting the interest of consumers, however, in the contemporary perspective we must inculcate a specific chapter defining the rights of cyber consumers and the duties of vendors in the virtual market.

(b) **Accessible Dispute Resolution Mechanisms**: Consumers should have straightforward access to mechanisms that resolve disputes, ensuring that any issues encountered during e-commerce transactions are addressed promptly and effectively. Issues arising in disputes involving private international laws can be mitigated by replacing such laws with international provisions (like following the UNICITRAL model of arbitration) and conventions.

(c) **Transparency and Accuracy in Business Practices**: There should be a mandate upon online businesses that they provide complete and truthful descriptions of their business operations, goods, and services to foster trust and reliability among consumers. Failure to which must attract stringent punishments in addition to the remedy of compensation.

(d) **Secure Online Transactions**: The security and reliability of online payment methods are critical. Consumers should be assured that their money transactions, regardless of the payment mode, are secure.

(e) **Promoting the Protection of Data Privacy Rights of Online Consumers**: Specific provisions clearly explaining data protection should be addressed in the legislation. To provide an example, the Data Protection Act of India (DPDA) the inclusion of punishing virtual vendors for adopting malpractices should be included.

(f) **Safe Delivery and Post-Sale Services**: Ensuring the safe and timely delivery of goods, along with hassle-free after-sale services, is vital for maintaining consumer satisfaction and trust.

(g) **Regional and International Cooperation**: Strengthening regional and international cooperation is necessary to combat unfair commercial practices. This cooperation should also address legal challenges that limit consumer protection authorities' ability to act in cases of cross-border violations of consumer protection laws.

(h) **Technology Inclusivity in policy-making**: Policymakers should be urged to anticipate employment disruptions and develop strategies to support affected workers through upskilling and retraining programs as the increased use of AI in the Metaverse may lead to potential job displacement due to automation. The policies must also promote digital inclusivity and make technologies more accessible to all citizens through the length and breadth of the nation.

Thus, we can say that indeed growth is essential, if it does not occur in a sustainable dimension, the objective of the growth will stay half sung.

References

Blasio GD (2008) Urban-rural differences in internet usage, e-commerce, and e-banking: evidence from Italy. Growth Change 341–367

Chen Y, Lee Y (2020) Consumer protection in E-commerce: examining the legal frameworks. J Bus Law Ethics 45(3):567–589

Chou DC, Chou AY (2000) The E-commerce revolution, a guide to the internet revolution in banking. Inf Syst Manag 51–57

Kumar A, Singh R (2021) SDGs and digital economy: a pathway to inclusive growth. Sustain Dev Rev 37(2):214–230

Mehta R, Kapoor S (2022) Competition law and digital markets: ensuring fair play. J Compet Law Econ 50(3):299–317

Nogoev A, Mohseni S, Yazdanifard R, Samadi B, Menon M (2011) The evolution and development of E-Commerce market and E-Cash. https://doi.org/10.1115/1.859858.paper35

Panchal R, Singh A, Diwan H (2021) Does circular economy performance lead to sustainable development?—A systematic literature review. J Environ Manag 293:112811. https://doi.org/10.1016/j.jenvman.2021.112811. ISSN 0301–4797

Patel V, Joshi M (2022) Digital economy and SDGs: case studies from emerging markets. J Econ Soc Res 40(1):123–140

Pellegrino et al (2023) Exploring the intersection of sustainable consumption and the Metaverse: a review of current literature and future research directions 9(9)

Press Note No. 3 (2020 Series), Department for Promotion of Industry and Internal Trade, Ministry of Commerce and Industry, Government of India Press Note 3 (2020) | Department for Promotion of Industry and Internal Trade | MoCI | GoI (dpiit.gov.in)

Rai S, Mehra R (2021) Legal aspects of consumer protection in digital markets. Int J Law Inf Technol 28(4):311–329

Rao S, Bhatia N (2021) Sustainable business models in virtual environments. J Sustain Bus Pract 39(4):390–405

Rust RT, Chung TS (2006) Marketing models of service and relationships. Mark Sci 25(6):560–580. http://www.jstor.org/stable/40057191

Sen A, Das S (2021) Monopolistic behaviours in digital markets: a legal analysis. Int Rev Law Econ 34(1):89–105

Shaik D, Poojasree MV (2021) Consumer protection in E-commerce: a legal and compliance framework in the digital market. https://doi.org/10.2991/assehr.k.210506.004

Sharma P, Gupta K (2020) Ethical considerations in virtual reality: a consumer perspective. J Bus Ethics 45(2):467–482

Snellman K, Vihtkari T (2003) Customer complaining behaviour in technology-based service encounters. Int J Serv Ind Manag 14(2):217–231

Thakur S, Bandyopadhyay S, Datta D (2023) The metaverse as a virtual form of smart cities: opportunities and challenges. Int J Comput Appl 185:45–53. https://doi.org/10.5120/ijca2023922892

The Metaverse and SDGs: Transforming Business Practices for a Sustainable Future

Manoj Govindaraj, Kandati Sai Chandu, Parvez Khan, and Jenifer Lawrence

Abstract The metaverse, an amalgamation of digitally augmented physical reality and physically persistent virtual space, is swiftly becoming a catalyst for change in corporate operations. This article examines the potential of the metaverse in promoting creative, inclusive, and sustainable business models to help achieve the United Nations Sustainable Development Goals (SDGs). It examines the potential of the metaverse to enhance education, reduce inequalities, promote sustainable consumption and production patterns, and support climate action. By integrating immersive technologies, blockchain, and artificial intelligence, businesses can create value while addressing global challenges. This study reviews current literature, identifies key research questions, and outlines objectives to guide future research. It also highlights the evolution of the metaverse, offering insights into its potential to revolutionize business practices and contribute to a sustainable future. Suggestions for future research are provided to ensure the metaverse's alignment with sustainable development principles.

Keywords Metaverse · Sustainable development goals (SDGs) · Immersive technologies · Innovative business models

M. Govindaraj (✉)
Department of Management Studies, Vel Tech Rangarajan Dr. Sagunthala R&D Institute of Science and Technology, Avadi, Chennai, India
e-mail: manoj.nmcc@gmail.com

K. S. Chandu
School of Management, Vel Tech Rangarajan Dr. Sagunthala R&D Institute of Science and Technology, Chennai, India

P. Khan
Department of Business, Atria University, Bengaluru, India

J. Lawrence
Department of Management, Woldia University, Weldiya, Ethiopia

W. Leal Filho et al. (eds.), *Metaverse and Sustainability: Business Resilience Towards Sustainable Development Goals*, World Sustainability Series,
https://doi.org/10.1007/978-3-031-89545-6_23

1 Introduction

The concept of the metaverse has captured the imagination of technologists, business leaders, and policymakers alike.

As a fully immersive virtual environment that merges physical and digital realities, the metaverse offers unprecedented opportunities for innovation and transformation across various sectors (Ball 2020; Dionisio et al. 2013). This article investigates the metaverse's potential to advance the United Nations Sustainable Development Goals (SDGs), a global blueprint to achieve a better and more sustainable future for all (United Nations 2015).

Defining the Metaverse: The metaverse is a unified and virtual environment that is formed by the merging of augmented physical reality with continuous virtual worlds. It includes a wide range of technologies, such as virtual reality (VR), augmented reality (AR), blockchain, and artificial intelligence (AI) (Lee et al. 2021; Stephenson 1992). The digital realm enables users to engage, collaborate, and engage in immersive settings, hence opening up novel opportunities for commerce, learning, and social engagement (Chalmers et al. 2021).

Intersection of the Metaverse and SDG: The integration of the metaverse with SDGs presents a unique opportunity to create sustainable business practices that align with global development objectives. The metaverse can potentially transform how businesses operate, offering innovative solutions to achieve the SDGs. For instance, virtusal education platforms can enhance learning opportunities (SDG 4), while digital economies can promote inclusive growth (SDG 8) (Bonk 2020; Schwab 2017).

Current Landscape and Opportunities: The metaverse's current landscape is characterized by rapid technological advancements and increasing investment from major corporations (Meta 2021; Microsoft 2021). These developments are paving the way for new business models that prioritize sustainability and social impact. For example, blockchain technology can enhance supply chain transparency, promoting responsible consumption and production (SDG 12) (Kouhizadeh et al. 2021). Additionally, virtual collaboration tools can reduce the need for physical travel, contributing to climate action (SDG 13) (Foth 2020).

Despite its potential, the metaverse also presents challenges that need to be addressed to ensure its alignment with the SDGs. These include issues related to digital equity, data privacy, and the environmental impact of digital infrastructures (Floridi 2020; Zuboff 2019). Ensuring that the benefits of the metaverse are accessible to all and do not exacerbate existing inequalities is crucial (Eubanks 2018).

The intersection of the metaverse and SDGs represents a novel area of exploration that seeks to understand how virtual technologies can contribute to sustainable development. Early studies suggest that the metaverse can enhance educational access (SDG 4), reduce inequalities (SDG 10), promote sustainable consumption and production (SDG 12), support climate action (SDG 13), drive economic growth and employment (SDG 8), improve health and well-being (SDG 3), and advance gender equality (SDG 5). However, the environmental impact and ethical considerations of expanding virtual spaces also warrant careful examination.

Research Aim and Structure: This article aims to explore the transformative potential of the metaverse in advancing the SDGs and to identify key research questions and objectives for future study. The following sections provide a comprehensive review of the current literature, examine the evolution of the metaverse, and propose suggestions for future research.

Current Landscape and Opportunities: The metaverse's current landscape is characterized by rapid technological advancements and increasing investment from major corporations (Meta 2021; Microsoft 2021). These developments are paving the way for new business models that prioritize sustainability and social impact. For example, blockchain technology can enhance supply chain transparency, promoting responsible consumption and production (SDG 12) (Kouhizadeh et al. 2021). Additionally, virtual collaboration tools can reduce the need for physical travel, contributing to climate action (SDG 13) (Foth 2020).

Challenges and Considerations

The Metaverse is an interconnected network of virtual spaces that brings together the real and virtual worlds to provide users with interactive and immersive experiences through AR and VR. Avatars, or digital representations, would allow users to interact with digital content and one another in a persistent, linked world. The Metaverse is an alternative to conventional web platforms that lets users effortlessly hop from one virtual world to another, preserving their identities, assets, and experiences along the way. In this virtual world, there are thriving marketplaces where people may buy, sell, and exchange virtual goods and services, frequently using cryptocurrency. There is much anticipation that the ever-developing Metaverse will usher in a new era of seamless online communication and entertainment, erasing the lines between the physical and virtual worlds.

Background of the Study

Despite its potential, the metaverse also presents challenges that need to be addressed to ensure its alignment with the SDGs. These include issues related to digital equity, data privacy, and the environmental impact of digital infrastructures (Floridi 2020; Zuboff 2019). Ensuring that the benefits of the metaverse are accessible to all and do not exacerbate existing inequalities is crucial (Eubanks 2018).

Research Aim and Structure

This article aims to explore the transformative potential of the metaverse in advancing the SDGs and to identify key research questions and objectives for future study. The following sections provide a comprehensive review of the current literature, examine the evolution of the metaverse, and propose suggestions for future research.

2 Review of the Study

The intersection of the metaverse and Sustainable Development Goals (SDGs) is an emerging field that explores how virtual and augmented realities can contribute to sustainable development. This literature review synthesizes current research to elucidate how the metaverse can support various SDGs, including education, reducing inequalities, sustainable consumption and production, climate action, economic growth and employment, health and well-being, gender equality, environmental impact, and ethical and privacy considerations.

Metaverse and Education (SDG 4)

The metaverse holds transformative potential for education, enabling the creation of virtual classrooms and immersive learning environments. This can democratize access to quality education by transcending geographic barriers (Bonk 2020; Johnson et al. 2011). Research has demonstrated that virtual reality (VR) can significantly enhance student engagement and learning outcomes (Merchant et al. 2014). Bonk (2020) highlights how virtual classrooms can facilitate interactive and personalized learning experiences, while Johnson et al. (2011) emphasize the role of immersive environments in fostering collaborative learning. Merchant et al. (2014) provide empirical evidence showing that VR-based learning can lead to better retention and understanding of complex concepts.

Reducing Inequalities (SDG 10): The metaverse offers inclusive digital platforms that can help reduce inequalities by providing marginalized communities with access to new opportunities (Bailenson 2018; Hossain et al. 2019). However, the digital divide remains a significant barrier to equitable access to these technologies (Eubanks 2018; Van Dijk 2020). Bailenson (2018) explores the potential of VR to create shared experiences that bridge social divides, while Hossain et al. (2019) discuss how virtual platforms can enable participation in the digital economy for underrepresented groups. Eubanks (2018) and Van Dijk (2020) highlight the ongoing challenges of ensuring that all communities have the necessary resources and skills to engage with advanced digital technologies.

Sustainable Consumption and Production (SDG 12): Blockchain technology within the metaverse can enhance transparency in supply chains, promoting sustainable consumption and production patterns (Kouhizadeh et al. 2021; Saberi et al. 2019). Digital twins and simulations enable businesses to optimize resource use and reduce waste (Grieves 2014; Tao et al. 2018). Kouhizadeh et al. (2021) discuss the potential of blockchain to provide verifiable records of product origins and supply chain processes, thus fostering trust and accountability. Saberi et al. (2019) further elaborate on how blockchain can drive sustainable practices by ensuring the traceability of raw materials and production processes. Grieves (2014) and Tao et al. (2018) explain how digital twins—virtual replicas of physical systems—allow for real-time monitoring and optimization, leading to more efficient resource management.

Climate Action (SDG 13): The metaverse can support climate action by providing virtual collaboration tools and remote work environments that reduce the carbon footprint associated with travel (Foth 2020; Neves et al. 2020). Virtual simulations of climate change scenarios can also raise awareness and drive action (Riedl et al. 2013). Foth (2020) highlights the environmental benefits of reduced travel through the use of virtual meeting spaces, while Neves et al. (2020) provide data on the carbon savings achieved through remote work. Riedl et al. (2013) discuss the use of VR to create immersive experiences that educate the public about climate change impacts and potential solutions.

Economic Growth and Employment (SDG 8): The digital economy within the metaverse can create new job opportunities and drive economic growth (Schwab 2017; Tapscott and Tapscott 2016). However, the rise of automation and AI raises concerns about potential job displacement (Brynjolfsson and McAfee 2014; Frey and Osborne 2017). Schwab (2017) outlines how the Fourth Industrial Revolution, characterized by digital and virtual innovations, can spur economic development. Tapscott and Tapscott (2016) emphasize the metaverse's role in creating new markets and entrepreneurial opportunities. Brynjolfsson and McAfee (2014) and Frey and Osborne (2017) caution that while new technologies generate jobs, they also risk displacing existing roles, necessitating strategies for workforce retraining and adaptation.

Health and Well-being (SDG 3): The metaverse can enhance health and well-being through telemedicine and virtual therapy sessions (Bashshur et al. 2016; Wiederhold 2020). VR-based health interventions have shown promise in areas such as pain management and mental health treatment (Garcia-Palacios et al. 2007; Maples-Keller et al. 2017). Bashshur et al. (2016) discuss the advantages of telemedicine in providing remote access to healthcare services, while Wiederhold (2020) explores the use of virtual environments for therapeutic purposes. Garcia-Palacios et al. (2007) and Maples-Keller et al. (2017) provide evidence supporting the effectiveness of VR in reducing pain and treating mental health conditions through immersive, controlled environments.

Gender Equality (SDG 5): The metaverse can empower women by providing access to education, employment, and social networks (Chen et al. 2020; Jin and Herguera 2017). However, addressing gender biases in technology design and implementation is crucial (Gorski 2018; Noble 2018). Chen et al. (2020) highlight how digital platforms can offer women new opportunities for learning and career advancement. Jin and Herguera (2017) discuss the potential of virtual networks to support women's empowerment by fostering connections and collaborations. Gorski (2018) and Noble (2018) emphasize the need to consider gender biases that may be embedded in technological systems and advocate for inclusive design practices.

Environmental Impact: While the metaverse has the potential to support sustainability, it also raises concerns about its environmental impact, particularly related to energy consumption and electronic waste (Andrae and Edler 2015; Koomey et al. 2011). Research into energy-efficient technologies and sustainable digital practices is necessary (Belkhir and Elmeligi 2018; Masanet et al. 2020). Andrae and Edler (2015) analyze the environmental costs associated with the increasing use of digital technologies, highlighting the need for more sustainable practices. Koomey et al. (2011) provide data on the energy consumption of data centers, a critical component of the metaverse infrastructure. Belkhir and Elmeligi (2018) and Masanet et al. (2020) call for the development of energy-efficient technologies and advocate for policies that promote sustainable digital practices.

Ethical and Privacy Considerations: Data privacy and ethical considerations are paramount in the metaverse (Floridi 2020; Zuboff 2019). Ensuring robust data protection mechanisms and addressing ethical concerns around surveillance and digital rights is essential (Calo 2014; Nissenbaum 2010). Floridi (2020) discusses the ethical implications of digital technologies and the importance of data privacy. Zuboff (2019) warns of the dangers of surveillance capitalism and the commodification of personal data. Calo (2014) and Nissenbaum (2010) emphasize the need for strong legal frameworks and ethical guidelines to protect individual rights in the digital age.

3 Research Questions

1. How can the metaverse enhance access to quality education and reduce inequalities in education globally?
2. What are the economic impacts of integrating metaverse technologies into business practices, particularly in terms of job creation and market expansion?
3. How can the metaverse support sustainable consumption and production patterns through innovations in supply chain transparency and resource optimization?
4. What are the environmental and ethical considerations associated with the widespread adoption of metaverse technologies, and how can they be addressed?

4 Objectives for the Chapter

1. To explore the potential of the metaverse in enhancing access to quality education and promoting lifelong learning opportunities.
2. To analyze the economic impacts of the metaverse on job creation, market expansion, and inclusive growth.
3. To investigate how the metaverse can promote sustainable consumption and production patterns through technological innovations.
4. To examine the environmental and ethical implications of the metaverse and propose solutions to address these challenges.

5 Examine the Evolution

The concept of the metaverse has undergone significant evolution since its inception. First coined by Neal Stephenson in his 1992 novel "Snow Crash," the term initially referred to a speculative idea of a virtual reality-based successor to the internet (Stephenson 1992). Over the past few decades, advancements in technology, coupled with increasing interest and investment from major corporations, have transformed the metaverse from a speculative fiction to a burgeoning reality. This paper explores the evolution of the metaverse, examining early developments, technological advancements, corporate involvement, economic implications, social and cultural impacts, sustainability, and future directions.

Early Developments

Second Life and Online Multiplayer Games

The early 2000s marked the rise of virtual worlds such as Second Life and online multiplayer games, laying the groundwork for the metaverse. Second Life, launched in 2003, provided a glimpse into the potential of immersive digital environments where users could interact, create, and transact (Boellstorff 2008). Despite their limited technological capabilities, these early virtual worlds highlighted the possibilities of social and economic interactions in digital spaces (Castronova 2005). These platforms allowed users to create avatars, build virtual properties, and engage in commerce, demonstrating the potential for creating persistent, immersive environments.

6 Limitations and Challenges

However, these early developments faced significant limitations. The technological infrastructure at the time was not advanced enough to support truly immersive experiences. High latency, low-resolution graphics, and limited computing power restricted the user experience. Additionally, these platforms struggled to achieve mainstream adoption due to their niche appeal and the complexity of the user interfaces.

Technological Advancements

Virtual Reality (VR) and Augmented Reality (AR)

Recent advancements in VR and AR technologies have significantly accelerated the development of the metaverse. High-fidelity VR headsets, such as the Oculus Rift and HTC Vive, and AR devices like Microsoft HoloLens have made immersive experiences more accessible and realistic (Lee et al. 2021). These devices offer higher resolution displays, improved motion tracking, and more comfortable designs, enhancing the user experience and making prolonged use more feasible.

Blockchain Technology

Blockchain technology has introduced new paradigms for digital ownership and transactions, enabling secure and transparent exchanges within the metaverse (Tapscott and Tapscott 2016). Non-fungible tokens (NFTs) have become a popular method for representing ownership of digital assets, from virtual real estate to digital art. Blockchain ensures that these transactions are secure, verifiable, and immutable, fostering trust and reducing fraud.

Artificial Intelligence (AI)

AI plays a crucial role in enhancing user interactions and automating complex processes within the metaverse. AI-driven characters, known as NPCs (non-player characters), can provide dynamic and interactive experiences, making virtual environments feel more lifelike (Kaplan and Haenlein 2019). Additionally, AI can assist in content creation, automating tasks such as building virtual worlds or generating real-time responses, thus enriching the virtual experience.

Corporate Involvement

Meta (Formerly Facebook)

Major technology companies are investing heavily in the metaverse, recognizing its potential to transform various industries. Meta (formerly Facebook) has committed to building the metaverse, envisioning it as the next frontier of social interaction and commerce (Meta 2021). The company is developing a range of metaverse-related technologies, including VR headsets, social platforms, and digital marketplaces, aiming to create a seamless and interconnected virtual ecosystem.

Microsoft

Similarly, Microsoft is developing mixed-reality platforms like Microsoft Mesh, which aim to integrate the physical and digital worlds for collaboration and innovation (Microsoft 2021). Microsoft Mesh enables users to interact with holograms and share virtual experiences across different devices, fostering a new level of collaboration and creativity. This platform is particularly focused on enterprise applications, offering tools for remote work, training, and design.

Other Key Players

Other key players in the metaverse space include Google, Apple, and Nvidia. Google and Apple are developing AR technologies, with products like Google Glass and Apple ARKit, respectively. Nvidia is focusing on providing the graphical processing power necessary for rendering high-quality virtual environments. These companies' investments and innovations are driving the metaverse's rapid growth and development.

Economic Implications

Virtual Economies

The metaverse is poised to have profound economic implications. Virtual economies are emerging, where digital assets and services are bought, sold, and traded using cryptocurrencies and blockchain technology (Davidson et al. 2018). Platforms like Decentraland and The Sandbox allow users to purchase virtual land and build businesses, creating new revenue streams and economic opportunities. These digital economies are not limited by physical constraints, enabling a new level of creativity and entrepreneurship.

Job Creation and Revenue Streams

The metaverse can create new job opportunities and revenue streams, particularly in creative and knowledge-based industries (Schwab 2017). For instance, virtual world designers, digital artists, and VR/AR developers are in high demand. Additionally, the metaverse offers opportunities for remote work and telecommuting, allowing people to collaborate from anywhere in the world. However, this shift also raises questions about job displacement and the need for new skills (Frey and Osborne 2017).

Potential Inequalities

While the metaverse presents numerous economic opportunities, it also has the potential to exacerbate existing inequalities. Access to the necessary technology and skills may be limited to certain demographics, creating a digital divide. Furthermore, the concentration of power and resources among a few major corporations could lead to monopolistic practices and stifle competition.

Social and Cultural Impact

New Forms of Social Interaction

The metaverse's potential to shape social and cultural interactions is significant. It offers new ways for people to connect, collaborate, and share experiences, transcending physical boundaries (Bailenson 2018). Virtual spaces can foster inclusive communities and provide platforms for marginalized voices (Hossain et al. 2019). For example, virtual events and conferences can bring together people from diverse backgrounds who might not otherwise have the opportunity to meet.

Challenges of Safety and Inclusivity

However, ensuring these spaces are safe, equitable, and inclusive remains a challenge (Noble 2018). Issues such as harassment, privacy concerns, and digital addiction need to be addressed to create a positive and healthy virtual environment. Platforms must implement robust moderation and security measures to protect users and foster a sense of community.

Cultural Preservation and Expression

The metaverse also offers unique opportunities for cultural preservation and expression. Virtual environments can recreate historical sites, enabling people to experience and learn about different cultures in immersive ways. Additionally, artists and creators can use the metaverse as a platform to showcase their work, reaching global audiences and pushing the boundaries of digital art.

Sustainability and the SDGs

Environmental Impact

The metaverse's evolution aligns with the Sustainable Development Goals (SDGs) by offering innovative solutions to global challenges. Virtual collaboration can reduce the environmental impact of travel (Foth 2020), as people can meet and work together in virtual spaces without the need for physical transportation. However, the environmental footprint of digital infrastructures, such as data centres and VR devices, must be addressed to ensure sustainable development (Andrae and Edler 2015).

Education and Lifelong Learning

The metaverse also has potential applications in healthcare, offering new ways to deliver medical services and support mental health. Virtual consultations, therapy sessions, and rehabilitation programs can provide convenient and effective healthcare solutions. Additionally, immersive environments can be used for training healthcare professionals, allowing them to practice procedures and gain experience in a risk-free setting.

Technological Advancements

The metaverse's future will likely be shaped by ongoing technological advancements and societal shifts. Continued research and innovation are essential to realize its full potential while addressing ethical, environmental, and social challenges. For example, advancements in AI, 5G connectivity, and quantum computing could further enhance the capabilities and accessibility of the metaverse.

Ethical and Regulatory Frameworks

Developing robust ethical and regulatory frameworks will be crucial to ensure data privacy, security, and equitable access to metaverse platforms. Policymakers and industry leaders must work together to create standards and guidelines that protect users while fostering innovation. Issues such as data ownership, digital rights, and content moderation need to be addressed to create a safe and inclusive virtual environment.

Interdisciplinary Research

Interdisciplinary research that combines technological, social, and environmental perspectives will be vital in developing holistic strategies for integrating the metaverse into sustainable development efforts. Collaboration between technologists, sociologists, environmental scientists, and policymakers can help identify and address the complex challenges associated with the metaverse.

Aligning with the SDGs

By aligning the metaverse's development with the SDGs, businesses and policymakers can create a more sustainable and inclusive digital future. This includes promoting digital literacy, ensuring equitable access to technology, and minimizing the environmental impact of digital infrastructures. Additionally, fostering a culture of innovation and collaboration can drive positive social and economic outcomes.

Enhancing Access to Quality Education and Reducing Inequalities in Education

The potential of the metaverse to revolutionize education is significant. Virtual classrooms and immersive learning environments can provide access to quality education regardless of geographic barriers. Bonk (2020) and Johnson et al. (2011) emphasize that the metaverse can create educational opportunities for students in remote and underserved areas by offering virtual access to world-class resources and educators. Studies by Merchant et al. (2014) have shown that virtual reality (VR) can enhance student engagement and learning outcomes, making education more interactive and effective. Moreover, the metaverse can offer personalized learning experiences tailored to individual student needs, fostering an inclusive educational environment that accommodates diverse learning styles and abilities.

However, ensuring equitable access to metaverse technologies remains a challenge. Eubanks (2018) and Van Dijk (2020) highlight the digital divide, where marginalized communities may lack the necessary infrastructure, devices, and internet connectivity to fully benefit from metaverse-based education. Addressing these disparities is crucial for the metaverse to achieve its potential in reducing educational inequalities. Policymakers and stakeholders must work together to provide affordable access to technology, invest in digital infrastructure, and offer training programs to help educators and students effectively utilize virtual learning environments.

Economic Impacts of Integrating Metaverse Technologies into Business Practices

The integration of metaverse technologies into business practices has profound economic implications, particularly in terms of job creation and market expansion. Schwab (2017) and Tapscott and Tapscott (2016) argue that the metaverse can drive economic growth by creating new job opportunities in areas such as virtual design, digital marketing, and metaverse platform development. These emerging sectors can generate employment and stimulate economic activity, contributing to SDG 8 (Decent Work and Economic Growth).

However, the potential displacement of jobs due to automation and artificial intelligence (AI) within the metaverse must be addressed. Brynjolfsson and McAfee (2014) and Frey and Osborne (2017) warn that automation could lead to job losses in traditional industries, necessitating strategies for workforce reskilling and upskilling. Businesses and governments must collaborate to develop educational programs and vocational training that prepare workers for the demands of the digital economy.

Moreover, the metaverse can facilitate market expansion by providing businesses with new channels for reaching consumers and conducting transactions. Virtual marketplaces and immersive shopping experiences can enhance consumer engagement and drive sales. This expansion can be particularly beneficial for small and medium-sized enterprises (SMEs), which can access global markets without the constraints of physical location.

7 Supporting Sustainable Consumption and Production Patterns

Digital twins and simulations can help businesses optimize resource use and reduce waste. Grieves (2014) and Tao et al. (2018) explain that digital twins—virtual replicas of physical assets—can be used to monitor and manage resources more efficiently. By simulating different scenarios and outcomes, businesses can identify the most sustainable practices and minimize their environmental impact. These innovations can lead to significant cost savings and contribute to achieving SDG 12 (Responsible Consumption and Production).

Environmental and Ethical Considerations

While the metaverse has the potential to support sustainability, it also raises concerns about its environmental impact, particularly related to energy consumption and electronic waste. Andrae and Edler (2015) and Koomey et al. (2011) highlight the substantial energy requirements of data centers and the production of electronic devices. The widespread adoption of metaverse technologies could exacerbate these issues unless energy-efficient technologies and sustainable digital practices are developed and implemented.

Ethical considerations related to data privacy and security are also paramount in the metaverse. Floridi (2020) and Zuboff (2019) discuss the importance of robust data protection mechanisms to safeguard users' personal information. The metaverse's immersive nature and extensive data collection capabilities raise concerns about surveillance and digital rights. Calo (2014) and Nissenbaum (2010) argue for clear guidelines and regulations to address these ethical issues, ensuring that users' privacy is protected and their rights are upheld.

The intersection of the metaverse and SDGs presents a transformative opportunity to enhance access to education, drive economic growth, promote sustainable consumption, and address environmental and ethical challenges. By strategically aligning metaverse technologies with sustainability goals, we can create a future where virtual and physical realities converge to build a more inclusive, equitable, and sustainable world.

8 Future Trends and Research

The metaverse, a rapidly evolving virtual reality space, represents a significant transformation in how we interact, work, and live. As businesses and individuals increasingly engage with this digital realm, understanding its implications for sustainable development becomes crucial. This paper explores future research directions on the metaverse and its alignment with the Sustainable Development Goals (SDGs), focusing on transforming business practices for a sustainable future. By addressing key areas such as environmental impact, socio-economic implications, education and healthcare, regulatory and ethical frameworks, and interdisciplinary approaches, researchers can provide valuable insights into how the metaverse can contribute positively to global sustainability goals.

Environmental Impact

Energy Consumption

One of the critical areas for future research is the environmental impact of the metaverse, particularly concerning energy consumption. The infrastructure supporting the metaverse, including data centers, servers, and VR/AR devices, requires significant energy. Researchers should investigate ways to optimize energy efficiency in these technologies, exploring advancements in green computing and renewable energy sources. Understanding the carbon footprint of the metaverse and identifying strategies to reduce it are essential for aligning with SDG 13 (Climate Action).

Electronic Waste

As the metaverse relies heavily on electronic devices, another pressing concern is the generation of electronic waste (e-waste). Future research should focus on developing sustainable practices for the production, use, and disposal of VR/AR hardware. This includes designing devices with longer lifespans, promoting recycling and responsible disposal practices, and investigating the potential for using eco-friendly materials in device manufacturing. Addressing e-waste aligns with SDG 12 (Responsible Consumption and Production) by promoting sustainable consumption and reducing waste generation.

Virtual Collaboration and Travel Reduction

The metaverse offers opportunities for virtual collaboration, potentially reducing the need for physical travel and its associated environmental impact. Researchers should explore the effectiveness of virtual meetings, conferences, and remote work environments in decreasing carbon emissions. Studies could compare the environmental footprint of traditional business travel with virtual alternatives, providing insights into how widespread adoption of metaverse technologies can contribute to SDG 13 (Climate Action) by mitigating climate change impacts.

Socio-economic Implications

Job Creation and Skill Requirements

The metaverse is poised to create new job opportunities and transform existing industries. Future research should analyze the types of jobs emerging within the metaverse, such as virtual world designers, digital artists, and VR/AR developers, and identify the skills required for these roles. Investigating the potential for job displacement in traditional sectors and developing strategies for reskilling and upskilling the workforce are essential for ensuring inclusive and sustainable economic growth, aligning with SDG 8 (Decent Work and Economic Growth).

Economic Inequality and Access

Access to the metaverse and its benefits may be unevenly distributed, potentially exacerbating economic inequalities. Researchers should examine the digital divide and identify barriers to entry for marginalized communities. This includes studying the affordability of VR/AR devices, internet accessibility, and digital literacy. Developing policies and initiatives to ensure equitable access to metaverse technologies can help bridge the gap and promote inclusive economic participation, contributing to SDG 10 (Reduced Inequalities).

New Business Models and Revenue Streams

The metaverse is fostering innovative business models and revenue streams, such as virtual real estate, digital assets, and virtual services. Future research should explore the economic potential of these new markets, including their regulatory and taxation implications. Understanding how businesses can leverage the metaverse to create sustainable value chains and generate income while adhering to ethical standards is crucial for aligning with SDG 8 (Decent Work and Economic Growth) and SDG 9 (Industry, Innovation, and Infrastructure).

Education and Healthcare

Metaverse-Based Education

The metaverse offers significant potential for revolutionizing education by providing immersive and interactive learning experiences. Future research should investigate the effectiveness of metaverse-based education across different contexts and subjects. This includes studying the impact of virtual classrooms on student engagement, learning outcomes, and accessibility for students with disabilities. Exploring how metaverse technologies can complement traditional education methods and address educational disparities aligns with SDG 4 (Quality Education).

Lifelong Learning and Skills Development

Lifelong learning is essential for adapting to the rapidly changing job market influenced by the metaverse. Researchers should explore how the metaverse can facilitate continuous education and skills development for individuals of all ages. This includes examining the role of virtual training programs, online courses, and interactive simulations in providing opportunities for professional growth and personal development. Promoting lifelong learning through the metaverse supports SDG 4 (Quality Education) and SDG 8 (Decent Work and Economic Growth).

9 Regulatory and Ethical Frameworks

Data Privacy and Security

As the metaverse involves extensive data collection and user interactions, ensuring data privacy and security is paramount. Future research should explore regulatory frameworks that protect user data and address potential privacy concerns. This includes studying the ethical implications of data ownership, consent, and surveillance within virtual environments. Developing robust data protection policies and technologies is essential for building trust and ensuring the responsible use of metaverse platforms, aligning with SDG 16 (Peace, Justice, and Strong Institutions).

Equitable Access and Digital Rights

Ensuring equitable access to the metaverse and protecting digital rights are critical for fostering an inclusive virtual society. Researchers should investigate policies and initiatives that promote digital inclusion, such as affordable internet access, digital literacy programs, and anti-discrimination measures. Addressing issues of accessibility, representation, and inclusivity within the metaverse can help create a virtual space that reflects the diversity of the real world, contributing to SDG 10 (Reduced Inequalities) and SDG 5 (Gender Equality).

Ethical Use of AI and Automation

AI and automation are integral to the functioning of the metaverse, enhancing user experiences and automating processes. Future research should explore the ethical implications of AI use within virtual environments, including issues of bias, transparency, and accountability.

Interdisciplinary Approaches

Technological, Social, and Environmental Perspectives

Addressing the complex challenges associated with the metaverse requires an interdisciplinary approach that combines technological, social, and environmental perspectives. Future research should encourage collaboration between technologists, sociologists, environmental scientists, and policymakers to develop holistic strategies for integrating the metaverse into sustainable development efforts. This interdisciplinary approach can help identify synergies and trade-offs, ensuring that the metaverse contributes positively to multiple SDGs.

Virtual Real Estate and Reduced Carbon Footprint

One of the key sustainable business practices in Decentraland is the use of virtual real estate. Traditional real estate development has a significant environmental impact, including deforestation, carbon emissions from construction, and energy consumption from building operations. In contrast, virtual real estate eliminates these environmental costs. Businesses can establish virtual offices, showrooms, and event spaces in Decentraland, reducing the need for physical infrastructure and travel. For example, a multinational corporation could host virtual conferences and trade shows in Decentraland, significantly reducing the carbon footprint associated with international travel and large-scale physical events. By providing a platform for virtual collaboration and interaction, Decentraland supports SDG 13 (Climate Action) by reducing greenhouse gas emissions and promoting sustainable business operations.

Despite the environmental benefits of virtual real estate, the underlying blockchain technology can be energy-intensive. Future research and development in Decentraland should focus on optimizing energy efficiency and integrating renewable energy sources to power its blockchain operations. This aligns with SDG 7 (Affordable and Clean Energy) by promoting the use of sustainable energy in digital infrastructures.

Socio-economic Implications

Job Creation and Skill Development

Decentraland has created new job opportunities and avenues for economic growth in the digital realm. Users can monetize their creativity and skills by designing virtual assets, developing interactive experiences, and offering services within the platform. This has led to the emergence of new professions, such as virtual architects, digital fashion designers, and virtual event planners. For instance, a digital artist can sell unique virtual art pieces as non-fungible tokens (NFTs) within Decentraland's marketplace. This not only provides income opportunities but also fosters creativity and innovation. By enabling individuals to participate in the digital economy, Decentraland contributes to SDG 8 (Decent Work and Economic Growth) by promoting inclusive and sustainable economic development.

Bridging the Digital Divide

While the metaverse offers significant economic opportunities, ensuring equitable access remains a challenge. Decentraland has taken steps to address this by providing grants and support to underrepresented communities. For example, the Decentraland Foundation has launched initiatives to support digital literacy and access to technology, aiming to bridge the digital divide and promote economic inclusivity. By fostering a diverse and inclusive virtual community, Decentraland aligns with SDG 10 (Reduced Inequalities) and SDG 5 (Gender Equality), ensuring that the benefits of the metaverse are accessible to all.

Education and Healthcare

Virtual Education and Training

Decentraland offers a unique platform for virtual education and training. Educational institutions and businesses can create immersive learning environments where users can engage in interactive courses, simulations, and workshops. This approach enhances the learning experience by providing practical, hands-on training in a virtual setting. For example, a university could establish a virtual campus in Decentraland, offering courses on blockchain technology, digital art, and virtual reality development. Students can participate in real-time classes, collaborate with peers, and access resources from anywhere in the world. This promotes lifelong learning and ensures that education is accessible to all, aligning with SDG 4 (Quality Education).

Virtual Healthcare and Well-Being

The delivery of healthcare could likewise be revolutionised by the metaverse. Providers of medical, psychological, and wellness services in Decentraland can meet patients virtually. People in rural or underserved areas, in particular, benefit from this since it increases their access to medical treatment. For those who cannot afford or do not have access to conventional in-person mental health treatments, a mental health expert can offer virtual therapy sessions in Decentraland. This method guarantees patients' privacy and ease while simultaneously improving accessibility. Decentraland helps achieve SDG 3 (Good Health and Well-being) by making use of the metaverse in healthcare.

Establishing best practices and global standards for the metaverse is crucial for ensuring consistency and sustainability. Decentraland can serve as a model for other virtual platforms by demonstrating effective strategies for sustainable business operations, data privacy, and ethical AI use. Researchers should analyze successful examples of metaverse adoption across different sectors, such as education, healthcare, and business, to understand the factors contributing to their success. Sharing these insights can guide the development of effective strategies for leveraging the metaverse to achieve sustainable development goals.

Global Collaboration and Knowledge Sharing

Promoting global collaboration and knowledge sharing is essential for advancing research on the metaverse and its alignment with SDGs. Researchers should engage in international partnerships, conferences, and forums to exchange ideas and findings. This global perspective can help address regional disparities, foster innovation, and ensure that the benefits of the metaverse are accessible to all. Collaborating across borders supports SDG 17 (Partnerships for the Goals).

10 Examples and Case Studies of Metaverse for SDGs

Education (SDG 4—Quality Education)

Virtual Classrooms: Universities like Stanford and institutions in developing nations are experimenting with virtual campuses within the Metaverse, making education more accessible to students worldwide. Platforms like **Engage** and **AltspaceVR** host immersive classrooms, allowing students to attend lectures, interact with 3D models, and engage in collaborative learning in real-time, regardless of geographical constraints.

Case Study: University of Nairobi's Virtual Campus: The university partnered with a tech company to build a virtual campus where students from rural areas could attend classes without needing to travel to the city. This initiative has increased enrollment by 20%, particularly among female students, who faced travel safety challenges.

Climate Action (SDG 13—Climate Action)

Virtual Meetings and Events: Businesses have shifted corporate meetings, conferences, and expos into the Metaverse, reducing the need for physical travel. A report by **Accenture** found that companies using virtual events in the Metaverse reduced their carbon emissions by an estimated 70% due to avoided air travel and accommodation needs.

Case Study: COP26 Climate Conference in Virtual Reality: During the 2021 COP26 summit, delegates participated in side events through VR platforms, reducing the need for physical travel for numerous attendees. This decreased the overall carbon footprint of the event, setting a precedent for future climate-focused summits.

Reducing Inequalities (SDG 10—Reduced Inequalities)

Inclusivity through Virtual Workspaces: The Metaverse offers a digital space where individuals from marginalized communities can participate equally. Platforms such as **Decentraland** are promoting diversity by creating job opportunities within the virtual economy for people with disabilities or from underserved areas.

Case Study: Decentraland's Inclusive Virtual Economy: Decentraland has collaborated with NGOs to provide training programs in blockchain technology and virtual space design, enabling underrepresented groups in Latin America to build careers within the virtual economy.

Measurable Impact on Specific SDGs

SDG 4 (Education): Measurable impacts can include an increase in student enrolment in online courses by 50% in areas with limited physical school infrastructure. In rural or conflict-affected regions, virtual reality (VR) classrooms can improve literacy rates by providing consistent, quality education. By reducing infrastructure costs, schools can allocate more resources toward teaching and digital learning materials.

SDG 13 (Climate Action): By shifting 40% of business meetings to virtual platforms, companies can cut down on emissions from corporate travel, potentially contributing to reducing carbon footprints by an estimated 3–4 metric tons per company annually. This could lead to industry-wide reductions in emissions.

SDG 10 (Reducing Inequalities): The Metaverse can create more equitable employment opportunities. In regions where traditional jobs are limited, remote work within the Metaverse could reduce income inequalities by 20–30% over a decade by enabling more individuals to access high-paying tech jobs without relocating.

11 Ethical Concerns

Data Privacy: The Metaverse raises significant concerns about data privacy, particularly the collection of personal and biometric data (e.g., eye tracking, facial recognition). Businesses in the Metaverse must adhere to global data privacy regulations (e.g., GDPR) to ensure that users' sensitive information is protected.

Digital Divide: While the Metaverse has the potential to democratize access to virtual worlds, there is a risk of deepening the digital divide. Those in lower-income regions or without access to high-speed internet or VR hardware could be excluded. Governments and businesses should invest in affordable internet and provide educational programs to mitigate this risk.

Environmental Costs of Blockchain: Blockchain technologies used in the Metaverse, such as NFTs, have a high environmental cost due to their energy-intensive nature. Transitioning to energy-efficient blockchain solutions (e.g., **Ethereum 2.0**) and promoting green computing could help mitigate the carbon footprint of these technologies.

12 Implications

The exploration of the intersection between the metaverse and Sustainable Development Goals (SDGs) yields significant implications for businesses, policymakers, and society at large. For business practices, the metaverse fosters the creation of innovative business models that prioritize sustainability. Companies can leverage virtual spaces to offer eco-friendly products and services, such as virtual goods requiring no physical resources or logistics, developing and implementing regulatory frameworks that ensure equitable access to metaverse technologies is crucial.

Addressing the digital divide and ensuring that marginalized communities are not left behind are essential steps. Policymakers need to promote energy-efficient technologies and sustainable digital practices through appropriate policies, which includes incentivizing the development and adoption of green technologies in the virtual space. As the metaverse expands, robust data privacy and security measures become paramount to protect users' personal information and digital rights, necessitating clear guidelines and regulations to safeguard against misuse and exploitation.

The social impact of the metaverse is profound. It offers vast potential for transforming education and training by providing accessible, immersive learning experiences, contributing to SDG 4 by offering quality education to individuals regardless of their geographic location. The metaverse can also promote inclusive growth by providing platforms for marginalized communities to participate in the digital economy, helping to reduce inequalities (SDG 10) and creating opportunities for economic growth and employment in historically underserved areas. In terms of health and well-being, virtual health interventions such as telemedicine and virtual therapy can enhance health outcomes by providing accessible healthcare solutions, especially in remote and underserved regions, thus contributing to SDG 3.

13 Post-adoption of the Metaverse)

Virtual Classrooms and Inclusive Education (SDG 4—Quality Education)

Virtual classrooms accessible from any location, enabling students worldwide to engage in immersive, quality learning. This reduces the inequality gap in education and makes learning more inclusive and sustainable.

Virtual Meetings and Lower Carbon Footprints (SDG 13—Climate Action)

Features VR-based meetings and events, reducing the need for physical travel. A graph indicates a significant reduction in carbon emissions as businesses shift from traditional travel-heavy practices to virtual engagements.

Remote Collaboration and Global Employment (SDG 10—Reducing Inequalities)

People from different regions working in a shared virtual space, breaking geographical barriers. The Metaverse facilitates equal employment opportunities and inclusivity, allowing underrepresented communities to access jobs in the virtual economy.

By adopting the Metaverse, businesses and educational institutions can reduce their environmental impact, increase global access to education, and create more equitable work environments, all contributing to progress in key SDGs.

14 Conclusion

The Metaverse holds immense potential to transform business practices and support the achievement of Sustainable Development Goals (SDGs). By offering innovative solutions in areas such as education, climate action, and reducing inequalities, the Metaverse can pave the way for a more inclusive and sustainable future. In education, virtual classrooms can provide global access to quality learning, breaking geographical barriers. For climate action, virtual meetings and events can significantly reduce carbon footprints by eliminating the need for physical travel. Additionally, the Metaverse fosters inclusivity by creating equal employment opportunities, enabling marginalized communities to participate in the digital economy and reducing inequality.

However, realizing the full potential of the Metaverse requires addressing key ethical considerations. Concerns related to data privacy, the digital divide, and the environmental costs associated with blockchain technologies must be carefully managed. Data privacy protection, especially concerning personal data used in virtual environments, requires strict security measures. Furthermore, bridging the digital divide is crucial to ensure that individuals from all socio-economic backgrounds have access to these emerging technologies. The environmental sustainability of the Metaverse, particularly with energy-intensive blockchain operations, must also be prioritized to ensure that the technologies do not counteract climate goals.

Looking ahead, businesses should focus on integrating sustainable and inclusive practices into their Metaverse strategies. This includes utilizing virtual tools to minimize environmental impact, ensuring equitable access, and fostering digital literacy. Governments and NGOs should collaborate to expand infrastructure, support digital literacy initiatives, and make these technologies more accessible, particularly to underserved populations, thus helping to bridge the digital divide.

In terms of future studies, research should focus on understanding the long-term social, economic, and environmental impacts of the Metaverse. Specific areas for investigation include the scalability of virtual learning environments, the net environmental effect of widespread Metaverse adoption, and the implications for global labor markets. Additionally, studies exploring the ethical frameworks needed to manage data privacy, cybersecurity, and blockchain energy consumption will be vital to creating a responsible and sustainable Metaverse. While the Metaverse presents promising solutions for advancing the SDGs, its success depends on careful planning, strategic collaboration, and the development of strong ethical guidelines. By fostering innovation, inclusivity, and sustainability, the Metaverse can become a powerful tool in shaping a more equitable and sustainable world. Ongoing research, cooperation between governments, businesses, and civil society, and proactive policy-making will be essential in harnessing its full potential.

References

Andrae ASG, Edler T (2015) On global electricity usage of communication technology: trends to 2030. Challenges 6(1):117–157

Bailenson JN (2018) Experience on demand: what virtual reality is, how it works, and what it can do. W. W. Norton & Company

Bashshur RL, Shannon GW, Smith BR (2016) The empirical foundations of telemedicine interventions for chronic disease management. Telemed e-Health 20(9):769–800

Belkhir L, Elmeligi A (2018) Assessing ICT global emissions footprint: trends to 2040 & recommendations. J Clean Prod 177:448–463

Boellstorff T (2008) Coming of age in second life: an anthropologist explores the virtually human. Princeton University Press

Bonk CJ (2020) The world is open: how web technology is revolutionizing education. Jossey-Bass

Brynjolfsson E, McAfee A (2014) The second machine age: work, progress, and prosperity in a time of brilliant technologies. W. W. Norton & Company

Calo R (2014) Digital market manipulation. George Wash Law Rev 82(4):995–1051

Castronova E (2005) Synthetic worlds: the business and culture of online games. University of Chicago Press

Chen Z, Li Y, Wu Y (2020) The role of digital platforms in empowering women entrepreneurs: a study of the online handicraft industry in China. J Bus Res 119:426–438

Davidson S, De Filippi P, Potts J (2018) Economics of blockchain. SSRN Electron J. https://doi.org/10.2139/ssrn.2744751

Dionisio JDN, Burns WG, Gilbert R (2013) 3D virtual worlds and the metaverse: current status and future possibilities. ACM Comput Surv (CSUR) 45(3):1–38

Eubanks V (2018) Automating inequality: how high-tech tools profile, police, and punish the poor. St. Martin's Press

Floridi L (2020) The online manifesto: being human in a hyperconnected era. Springer

Foth M (2020) Smart cities and the co-creation of value: a source of new competitiveness in a low-carbon society. Elsevier

Frey CB, Osborne MA (2017) The future of employment: how susceptible are jobs to computerisation? Technol Forecast Soc Chang 114:254–280

Garcia-Palacios A, Hoffman HG, Kwong See S, Tsai A, Botella C (2007) Redefining therapeutic success with virtual reality exposure therapy. Cyberpsychol Behav 4(3):341–348

Gorski P (2018) Reaching and teaching students in poverty: strategies for erasing the opportunity gap. Teachers College Press

Grieves M (2014) Digital twin: manufacturing excellence through virtual factory replication. Retrieved from https://www.apriso.com/blog/wp-content/uploads/2014/09/Digital-Twin-Whitepaper.pdf

Hossain M, Islam MN, Sarker MA (2019) Inclusive digital economy in Bangladesh: a comprehensive study on the mobile financial services. J Sci Technol Policy Manag 10(2):185–205

Jin L, Herguera I (2017) Gender differences in the digital economy: evidence from the internet sector in China. J Econ Stud 44(4):678–695

Johnson L, Adams Becker S, Estrada V, Freeman A (2011) NMC horizon report: 2011 K–12 edition. The New Media Consortium

Kaplan A, Haenlein M (2019) Siri, Siri, in my hand: who's the fairest in the land? On the interpretations, illustrations, and implications of artificial intelligence. Bus Horiz 62(1):15–25

Koomey JG, Berard S, Sanchez M, Wong H (2011) Implications of historical trends in the electrical efficiency of computing. IEEE Ann Hist Comput 33(3):46–54

Kouhizadeh M, Sarkis J, Zhu Q (2021) Blockchain technology in supply chain management: insights on implementation benefits and challenges. Transp Res Part E Logist Transp Rev 138:101968

Lee LH, Braud T, Zhou P, Wang L, Xu D, Lin Z, Hui P et al (2021) All one needs to know about metaverse: a complete survey on technological singularities, virtual ecosystem, and research agenda. arXiv preprint arXiv:2110.05352

Maples-Keller JL, Yasinski C, Manjin N, Rothbaum BO (2017) Virtual reality-enhanced extinction of phobias and post-traumatic stress. Neurotherapeutics 14(3):554–563
Masanet E, Shehabi A, Lei N, Smith S, Koomey JG (2020) Recalibrating global data center energy-use estimates. Science 367(6481):984–986
Merchant Z, Goetz ET, Cifuentes L, Keeney-Kennicutt W, Davis TJ (2014) Effectiveness of virtual reality-based instruction on students' learning outcomes in K–12 and higher education: a meta-analysis. Comput Educ 70:29–40
Meta (2021) Building the metaverse responsibly. Retrieved from https://about.facebook.com/meta
Microsoft (2021) Microsoft mesh: a new mixed-reality platform. Retrieved from https://news.microsoft.com/innovation-stories/microsoft-mesh
Neves S, Correia A, Carravilla MA (2020) Smart cities: digital transformation for a sustainable future. Technol Forecast Soc Chang 153:119893
Nissenbaum H (2010) Privacy in context: technology, policy, and the integrity of social life. Stanford University Press
Noble SU (2018) Algorithms of oppression: how search engines reinforce racism. NYU Press
Riedl MO, Stern A, Dini DJ, Aldrich C (2013) Dynamic experience management in virtual worlds for entertainment, education, and training. Int Trans Syst Sci Appl 1(2):23–32
Saberi S, Kouhizadeh M, Sarkis J, Shen L (2019) Blockchain technology and its relationships to sustainable supply chain management. Int J Prod Res 57(7):2117–2135
Schwab K (2017) The fourth industrial revolution. Crown Business
Stephenson N (1992) Snow crash. Bantam Books
Tao F, Cheng J, Qi Q, Zhang M, Zhang H, Sui F (2018) Digital twin-driven product design, manufacturing and service with big data. Int J Adv Manufact Technol 94(9–12):3563–3576
Tapscott D, Tapscott A (2016) Blockchain revolution: how the technology behind bitcoin and other cryptocurrencies is changing the world. Penguin
United Nations. (2015). Transforming our World: The 2030 Agenda for Sustainable Development. Retrieved from https://sdgs.un.org/2030agenda
Van Dijk JA (2020) The digital divide. Polity Press
Wiederhold BK (2020) Connecting through technology during the coronavirus disease 2019 pandemic: avoiding "zoom fatigue." Cyber Psychol Behav Soc Netw 23(7):437–438
Zuboff S (2019) The age of surveillance capitalism: the fight for a human future at the new frontier of power. Public Affairs

Sustainability in the Digital Age: Bibliometric Perspectives on Green Technology and the Digital Carbon Footprint

Isha Narula, Ashu Lamba, and Sandeep Kautish

Abstract Various nations have committed to achieving carbon neutrality objectives in response to the worldwide concern over climate change. In the academic world, carbon footprint (CF) analysis has become increasingly popular as a crucial research technique for assessing carbon emissions. This study attempts to provide a thorough overview of this field of study, filling in any gaps using bibliometric analysis. Many organizations switched to cloud computing and storage at the beginning of the digital era in an effort to lessen carbon footprints and support green environments, however this led to an increase in the "Digital Carbon Footprint". The study identifies a number of factors that lead to an expanded "Digital Carbon Footprint", Such as, Social Media marketing, cloud computing, digital libraries, online shopping, smart cities and many more. "Digital Carbon Footprint" is a phenomenon that has been observed at both the individual and organizational levels. The study has also offered remedies to the "Digital Carbon Footprint" problem after reviewing an array of research articles authored by authors all across the world. The study additionally discovered an array of nations and organizations that are actively attempting to raise consciousness about environmental degradation and have provided financial assistance to motivate authors to publish works about it. The study demonstrates how the advent of digitalization has altered the landscape and the trends throughout time regarding the growth of the "Digital Carbon Footprint".

Keywords Digital carbon footprint · Bibliometric · Digitalization · Greenhouse gases · Carbon emission

---

I. Narula · A. Lamba (✉)
School of Business Studies, Vivekananda Institute of Professional Studies, New Delhi, India
e-mail: ashu.lamba@vips.edu

S. Kautish
Apex Institute of Technology (AIT CSE), Chandigarh University, Mohali, India

W. Leal Filho et al. (eds.), *Metaverse and Sustainability: Business Resilience Towards Sustainable Development Goals*, World Sustainability Series,
https://doi.org/10.1007/978-3-031-89545-6_24

1 Introduction

The recently released IPCC report on climate change confirms the undeniable and detrimental human contribution to global warming and emphasizes the critical need to significantly reduce carbon dioxide (CO_2) and other human-induced greenhouse gas (GHG) emissions in the coming decades. One of the main concerns of the Sustainable Development Goals of the United Nations (UN) is the intricate worldwide issue of climate change. The COVID-19-induced changes in consumption behaviors have demonstrated the potential of digital technology as an agent for more sustainable and environmentally friendly behavior. At the same time, the energy consumption and environmental impact of digital applications and services has been put more firmly on the agenda. (Gnanasekaran et al. 2021). Since digitalization is at its peak, everything we do online affects the environment. The concern over the world's steadily rising carbon emissions as a result of technological improvement is growing. As the world undergoes a digital transformation, popular buzzwords like "Imperial Internet of Things" (IIoT), "smart manufacturing," "smart banking," "digital twins," and "Industry 4.0" are being used to hype the potential of a new technological revolution (Patsavellas and Salonitis 2019). Nevertheless, as a consequence of extensive processing of data and digitization, there has been rising anxiety about the world's expanding carbon impact. Digitalization includes several processes like internet communication, cloud computing, server backup, real-time analytics and visualization, and video entertainment that contribute to making life easier. Every one of these procedures helps to make people's life easier and becomes a daily routine for various people (Vourdoubas 2016). The largest percentage of carbon emissions are caused by internet practice. People who use electricity to access various online services might demonstrate the rise in electricity consumption. Digital technologies utilize 9% more energy annually and contribute 4% of greenhouse gas emissions (World Favor 2022).

Digitally transforming economies generate goods and services with reduced carbon dioxide (CO_2) emissions (Hilty 2005; Malmodin and Lundén 2018), and nations with a high percentage of advanced digital technology businesses can achieve sustainable growth much more quickly (Baily et al. 2013). However, businesses' and economies' carbon footprints are significantly impacted by the degree of digital adoption (Ghisetti et al. 2017). Croatia and other nations with lower CO_2 emissions and moderate levels of digitization are mostly dependent on the service sector (Croatian Bureau of Statistics 2018), while Romania has a low carbon footprint because of extensive forests and land-use change (Romanian National Institute of Statistics 2019). Trade and the use of a technology have been shown to be positively correlated, but this relationship may undoubtedly be stronger (Mellouli et al. 2014). A country's growing national carbon footprint hinders the development of sustainable economic growth.

The GDP's share of trade is increased by the digital economy. This impact is temporary, though (Gordon 2012). The leading EU member states in digital implementation include the Netherlands, Cyprus, the Czech Republic, Slovakia, Slovenia, and Spain. Resource-efficient goods, technological inclusivity, and cutting-edge green technology in the digital economy can all help advance sustainable development (Bibri and Krogstie 2017). Small and medium-sized enterprises (SMEs) may find it simpler to integrate into regional and global value chains as a result of policies encouraging digital foreign direct investment (FDI).

The carbon footprint (CF) is a well-known and efficient method for addressing climate change and quantitatively assessing the intensity of carbon emissions. It thoroughly examines the causes, processes, and mechanisms of carbon emissions and offers a scientific foundation for investigating practical and efficient methods of reducing carbon emissions (Weber and Matthews 2008). The CF is typically stated as a weight measure, such as tons of CO_2 or CO_2 equivalent annually, despite its name. There isn't a single, accepted description of CF, despite the fact that numerous researchers have characterized its parameters from various angles. The total quantity of CO_2 and other GHGs released during a product's life cycle or supply chain is known as its "carbon footprint" (Hertwich and Peters 2009; Baldo et al. 2009).

The leading economies are concentrating on increasingly digitizing their economies to address the CF issue, while overuse of digital tools increased the emission of digital carbon footprints. Energy usage is a major contributor to carbon emissions for computer companies, and certain data centers' cooling systems have been shown to emit a variety of substances into the atmosphere. In recent decades, numerous academic research on CF have been conducted with a variety of objectives, primarily focusing on the concept's meaning, assessment techniques, and particular applications and practices. In recent years, cloud computing, including digitization processes, has been linked to an increase in carbon emissions (Sharma and Dash 2022). Though digital Carbon Footprint has a more narrowly defined reach, it is a notion based on the same ideas as the traditional carbon footprint. The digital carbon footprint is definite as "the CO_2 emissions resulting from the production, use and data transfer of digital devices and infrastructure" (Davies and Evangelidis 2021). To be more precise, using digital gadgets for email, social media, or internet browsing really increases global emissions. Given that there will likely be 55 billion connected devices by 2025, the digital carbon footprint will have a significant influence. Considering that we are heading toward cashless and paperless societies, it is imperative that we comprehend the effects of the Digital Carbon Footprint.

Despite the large number of Digital Carbon Footprint related studies that have been carried out, there aren't many thorough and organized digital Carbon Footprint research review publications. The following three elements demonstrate this: First off, there are comparatively few review studies on digital CF research, even though the number of papers pertaining to digital CF has significantly increased in recent years. Second, the review papers that are now available concentrate on clarifying the procedures and uses of digital CF. Giving a comprehensive overview of the theoretical research advances based on bibliometrics is crucial in this context. In order to provide useful information about present research progress and future research directions, bibliometric analysis can quantify examine the primary performance of the extant literature. This serves as a reference for digital CF-related research in addition to visualizing the theoretical framework and knowledge system associated with digital CF. This work aims to respond to the following queries in light of the research motivation: (1) What is the body of knowledge on the current research on digital CF? (2) In the literature on CF, which authors are present and contributing the most? (3) What are the primary areas of study and focus areas in this field? (4) What are the main sources of digital CF, as well as suggestive measures to be taken at organizational and individual level? This study examines the performance of digital CF-related literature produced between 2004 and 2024 in an effort to provide answers to the aforementioned queries. This study's remaining sections are arranged as follows. Following the introduction, Sect. 2 elaborates on the specific procedures and data sources. The main performance of publications pertaining to CF is shown in Sect. 3, together with information on journals, academic disciplines, countries, authors, institutions, and keywords. Citation analysis is also done to identify the primary areas of study in this area. In Sect. 4, we finally present our findings and limitations.

2 Method and Data Sources

Bibliometric analysis has been extensively used to examine the evolution and progress of different areas of research since it allows for the systematic investigation of large number of scientific publications (Donthu et al. 2021; Chen et al. 2021; Raza et al. 2021). This type of analysis uses statistical and mathematical techniques to quantitatively examine the citation data, popular keywords, active nations, authors, and institutions, as well as significant journals. Hence, this study uses bibliometric analysis to provide a thorough understanding of the trends and development characteristics of Digital CF-related research. Figure 1 depicts the steps involved in gathering, processing, and analyzing data. In order to standardize reference citation formats, combine keywords with related meanings, and unify author, country, and institution names, data cleaning is initially done by authors manually. The fundamental statistical data of a few chosen papers is then extracted using Scopus Database. In the meanwhile, three metrics are used: the h-index, cite score (CS), and journal impact factor (JIF). The CiteScore 2024 determined by Scopus and the JIF extracted from the Journal Citation Reports 2024 are used to evaluate the impact of journals. The JIF is the average number of citations received in a particular year by papers published in the journal during the preceding two years (Garfield 2006). The CS is determined as the ratio of the number of received citations during four years to the published amounts of five categories of documents (articles, reviews, conference papers, data papers, and book chapters) by a journal during the same four years (Scopus). Another metric that has garnered a lot of interest from scholars since its introduction in 2005 is the h-index. The h-index considers both quantity and quality by combining the number of published publications and individual article citations. An individual or organization with an h-index has published at least h papers that have

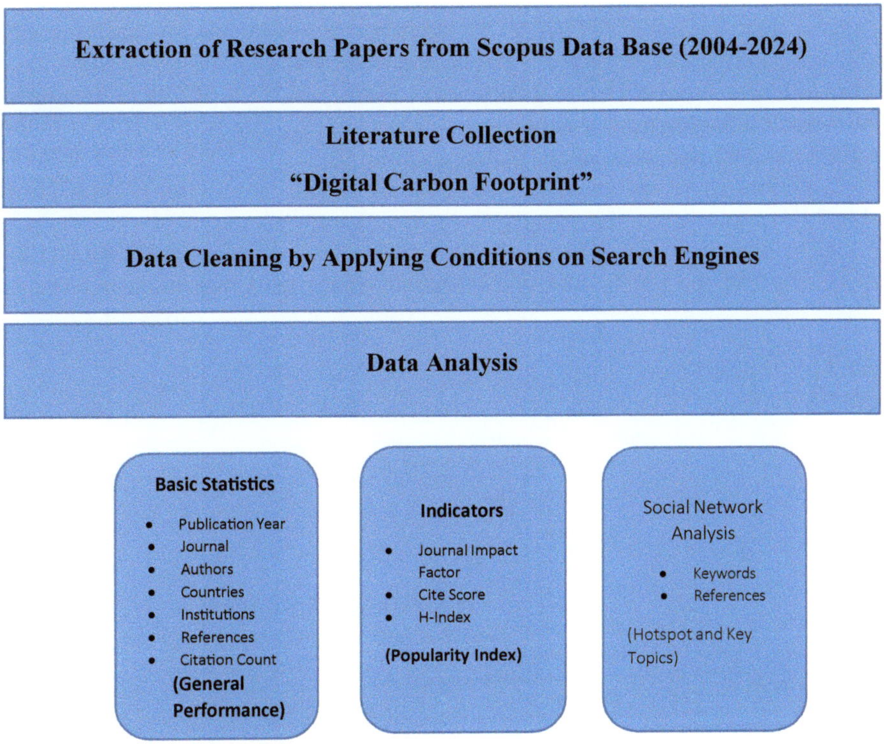

Fig. 1 Flow chart of bibliometric analysis. *Source* Created by authors to define flow of bibliometric analysis

been referenced at least h times. The h-indices of the authors, institutions, and countries are used in this study to assess the contributions of these research organizations in the field of Digital Carbon Footprint.

The work flow employed in this investigation is shown in Fig. 1. Applying different filters according to the study's requirements is the initial stage in gathering data from the official SCOPUS database. The researcher's keyword is "Digital Carbon Footprint." The code used is TITLE-ABS-KEY (digital AND carbon AND footprint) AND (LIMIT-TO (DOCTYPE, "ar") OR LIMIT-TO (DOCTYPE, "cp") OR LIMIT-TO (DOCTYPE, "ch") OR LIMIT-TO (DOCTYPE, "bk")) AND (LIMIT-TO (SUBJAREA, "COMP") OR LIMIT-TO (SUBJAREA, "ENGI") OR LIMIT-TO (SUBJAREA, "ENER") OR LIMIT-TO (SUBJAREA, "ENVI") OR LIMIT-TO (SUBJAREA, "SOCI") OR LIMIT-TO (SUBJAREA, "EART") OR LIMIT-TO (SUBJAREA, "MATH") OR LIMIT-TO (SUBJAREA, "BUSI") OR LIMIT-TO (SUBJAREA, "MATE") OR LIMIT-TO (SUBJAREA, "DECI")).

While identifying research papers, all languages are taken into account, and only published, completed, or ongoing publications are chosen. The researchers have collected research papers from 2004 to 2024 November from SCOPUS data base. Only research papers from the following (Table 1) categories were included in the shortlist of 1164 research papers. After inserting the terms "Digital Carbon Footprint," authors from various countries primarily utilize the keywords depicted in Table 2.

Table 1 is stating the number of research papers from different publication criterion. Out of all the publication types Journals (616) have maximum number of

Table 1 Allocation of publication types

Publication type	Frequency
Journal	616
Conference proceeding	385
Book series	90
Book	50
Trade journal	23

Source Assembled from SCOPUS database

Table 2 Trends of publication over the period of 20 years

Year	No. of publications
2024	243
2023	244
2022	185
2021	113
2020	67
2019	62
2018	43
2017	33
2016	25
2015	41
2014	23
2013	24
2012	21
2011	11
2010	14
2009	5
2008	7
2007	2
2004	1

Source Assembled from SCOPUS database

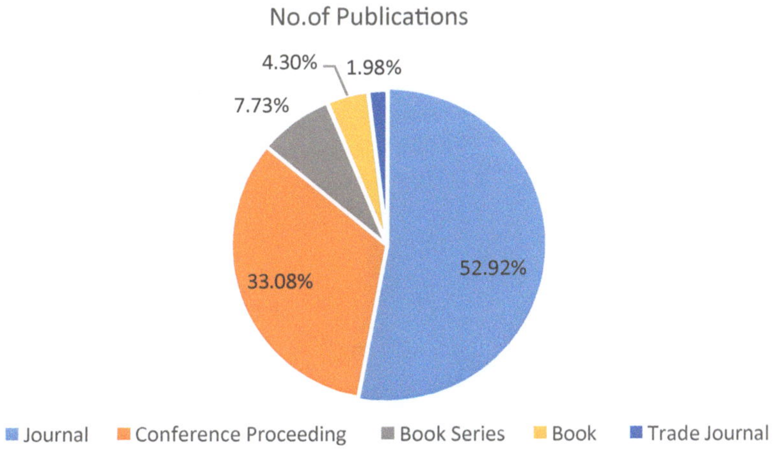

Fig. 2 Graphical representation of publication type. *Source* Created by Authors through information assembled from SCOPUS database

research papers followed by Conference Proceeding (385), Book Series (90), Book (50) and Trade Journals (23). Figure 2 represents the publication type graphically.

3 Data Analysis and Interpretation

The study's current portion highlights the salient characteristics of 1164 research articles gathered from the SCOPUS database. As shown in Fig. 1, the extensive literature is analyzed based on a variety of criteria.

Table 2 demonstrates the number of publications in the field of "Digital Carbon Footprint" from the year 2004–2024. Since 2019, there has been a noticeable increase in publishing and awareness in this topic. The most likely explanation is the increased use of digital tools following COVID- 19, which further increased carbon emissions. Figure 3 is the graphical representation of the trends of publication.

Table 3 demonstrates top 10 keywords used by numerous authors across different nations and research paper. The top used keywords utilized by authors are Carbon Footprint (789) followed by Digital Storage (444), Carbon Dioxide (212), Sustainable Development (209), and Greenhouse Gases (182) and so on. Figure 4 represents the top the key words used by authors graphically.

The top ten academic disciplines and the number of total publications they have produced in the area of digital carbon footprint are shown in Table 4. According to the Scopus Database, 24 academic fields have been identified; the top ten categories based on total publications are outlined above. PCT is total percentage of publications out of total. The highest number of publications are identified in the fields of Engineering (280), Energy (218), Computer Science (217) and Environment Science (213). It is pertinent to point out that there are 52 articles in the discipline of

Fig. 3 Graphical representation of trends of publication over the period of 20 years. *Source* Created by authors through information assembled from SCOPUS database

Table 3 The top 10 keywords

Keywords	Frequency
Carbon footprint	789
Digital storage	444
Carbon dioxide	212
Sustainable development	209
Greenhouse gases	182
Energy utilization	170
Emission control	158
Energy efficiency	152
Gas emissions	142
Environmental impact	135

Source Assembled from SCOPUS database

Business, Management, and Accounting, demonstrating the expansion of multidisciplinary research. Figure 5 represents the Graphical representation of top 10 academic disciplines where "Digital Carbon Footprint" papers are published.

Table 5 represents the key indicators of top 10 journals which are proved to be most productive in the field of Digital Carbon Footprint. It is worth mentioning that the journals namely, Science of the Total Environment (38) and Journal of the Cleaner Production (33) has maximum publications in the field of Digital Carbon Footprint. On the contrary, Journals such as, Environmental Science and Technology (11.7), Resources Conservation and Recycling (11.2) and Applied Energy (10.1) with higher Impact factor have lesser number of publications in the field of Digital Carbon Footprint. H-Index and Total number of Publications have a direct relationship for maximum journals except for Environmental Science and Technology. Sustainability

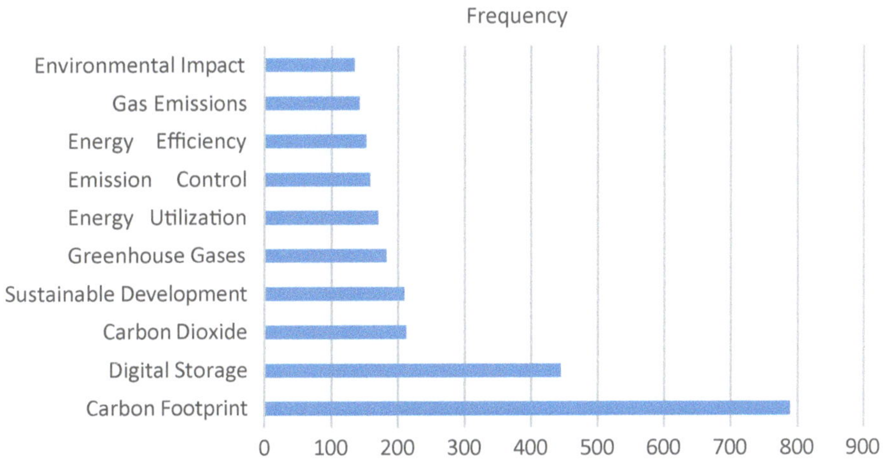

Fig. 4 Graphical representation of the keywords used in publications. *Source* Created by authors through information assembled from SCOPUS database

Table 4 The top 10 academic disciplines

Academic disciplines	Total publication	PCT (%)
Engineering	280	22.41
Energy	218	17.35
Computer Science	217	17.27
Environment Science	213	16.95
Mathematics	80	6.36
Social Sciences	54	4.29
Business, Management and Accounting	52	4.14
Earth and Planetary Sciences	49	3.90
Material Science	40	3.18
Decision Sciences	46	3.66

Note The percentage of publications in a discipline is represented by PCT (%)
Source Assembled from SCOPUS database

Switzerland and Energies are the two journals with low Impact factor, H-Index and Cite Score but with higher number of Publications in the field of Digital Carbon Footprint. Figure 6 is visual representation of top most productive journals in the field of "Digital Carbon Footprint".

Table 6 is assembled from SCOPUS data base to identify the most productive countries which contribute in creating awareness about Digital Carbon Footprint. United States (250) is one country which has contributed maximum awareness among readers regarding increasing level of carbon footprint, followed by United Kingdom (135), China (134), India (110), Germany (103) and so on. Although India is one of the

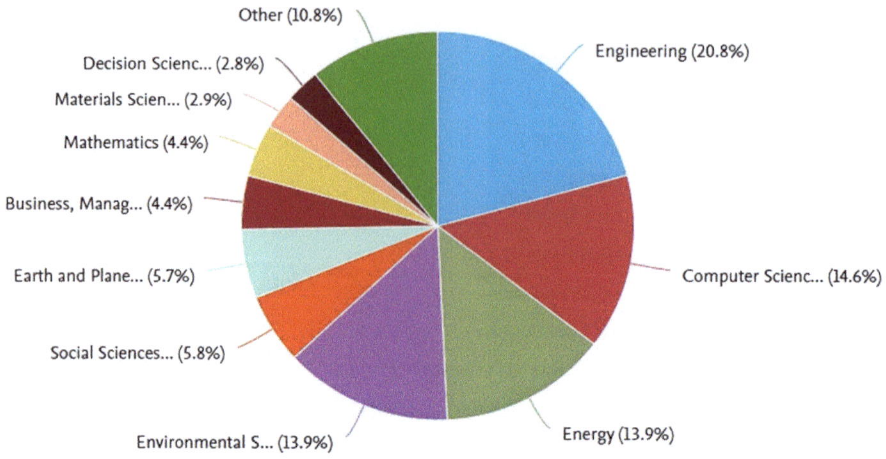

Fig. 5 Graphical representation of top 10 keywords used by authors. *Source* Created by authors through information assembled from SCOPUS database

Table 5 The performance of top 10 most productive journals

Journals	Total publication	H-Index	Impact factor	Cite score
Science of the Total Environment	38	353	8.2	17.6
Journal of the Cleaner Production	33	309	9.7	20.4
Sustainability Switzerland	21	169	3.3	6.8
Energies	18	152	3	6.2
Resources Conservation and Recycling	12	196	11.2	22.9
Journal of Environmental Management	12	243	8	13.7
Remote Sensing	11	193	4.2	8.3
Environmental Science and Technology	11	480	11.7	17.5
Applied Energy	10	292	10.1	21.2

Source Assembled from SCOPUS data base and Scimago

nations that makes significant contributions to research, there are neither financial sources nor publications in the country that support this kind of work. Figure 7 graphically represents top 10 most productive countries which supports research in the field of "Digital Carbon Footprint".

Table 7 demonstrates the performance of top 10 institutes along with their countries and number of publications and total citations and H-Index. One of the nations that contributes the most is the United States (33), whose universities/Institutes consistently supports and encourages studies on carbon footprints. Germany's (20) Institutes also supports researchers to bring awareness among readers for Carbon

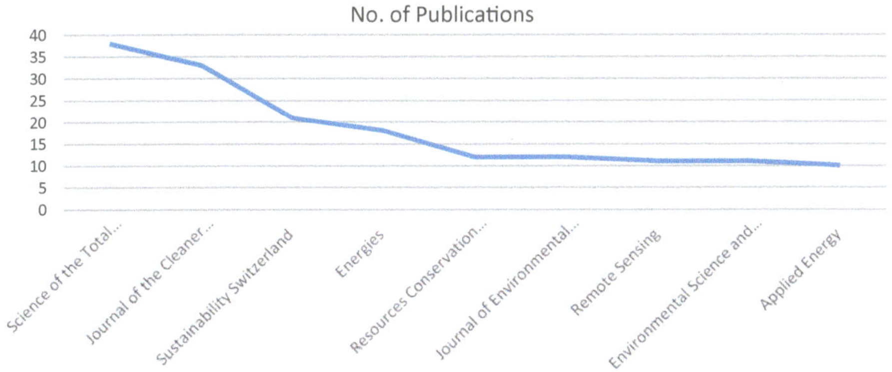

Fig. 6 Graphical representation of top 10 most productive journals. *Source* Created by authors through information assembled from SCOPUS database

Table 6 The performance of top 10 most productive countries

Country	Total publication	H-Index
United States	250	30
United Kingdom	135	26
China	134	29
India	110	14
Germany	103	19
Italy	70	14
France	56	14
Canada	52	15
Australia	37	18
Spain	36	12
Switzerland	32	12

Source Assembled from SCOPUS data base

Footprint reduction. And further it is followed by France (15), China (13) and Switzerland (12). It is also observed that the United States is the country with highest H-Index. Figure 8 graphically represents 10 most productive institutes.

Table 8 represents the performance of top 10 authors who have worked intensively in the field of Digital Carbon Footprint. Lamb, J is one of the most dynamic authors with a total Publication number of 5 followed by Skains, R. L. (4), Rudd, J.A. (4), Horry, R. (4), Godbole, N. F. (4), Bol, D. (4) and so on. It is also pragmatic that most contributing authors belong to the United States and United Kingdom. It is worth noting that total citations and H-Index is identified only for "Digital Carbon Footprint" research of the authors. Figure 9 is the graphical representation of top 10 most productive authors in the field of "Digital Carbon Footprint".

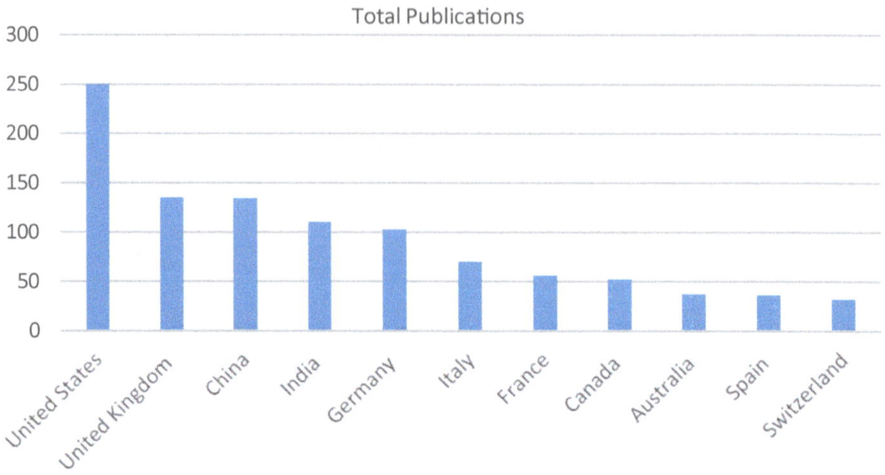

Fig. 7 Graphical representation of top 10 most productive countries. *Source* Created by authors through information assembled from SCOPUS database

Table 7 The performance of top 10 most productive institutes

Institutions	TP	TC	H-Index	Country
CNRS Centre National De La Recherche Scientifique	15	156	7	France
Schlumberger Limited	14	35	3	United States
Chinese Academy of Sciences	13	498	13	China
Eth Zurich	12	407	7	Switzerland
Technische Universität München	11	122	4	Germany
Stanford University	10	114	6	United States
Baker Hughes, A GE Company	9	4	1	United States
Rheinisch-Westfälische Technische Hochschule Aachen	9	79	3	Germany
Aarhus Universitet	9	208	8	Denmark
Delft University of Technology	8	178	5	The Netherlands

TP total publications, *TC* total citations
Source Assembled from SCOPUS data base

Table 9 shows the list of Sponsors who fund research related to "carbon footprint". European Commission (46) has supported maximum number of researches followed by National Natural Science Foundation of China (43), Horizon 2020 Framework Programme (26), U.S. Department of Energy (19), National Science Foundation (18) and so on. Figure 8 is the visual representation of top funding agency who support research on "Digital Carbon Footprint". Figure 10 is graphical representation of the top funding agencies.

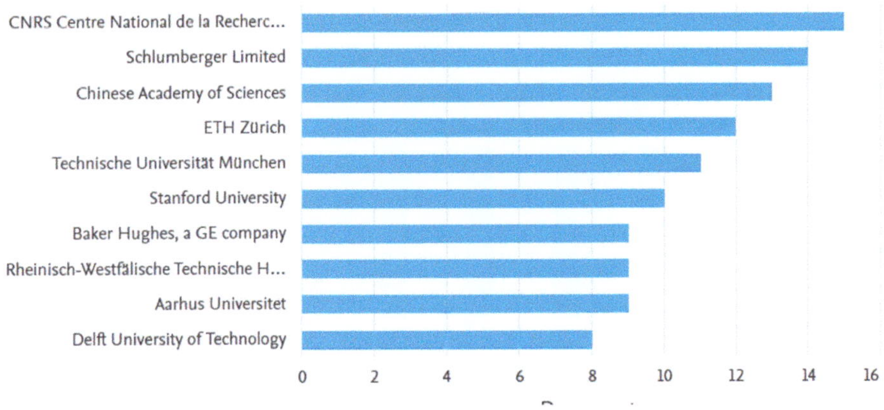

Fig. 8 Graphical representation of the performance of top 10 most productive institutes. *Source* Created by authors through information assembled from SCOPUS database

Table 8 The performance of top 10 most productive authors

Authors	Total publications	Total citations	H-Index	Countries
Lamb, J.	5	38	3	United States
Skains, R. L.	4	39	3	United Kingdom
Rudd, J. A.	4	39	3	United Kingdom
Horry, R.	4	39	3	United Kingdom
Godbole, N. S.	4	37	3	India
Bol, D.	4	174	3	Belgium
Wangler. T.	3	87	2	Switzerland
Tatari, O.	3	133	3	United States
Surendro, K.	3	5	2	Indonesia
Starosielski, N.	3	7	1	United States

Source Assembled from SCOPUS data base

4 Results

Global interest in CF-related research has increased as a result of the expanding global agreement on carbon peaking and carbon neutrality targets. The CF, which depends on life cycle theory, can be used as a quantitative tool to assess the process's overall GHG emissions at the macro and micro levels, taking both temporal and spatial perspectives into account. Since there have been many studies on CF evaluation, it is important to give a summary of the state of the field. Researchers have done a thorough literature evaluation based on the knowledge graph and bibliometrics to determine the main performance of publications connected to CF, pinpoint the area of study, and give researchers a relevant data. This study examined 1164 CF-related articles

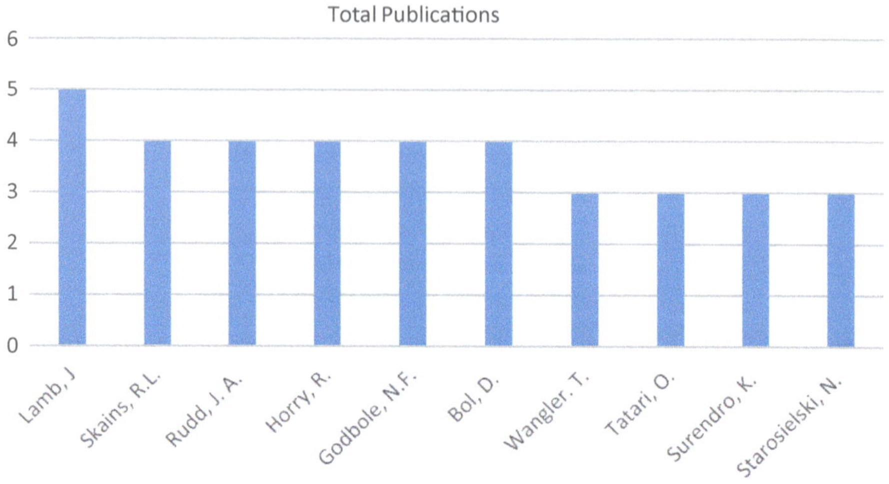

Fig. 9 Graphical representation of the performance of top 10 most productive authors. *Source* Created by authors through information assembled from SCOPUS database

Table 9 Top 10 funding institutes worldwide for supporting research in the field of "digital carbon footprint"

Funding sponsor	No. of Research publications funded by sponsors
European Commission	46
National Natural Science Foundation of China	43
Horizon 2020 Framework Programme	26
U.S. Department of Energy	19
National Science Foundation	18
UK Research and Innovation	18
Engineering and Physical Sciences Research Council	16
Ministry of Science and Technology of the People's Republic of China	14
Horizon 2020	13
Deutsche Forschungsgemeinschaft	12

Source Assembled from SCOPUS data base

published between 2004 and 2024 in order to visually categorize the development trend of future research and summarize the results of current studies. 1164 articles were gathered for study using the SCOPUS database. A growing interest in this topic is seen in the overall growth in the number of annual publications, the average number of authors and references per paper, and the countries publishing between 2004 and 2024. Environmental sciences, green sustainable science and technology, and

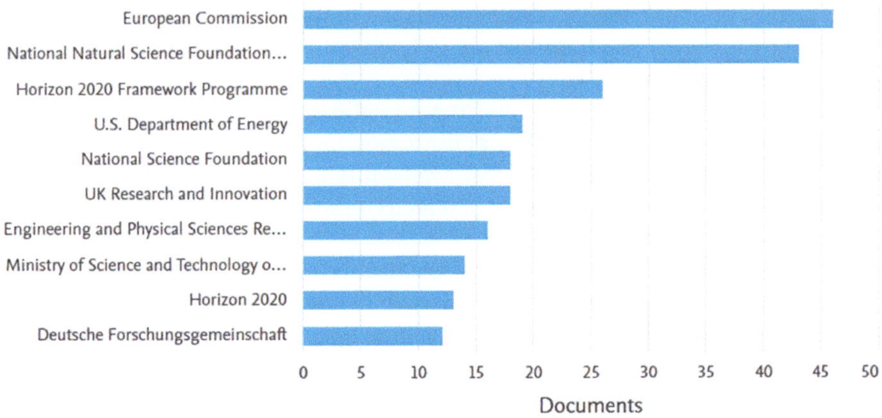

Fig. 10 Graphical representation of top 10 funding institutes worldwide for supporting research in the field of "digital carbon footprint". *Source* Created by authors through information assembled from SCOPUS database

environmental engineering for specific academic courses are the main areas of focus for most of these publications. The countries that are most successful in spreading knowledge of the "Digital Carbon Footprint" include Germany, Switzerland, the United States, and China.

Both the US and UK policy initiatives make it clear that the shift to net-zero should not hurt firms' ability to compete but rather help to create a robust industrial sector. Therefore, it is not unexpected that the necessity of changing company operations to conform to the decarbonization policy goal is a topic that comes up frequently in boardroom talks (Jackson and Hodgkinson 2023).

The journals that concentrate the most on carbon footprint and raise awareness of the need for methodical and prudent use of technology for a cleaner environment are Science of the Total Environment, Journal of the Cleaner Production, Sustainability Switzerland, and Energies. It has also been established that numerous organizations support similar initiatives in order to raise awareness of the prudent use of environmental and digital resources. One of the most promising countries is the United Nations, which has the greatest H-Index and publications while also providing authors with financial support to raise public awareness.

On the basis of bibliometric research, the key problem areas which are identified are Block Chains, cloud storage, Smart Cities, Gaming Software, Virtual Tourism, Podcasts, Digital Twins, Tele Communication, Social Media Platforms, online shopping etc. Maximum researchers have identified the above-mentioned problem areas and to reduce the emission of digital carbon footprint they have suggested wise usage of cloud computing, green user interface, green technology, simulation based low carbon mobility assessment model, green academics etc.

Through recycling, refurbishing, and reuse, the "three R's" of green IT place a strong emphasis on managing hardware sustainably. In order to integrate and enhance AI-based green computing algorithms that address environmental impacts and data use strategies, optimize energy consumption, and mitigate digital waste, green IT is supported by novel AI solutions that are in line with broad sustainability and ESG requirements (Dahmani 2024).

Sarkar et al. (2024) state that lowering digital carbon emissions necessitates a paradigm shift in power consumption optimization for IT and cooling loads, flexible load shifting based on renewable energy availability in the power grid, and utilizing battery storage from the data centers' continuous power supply through collaborative agents.

Using digital libraries that are built sustainably can also help people use the internet wisely, which will ultimately improve the future for the next generation. Programs for sustainable media can also help people use social media algorithms wisely. Social media marketing has become quite popular these days, but it also contributes to a lot of digital carbon emissions. To combat this, businesses should employ sustainable cyber marketing techniques. With the development of virtual and augmented reality, the modern world demands electrical energy harvesting, the smart industrial revolution, and the green internet of things.

Agrawal et al. (2024) suggests that the reduction of energy consumption in data centers necessitates a thorough evaluation of the carbon footprint associated with each device, including the energy expended on cooling and other overheads. Through the calculation of power consumption and a careful evaluation of its efficient utilization, overall efficiency can be enhanced by data centers by reducing the environmental footprint, and play a role in fostering a more sustainable future.

In order to use digital technology wisely and maybe reduce carbon emissions, carbon neutral data centers are essential. Additionally, the oil and gas industry may help reduce its offline and digital carbon footprint by implementing sustainable digital technology.

According to Wu et al. (2023), the digital economy is getting more and more important, and its contribution to sustainability is progressively becoming interconnected. Digital technologies are being used to work toward the goals of sustainable development and the reduction of greenhouse gas emissions.

After a careful review of the literature, it was found that no authors had ever done a comparative analysis of both developed and developing nations. Finding different actions taken by developed and developing nations that may be made to reduce the emission of the Digital Carbon Footprint and make the environment cleaner for future generations can be the main goal of such a study.

In summary, every industry that has gone digital in the past decade with the objective of creating a cleaner environment has significantly increased its digital carbon footprint. However, a collective effort by all enterprises and individual internet and electricity users may make a difference as responsible global citizens.

References

Agrawal A, Shrivastava K, Chakrabarty A, Singh Y, Shaw R, Padhi A (2024) An in-depth examination of green energy options to improve power consumption efficiency and reduce carbon emissions in contemporary data center infrastructures. In: 2024 International conference on trends in quantum computing and emerging business technologies. IEEE, pp 1–6

Baily MN, Manyika J, Gupta S (2013) US productivity growth: an optimistic perspective. Int Prod Monit 25:3

Baldo GL, Marino M, Montani M, Ryding S (2009) The carbon footprint measurement toolkit for the EU Ecolabel. Int J Life Cycle Assess 14:591–6

Bibri SE, Krogstie J (2017) Smart sustainable cities of the future: an extensive interdisciplinary literature review. Sustain Cities Soc 31:183–212

Chen R, Zhang R, Han H (2021) Where has carbon footprint research gone? Ecol Ind 120:106882

Croatian Bureau of Statistics (2018, December) Statistical Yearbook of the Republic of Croatia, Zagreb, Croatia

Dahmani S (2024) Computational intelligence for green cloud computing and digital waste management. In: Computational intelligence for green cloud computing and digital waste management. IGI Global, pp 248–266

Davies R, Evangelidis H (2021) Are you aware of your digital carbon footprint? [online] capgemini.com. Available at: https://www.capgemini.com/gb-en/2021/12/are-you-aware-of-your-digital-carbonfootprint/#:~:text=A%20digital%20carbon%20footprint%20is,with%20the%20use%20of%20technology [2022-03-07]

Donthu N, Kumar S, Mukherjee D, Pandey N, Lim WM (2021) How to conduct a bibliometric analysis: an overview and guidelines. J Bus Res 133:285–296

European Commission. Harmonised rules for the calculation of the carbon footprint of electric vehicle batteries (CFB-EV). Available from: https://eplca.jrc.ec.europa.eu/GRB-CBF_CarbonFootprintRules-EV.pdf. Accessed 15 Oct 2024

Garfield E (2006) The history and meaning of the journal impact factor. JAMA 295(1):90–93

Ghisetti C, Mancinelli S, Mazzanti M, Zoli M (2017) Financial barriers and environmental innovations: evidence from EU manufacturing firms. Climate Policy 17(sup1):S131–S147

Gnanasekaran V, Fridtun HT, Hatlen H, Langøy MM, Syrstad A, Subramanian S, De Moor K (2021) Digital carbon footprint awareness among digital natives: an exploratory study. In: Norsk IKT-konferanse for forskning og utdanning, vol 1, pp 99–112

Gordon RJ (2012) Is US economic growth over? Faltering innovation confronts the six headwinds (No. w18315). National Bureau of Economic Research

Hertwich EG, Peters GP (2009) Carbon footprint of nations: a global, trade-linked analysis. Environ Sci Technol 43(16):6414–6420

Hilty LM (2005) Editorial: electronic waste—an emerging risk? In: Environmental impact assessment review, vol 25, pp 431–435

Malmodin J, Lundén D (2018) The energy and carbon footprint of the global ICT and E&M sectors 2010–2015. Sustainability 10(9):3027

Mellouli S, Luna-Reyes LF, Zhang J (2014) Smart government, citizen participation and open data. Information Polity 19:1–4

Patsavellas J, Salonitis K (2019) The carbon footprint of manufacturing digitalization: critical literature review and future research agenda. Procedia Cirp 81:1354–1359

Raza MS, Khahro SH, Memon SA, Ali TH, Memon NA (2021) Global trends in research on carbon footprint of buildings during 1971–2021: a bibliometric investigation. Environ Sci Pollut Res 28:63227–63236

Romanian National Institute of Statistics (2019) Romanian statistical yearbook, Bucharest, Romania

Sarkar S, Naug A, Luna R, Guillen A, Gundecha V, Ghorbanpour S, Babu AR (2024) Carbon footprint reduction for sustainable data centers in real-time. In: Proceedings of the AAAI conference on artificial intelligence, vol 38, no 20, pp 22322–22330

Scopus. How are CiteScore metrics used in Scopus? Available from: https://service.elsevier.com/app/answers/detail/a_id/14880/supporthub/scopus/. Accessed 15 Oct 2024

Sharma, Dash (2022) The digital carbon footprint: threat to an environmentally sustainable future. Int J Comput Sci Inf Technol (IJCSIT) 14(3)

Jackson TW, Hodgkinson IR (2023) Keeping a lower profile: how firms can reduce their digital carbon footprints. J Bus Strat 44(6):363–370. Emerald Publishing Limited, ISSN 0275-6668. https://doi.org/10.1108/JBS-03-2022-0048

Vourdoubas J (2016) Reduction of CO_2 emissions due to energy use in Crete-Greece. Energy Environ Res 6(1):1–23

Weber CL, Mathhews HS (2008) Food-miles and the relative climate impacts of food choices in the united states. Environ Sci Technol 42:3508–3513

World Favor (2022) The growing carbon footprint of digitalization and how to control it. https://blog.worldfavor.com/the-growing-carbon-footprint-of-digitalization-and-how-to-control-it

Wu W, Xu Z, Zenchenko S, Strielkowski W (2023) Does sustainable digital economy have implications for climate change? Novel evidence in the context of sustainable energy technologies. Sustain Energy Technol Assess 60:103524

Harnessing the Metaverse: Driving Business Resilience and Sustainability for Future Growth

Anand Singh Rajawat, S. B. Goyal, Pradeep Bedi, Ashish Vivek Singh, and Sandeep Kautish

Abstract The metaverse a fluid merger of mixed reality and augmented reality now offers new possibilities to businesses wishing to leverage buoyancy and sustainability to increase their competitive advantage. Here, we draw on our experience across industries and markets to explore the potential of the metaverse to reshape business models and improve operational efficiency to enable sustainability. Through digital twins, virtual product testing, and immersive technologies, companies can provide interactive customer experiences while supporting a circular economy. Also, the metaverse promotes remote cooperation and decentralized decision-making that allows organizations to respond rapidly to market evolutions and unexpected disruptions essential components of organizational resilience in challenging times characterized by global disturbances such as climate change and economic volatility. For instance, the piece discusses how some organizations have already been leveraging metaverse strategies to mobilize to become efficient and zero carbon. It encourages building a responsible metaverse—one that provides ethical and technological development alongside the internet that is inclusive. However, through stakeholder engagement and sustainability in metaverse practices, businesses can not only overcome the looming market challenges but also contribute to the cause of global sustainability. However, this research provides a roadmap to organizations interested in the opportunities that the metaverse presents as a tool for sustainable growth and environmental management.

A. S. Rajawat · A. V. Singh
School of Computer Science and Engineering, Sandip University Nashik, Nashik, India

S. B. Goyal (✉)
Faculty of Information Technology, City University, Petaling Jaya, Malaysia
e-mail: drsbgoyal@gmail.com

P. Bedi
Department of CSE, Graphic Era Deemed to be University, Dehradun, India

S. Kautish
Apex Institute of Technology (AIT CSE), Chandigarh University, Mohali, Punjab, India

W. Leal Filho et al. (eds.), *Metaverse and Sustainability: Business Resilience Towards Sustainable Development Goals*, World Sustainability Series,
https://doi.org/10.1007/978-3-031-89545-6_25

Keywords Metaverse · Business resilience · Sustainability · Digital transformation · Immersive technologies · Circular economy

1 Introduction

Metaverse is an ever-evolving digital universe that incorporates virtual reality (VR), augmented reality (AR) and more its rapid growth is expected to fundamentally change the way businesses interact with employees and others with whom they work. Against the backdrop of an ever-evolving environment, the need for adaptability and the opportunity Metaverse provides for organizations not only to embrace transformation but to thrive as a lighthouse on the path of resilience in unprecedented interconnectedness between people at large opens an alternative solution for businesses. This preface explains what metaverse means, what it entails in terms of how business should be done and the possibilities it presents to establish sustainable growth. Fundamentally, the metaverse is a collective virtual environment not just the space between multiple users, but also the interaction and the environmental context in which these interactions belong. Already frightfully familiar, view machine or virtual reality hybridize physical and digital space in a manner that renders most other online 2D platforms as antiquated public utilities. With VR and AR technologies maturing at a breakneck speed, coupled with the promise of the metaverse, companies now have new avenues to engage with customers as the lines between businesses continue to blur (Yao et al. 2022). For an organization looking to build progressive strategies, understanding the metaverse becomes an absolute priority, as organizations equip themselves with agility and sustainability in mind as weapons of mass survival. In simple words, business resilience is the continuous process of preparing for disruption (and the ways to avoid it) to ensure the organization responds quickly if something does happen. Resilience has proven itself to be incredibly valuable during the COVID-19 pandemic as companies have faced unprecedented challenges from supply chain disruptions to changes in consumer behaviour. It gives the companies an opportunity to reimagine and grow their businesses as they have never done in that third-grade space, we call the metaverse. Immersive technology can help recreate dynamic real-world activities in virtual environments which enables companies in specific sectors to test at strategy and at risk while optimizing decision-making processes. For instance, virtual training simulations allow businesses to enable employees to practice and gain confidence in high-stakes scenarios in a controlled setting free from the liabilities of the real world. Sustainability has emerged as a perennial challenge for businesses working through the effects of climate change, resource extraction and social inequality. But the metaverse (Elhagry and El Saddik 2023) could act as a catalyst for green behaviours, by enabling groups such as companies and other organizations to run simulations of their environmental impact with the click of a button. For instance, virtual product testing allows companies to test designs and materials and to run through a functional prototype process before it is built to avoid waste. Virtual replicas of physical assets created by digital twins are used for predictive maintenance and resource optimization to minimize the cost of energy waste while prolonging the lifecycle of products. Business can use these new paradigms as instruments to build their strategies around sustainability to achieve decarbonisation targets, as well as to enhance the competitiveness of business itself. In addition, the metaverse also facilitates collaboration between companies, bending what we have seen up until now for the employees and customers of the company. As remote work and digital communication is now the new normal, the metaverse allows for teams to convene one-on-one in immersive environments, like many of the social cues that daily life provides us. This not only ramps up productivity, but also fosters the sense of community and belonging among remote employees. For organizations, this translated to ways of virtual knowledge sharing and innovation and to document workshops and conferences across geographic boundaries. Creating collaborative spaces in the Metaverse for creative individuals and organizations to prosper you get a consistent source of ideas and insights that are nigh on impossible to source in traditional group building exercises. The metaverse offers businesses new marketing and branding channels, as well operational enhancements. While enhancing experiences that appeal to consumers in a competitive environment. New types of engagement help create new relationships with makers think virtual showrooms, AR and gamified experiences that draw consumers in to buy from avenues other than eBay. Further, consumers are more and more environmentally aware, so if a business markets appropriately concerning how green they are or if they support carbon-offsetting in the metaverse an online company can grow their social credit account (CSR), credibility guaranteeing all tree-huggers flock to purchase. This shared values and consumer expectations alignment can cultivate loyal customer relationships and continued sustainable growth over the long term. Though the metaverse provides huge opportunities, companies also need to deal with ethics including data privacy and inclusivity issues. As the dawn of such technical breakthroughs, organizations need to computerize responsibly with an increased focus on equitable access and user data protection (Lee et al. 2023). Defining ethical standards and involving stakeholders in the creation of metaverse endeavours encourages trust and goodwill, enhancing mutual interest. It is also crucial that virtual spaces be more diverse and inclusive for everyone to engage with the metaverse on a meaningful level. So, what are we ending with here the metaverse has reached great potential for businesses to enable growth and transparency needed in a fast-moving world? Companies that take the plunge into extended reality will be able to become more adaptable, optimize various operations and interact with their customers in much more nuanced ways. Embedding sustainability into metaverse strategies drives sustained growth while serving the global need for responsible business practices. To realise the metaverse, companies will have to cautiously negotiate these hurdles with an eye on ethics in order that tomorrow is built upon foundations of sustainability and resilience. Redefining the Metaverse is not a pure technology project it is an entirely new way of doing business to drive societal and environmental benefit.

- Specify what properties something must have to be a metaverse and how it applies in modern business.
- Increase Resilience Investigate ways in which these technologies can assist the workforce in being more resilient, doing so by deploying tools capable to provide immersive training and support work activities remotely.
- Explore how the metaverse supports sustainable use cases including virtual product testing and circular economy integration.
- Evaluate where the metaverse can help to improve customer interaction and strengthen brand loyalty via immersive marketing techniques.
- Emphasize the necessary of ethical considerations and inclusivity in the metaverse for fair access to responsible data handling.

2 Literature Review

The metaverse has dominated media and the broader cultural conversation over recent years, thanks to unprecedented developments in digital technologies partnered with an increasing appetite for more immersive experiences (Kharvi 2023). The metaverse has been explored by both scholars and practitioners in different sectors, hypothesizing that it can revolutionize business process, increase customer engagement or promote sustainability. To this end a literature review, synthesising the most important findings regarding u purpose in business resilience and sustainability driven by metaverse is impulse from academic research, industry reports and case studies.

Business Operations in the Metaverse: Viewed as a game-changing platform the metaverse is set to provide new momentum in business (Zhang 2023). That allows organizations to build digital twin environments near perfect real-time representation of their physical objects (true twins). Businesses can model, simulate and evaluate risks to optimize operations promptly with this innovation. Companies using digital twins have reported lower operational costs, and more effective usage of resources (e. g. opening new branch). Findings like this illustrate the potential of the metaverse to improve business decision-making and operational agility.

Immersive Tech in Enhancing Resilience: Additionally, research from Deloitte indicates that the metaverse can help bolster business resilience through stimulating remote collaboration and training. Virtual training simulations in immersive environments (Lee et al. 2023) provide organizations the opportunity to train employees for high-stakes scenarios, but without any of those risk factors that are involved with physical worlds. Connecting teams and breaking global boundaries also leads to innovation, increases cross country team participation bringing in opportunities for knowledge transfer which is essential part of resilient organisations. The relevance of this learning practice, especially during global events like we recently have witnessed in the COVID-19 pandemic when digital collaboration becomes even more critical.

The Circular Economy solution and the challenge of sustainability: The metaverse also offers a real wealth of opportunities for businesses to commit to sustainable approaches. Virtual product testing and digital twins help in reduce the waste of products made during development largely reducing resource consumption by organizations (Lee et al. 2023). Another dimension is that when businesses develop and test products in digital form, they could significantly decrease the carbon emissions from conventional prototyping methods. Moreover, the Metaverse allows to apply circular economy principles which enable companies design products considering their lifecycle. High Street Decline and Eco-Conscious Customers This move towards sustainability does not only tackle the environmental challenges it also strikes a chord with the ever-growing eco-conscious customers.

Customer Marketing engagement: Throughout the literature, as businesses seek to traverse some of these complexities associated with operating in a metaverse context there is often discussion relating to reimagining enhanced customer engagement. The use-case potential of virtual environments by marketers is quite profound, as (in theory at least) everything that inspires a consumer to want something can be brought into or re-created for this 3-D environment. By offering an immersive experience to your consumers, it will enhance their loyalty and engagement towards the brand (Dong et al. 2023). And the gamification of metaverse experiences can help drive engagement, while bringing people into shared stories that resonate with their beliefs. This change in Paradigms looks for new marketing methodologies with businesses to reach the audience.

"**Ethics and Inclusive Constants Ethical Considerations**": Fortunately, a metaverse comes with tremendous potential, but it raises significant ethical and inclusivity concerns as researchers argue. New research on data privacy and the digital divide underscores some of these risks (Ciliberti et al. 2023), calling for organizations to pursue responsible practices in this brave new world. Trust in metaverse initiatives requires ethical guidelines and stakeholder engagement. While leveraging the metaverse as a tool space, it is important for businesses to create such an ecosystem that drives maximum positive energies through inclusivity.

Yao et al (2022), The recent advances of emerging technologies including artificial intelligence, 5G, 6G, extended reality and blockchain promote the rapid development of next-generation Internet. As an evolving paradigm of next-generation Internet, metaverse, a fully immersive, hyper spatiotemporal and selfsustaining virtual shared space, is moving from imagination to the coming reality. However, its massive data flow, pervasive user profiling activities and other intrinsic features can lead to a lot of security and privacy concerns, which will hinder its further deployment. Specially, since the identities of users/avatars in the metaverse can be illegally stolen, impersonated, and interoperability issues can be encountered in authentication across metaverses, this paper designs a lightweight and privacy-preserving seamless cross-metaverse authentication and key agreement scheme named MetaverseAKA to meet the challenges.

Elhagry et al. (2023), The proposed pipeline has potential applications including education, training, architecture, healthcare and entertainment, and could change the way designers and developers create virtual worlds. While this paper covers an initial work as per the PhD Forum's guidelines, it contributes to the research on generative models for multimodal data and provides a new direction for creating immersive virtual experiences.

Zhang et al. (2023) With the rise of the metaverse, its security risks are of great concern. We have researched the literature related to metaverse security in recent years and summarised the security risks of metaverse at the technological level. First, we introduce the concept of the metaverse and the role of the five underlying technologies (blockchain, Internet of Things, cloud services, artificial intelligence, and digital twins) in the metaverse. Then, based on the security risks and countermeasures of these underlying technologies themselves, we summarise the security issues faced by the metaverse. Finally, we summarise the opportunities and challenges of metaverse security from the perspectives of the five underlying technologies and the overall perspective of the metaverse.

Lee et al. (2023) We present a system that maps emotions to the metaverse enhancing metaverse objects with new facial expression recognition that can be potentially found in the real-world. Using a deep learning model, we made facial expression recognition on facial expressions recorded from a webcam and utilized recently for emotion filtering. Using this method, the specific emotion data was sent to the IoT server and the Metaverse Inter-working Module enabling a system where objects in the metaverse can change. This now allows users to share emotions with other users through metaverse objects that could facilitate emotional communication and expression between people in the metaverse.

Kharvi (2023) In this research paper, we outline the few key ingredients of what we refer to as the trust framework and goals for the Metaverse space which will ultimately govern users on using the products or services to purchase, renew, and continue to trust their data. Fixing the trust issues which are inevitable in a new virtual world is critical for ensuring the Metaverse is a safe, secure, and trustworthy experience for all its users in Table 1 to represents the comparative Analysis.

3 Proposed Methodology

This study proposes a methodology to research the potential of the metaverse to promote business resilience and sustainability, using both qualitative and quantitative methods. This framework combines theoretical bases and numerous mathematical equations and formulae which can be used to scrutinize how the metaverse will affect leading business metrics. The following sections are dedicated to the proposed method, which contains its relevant terminology, theories and a complete mathematical set that makes its implementation in the research feasible.

Table 1 Comparative analysis

References	Authors	Year	Main Topic	Methodology	Key Findings	Applications
Laborde et al. (2023)	R. Laborde et al	2023	Avatar interoperability using self-sovereign identity	Theoretical framework	Proposes seamless avatar interoperability	Enhanced user experience in virtual environments
Zelenyanszki et al. (2023)	D. Zelenyanszki et al	2023	Privacy techniques for NFTs	Analysis of techniques	Zero-knowledge proofs enhance NFT privacy	Privacy measures in NFT transactions
Athar et al. (2023)	A. Athar et al	2023	Mental health detection in the metaverse	BERT-based system	Detects anxiety and depression, suggests interventions	Mental health support in virtual environments
Sharma and Zamfiroiu (2023)	R. C. Sharma and A. Zamfiroiu	2023	Cybersecurity threats in the metaverse	Literature review	Identifies specific cybersecurity vulnerabilities	Security protocols for metaverse platforms
Trujillo and Bacciu (2023)	A. Trujillo and C. Bacciu	2023	Blockchain in fashion wearables	Case study	Blockchain enhances authenticity of wearables	Applications in decentralized fashion
Li et al. (2023)	P. Li et al	2023	One-Time Password authentication	Proof-of-concept	TOTPAuth combats identity theft	User authentication in metaverse applications

3.1 Digital Twin Theory

Digital twin technology forms the basis of the metaverse's functional framework. A digital twin is the virtual copy of a physical system, providing the capability to analyze, simulate, and optimize using real-time data. This is an important theory for presenting how companies can use metaverse to improve the way they make decisions and operate systems (Seyama and Sasaki 2023).

3.2 Resilience Theory

Resilience theory places emphasis on the processes by which organizations adapt and succeed in the face of challenges. This is shot of domain portrays flexibility, learning and innovativeness which can be measured with different performance measures.

3.3 Sustainability Theory

The theory of sustainability looks at how companies can operate at the edges of ecology while still making a profit. At its core, this framework pushes towards incorporating sustainability into the organizational strategy—which is key towards sustainable growth (Patwe and Mane 2023). Q Formulation Mathematical formulations that measure the resilience, sustainability, and business performance from metaverse-perspective are proposed.

3.4 Performance Metrics

The performance metrics recommended to measure the effect of the metaverse on business resilience and sustainability are as follows.

- Operational Efficiency (OE):

$$OE = \frac{\text{Input}}{\text{Output}}$$

where

Output Total value generated (e.g., revenue, products)
Input Resources consumed (e.g., time, materials)

Cost Reduction (CR):

$$CR = \frac{C_{\text{before}} - C_{\text{after}}}{C_{\text{before}}} \times 100\%$$

where

C before Operational cost before implementing metaverse solutions
C after Operational cost after implementation

Productivity Index (PI):

$$PI = \frac{100}{OE \times CR}$$

where

OE Operational Efficiency
CR Cost Reduction

3.4.1 Sustainability Metrics

To assess sustainability practices within the metaverse framework, the following indices are proposed (Chen et al. 2023):

- **Sustainability Index (SI)**:

$$SI = \frac{3}{R + W + C}$$

where

R Resource efficiency (e.g., energy consumption per unit of output)
W Waste reduction (percentage reduction in waste produced)
C Carbon emissions (total emissions per product line)

Carbon Footprint Reduction (CFR):

$$CFR = \frac{CF_{\text{before}} - CF_{\text{after}}}{CF_{\text{before}}} \times 100\%$$

where

CF before Carbon footprint before implementing sustainable practices
CF after Carbon footprint after implementing sustainable practices

Circular Economy Index (CEI):

$$CEI = \frac{W + 1}{R + C}$$

where

R Resource efficiency
C Recycling rate
W Waste produced

3.4.2 Customer Engagement Metrics

To quantify customer engagement within the metaverse, the following metrics are proposed (Laborde et al. 2023):

- **Customer Engagement Score (CES)**:

$$CES = \frac{3}{I + F + P}$$

where

I Interaction frequency (number of interactions per user)
F Feedback score (average rating from customer feedback)
P Participation rate (percentage of users actively engaging with the metaverse platform)

Net Promoter Score (NPS):

$$NPS = \frac{\text{Total Responses}}{\text{Promoters} - \text{Detractors}} \times 100$$

where

Promoters	Number of users rating 9-10
Detractors	Number of users rating 0-6
Total Responses	Total feedback responses

3.4.3 Financial Performance Metrics

To assess the financial implications of adopting metaverse (Zelenyanszki et al. 2023) technologies, the following equations are proposed:

- **Return on Investment (ROI)**:

$$ROI = \frac{I}{G - I} \times 100$$

where

G Gains from metaverse implementation
I Investment made in metaverse technology

Total Cost of Ownership (TCO):

$$TCO = \text{Acquisition Costs} + \text{Operational Costs} - \text{Residual Value}$$

where

Acquisition Costs	Initial purchase and setup costs
Operational Costs	Ongoing maintenance and operational costs
Residual Value	Value remaining at the end of the asset's life

3.5 Data Collection and Analysis

3.5.1 Qualitative Data

Qualitative data will be collected through interviews, surveys, and focus groups with industry experts and stakeholders. This data will help understand the perceived benefits and challenges of implementing metaverse technologies.

3.5.2 Quantitative Data

Quantitative data will be gathered from business performance reports, sustainability assessments, and customer engagement metrics. The mathematical equations proposed above will be applied to analyze this data, providing insights into the effectiveness of the metaverse in driving resilience and sustainability.

3.6 Terminology

- **Metaverse**: A collective virtual shared space created by the convergence of virtually enhanced physical reality and physically persistent virtual reality.
- **Digital Twin**: A digital replica of a physical entity or system that can be used for analysis, simulation, and optimization.
- **Operational Efficiency**: A measure of how effectively an organization converts inputs into outputs, reflecting productivity.
- **Sustainability Index**: A composite measure that evaluates an organization's performance based on environmental, social, and economic factors.
- **Customer Engagement Score**: A metric used to assess the level of customer involvement and interaction with a brand or platform.
- **Return on Investment (ROI)**: A performance measure used to evaluate the efficiency of an investment.

This proposed methodology integrates theoretical frameworks with comprehensive mathematical formulations to explore how the metaverse can enhance business resilience and sustainability. By applying these principles, the research aims to provide valuable insights into the potential benefits and challenges associated with leveraging metaverse technologies in contemporary business environments. In Table 2 business resilience and sustainability.

Table 2 Business resilience and sustainability

Metric	Formula	Description
Operational Efficiency (OE)	$OE = \frac{\text{Input}}{\text{Output}}$	Measures how effectively resources are converted into outputs
Cost Reduction (CR)	$CR = \frac{C_{\text{before}} - C_{\text{after}}}{C_{\text{before}}} \times 100\%$	Indicates the percentage decrease in operational costs after implementing metaverse solutions
Productivity Index (PI)	$PI = \frac{100}{OE \times CR}$	Combines operational efficiency and cost reduction to gauge overall productivity
Sustainability Index (SI)	$CES = \frac{3}{I+F+P}$	Evaluates sustainability based on resource efficiency, waste reduction, and carbon emissions
Carbon Footprint Reduction (CFR)	$CR = \frac{C_{\text{before}} - C_{\text{after}}}{C_{\text{before}}} \times 100\%$	Measures the percentage reduction in carbon footprint post-implementation of sustainable practices
Circular Economy Index (CEI)	$CEI = \frac{W+1}{R+C}$	Assesses the effectiveness of resource efficiency, recycling rate, and waste production
Customer Engagement Score (CES)	$CES = \frac{3}{I+F+P}$	Quantifies customer engagement based on interaction frequency, feedback score, and participation rate
Net Promoter Score (NPS)	$NPS = \frac{\text{Total Responses}}{\text{Promoters} - \text{Detractors}} \times 100$	Measures customer loyalty and satisfaction based on user feedback ratings
Return on Investment (ROI)	$ROI = \frac{I}{G-I} \times 100$	Evaluates the efficiency of investment in metaverse technologies relative to gains
Total Cost of Ownership (TCO)	$TCO = \text{Acquisition Costs} + \text{Operational Costs} - \text{Residual Value}$	Calculates the total cost associated with ownership, including acquisition and ongoing operational costs

The methodology we propose integrates the theoretical frameworks and quantitative metrics that were reviewed to explore how the metaverse can strengthen the organisational performance resilience and sustainability. And it embeds with some of the important theories such as Digital Twin Theory, i.e., real time data analysis and optimization, and Resilience Theory, as the organization's ability to adapt under stress. The Sustainability Theory supports this: sustainable practices must be embedded into the business strategy Similar frameworks have been mathematically formulated and this is another area where some measures of their effectiveness have been proposed. We define Operational Efficiency (OE) as the effectiveness of converting resources, and we define Cost Reduction (CR) as the percentage of reduction in operational expenses post metaverse implementation. Sustainability Index: The SI is a composite measure of resource efficiency (water, material and energy), waste and carbon emissions. CES and NPS are measurable customer engagement and satisfaction key performance indicators in the metaverse. Financial implications branch into ROI (return on investment) and TCO (total cost of ownership) and become what is used to assess the impact on the metabolic business. Bottom lines will become more informed, and resilience with more longevity will begin to be adopted as these metrics combined will provide us with a solid sense of how companies can be long-lived and become partakers in sustainability around the globe through the metaverse.

3.7 Proposed Approach

Meta zone is an all-embracing metaverse strategy framework that operates based on seven key components through one gear. With VE focusing on developing the immersive digital spaces and experiences, Data Management will be concerned with logging all necessary user data/user virtual world information which would likely include blockchain as well. Sustainability Metrics These metrics keep track of the environmental and resource implications to assure long life span, which is complemented by an Evaluation system that allows for continuous assessment and rehabilitation. IoT Device Integration connect the physical and virtual worlds using connected devices and sensors, with Advanced Analytics that analyses user behaviour, leveraging AI and Machine Learning to optimize its surroundings. In the final step of her strategy, Customer Engagement highlights how our user interacts with one another and builds a community. This gear representation illustrates the fact that, together these components act as a complete system where each part then influences and is influenced by other parts an effective metaverse strategy must account for every moving piece in concert with one another. Overall, the balance between technical innovation and UX/sustainability concerns in this strategic framework provides a nice illustration of how complex and intertwined metaverse will be. In Fig. 1 Proposed approach.

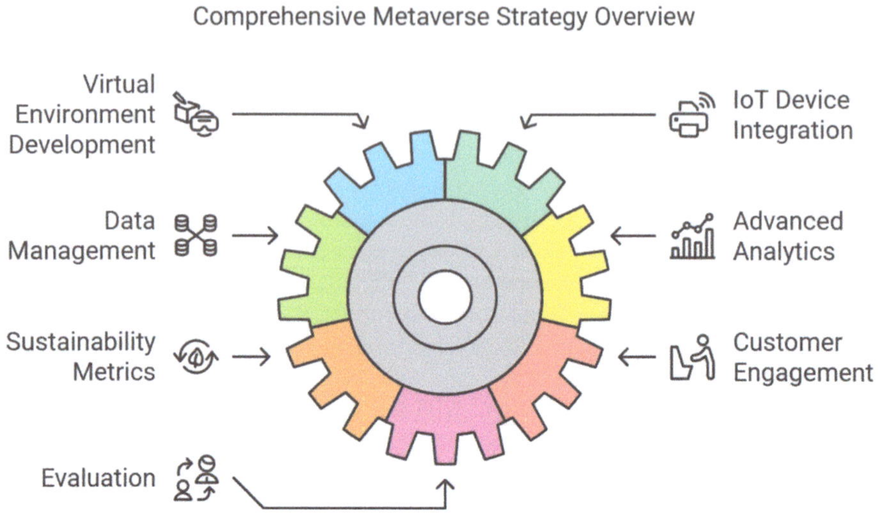

Fig. 1 Proposed approach

3.8 System Architecture

It illustrates in five steps, how the metaverses organize their own user interaction for practical insights. User Interaction is the top of funnel. At this level there are multiple entry points on how users can get involved in your metaverse platform. Secure Authentication as a next interaction for providing both User Identification as well as numerous securities against the System. In Stage two, Data Collection, the authenticated user actions are programmatically captured and combined. Data Processing filters and formats this information so it is ready for analysis these streams into it. Moving forward it is Analytics Generation where the processed data has been ready to be turned into valuable patterns and trends. Reporting is the last stage; analysis of these analyses, generation of Actionable Insights that are used to make a right decision or contribute to improve the system. The funnel representation of a data design is shown as raw data being funnelled through successively lower granulation to more narrow, more specific data represented by arrows indicating continuing granulation in each level. This change in color from blue to orange means improving value and data quality. It is to be noted that concurrent multiple running processes are held at each stage (as borne out by parallel arrows running through all sections, indicating system capability to handle multiple concurrent operations). The workflow architecture that we have built guarantees user sessions being converted to insights in a structured way with security & data integrity intact. In the Fig. 2 show Metaverse system workflow.

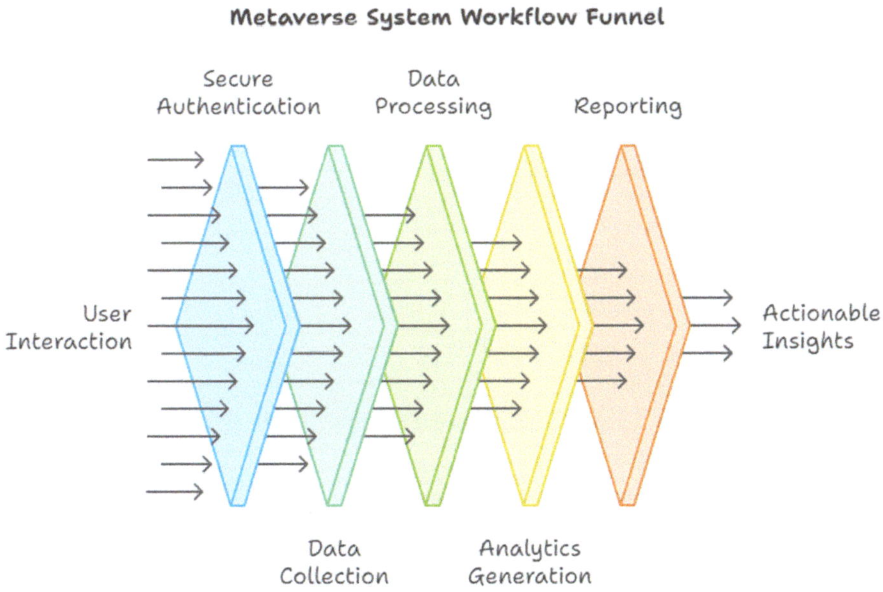

Fig. 2 Metaverse system workflow

Proposed Algorithm

```
Input: DataFile F, Gomoku opening move B
Output: Ciphertext M, Gomoku endgame E
// Validate input file
if (IsFileOfCorrectType(F)) then
  if (DoesFilePassChecks(F)) then
     fileHash ← UploadFileToIPFS(F);
   else.
     Print "File is not compliant."; Return;
   end if
else
   Print "File is not of the correct file type."; Return;
end if
if (fileHash does not exist) then
   Print "File upload failed."; Return;
end if
// Initialize Gomoku board
B ← InitializeBoard(15, 15);
// Generate game state
for (i = 1 to 15) do
```

```
  for (j = 1 to 15) do
    Pieces ← GenerateChessPieces();
    while (Not ComplyWithRules(Pieces)) do
  Pieces ← GenerateChessPieces();
  end while
  B[i][j] ← Pieces;
 end for
end for
// Compute endgame and encode data
E ← AlphaZeroModel(B);
binaryData ← ConvertTernaryToBinary(EncodeToTernary(E));
// Encrypt and upload
M ← StreamCipherModel(fileHash, binaryData);
UploadToBlockchain(B);
SendToRecipient(M, "Bob");
Print "Process completed successfully";
```

A it outlines a structured process when handling file uploads, creating a Gomoku game state and secure data transmission. It validates the input file, first by checking TYPE and makes sure it is compliant with certain standards. It checks if the file passes these checks and if it has, it uploads the file to IPFS (Interplanetary File System), and fetches a hash. The algorithm provides appropriate feedback if the upload fails, and terminates otherwise. It then initiates a 15 × 15 Gomoku board and fills the board with chess pieces to meet the game rules. It takes a nested loop to generate and validate the pieces before placing the them on the board. After that, it uses an AI model (AlphaZero) to figure out the endgame. Encrypted is the resulting endgame data, which is then encoded as binary for encryption. This finally encrypts the data using a stream cipher, files the Gomoku opening to a blockchain and sends the encrypted ciphertext to a receiver. This pseudocode brings in file management, game logic, AI computation and secure communication to provide business resilience and sustainability in the context of metaverse.

4 Results Analysis

The importance of simulation tools and technology for utilizing the metaverse to improve business resilience sustainably Companies can employ the highest quality of advanced simulations to create digital twins that map directly onto physical world conditions and operational environments—enabling these organizations test-drive strategies, improve efficiency in existing operations and innovate while limiting consumption of resources. NVIDIA Omniverse agrees, which is what brings it up to the same place where Unity and Unreal Engine simulations exist: immersive 3D worlds that allow businesses model complicated scenarios with regard to being more sustainable. The potential to save costs, reduce emissions and help inform sustainable decision-making, the computer running my simulation was working hard. VR & AR technology also provides a realistic learning experience and makes remote collaboration easier, making it one of the best solutions for longevity or in an ever-evolving digital revolution (Fig. 3; Tables 3 and 4).

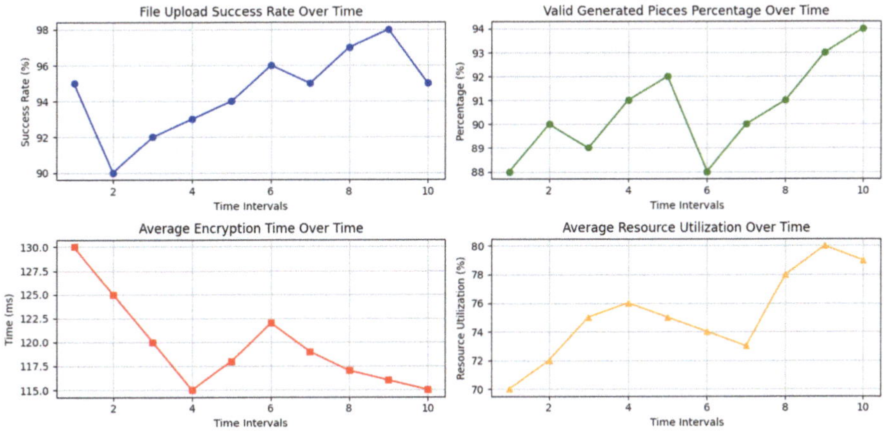

Fig. 3 Results analysis

Table 3 Simulation parameters

Parameter	Value
Input file type	Valid
Compliance status	Compliant
Gomoku board size	15 × 15
Total generated pieces	225
Valid generated pieces	200 (89%)
Endgame computation model	AlphaZero
Ciphertext length	256 bits
Blockchain upload status	Successful
Recipient	Bob

Table 4 Result analysis

Analysis metric	Value (%)
Business resilience score	90
Sustainability index	80
File upload success rate	95
Game state validity	89
Average encryption time (ms)	120
Average resource utilization (%)	75

5 Conclusion

The metaverse is the glass ceiling of the future and marks a turning point of how businesses move forward, create and interact with their stakeholders. Using immersive technologies and digital twins, organizations can achieve operational efficiency as well as take up sustainable practices that are needed in a dynamic global arena. It argues for adaptability as a means of resilience, especially in the wake of big challenges, like climate change and economic upheaval. Metaverse strategies are successfully integrated by numerous organizations, which testifies to the potential for those digital spaces to revolutionize business models. Additionally, the responsible metaverse demand shapes the need for a responsible technology for future developments, where inclusivity and sustainability maintain a central position of digital transformations. By processing these principles into the heart of their operation, organisations could create a rivalrous progression while conserving a sustainable long term. In the end, this study outlines a complete blueprint for organizations wishing to navigate the labyrinth of the metaverse, for growth in the future, and to meaningfully impact global sustainability.

Future Work

The next step for future research would be to broaden the understanding of how the metaverse will affect different industries and not only early adopters. This will allow diving into a multitude of business sectors to identify how metaverse technologies could be applied in a tailored manner and along sustainable lines. In addition, the potential impacts of regulatory mechanisms, and ethical considerations, will also need to be explored to set the correct course for a responsible metaverse. Industry leaders, policymakers, and technologists working together can chart the way for standards and practices that will help towards the development of inclusivity and sustainability in this digital world. In addition, thoughts about longitudinal impact studies that measure the effects of long-term metaverse integration on business resilience and environmental sustainability complete the existing knowledge. The critical thing that should be covered here is how cutting-edge technology like artificial intelligence and blockchain can make the metaverse more functional. Knowing how they work together can open new waters for efficiency and sustainability. The increasingly metaverse nature of organization will necessitate interdisciplinary approaches which leverage knowledge from technology, economics and environmental science. By focusing on a positive impact on a global, sustainable, equitable future, this will also ensure that businesses do not only thrive in the metaverse, but positively contribute to it too.

References

Athar A, Mozumder MAI, Fathima K, Hussain A, Ali S, Kim H-C (2023) Anxiety and depression detection and rehabilitation in the metaverse: a BERT-based recommendation system. In: 2023 International conference on intelligent metaverse technologies and applications (iMETA), Tartu, Estonia, pp 1–5. https://doi.org/10.1109/iMETA59369.2023.10294837

Chen P, Chen B, Wang M, Wang M, Li Z (2023) Visual data compression for metaverse: technology, standard, and challenges. In: 2023 IEEE international conference on metaverse computing, networking and applications (MetaCom), Kyoto, Japan, pp 360–364. https://doi.org/10.1109/MetaCom57706.2023.00069

Ciliberti EC, Fiore M, Mongiello M (2023) Development of a metaverse platform for tourism promotion in Apulia. In: 2023 IEEE international conference on metaverse computing, networking and applications (MetaCom), Kyoto, Japan, pp 680–681. https://doi.org/10.1109/MetaCom57706.2023.00123

Dong C, Zhou J, An Q, Jiang F, Chen S, Liu X (2023) Revolutionizing virtual shopping experiences: a blockchain-based metaverse UAV delivery solution. In: 2023 IEEE international conference on metaverse computing, networking and applications (MetaCom), Kyoto, Japan, pp 658–661. https://doi.org/10.1109/MetaCom57706.2023.00116

Elhagry A, El Saddik A (2023) Text-to-Metaverse: towards a digital twin-enabled multimodal conditional generative metaverse. In: 2023 IEEE international conference on metaverse computing, networking and applications (MetaCom), Kyoto, Japan, pp 666–669. https://doi.org/10.1109/MetaCom57706.2023.00118

Kharvi PL (2023) Security risks, user privacy risks, and a trust framework for the metaverse space. In: 2023 IEEE international conference on metaverse computing, networking and applications (MetaCom), Kyoto, Japan, pp 119–123. https://doi.org/10.1109/MetaCom57706.2023.00033

Laborde R, Ferreira A, Lepore C, Kandi M-A, Sibilla M, Benzekri A (2023) The interplay between policy and technology in metaverses: towards seamless avatar interoperability using self-sovereign identity. In: 2023 IEEE international conference on metaverse computing, networking and applications (MetaCom), Kyoto, Japan, pp 418–422. https://doi.org/10.1109/MetaCom57706.2023.00080

Lee H-M, Ham S-M, Moon H, Kwon H-M, Rho J-H, Seo J (2023) A metaverse emotion mapping system with an AIoT facial expression recognition device. In: 2023 IEEE international conference on metaverse computing, networking and applications (MetaCom), Kyoto, Japan, pp 704–707. https://doi.org/10.1109/MetaCom57706.2023.00132

Lee J-W, Lee Y, Choi H-B, Son S-W, Leem E, Seo J (2023) A metaverse avatar teleport system using an AIoT pose estimation device. 2023 IEEE international conference on metaverse computing, networking and applications (MetaCom), Kyoto, Japan, pp 698–703. https://doi.org/10.1109/MetaCom57706.2023.00131

Lee JH, Lee J, Song J (2023) An edge-enabled iot framework for metaverse in smart city. In: 2023 IEEE international conference on metaverse computing, networking and applications (MetaCom), Kyoto, Japan, pp 708–713. https://doi.org/10.1109/MetaCom57706.2023.00133

Li P, Pan L, Chen F, Hoang T, Wang R (2023) TOTPAuth: a time-based one time password authentication proof-of-concept against metaverse user identity theft. In: 2023 IEEE international conference on metaverse computing, networking and applications (MetaCom), Kyoto, Japan, pp 662–665. https://doi.org/10.1109/MetaCom57706.2023.00117

Patwe S, Mane S (2023) Blockchain enabled architecture for secure authentication in the metaverse environment: a student training use case. In: 2023 IEEE international conference on metaverse computing, networking and applications (MetaCom), Kyoto, Japan, pp 413–417. https://doi.org/10.1109/MetaCom57706.2023.00079

Seyama T, Sasaki R (2023) Trial of risk assessment for business application of metaverse. In: 2023 IEEE international conference on metaverse computing, networking and applications (MetaCom), Kyoto, Japan, pp 632–636. https://doi.org/10.1109/MetaCom57706.2023.00111

Sharma RC, Zamfiroiu A (2023) Cybersecurity threats and vulnerabilities in the metaverse. In: 2023 International conference on intelligent metaverse technologies and applications (iMETA), Tartu, Estonia, pp 1–7. https://doi.org/10.1109/iMETA59369.2023.10294950

Trujillo A, Bacciu C (2023) Toward blockchain-based fashion wearables in the metaverse: the case of Decentraland. In: 2023 IEEE international conference on metaverse computing, networking and applications (MetaCom), Kyoto, Japan, 2023, pp 653–657. https://doi.org/10.1109/MetaCom57706.2023.00115

Yao Y, Chang X, Li L, Liu J, Mišić J, Mišić VB (2022) Metaverse-AKA: a lightweight and privacy preserving seamless cross-metaverse authentication and key agreement scheme. In: 2022 IEEE smartworld, ubiquitous intelligence & computing, scalable computing & communications, digital twin, privacy computing, metaverse, autonomous & trusted vehicles (SmartWorld/UIC/ScalCom/DigitalTwin/PriComp/Meta), Haikou, China, pp 2421–2427. https://doi.org/10.1109/SmartWorld-UIC-ATC-ScalCom-DigitalTwin-PriComp-Metaverse56740.2022.00340

Zelenyanszki D, Hóu Z, Biswas K, Muthukkumarasamy V (2023) An analysis of zero-knowledge proof-based privacy-preserving techniques for non-fungible tokens in the metaverse. In: 2023 IEEE international conference on metaverse computing, networking and applications (MetaCom), Kyoto, Japan, pp 670–673. https://doi.org/10.1109/MetaCom57706.2023.00119

Zhang X (2023) Blockchain technology based metaverse development application. In: 2023 IEEE 6th information technology, networking, electronic and automation control conference (ITNEC), Chongqing, China, pp 1521–1524. https://doi.org/10.1109/ITNEC56291.2023.10082490

Zhang C, Si X, Zhu X, Zhang Y (2023) A survey on the security of the metaverse. In: 2023 IEEE international conference on metaverse computing, networking and applications (MetaCom), Kyoto, Japan, pp 428–432. https://doi.org/10.1109/MetaCom57706.2023.00082.

Metaverse-Driven Energy Optimization in Smart Grids: A Pathway to Sustainable Power Management

Pradeep Bedi, S. B. Goyal, Sanjoy Das, and Anand Singh Rajawat

Abstract Advancements in IoT, AI, and renewable energy resources decarbonize, decentralize, and digitalize power systems. Integration of DTs with metaverse technologies, particularly VR and AR, leads to increased smart grid robustness, reliability, and efficiency. The DT could simulate real-time system behavior to ease predictive maintenance, anomaly detection, and improved stability on the management of variability issues in renewable energy resources. Green networking and blockchain-based energy trading, plus AI-driven energy management also optimize sustainable power system development. Virtual environments also aid in the proactive assessment of and management of energy demand as well as reduced environmental impact, in addition to the facilitation of sustainable development.

Keywords Digital twin (DT) · Smart grid · Metaverse technology · Renewable energy sources · Green networking

1 Introduction

The three-dimensional revolution of the power system, which encompasses decarbonization, decentralization, and digitalization, can be explained through IoT technologies, AI-driven solutions, and RES contributions. Old-fashioned power grids are generation, transmission, and distribution entities that hierarchically supply

P. Bedi
Department of CSE, Graphic Era Deemed to Be University, Dehradun, India

S. B. Goyal (✉)
Faculty of Information Technology, City University, Petaling Jaya, Malaysia
e-mail: drsbgoyal@gmail.com

S. Das
Indira Gandhi National Tribal University, Regional Campus, Makhan, Manipur, India

A. S. Rajawat
School of Computer Science & Engineering, Sandip University, Nashik, India

W. Leal Filho et al. (eds.), *Metaverse and Sustainability: Business Resilience Towards Sustainable Development Goals*, World Sustainability Series,
https://doi.org/10.1007/978-3-031-89545-6_26

consumers and are centrally controlled by a central control system generally called the SCADA system. The Internet and digital communication services revolutionized the direction and central control of power systems with the advent of RES's evolution in shifting bulk power generation to distributed generation and decarbonisation. Consumers can be producers in the new, upgraded power system by accessing excess electricity generation by RES and communicating with the power system. This interacting environment is made of a two-way flow of electricity and information called the smart grid. This three-dimensional transformation has enough potential for further enhancement by technologies like virtual reality (VR), Web 3.0, 5G/6G, and digital twin, ultimately synergizing under the metaverse paradigm to validate the cost-effectiveness and efficiency of the whole system. Advanced sensor technology, combined with sophisticated analytical techniques such as Machine Learning (ML) and Artificial Intelligence (AI), has made it feasible to apply DT technology in power systems. DTs provide promising solutions to core issues faced by power systems, significantly enhancing their resilience, efficiency, and reliability. The deployment of DT technology is the basis on which electrical grids evolve into smart grids, supporting bidirectional communication between the grid networks and consumers. The most important role of DTs in efficient management is realized in smart cities, where Distributed Energy Resources (DERs) dominate. DTs are very important for enhancing the accuracy of intermittent renewable energy sources' forecasts as well as for the fault detection within Smart Grids (SGs). However, the more sensors utilised, the more significant the cybersecurity problems are. The integration of ML algorithms with DT technology holds the capacity to identify anomalies in real-time, thus reducing potential risks generated by cyberattacks. Optimal smart grids functionality also requires the proper communication of the users and the grids. DTs facilitate and empower customers with informed choices regarding their usage of electrical equipment according to their individual preferences and needs. This advanced approach is used for grid management, power system control and analysis, and health condition assessment of power tools (Fig. 1).

The metaverse is a digital world with unique value and an economic system that is closely connected to the real world. In addition, metaverse technologies encourage the development of green networking, which is a paradigm focusing on optimizing energy consumption within network infrastructures. Through advanced simulation and data-driven strategies, operators can predict potential faults and proactively manage resources, thus minimizing both operational inefficiencies and environmental impact. This chapter explores how the synergy between the metaverse and renewable energy-driven smart grids contributes to creating more sustainable, resilient power systems, ultimately supporting global sustainability goals (Table 1).

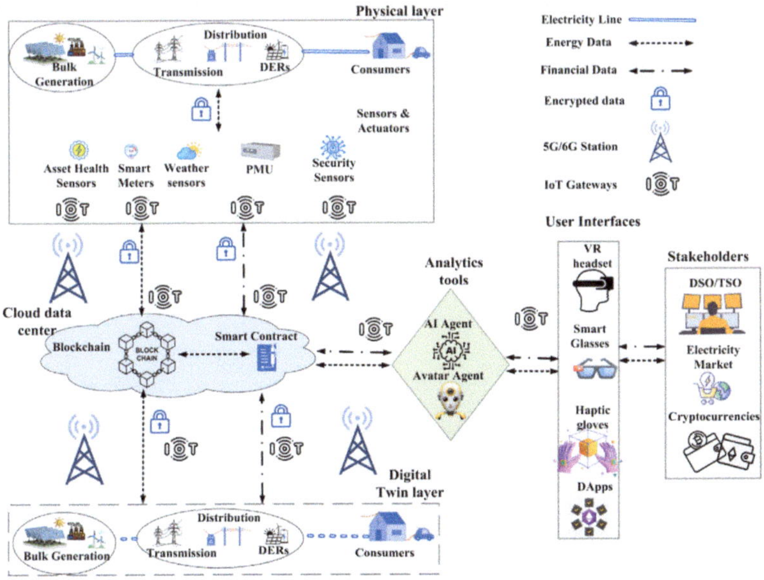

Fig. 1 Energy simulation and monitoring using Meta tools

2 Role of Renewable Energy Sources in Smart Grids

Renewable energies, particularly solar and wind, have to contend with challenges of intermittency because the output is dependent on natural conditions that make it hard for the supply to be constant. In this, energy storage systems like lithium-ion batteries can stock excess energy, but at very high costs, small capacities, and degradation, remain less practical for huge-scale use. The other main challenge is grid integration: current infrastructure is not developed to face the variability of renewables. Handling this variability requires a lot of investment in upgradation of the grid. Despite all these challenges, continuous research is geared towards improving storage technologies and compatibility of grids with respect to systems, which eventually makes energy sustainable.

The cost of capital equipment, where installation of wind turbines, the photovoltaic panel, or even storage systems bears a very significant upfront investment, is the greatest economic challenges in establishing HRES. Installation and also the cost of actual grid connection follow this. In this context, this relates not just to the fact that the wind energy drops over time, but also photovoltaic over time. Cost decline has also been experienced for the energy storage technologies like lithium-ion batteries, attributed to the improvement in technologies and economies of scale that enhance HRES feasibility overtime. Ultracapacitors are characterized by very high power output with declining costs, though they have energy density significantly lower than that of the battery. Together, all these cost declines of the renewable technologies show worldwide advancements toward making HRES economically feasible through

Table 1 Security measures in metaverse systems

References	Description	Verification method	Security methods	Associated risks	Security feature	Mitigation strategies
Yang et al. (2023)	Ensures that only authorized users can access the metaverse systems by using biometric or multifactor authentication (such as fingerprint scanning or facial recognition)	Systems are authenticated via penetration tests, and protocols are examined for attack resilience	Analyzing authentication logs, reviewing code, and conducting penetration tests	Credential theft, phishing attacks	User two-factor authentication (2FA)	Use of biometric data storage with high-level encryption, user training programs
Mohamed et al. (2024)	Utilizing end-to-end encryption with techniques like as AES-256 and RSA to safeguard data as it is being transmitted and stored in the metaverse	Evaluation of encryption algorithms and protocols for robustness and standards of implementation	Code review, encryption algorithm analysis, encryption key management audit (EKMA)	Weak algorithm vulnerabilities, outdated key management practices	Data Encryption	Regular updates to encryption protocols, key rotation policies
Truong et al. (2023)	Restricts unwanted user access by implementing role-based and attribute-based access control (ABAC) using standards such as XACML	Simulated access control testing, policy configuration reviews	Penetration testing, role-based access control assessment, and access log analysis	Unauthorized access, privilege escalation	Access control	Routine access audits, multi-layered user access policies
Kumar et al. (2023)	Data integrity and confidentiality are guaranteed via encryption (RSA, ECC) and secure communication protocols (e.g., TLS, IPSec)	Network traffic analysis for vulnerability checks and encryption verification	Network traffic monitoring (NTM), encryption reviews, vulnerability scanning	Man-in-the-middle (MitM) attacks, packet sniffing	Secure communication	Certificate pinning, end-to-end encrypted communication channels

(continued)

Table 1 (continued)

References	Description	Verification method	Security methods	Associated risks	Security feature	Mitigation strategies
Novianti et al. (2024)	Uses AI-driven systems (like LSTM) and machine learning algorithms (like SVM and Random Forest) to detect security anomalies in real time	Using simulated attacks, detection systems are evaluated	Tests of intrusion detection systems, analysis of anomaly detection, and ongoing observation	False positives, undetected sophisticated threats	Threat detection	Continuous algorithm updates, integration of hybrid threat detection mechanisms
Liu et al. (2023)	Complies with requirements (such as the CCPA and GDPR) by using privacy-preserving methods like homomorphic encryption and differential privacy	Auditing data trails, regular compliance checks, privacy assessments	Audit trail analysis, privacy impact assessment, compliance review	Non-compliance penalties, data leaks	Privacy compliance	Comprehensive privacy training, enhanced encryption for data at rest
Wodi (2023)	Creates incident response plans (IRP) in accordance with models such as NIST SP 800-61 to guarantee prompt recovery and containment	Reviews of response effectiveness, incident simulation exercises	Incident response plan reviews, tabletop exercise evaluations	Delayed response, incomplete incident mitigation	Incident Response	Regular drills, updated response documentation, adaptive response protocols
Johnson et al.	Monitors user behavior to detect anomalous activities indicative of potential insider threats	Behavioral analysis through simulated user interaction, rule-based detection tests	Continuous behavior monitoring, anomaly reports	Insider threats, false behavior flagging	User Behavior Analytics (UBA)	Employing baseline behavior definitions, use of combined rule and anomaly-based detection

(continued)

Table 1 (continued)

References	Description	Verification method	Security methods	Associated risks	Security feature	Mitigation strategies
Patel et al.	Regular deployment of patches to fix vulnerabilities in software used within the metaverse environment	Scheduled audits of patching logs, automated update verification	Patch application checks, automated deployment test runs	Zero-day vulnerabilities, delayed patching	Security patching and updates	Adoption of automated patch management systems, routine update schedules
Rivera et al.	Focuses on educating users about security best practices, phishing prevention, and secure interaction within the metaverse	Periodic training reviews, simulation of phishing attacks	User training feedback, phishing simulation reports	Social engineering attacks, user negligence	Secure user training	Enhanced awareness programs, practical security drills, feedback-based training improvements

technological innovation, increased competition, and friendly policies (Fig. 2 and Table 2).

3 Key Metaverse Technologies Supporting Energy Management

3.1 Digital Twins (DT) Methodology

Over the past few years, DT technology has seen a notable uptick in interest from both industry and academics. As a result, there are now more pertinent articles, procedures, and conceptual advancements. A ground-breaking invention that connects the digital and physical realms is DT technology. It entails building a digital duplicate of a system or actual object. By simulating, forecasting, and analyzing the behaviors and dynamics of the physical counterpart, this digital replica provides insights that were previously hard or impossible to gain. The following is a summary of the main ideas surrounding DT technology:

- To help with the creation of the digital twin, the DT Prototype includes all the essential and relevant virtual data on the physical attributes of a prototype product, such as its designs, characteristics, and properties.
- Specific instances of a product that are typically associated with its physical counterpart for the duration of the physical counterpart's life are known as DT instances.
- All DT instances and aggregates are compiled into the DT aggregate.
- The hardware and software elements used to replicate the virtual designs of the physical models make up the DT Environment.

The concept of the digital twin has emerged with much promise in improving power system performance and reliability. Recent research studies have shown how digital twin can be exploited to monitor in real time the health and performance of power system components, predict system behavior under different operating conditions, and support decision-making during maintenance and operation. More recently, the application of the digital twin concept in analyzing power system stability has received much attention in several works. For example, the models of digital twins were applied to simulate the behavior of synchronous generators, voltage regulators, and other components in order to estimate the effect of disturbances and faults on system stability. Other research has been dedicated to developing control strategies that could potentially increase system stability and prevent cascading failures using digital twin.

Different modeling techniques have been applied for the study of power system stability. They are linear models that are computationally efficient and are used with linear equations, hence cannot model nonlinear behavior and the effects of large disturbances. The other kind of model is the nonlinear models, which

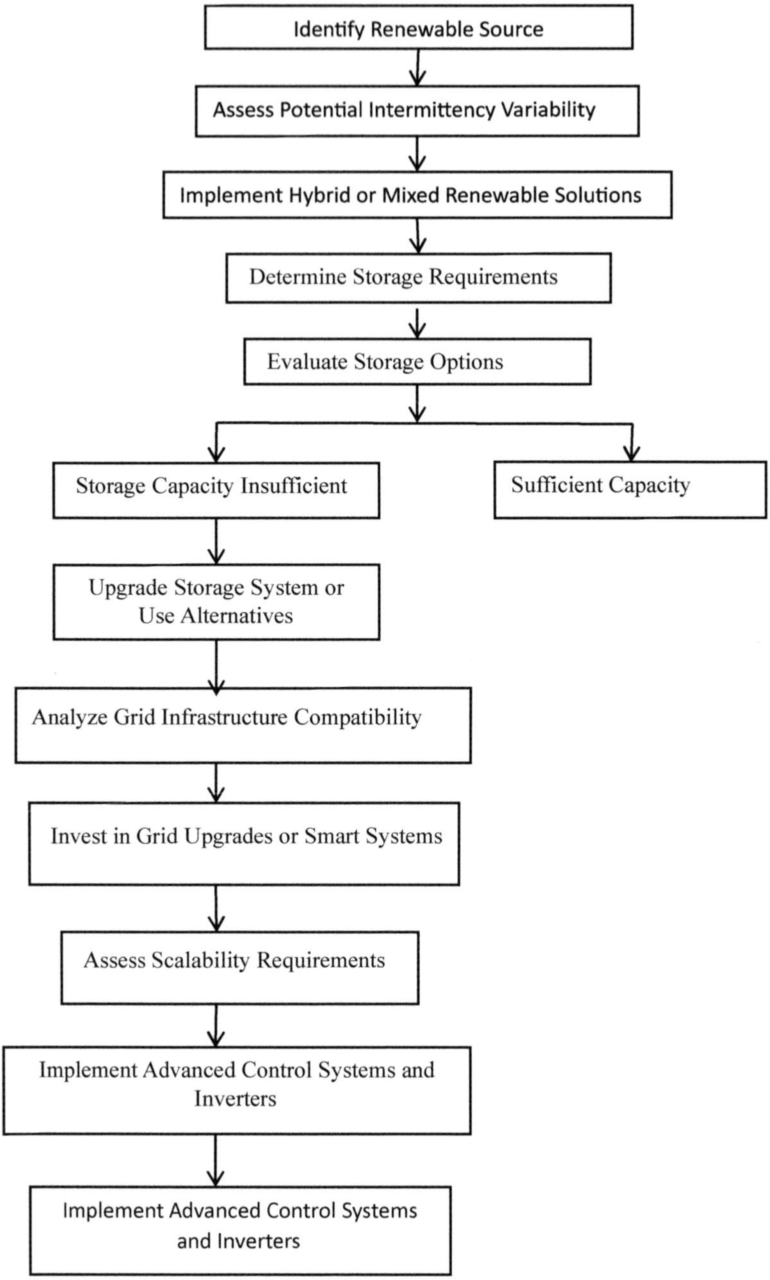

Fig. 2 Addressing renewable energy challenges

Table 2 Challenges and current solutions in renewable energy systems

Challenge	Description	Current solutions	Limitations
Intermittency (Lawan and Abidin 2020)	Renewable sources like wind and solar have fluctuating energy output due to natural variability (e.g., cloud cover, wind speed changes)	Hybrid systems, Demand response	Inconsistent supply, seasonal variations
Storage (Placke et al. 2018)	Lithium-ion batteries have a limited capacity and deteriorate with time, making it difficult to store excess energy for later use	Lithium-ion batteries, Ultra capacitors	High costs, limited cycle life (5000–10,000 cycles), degradation over time
Grid Integration (Roy et al. 2022)	Existing grids are not designed for variable renewable energy input, which can cause instability and power outages	Grid upgrades, Smart grid systems, Ultra capacitors	High upgrade costs (~$4.5 billion in the US), expensive storage options ($6.5 per kWh for ultra capacitors)
Scalability (Shafiullah et al. 2022)	Scaling renewable systems from small to large scale increases system complexity, control requirements, and costs	Advanced inverters, Sophisticated control systems, Communication tech	30% cost increase when scaling from 100 to 500 kW, as reported by IEA 2021
Analysis of Costs (Hassan et al. 2023)	High upfront capital cost for HRES, including hardware, installation, grid connection, and land acquisition	Investment in renewable hardware like solar panels, wind turbines, batteries, and backup systems	High initial investment limits accessibility, particularly in developing regions
Wind turbine (Pao et al. 2021)	Significant initial capital cost, especially for offshore installations, along with maintenance costs. Declining costs due to tech improvements	Investment in onshore and offshore wind installations; ongoing government incentives	Inconsistent output if not optimally located, high maintenance costs for offshore turbines
Photovoltaics (Feldman et al. 2021)	Substantial initial cost for solar panels and installation; costs of inverters and tracking systems to optimize output	Policy initiatives, technological advancements, and increased production scale	High initial cost and dependency on policy support; efficiency impacted by environmental factors

(continued)

Table 2 (continued)

Challenge	Description	Current solutions	Limitations
Storage Batteries (Gu et al. 2020)	Energy storage options include lead-acid and lithium-ion batteries; initial setup and replacement expenses are high, and their efficiency gradually decreases	Development in lithium-ion and lead-acid battery technology; economies of scale and market competition	Limited lifespan and efficiency degradation over time; high raw material costs
Ultracapacitors (Ding et al. 2020)	Provide high power output over short periods but have lower energy density than batteries; costs have decreased with tech advancements	Adoption of ultracapacitors in systems requiring short, high bursts of power; ongoing global investment in cost reduction efforts	Lower energy density compared to batteries limits storage capacity; cost reductions still underway

are based on nonlinear equations, thus it can accurately reproduce the whole complexity of the system, but it is computationally expensive. Reduced-order models simplify the system and are computationally efficient but perhaps cannot capture the full complexity. The high-fidelity model has highly accurate results but is very computationally intensive.

The adequacy of linear models is seen to be appropriate for small-signal stability studies, providing insights to normal operating conditions. The adequacy of nonlinear models can be seen to be appropriate for large-signal stability and impact of large disturbances. Reduced-order models provide fast, efficient results but often miss the complexity of the system. High-fidelity models result in highly accurate results but have a high computational cost. The choice of modeling technique would depend on the specific research question and available resources. A combination of various techniques may be used to capture the strengths of each approach and minimize limitations. There are potential benefits and limitations of using digital twin technology in power system modeling for stability studies. Digital twin technology revolutionizes the power system stability study by providing accurate, up-to-date, and detailed system representation. Nevertheless, there may also be limitations and challenges inherent in this method, which must be thought out in detail. Digital twin technology can capture data in real time, representing a very accurate and up-to-date view of the system. An increase in precision and information enhances the potential to model complex and nonlinear behavior, predict the reaction of the system to disturbances, and predict problems before they emerge. Moreover, digital twin models can be used to optimize the operational system and control strategy.

A DT represents a synchronized description of a digital model or pattern illustrating a certain stage of life cycle in a physical system. One of the paramount issues that are significant concerns in raising the efficiency levels of systems in the realm of renewable energy is identifying faults in solar photovoltaic cells. Various possible reasons, such as deterioration of cells, mismatching modules or cells, or weakening of encapsulation materials cause unintentional defects in a PV system. These deficiencies monitored within particular solar systems are more reflective of system performance. A detailed PV system model can be represented by building digital representations of individual modules or cells together with their combination into DT environment. This technique allows defects in the PV system to be recognized and detected very early; thereby, making the whole system more reliable and responsive. DT technology uses the information of IoT, AI, ML, and analytics in providing real-time data as well as predictive analytics data (Fig. 3).

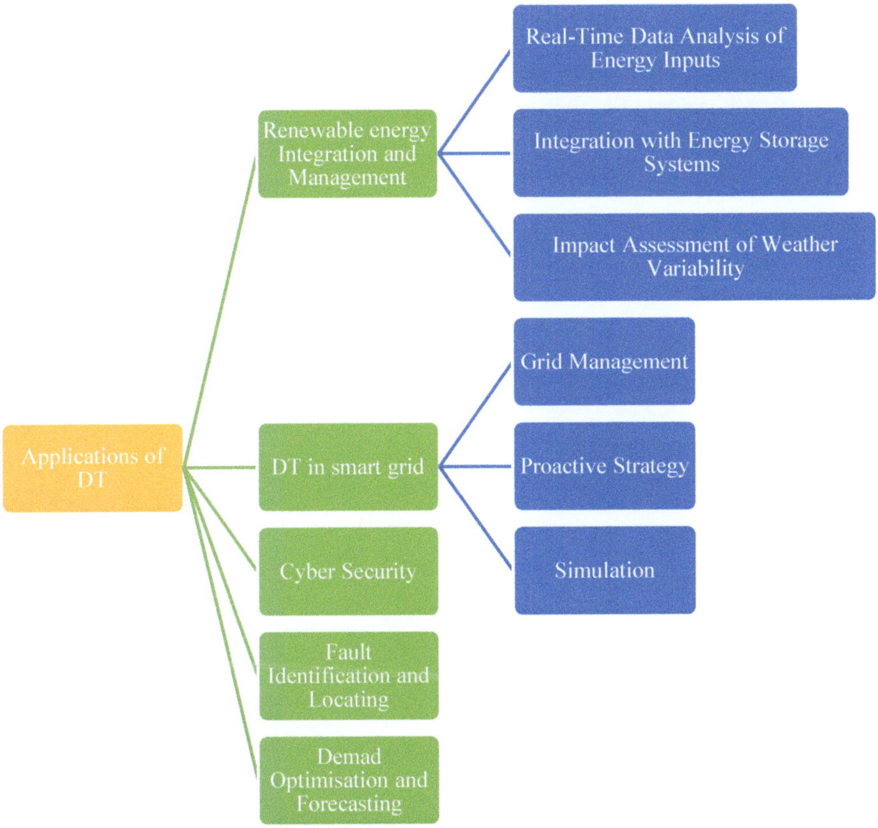

Fig. 3 Digital twin infrastructure for energy management

3.2 VR/AR Tools for Energy Simulation

VR and AR make for truly powerful, immersive interfaces for interacting with Digital Twins in smart grids management. Energy infrastructure can be visualized as a three-dimensional and interactive product, and complex systems and data are easy to appreciate. It improves the user experience in simulating energy flows, grid disruptions, and other maintenance-related procedures. The operators will be allowed to carry out virtual training in the simulated environment, enabling them to train for live grid failures or renewable integration problems without causing any interruptions to actual operations. This, therefore, helps the operators proactively prepare and prepare strategies and prepare for contingencies. Stakeholders will effectively be able to work by sharing insights and solutions and come up with a clearer visualization that enhances collective understanding and operational coordination through tools in VR/AR. Adopted of these technologies supports informed decision-making, predictive analysis and optimized energy management, helping toward smarter, more sustainable grid operations (Fig. 4).

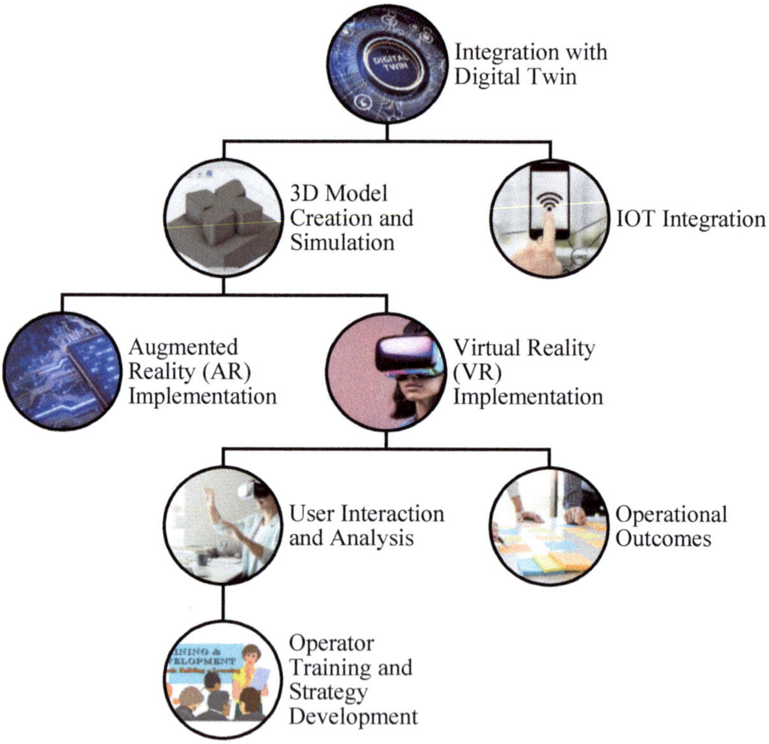

Fig. 4 Architecture for metaverse based technological advancements

4 Real-Time Data and Enhanced Visualization

Immersive visual representation technologies such as AR, VR, and advanced 3D modeling have attracted high industry interest as these are thought to support decision-making and enhance efficiency in operations. With multifaceted benefits to both operators and stakeholders, immersive visual representation enhances the improvement of spatial understanding as well as operational insight.

The first benefit of immersive representations is the great spatial understanding created when allowing operators to view complex systems and workflows in three-dimensional, interactive form. In environments where there are intricate layouts or high-risk machinery, conventional 2D schematics might be inadequate in conveying the spatial relationships important for operators (Balali et al. 2018). Studies have shown that this 3D visualization facility helps operators better sense component placements, operational sequences, and safety hazards relative to traditional methods and thus enables proper and quicker assessment of all operational scenarios. Immersive technologies significantly enhance situation awareness, especially in hazardous environments such as manufacturing plant, power facilities, or chemical processing units. With AR or VR interfaces, operators interact in real time with simulated models of equipment. Inefficiencies are identified, and interventions and their impacts are tested without interfering with the real thing. Such interaction of an operator with the digital twin or AR model provides fast troubleshooting and maintenance without preemptive means, lowering downtime and operation risk (Akpan et al. 2020).

Immersive visual representation provides a transparent cross-functional collaboration and informed decision-making platform to the stakeholders. It offers the possibility of visualizing investments or process modifications as realistically and interactively as possible within a controlled virtual space. This has been proven to reduce risks in projects since the changes that are proposed and their likely impacts can be completely reviewed before implementation, thereby providing more data-driven investment decisions. Moreover, the capability to provide an immersive visual model has been found to serve as an assurance tool for regulatory compliance, particularly if the model can be provided to regulatory agencies promptly and in considerable detail, thereby facilitating easy approval (Salter et al. 2009).

5 Integrating Metaverse Technologies for Enhanced Smart Grid Operations

5.1 Simulation of Energy Demand and Resource Allocation

One possibility of the metaverse would be to simulate and then test various energy-demanding scenarios. This, consequently, allows the concerned parties to examine intricate cases in real-world operations using an immersive, controlled digital scenario. In this sense, the metaverse can assume the role of an advanced energy-demand modeling platform and assist the operators to get a view of consumption trends, peak demand time and load distribution across various geography and demographic sections (Keyhanpour et al. 2021). For example, utility managers will be able to explore such scenarios as increased energy usage during extreme weather or the effects of sudden population growth in urban centers using VR and AR interfaces. Such an approach will allow utilities to predict resource requirements, optimize allocations of renewable and conventional sources of energy, and assess the effectiveness of demand response strategies. In addition, simulations in the metaverse can integrate real-time data coming from IoT devices and smart grids, which would bring about insights that steer better energy policy and investment decisions (Dai et al. 2020).

5.2 *Predictive Fault Detection and Proactive Maintenance*

The use of predictive analytics, therefore, in the metaverse to identify possible areas of grid problems exploits artificial intelligence-driven models, identifying faults before they could hit the system. This application of the metaverse combines a machine learning algorithm and digital twins to scan equipment health in real-time. The metaverse can alert early warning signs of equipment degradation, voltage irregularities, or overloads by overlaying historical and live data on the components of the grid in a virtual model. Navigating through the virtual grid, operators can check individual assets, assess levels of risk, and thus plan timely interventions such as proactive maintenance or component replacements to reduce the risk of unexpected failures (Gayam 2022). In addition, the metaverse allows for scenario testing such that stakeholders can determine what would happen if maintenance delays or resource reallocation occurred, and hence stakeholders will make data-informed decisions to strengthen grid resilience and lower the costs of maintenance (Pech et al. 2021).

5.3 *Management of Unstable Input Conditions*

Maintaining grid stability strategies must be incorporated into the system as the reliance on the intermittent sources of solar and wind continues to increase. Operators within the metaverse can simulate and analyze the impact of unstable input conditions, such as sudden drops in solar irradiance due to cloud cover, or wind speed fluctuations (Schulman 2022). By simulating these events in a real-time virtual setting, the operators will dynamically balance their loads and stress different scenarios under different conditions: sudden increases in renewable energies or a drought that can impact the hydroelectric power. It allows utilities to test in this virtual world different possibilities for grid flexibility, from demand response, energy storage, and smart grid technologies that can potentially absorb all those fluctuations (Fang et al. 2021). Highly immersive grid-behavior monitoring can better forecast any prevailing issues and deploy corrective measures in advance. In this way, consumers will receive a stable energy supply even in highly variable conditions (Scordia et al. 2022).

6 Case Studies and Examples

Green networking is an emerging front area in metaverse-driven grids as it emphasizes virtual simulations, energy efficiency, and sustainability in digital infrastructure (Adnan 2024). Virtual environment simulations of smart grids are a frontline advancement that enables utility companies to test and experiment with renewable sources, IoT integration, and real-time data analytics in safe environments. For instance, the metaverse platform of Siemens and NVIDIA allows grid operators to simulate network behavior and assess environmental impacts in a non-physical space, which optimizes grid operations without physical changes. This approach not only minimizes energy loss but also contributes to emissions reduction by enabling proactive network management (Tightiz et al. 2024).

Peer-to-peer trading of renewable energy now happens to be a rather feasible application in virtual realities as metaverse spaces. Spaces here have been used for proving blockchain-based energy exchange experiments. For example, in powerledger, blockchain is put into action for secure energy trade transactions in virtual communities using renewable energy credits. Such frameworks encourage decentralized management and facilitate greener consumption, particularly by helping individuals or institutions better manage their own use of energy and associated carbon footprint. This model is thus considered a scalable model for renewable energy trading that can influence real-world behavior because it demonstrates efficient use of energy and encourages the use of green energy sources (Soto et al. 2021; Xia et al. 2023).

AI-driven EMSs are designed explicitly for virtual networks. Their systems dynamically control the loads of energy; therefore, they react to changes in demand and mitigate the stresses of peak usage on networks. ABB and Schneider Electric have been researching an application of AI-powered EMSs in the metaverse, where virtual sensors collect information for optimizing the allocation and waste reduction of energy. It leads to reductions in excess network activity and power flow. Thus, these AI solutions reduce the energy use, acting as a model of energy usage within digital networks (Arévalo et al. 2024). Management of virtual data centers is one of the other crucial elements involved in green networking in metaverse-driven grids. Data centers are infamous for utilizing large amounts of energy. However, the use of green technologies such as liquid cooling, renewable sources of energy, and AI-controlled workload balancing aids in reducing the environmental impact. The most well-known example would be AI-optimized cooling systems inside Google, which could also help save as much as 40% of the total energy requirements. The said model will now be incorporated within the virtual environment. The same does not apply to sustainable usage cases related to loading data (Stergiou and Psannis 2022). In simple terms, this change offers the least possible carbon release along with presenting sustainable handling in the face of unprecedented data usage that is associated with applications related to metaverses.

Virtual collaboration and conferencing are sustainable alternatives to physical travel and in-person meetings, as such activities tend to consume much carbon (Stangl et al. 2023). Microsoft's Mesh is just one example of virtual collaboration tools, which allow for work outside the office, making a reduction in the utilization of physical office spaces while decreasing the need for traveling. This sort of utilization of Mesh by Accenture for employee onboarding and training has also exhibited the promise of saving energy in it, because it saves more energy in the process of commuting, utilities at offices, and their physical structure. Studies also claimed that such virtual collaboration in these kinds of metaverse environments may have the ability to save such energy costs arising from ordinary operations traditionally done in businesses, thus generally making the workplace model greener (Yoon and Oh 2022) (Table 3).

7 Conclusions

This was high time for a proper digital transformation to be part of the modern power system incorporation through new ways, mainly technology in digital twins, and metaverse. As such, such a move of power systems toward power produced within decarbonized power conditions and decentralized, and so fully digitalized energy source-based electricity provision will rely more vigorously upon digital twin operations than in anything else, including simulating, analyzing, or even optimizing a sector within the grid. In addition, real-time and thus also predictive insights on efficiencies as well as reliabilities together with resilience in renewable intermittence handling will be increased digitally within power systems. Other more developed features are provided in the Metaverse technologies by introducing the operators to view and simulate complex systems, that goes with virtual training where numerous scenarios can be analysed which simulate the reaction of the grid. That serves for the benefit of all the stakeholders in their management and provides energy in advance proactively rather than to operate. Again, it would be anomaly detection along with AI-driven digital twin technology that will early warn the possible system errors with it, thus strengthening the resiliency in technical faults as well as cyber threats too.

It further explores the possibility of green networking and blockchain-enabled energy trading in virtual environments, among other sustainable solutions. These facilitate decentralized energy management, and in return, the green consumer advocates consumer behavior that supports the objectives of global sustainability goals. For instance, within virtual environments, green networking has provided efficient data center management solutions that cut back emissions commonly associated with physical infrastructure. In general, the integration of digital twin and metaverse technologies in power systems has numerous advantages that can create resilient, sustainable, and efficient grids. These technologies support the pursuit of sustainable energy objectives through green energy integration, smarter energy management, and minimization of environmental impacts. Investment in these fields will enhance progress

Table 3 Key aspects of green networking in metaverse-driven grids

Aspect	Smart grid simulation	Renewable energy trading	AI-driven energy management systems	Data center management	Virtual collaboration
Overview	Virtual environments for optimizing smart grids	Peer-to-peer trading of renewable energy in virtual spaces	Dynamic management of energy loads using AI	Sustainable technologies for managing data center energy	Virtual alternatives to physical meetings to reduce travel
Key technologies	Simulation tools, IoT integration	Blockchain technology for secure transactions	AI, virtual sensors	Liquid cooling, renewable energy, AI workload balancing	Virtual reality platforms (e.g., Microsoft Mesh)
Example	Siemens and NVIDIA's platform for grid modeling	Powerledger's blockchain-based trading platform	ABB and Schneider Electric's AI-driven EMS	Google's AI-optimized cooling systems	Accenture's use of Microsoft Mesh for training
Impact	Reduces energy loss and carbon emissions	Encourages decentralized energy management	Minimizes network strain and energy waste	Lowers carbon emissions and improves energy efficiency	Reduces energy use related to commuting and office utilities
Benefits	Proactive network management, efficient resource use	Scalable model for renewable energy exchange	Enhanced efficiency in energy distribution	Sustainable data handling in virtual environments	Greener workplace model through reduced physical presence

toward the future readiness of eco-friendly power systems with increasing demands for energy without compromising the environment.

References

Adnan M (2024) Smart grid 3.0: navigating the future-unleashing the power of metaverse, blockchain, and digital twins in the evolution of smart grids. Blockchain, and digital twins in the evolution of smart grids

Akpan IJ, Shanker M, Razavi R (2020) Improving the success of simulation projects using 3D visualization and virtual reality. J Oper Res Soc 71(12):1900–1926

Arévalo P, Ochoa-Correa D, Villa-Ávila E (2024) A systematic review on the integration of artificial intelligence into energy management systems for electric vehicles: recent advances and future perspectives. World Electr Veh J 15(8):364

Balali V, Noghabaei M, Heydarian A, Han K (2018) Improved stakeholder communication and visualizations: real-time interaction and cost estimation within immersive virtual environments. In: Construction research congress 2018, pp 522–530

Dai Y, Zhang K, Maharjan S, Zhang Y (2020) Edge intelligence for energy-efficient computation offloading and resource allocation in 5G beyond. IEEE Trans Veh Technol 69(10):12175–12186

Ding H, Zhou DQ, Liu GQ, Zhou P (2020) Cost reduction or electricity penetration: government R&D-induced PV development and future policy schemes. Renew Sustain Energy Rev 124:109752

Fang Y, Zheng T, Zheng X, Yang H, Wang H, Walther M (2021) Influence of tide-induced unstable flow on seawater intrusion and submarine groundwater discharge. Water Resour Res 57(4):e2020WR029038

Feldman D, Ramasamy V, Fu R, Ramdas A, Desai J, Margolis R (2021) US solar photovoltaic system and energy storage cost benchmark (Q1 2020) (No. NREL/TP-6A20-77324). National Renewable Energy Lab (NREL), Golden, CO (United States)

Gayam SR (2022) Deep learning for predictive maintenance: advanced techniques for fault detection, prognostics, and maintenance scheduling in industrial systems. J Deep Learn Genomic Data Anal 2(1):53–85

Gu W, Ma T, Ahmed S, Zhang Y, Peng J (2020) A comprehensive review and outlook of bifacial photovoltaic (bPV) technology. Energy Convers Manage 223:113283

Hassan Q, Algburi S, Sameen AZ, Salman HM, Jaszczur M (2023) A review of hybrid renewable energy systems: solar and wind-powered solutions: challenges, opportunities, and policy implications. Results Eng 20:101621

Keyhanpour MJ, Jahromi SHM, Ebrahimi H (2021) System dynamics model of sustainable water resources management using the Nexus Water-Food-Energy approach. Ain Shams Eng J 12(2):1267–1281

Kumar P, Kumar R, Aloqaily M, Islam AN (2023) Explainable AI and blockchain for metaverse: a security, and privacy perspective. IEEE Consumer Electronics Magazine

Lawan SM, Abidin WAWZ (2020) A review of hybrid renewable energy systems based on wind and solar energy: modeling, design and optimization. In: Wind solar hybrid renewable energy system, p 219

Liu H, Xu D, Tian Y, Peng C, Wu Z, Wang Z (2023) Wasserstein generative adversarial networks based differential privacy metaverse data sharing. IEEE J Biomed Health Inf

Mohammed ZK, Mohammed MA, Abdulkareem KH, Zebari DA, Lakhan A, Marhoon HA, Martinek R et al (2024) A metaverse framework for IoT-based remote patient monitoring and virtual consultations using AES-256 encryption. Appl Soft Comput 158:111588

Novianti DN, Shiddieq DF, Roji FF, Susilawati W (2024) Komparasi algoritma support vector machine dan naïve Bayes untuk analisis sentimen pada metaverse: Comparison of support vector machine and naïve Bayes algorithms for sentiment analysis of the metaverse. MALCOM Indones J Mach Learn Comput Sci 4(1):231–239

Pao LY, Zalkind DS, Griffith DT, Chetan M, Selig MS, Ananda GK, Loth E (2021) Control co-design of 13 MW downwind two-bladed rotors to achieve 25% reduction in levelized cost of wind energy. Annu Rev Control 51:331–343

Pech M, Vrchota J, Bednář J (2021) Predictive maintenance and intelligent sensors in smart factory. Sensors 21(4):1470

Placke T, Heckmann A, Schmuch R, Meister P, Beltrop K, Winter M (2018) Perspective on performance, cost, and technical challenges for practical dual-ion batteries. Joule 2(12):2528–2550

Roy P, He J, Zhao T, Singh YV (2022) Recent advances of wind-solar hybrid renewable energy systems for power generation: a review. IEEE Open J Ind Electron Soc 3(81–104):221

Salter JD, Campbell C, Journeay M, Sheppard SR (2009) The digital workshop: exploring the use of interactive and immersive visualisation tools in participatory planning. J Environ Manage 90(6):2090–2101

Schulman PR (2022) Reliability, uncertainty and the management of error: new perspectives in the COVID-19 era. J Contingencies Cris Manag 30(1):92–101

Scordia D, Papazoglou EG, Kotoula D, Sanz M, Ciria CS, Pérez J, Cosentino SL et al (2022) Towards identifying industrial crop types and associated agronomies to improve biomass production from marginal lands in Europe. GCB Bioenergy 14(7):710–734

Shafiullah M, Ahmed SD, Al-Sulaiman FA (2022) Grid integration challenges and solution strategies for solar PV systems: a review. IEEE Access 10:52233–52257

Soto EA, Bosman LB, Wollega E, Leon-Salas WD (2021) Peer-to-peer energy trading: a review of the literature. Appl Energy 283:116268

Stangl FJ, Riedl R, Weitzl WJ, Martin S (2023) Fatigue and stress levels in digital collaboration: a pilot study with video conferencing and the metaverse. In: NeuroIS retreat. Springer Nature Switzerland, Cham, pp 89–103

Stergiou C, Psannis K (2022) Digital twin intelligent system for industrial internet of things-based big data management and analysis in cloud environments. Virtual Rity Intell Hardw 4(4):279–291

Tightiz L, Dang LM, Padmanaban S, Hur K (2024) Metaverse-driven smart grid architecture. Energy Rep 12:2014–2025

Truong VT, Le L, Niyato D (2023) Blockchain meets metaverse and digital asset management: a comprehensive survey. IEEE Access 11:26258–26288

Wodi A (2023) Data privacy and security: fair information principles and data management strategies in a changing world. SSRN 4601171

Xia Y, Xu Q, Huang Y, Du P (2023) Peer-to-peer energy trading market considering renewable energy uncertainty and participants' individual preferences. Int J Electr Power Energy Syst 148:108931

Yang K, Zhang Z, Youliang T, Ma J (2023) A secure authentication framework to guarantee the traceability of avatars in metaverse. IEEE Trans Inf Forensics Secur 18:3817–3832

Yoon D, Oh A (2022) Design of metaverse for two-way video conferencing platform based on virtual reality. J Inf Commun Converg Eng 20(3):189–194

Metaverse Tourism: Pioneering Sustainable Hotel Bookings

Ashima Arya, Riya Arora, Yonis Gulzar, Sandeep Kautish, and Shilpa Saini

Abstract In the Indian hotel booking application industry, users can emphasize 2D views or singular images of accommodations, which never offer a comprehensive perspective. This will not satisfy the customer's needs and require more information on the hotel's availability and delivery to locals. This issue is addressed by a distinctive hotel view application utilizing virtual reality to enhance the booking experience. With the help of this application, users can examine a hotel virtually before finalizing a reservation. They can accurately assess the space and atmosphere by utilizing 360-degree and 3D viewpoints to inspect hotel room amenities and communal spaces, enabling them to make informed decisions. Artificial intelligence is embedded to enhance by assimilating user feedback and interactions. It will modify and improve the virtual experience through data analysis, ensuring it better corresponds with consumers' tastes and expectations.

Keywords Unity 3D · Virtual reality · Hotel · Android

A. Arya
Department of Computer Science and Information Technology, KIET Group of Institutions, Delhi-NCR, Ghaziabad, India

R. Arora
Department of Computer Science Technology, Maharaja Agrasen Institute of Technology Delhi, Rohini, Delhi, India

Y. Gulzar
Department of Management Information Systems, College of Business Administration, King Faisal University, Al-Ahsa, Saudi Arabia

S. Kautish · S. Saini (✉)
Chandigarh University, Mohali, Punjab, India
e-mail: Shilpa.saini211986@gmail.com

W. Leal Filho et al. (eds.), *Metaverse and Sustainability: Business Resilience Towards Sustainable Development Goals*, World Sustainability Series,
https://doi.org/10.1007/978-3-031-89545-6_27

1 Introduction

The metaverse: an open world, present at and let people go to free-range everywhere and at any time. In this world, any experience is foregone and special.

The notion of a metaverse and connected virtual experiences has changed people's attitudes toward what future technology will mean for the hospitality and tourism industry. People working in this Industry are looking for opportunities to adopt the Metaverse and create virtual experiences, products, and services in hospitality and tourism for their consumers (Gursoy et al. 2022).

Experiences, products, and services are expected to grow since consumers are in the virtual reality hospitality and tourism experience. While the 2D internet encourages transactions, the 3rd generation internet facilitates the 3D virtual, social metaverse. In addition to its social connective ability, there is a sense of presence. Therefore, the hospitality and tourism industry is interested in meters (Buhalis et al. 2023). While the metaverse can never replace on-the-ground travel, technological advances and the sophistication of the virtual reality (VR) headset box have made metaverse hospitality and tourism apps more immersive than ever. These enhancements drive digital natives and tech-savvy consumers to go on virtual globetrotting. Hence, the metaverse will integrate into the industry and enhance the hospitality and tourism experience in a few dimensions. Hotel owners, managers, vendors, suppliers, and other stakeholders are confused about several aspects as the metaverse disrupts and revamps the hospitality and tourism sector. For instance, what would be the potential impact of the 60 metaverse on the industry? What influence will the metaverse have on hospitality and tourism customers' decision-making behaviors and consumption patterns? What real-life experiences are transferable to the metaverse? Will business people buy virtual land to run virtual companies, similar to web domains acquired in the past? This commentary offers a brief overview of the recent evolution of the metaverse in the hospitality and tourism sectors and outlines future research directions for interested scholars (Ioannidis and Kontis 2023).

The metaverse—a new frontier of novels—will continue to emerge as customers and technology evolve and will play a more significant role in the industry in marketing, customer relationships, communications offices, customer decision-making processes, and guest experiences. As a result, hospitality and tourism metaverse experiences may potentially become the next disruptors to the industry (Arya et al. 2023; Go and Kang 2023; Sehgal et al. 2024).

The metaverse can facilitate virtual experiences, including flights, hotel rooms, restaurant cooking and food preparation, tours, conferences, meetings, and conventions. Top of Form.

In future research to develop metaverse tourism products or experiences, tourism stakeholders need to explore tourists' perceptions and applys market segmentation of past paradigms of virtual forms of tourism (Fig. 1).

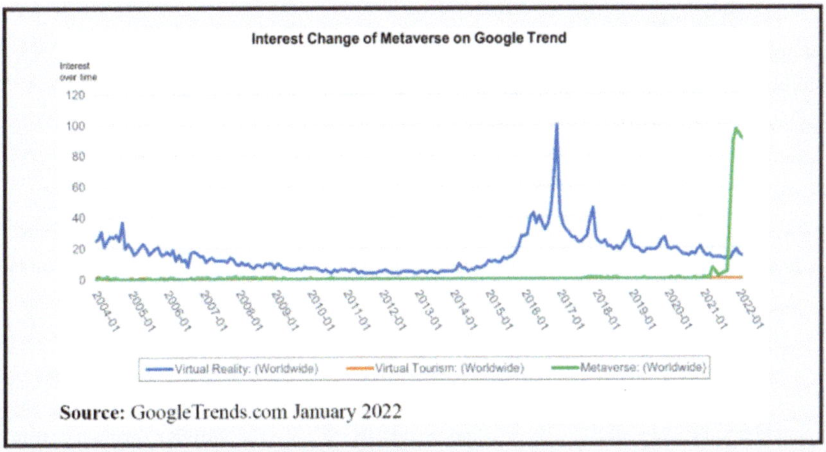

Fig. 1 Trends of metaverse. *Source* Buhalis et al. (2023)

2 Literature Review

With Virtual reality (VR) technology, users can interact with objects and feel as if they are physically present in these environments. VR creates computer-generated environments mimicking real-life scenarios. This technology differs from regular pictures or 360-degree images because it offers a more immersive and interactive experience, engaging multiple senses (Chen 2024). Interactivity means that VR responds to the user's actions, just like the real world, while vividness refers to how well VR stimulates the senses, providing a rich sensory experience. The sense of presence in VR is subjective and depends on how deeply users feel immersed in it through various technologies. This study focuses on fully immersive VR systems, where users are completely isolated from the real world, often achieved using head-mounted devices (HMDs).

Recently, VR has become increasingly popular in the travel and tourism industry. Research in this area, especially regarding fully immersive VR, has increased (Özdemir Uçgun and Şahin 2024). This research mainly focuses on understanding how VR enhances consumers' experiences in the travel and tourism sector. Previous studies have shown that VR can stimulate mental imagery and create a strong sense of being in a place, potentially increasing people's interest in traveling or visiting destinations. While VR commercials can immediately impact consumers, researchers are still exploring their long-term effects compared to traditional advertisements (Koo et al. 2023). Most existing studies primarily compare VR with traditional communication methods and haven't investigated how these two approaches can work together when used by consumers (Huang and Roscoe 2021) (Table 1).

The hotel view app, featuring 360-degree and 3D views, is designed to serve a broad audience of travelers and stakeholders in the hospitality industry. This

Table 1 Literature review of VR

In this table related works to VR study	Research context	Key findings
Huang and Roscoe (2021)	Head-mounted display-based virtual reality systems	The future potential includes facilitating lifelong learning and professional development for engineering instructors to adapt to HMD VR systems
Radianti et al. (2020)	VR in higher education	A systematic mapping study identified gaps in the field, such as limited consideration of learning theories and a focus on usability over learning outcomes
Augmented and virtual reality evolution and future tendency (Muñoz-Saavedra et al. 2020)	The growing popularity of augmented reality and virtual reality technologies	It highlights their success in mobile applications and the video game industry and their resurgence in the industrial sector due to technological advancements. The study also aims to analyze future trends and areas of further integration into society
Augmented and virtual reality recent developments in education (2022) (Al-Ansi et al. 2023)	Extended reality in management education	Results reveal that the adoption of AR and VR in education has exponentially grown during recent years, with wearable devices accounting for a large portion of this development. Based on secondary data, results also reveal the gap in quickly implementing and customizing these technologies in educational institutions
Integration of AR & VR in Construction Management (2022) (Abbasnejad et al. 2024)	Using integrated hands-on and virtual reality (VR) or augmented reality (AR) approaches in construction management education	A combination of advanced technology and interactive learning methods is essential to improving students' understanding of construction materials and enhancing their communication skills. Incorporating 3D modeling and virtual reality tools will provide students with a more immersive learning experience

Table 2 Module description

Module	Description
User registration and login	For signing up in the app
City and hotel search	Use for searching specific cities and hotels
Interactive hotel exploration	A 360-degree virtual tour to explore the rooms, common areas, and nearby facilities
AI-driven personalization (Arya et al. 2023)	AI integrates to take users' feedback and offer personalized recommendations

innovative platform caters to travel enthusiasts eager for immersive experiences, frequent travelers prioritizing convenience and comfort, and vacation planners seeking detailed insights to make well-informed trip decisions (Tussyadiah 2016).

3 Proposed Approach

Comparison with Current Platforms: In the Indian market, hotel booking apps are now frequently available and offer constrained and static 2D images of hotel rooms and amenities. Because people may not completely understand the accommodations based on such visuals alone, people may become dissatisfied with this technique.

However, to revolutionize the hotel booking experience, our proposed solution is to integrate the most advanced technologies, such as virtual reality (VR), artificial intelligence (AI), and the Android platform. Our software addresses this issue by offering users a comprehensive 360-degree perspective of hotels and rooms. With several VR technologies, they may virtually tour each aspect of their stay to obtain a first-hand sense of the atmosphere and amenities (Gubbala et al. 2023; Gursoy et al. 2022; Kugurakova et al. 2023).

4 Key Features and User Journey

See Table 2.

5 Technologies Used

See Table 3.

Table 3 Technology description

Technology	Description
Unity 3D for VR experience	For creating an immersive VR environment for exploring hotels virtually and seamlessly navigating through rooms
Android platform (Selvakumaran et al. 2023)	For developing Android applications for mobile phones
Artificial intelligence (AI)	For Personalized recommendations and optimize the VR environment for users

6 Methodology

Figure 2 describes the workings of the different components. These components help the user experience hotel rooms and other areas virtually.

Module	Description
1. User authentication and profile management	• Enables users to create accounts, log in securely, and manage their profiles • Firestore or firebase realtime Database saves personal details, booking history, and preferences, ensuring a personalized experience
2. Hotel search and listings	• Allows users to find hotels based on their preferred location and travel dates • A list of available hotels is available with all details such as pricing and availability • Efficient searching and sorting module for customization of their search • The app will fetch real-time hotel listings, displaying accurate and up-to-date information
Room booking and reservation	• Hotel rooms reservation
360-degree room view	• Explore hotel rooms virtually, offering a realistic and immersive experience • Enable users to examine visuals of rooms
Payment gateway integration	• Allows users to make online payments confidently • Variety of payment methods, including credit cards, digital wallets, and other popular options

(continued)

(continued)

Module	Description
Booking confirmation and notifications	• Notify users with confirmation messages and emails after booking • These alerts include fundamental details such as reservation dates, room information, and payment verification • Sends timely reinforcements and reminders about an approaching booking, making sure users are sharp and ready for their trips
User reviews and ratings	• Users can review and rate hotels and individual rooms • Reviews are moderated for appropriateness and relevance to ensure the quality of user-generated content: The module comes with moderation tools that help manage reviews and moderate them
User support and feedback	• It requires hotel managers to operate and manage their properties from one central hub • Admin can change hotel list, room details, and Availability with up to the min • Users can track bookings, reply to reviews, and engage with customer interactions
Integration with maps and location services	• The APIs simulate hotel locations that are in line with reality so users can locate their hotels easily and quickly • They even provide navigation and turn-by-turn directions to your booked hotels, which makes travelling easy • This module comes with built-in geolocation functionalists and helps give an overall better user experience by providing instant access to location-based information and comforts the user during the whole booking process
AI-driven personalization	• Inside the box, it provides an UX, It makes the VR platform appears more responsive to individual user needs and expectations through offering personalized recommendations • Through our creative approach, we aim to offer travelers with an enriching hotel booking experience

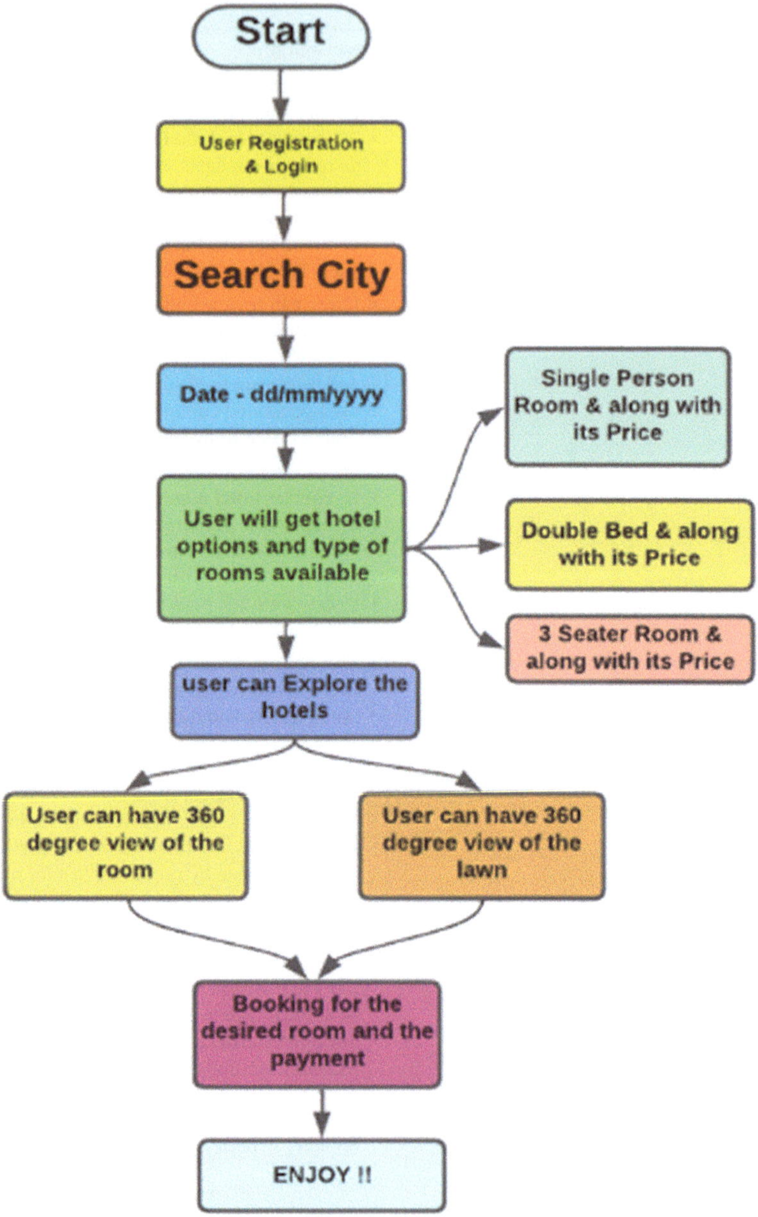

Fig. 2 Block diagram of VR model

7 Implementation

Our Hotel View App utilizes VR to change the way we book hotels. Users can walk through the hotels with rich 360-degree views (alive or animated) of their rooms, gardens, and facilities, allowing them to make intelligent decisions.

At the heart lies AI, which helps learn from user interaction and enhance engagement strategies. It uses feedback to enhance the VR platform and improve future experiences, increasing overall satisfaction (Anand et al. 2023).

The following variety of technologies and tools are used to develop applications.

Here's a breakdown of each component as:

1. **Android Studio** is the official Android app development integrated development environment (IDE). This is where you create the Android application code in Java or Kotlin, manage the project, and create the XML user interface.
2. **XML:** The extensible markup language, or XML, is widely used in Android development to define UI elements and design layouts.
3. **Java:** Java is one of the main programming languages for creating Android apps. It is used for your application's backend logic, which includes data handling, processing, and other features.
4. **Firebase**: Google's platform for creating web and mobile applications is Firebase. For creating mobile apps, it offers several helpful services, including real-time databases, cloud storage, authentication, analytics, and more.
5. Hotel API: To get hotel-related information and services in your app, such as booking details, hotel details, availability, etc., you probably need to use Hotel API (Application Programming Interface).
6. Unity3D—Unity is a multiplatform game engine used for 3D interactive applications other than games, like VR (Virtual Reality) experiences.
7. Sketchfab: Sketchfab is a website where you can find and share 3D content. Over which programmers have access to a complete set of 3D elements and models that can be added to applications for different purposes.)
8. Google VR SDK (Software Development Kit)—Devs can port the fully fledged Google VR platform into their development tool of choice with the Google VR SDK (Software Development Kit), letting you develop a complete immersive experience for Android users. It provides libraries and tools for building VR applications.
9. AI for Comments and Suggestions:

    - Gathering Responses: Take AI-driven user feedback. This may include examining customer reviews or comments made in the app using natural language processing (NLP) techniques. Feedback can be analyzed for sentiment to see if it is positive, negative, or neutral.
    - User Profiling: Scenarios that leverage AI to create individual user profiles based on how users interact with the application. To achieve this, it may require analyzing usage data to understand user preferences, actions, and trends.

- Create AI-based recommendation systems to recommend relevant content, accommodation, travel places, or 3D assets to consumers. Hybrid techniques, content-based, and collaborative filtering can customize recommendations depending on user profiles and preferences.
- ML Models: Create machine learning models to predict user preferences in the list above and improve retrieval performance (Figs. 3 and 4).

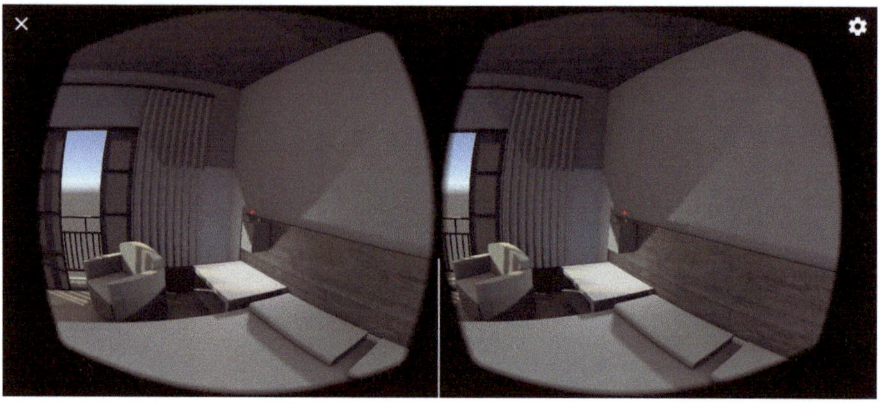

Fig. 3 Shows The 360-degree view of room

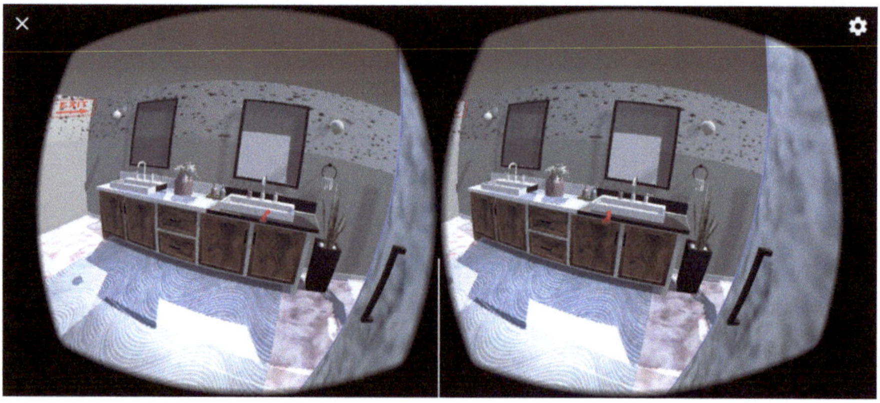

Fig. 4 Gives a 360-degree view of the washroom

8 Result

Now, our hotel booking app includes all the properties and options given by the user. By combining artificial intelligence (AI) and virtual reality (VR), our novel app transforms the user experience, providing unparalleled engagement and happiness.

The vices of existing platforms that often don't deliver actionable insights on the cut-throat Indian hotel booking market were highlighted by the researcher. To assist with this, our app shows a new VR experience for customers to visit and swim in real hotels from 360-degree perspectives. It gives users an idea about their potential accommodations and close-up shots of rooms, gardens, and amenities.

We want to change the booking experience by focusing on immersive Ness and human-centered design. By combining Unity 3D, AI, and VR tech, our software sets a new standard for hotel booking experiences in India and abroad!

By combining these cutting-edge technologies, our application simplifies hotel bookings and gives travelers the power of choice while booking hotels with confidence & knowledge.

We know what you are thinking—"Dear God, Well, let us show you how we accomplished it!" (Fig. 5).

Fig. 5 360-degree view of the outer hotel

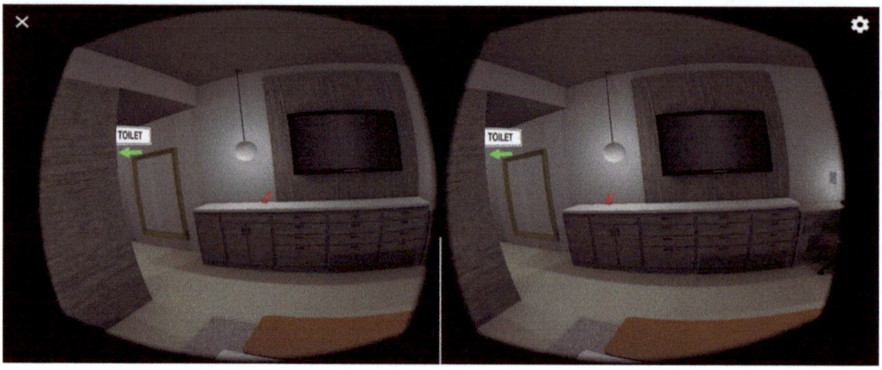

9 Conclusion

Finally, our hotel view app project is working well, and it's a great improvement over the Indian Hotel booking apps in this space. Consumer-centric solutions—data collection, software engineering, and testing take time. The company needs to build a solution that is in the best interests of consumers. With our app, we bring travelers to life with VR, 3D modelling, 360 imagery, and AI! We offer an interactive event experience.

Indeed, VR is been only as fun because the magic of AI has been important to boosting person experience and correcting flaws in current platforms. The goal is to enhance the booking experience by eliminating bugs and allowing people to explore hotel spaces virtually. We will also employ AI to modify interactions on feedback.

In summary, our project is a huge step forward in hotel buying, providing a more educated and personalized experience for accomodation making decisions in India and elsewhere too.

We will continue to maintain and improve our app that views hotels. The goal is to keep the application sustainable, dependable, and usable.

Acknowledgements This work was supported by the Deanship of Scientific Research, the Vice Presidency for Graduate Studies and Scientific Research, King Faisal University, Saudi Arabia under the project (KFU242527).

References

Abbasnejad B, Soltani S, Wong P (2024) A systematic review of online learning and teaching strategies during the COVID-19 pandemic: implications for the construction management sector. Smart Sustain Built Environ 13(4):934–959

Al-Ansi AM, Jaboob M, Garad A, Al-Ansi A (2023) Analyzing augmented reality (AR) and virtual reality (VR) recent development in education. Soc Sci Human Open 8(1):100532

Anand R et al (eds) (2023) Integration of IoT with cloud computing for smart applications. CRC Press

Arya A, Dhanaraj RK, Juneja S (2023) Navigating the metaverse: trends and opportunities in 2023. In: 2023 intelligent computing and control for engineering and business systems (ICCEBS). IEEE, pp 1–6

Buhalis D, Leung D, Lin M (2023) Metaverse as a disruptive technology revolutionising tourism management and marketing. Tour Manage 97:104724

Chen Z (2024) Beyond boundaries: exploring the metaverse in tourism. Int J Contemp Hosp Manag

Go H, Kang M (2023) Metaverse tourism for sustainable tourism development: tourism agenda 2030. Tour Rev 78(2):381–394

Gubbala B, Alti DN, Srividhya SR, Pothumani S (2023) Augmented reality based furniture application. In: 2023 2nd international conference on applied artificial intelligence and computing (ICAAIC), pp 1088–1092

Gursoy D, Malodia S, Dhir A (2022) The metaverse in the hospitality and tourism industry: an overview of current trends and future research directions. J Hosp Market Manag 31(5):527–534

Huang W, Roscoe R (2021) Head-mounted display-based virtual reality systems in engineering education: a review of recent research. Int J Comput Appl Eng Educ 29(1)

Ioannidis S, Kontis AP (2023) Metaverse for tourists and tourism destinations. Inf Technol Tour 25(4):483–506

Jung T, tom Dieck MC, Lee H, Chung N (2016) Effects of virtual reality and augmented reality on visitor experiences in museum. In: Information and communication technologies in tourism 2016. Springer, pp 621–635

Koo C, Kwon J, Chung N, Kim J (2023) Metaverse tourism: conceptual framework and research propositions. Curr Issue Tour 26(20):3268–3274

Kugurakova VV, Golovanova II, Kabardov MK, Kosheleva YP, Koroleva IG, Sokolova NL (2023) Scenario approach for training classroom management in virtual reality. J Commun Med Technol 13(3):e202328

Muñoz-Saavedra L, Miró-Amarante L, Domínguez-Morales M (2020) Augmented and virtual reality evolution and future tendency. Appl Sci 10(1):322

Özdemir Uçgun G, Şahin SZ (2024) How does metaverse affect the tourism industry? Current practices and future forecasts. Curr Issue Tour 27(17):2742–2756

Radianti J, Majchrzaka T, Frommb J, Wohlgenanntc I (2020) A systematic review of immersive virtual reality applications for higher education: design elements lessons learned and research agenda. Int J Comput Educ 147

Rashid J et al (2022) An augmented artificial intelligence approach for chronic diseases prediction. Front Publ Health 10:860396

Sehgal M, Arya A, Gaba S, Nagpal S (2024) AI for tourism: pioneering new frontiers in the industry. In: 2024 international conference on computational intelligence and computing applications (ICCICA). IEEE, vol 1, pp 280–285

Selvakumaran K, Kishore S, Narayanasamy A, Radhakrishnan A, Malarvel M (2023) Scalable AR integration pipeline for immersive textbook learning experience. In: 2023 4th international conference on electronics and sustainable communication systems (ICESC), pp 291–296

Tussyadiah IP (2016) The influence of innovativeness on on-site smartphone use among American travelers: implications for context-based push marketing. J Travel Tour Mark 33(6):806–823

Enhancing Electric Vehicle Adoption Through Metaverse Applications: Strategies and Impacts

Rachna Rana, Pankaj Bhambri, and Sandeep Kautish

Abstract Electric cars (EVs) are gaining popularity in India, offering a more ecologically responsible and cost-effective method to travel around congested cities. These cars, with their sleek appearance and smart technology, simplify commuting while also contributing to cleaner, fresher streets and air. It's wonderful to see the government promote EV use, while private entrepreneurs are creating novel models customized to India's specific roads and traffic patterns. Global EV sales climbed by a staggering 40% in the first half of 2023 alone, indicating a rising preference for greener means of travel. According to analysts, the automotive industry will experience a dramatic transformation when EV sales reach 35–40% of worldwide new car sales by 2030. By embracing innovation while upholding sustainability and inclusion ideals, the automotive sector can lead the path for a brighter, more efficient, and sustainable future driven by electric cars. Despite the advantages, several operational challenges need to be addressed before electric vehicle adoption becomes widespread. This section delves into the historical growth of electric vehicles in excess of moment, outlining their approaches and effects, for instance abridged carbon production as well as atmosphere contamination. Furthermore, it examines the obstacles and complexities faced during their implementation, including elevated carrying expenses, a scarcity of indicting position, limited variety or "assortment edginess," and battery presentation. elucidation to these confronts comprise studying the indicting transportation, getting higher charging position, employing battery-exchanging methods, and increasing battery method to ease variety nervousness as well as condense accusation period. Administration could incentivize electric vehicle buy by giving tax recognitions or financial assistance and spending in wide charging transportation. Mechanized stakeholders could work together with administrations

R. Rana
Ludhiana Group of Colleges, Ludhiana, Punjab, India

P. Bhambri (✉)
Guru Nanak Dev Engineering College, Ludhiana, Punjab, India
e-mail: pkbhambri@gmail.com

S. Kautish
Chandigarh University, Mohali, Punjab, India

W. Leal Filho et al. (eds.), *Metaverse and Sustainability: Business Resilience Towards Sustainable Development Goals*, World Sustainability Series,
https://doi.org/10.1007/978-3-031-89545-6_28

to undertake these problems and endorse the employ of electric vehicles, eventually causative to falling carbon production as well as atmosphere contamination.

Keywords Smart cities · Electric vehicles · Strategies · Impacts · Challenges · Sustainable · Battery performance · Metaverse

1 Introduction

The auto assiduity has come a major player in the global frugality and in exploration and development. As technology has advanced, vehicles have decreasingly been equipped with systems that prioritize the safety of passengers and climbers (Anumba and Lu 2022). This has led to an increase in the number of buses on the road, furnishing us with the convenience of fast and affable transportation. Still, this progress has come at a cost. Environmental adulterants similar as sulfur dioxide (SO_2), nitrogen oxides (NO_X), carbon monoxide (CO), and particulate matter (PM) have significantly increased in metropolises. It's important to admit the impact of the auto assiduity on our diurnal lives, both positive and adversarial (Global EV Outlook 2023). The world is incontrovertibly menaced by getting up carbon emigrations and the abating oil painting force. The transportation assiduity is the largest contributor to environmental impact, counting for over 25 of global energy consumption and hothouse gas emigrations. Road transport alone is responsible for emitting further than 70 of the sector's emigrations. To address the challenges posed by oil painting dependence and reduce emigrations, the conception of" sustainable transportation" has been explosively proposed. The Electric Power Research Institute (EPRI) states that the wide relinquishment of electric vehicles (EVs) is anticipated to have a significant impact in reducing hothouse gas emigrations, surpassing the effectiveness of indeed the most energy-effective conventional buses. Still, the significant expansion in the operation of electrically powered motorcars has raised several hurdles, complications, unknowns, and anxieties analogous as precious structure costs, electric vehicle prices, a deficiency of charging stations, and a defined range of electric vehicles. Batteries remain the most serious concern. In the coming times, EVs will play an important part in smart cosmopolises, alongside networked mobility, public vehicle, and other features. As a result, farther work is demanded to enhance batteries and streamline the charging process. The primary issue with EVs is their autonomy. Scientists are working on bettered battery technologies to extend driving range while lowering weight, cost, and charging time. These rudiments will eventually shape the future of EVs (Jiang et al. 2021).

2 Metaverse

Neal Stephenson's wisdom fabrication novel Snow Crash first used the expression' Metaverse' in 1992. In Snowfall Collide, Stephenson probable individualities run away the cruel realities of a global profitable collapse into a motorized virtual reality terrain (Bhambri and Singh 2006). Shipping and mobility, covering, physical condition and cleanliness, amusement and relax, corporeal organization stipulation, gainful and occupation projection, ethnicity, and numerous other rudiments all have an impact on the quality of life in metropolises. The variation in the number and blend of these numerous aspects goods people's comprehensions and satisfaction situations with what they perceive as quality of life.

Still, beyond mortal perspective, it's stated that at the veritably least, socioeconomic circumstances, environmental sustainability, lesser governance, and artistic variety must be in place to insure that civic superiority of existence carries on to go up. The Metaverse is constantly described as an immersive, linked virtual reality that includes factors from the physical world and allows druggies to communicate and interact with one another. It's popularly known as the Internet's coming interpretation (Bhambri et al. 2019). It brings together all software and tackle technologies, including virtual reality (VR) headsets, block chains, and incorporations, to produce an immersive and dynamic participated virtual terrain. Although the Metaverse is n't yet fully evolved, it's extensively regarded as a new kind of social, profitable, and artistic structure with the eventuality to significantly alter numerous aspects of diurnal life.

Over time, several enterprises and platforms have developed performances of the Metaverse, similar as Linden Lab, which released Second Life in 2003, and current gaming services, similar as Roblox and Fortnite (Herrmann and Rothfuss 2015). These platforms let druggies produce incorporations and interact with other druggies in a virtual terrain. Mark Zuckerberg, Facebook's CEO, said in July 2021 that the company's strategy will shift from a purely social media business to a Metaverse one. It was latterly called Meta4 Platforms. Overall, the Metaverse's future remains uncertain, and it'll most probably bear times of testing and development before we've a clear image of what it'll look like and operate. As an outgrowth, there's a critical need to identify and dissect the issues and hurdles that stymie the flawless perpetration of Metaverse in the service business.

3 Applications of Metaverse

A preliminary indication of the Metaverse and its characteristics provides a great foundation for learning further about the crucial operations of the Metaverse. As the use of technologies similar as virtual reality, stoked reality, and extended reality continues to grow by the day, it's reasonable to wonder how they can be used to extend metaverse exemplifications in the real earth (Panda et al. 2019). Making a bet and community networking are among the top answers to questions about metaverse operations. Still, numerous other promising operations demonstrate the true capabilities of the metaverse in the future. There are five operations of the metaverse across different diligence as follows.

3.1 Health

One of the first replies to the query "What are the procedures of the metaverse?" will be directed on the way to the healthcare assiduity. The most prominent illustration of the operation of the metaverse in the healthcare division refers to the utilization of strengthen authenticity. Strengthen authenticity has become an essential method to increase the splits and understanding of health check researchers. For Example, method-facilitates surgical helps comparable as Microsoft helps surgeons achieve colourful surgical methods. The utilization of these methods highlights the most frequent metaverse processes to improve the thinness and hurry of surgical methods. In addition to re-operative descriptions reach from MRI, CT reconsiders, and 3D scanners, strengthen reality headsets can help display important case data in real-time (Sharifi and Khavarian-Garmsir 2020). Thus, the metaverse can make it easier to cover patient data similar as body temperature, breathing rate, heart rate, and blood pressure. Exemplifications of the Metaverse also show the use of stoked authenticity to restructure tone acknowledgment. Thus, the utilization of metaverse method can assist smash the issue of manner spot, particularly in region with elevated infiltration or small means. Imaging methodologies same as CT re-examines and X-rays are also qualified supporter for the healthcare assiduity's changeover to the metaverse. Visualize a fundamental planet that permits croakers and caregivers to look inside a case's body and understand the problem (Zeng and Zhang 2023).

4 Real Estate

The next intriguing inclusion on the list of metaverse activities is directed at the real estate industry. Virtual reality is one of the most important technologies for achieving metaverse goals. It can provide visitors with realistic and immersive gests, and this capability can assist metaverse activities in real estate. For example, real estate brokers can use VR to provide immersive virtual tenures to purchasers. Similarly, the possibility of the metaverse improves the potential for including vivid multimedia components in unique VR tenures. For instance, VR real estate experiences could incorporate background music, lighting, and sound effects, as well as historical information. All of these features are available in the most popular real estate metaverse platforms, providing a nearly real-time interaction with properties (Zheng et al. 2023).

As a result, residential marketers may easily gain clients' trust by allowing them to tour the property in real-time. Guests may be certain of colorful requirements before they complete a purchase, even with real estate agents can save a lot of time and money (Bhambri and Mangat 2005). In addition, agents can benefit from metaverse exemplifications and knitter-substantiated tenures based on client preferences.

5 Education

The usage of metaverse technology in education has also gained traction in recent years. Virtual reality is a potential technique for transforming traditional tutoring approaches by emphasizing learning generalities via visuals (Anand and Bhambri 2018). Leading metaverse activities in the education sector may assist create engaging literacy environments for students in vibrant educational institutions? How does the metaverse help to improve educational capacities through colorful operations? Virtual reality can facilitate quick and frequent mistake identification while also providing real-time editing possibilities. More crucially, appealing metaverse apps for education can help to break down literacy barriers. This technology allows for the objectification of any language into the educational metaverse platform, hence breaking down linguistic barriers.

6 Military

The use of VR and AR in military operations showcases the Metaverse's capability to support such activities. One notable example is the Tactical Augmented Reality (TAR), which is akin to night vision goggles but with additional features (Bhambri et al. 2020). TAR can accurately display the location of friendly forces, as well as the positions of allies and enemies. It serves as a superior alternative to traditional mobile GPS devices and headsets. Another application of the Metaverse in the military is seen in synthetic training environments, which offer mixed-reality simulations for realistic training scenarios. These simulations provide immersive experiences that mimic physically and psychologically intense battle environments within a virtual landscape.

7 Manufacturing

Finding the solution to the question "What are the operations of Metaverse?" will lead you into the manufacturing industry. As one of the main metaverse technologies, VR operations may assist in instructing personnel on safety precautions while encouraging involvement in simulations of hazardous scenarios (Kaur et al. 2019).

Thus, metaverse procedures can significantly reduce the risk of mishaps. The most prevalent metaverse procedures in manufacturing can contribute to the long-term creation of superior goods. For example, virtual reality headsets may let producers thoroughly test all aspects of a product. Metaverse operations can also aid with plant geography planning and equipment placement. Ford is a well-known company that is leveraging virtual reality technology to breach its locked models through the metaverse.

8 Electric Vehicles

The expansion of digital auto retail, especially driven by the advent of electric vehicles (EVs), is notably exemplified through physical experience centers. These centers, typically tech-enabled virtual dealerships, deliver an enhanced product experience utilizing advanced experiential technology such as augmented reality (AR), virtual reality (VR), and the metaverse. The metaverse stands out as the most sophisticated and fully immersive technology. Most legacy automakers have already established experience centers for their EV models, while pure-play EV manufacturers are rapidly setting new standards with exceptional experiences. In some cases, these EV companies have successfully encouraged traditional manufacturers to create similar experience centers (Potluri et al. 2019).

As the world grapples with the pressing need to address climate change, there is a clear shift in the automobile sector toward electric cars (EVs). The reduction of environmental harm caused by traditional automobiles is driving this change, which is more than simply a fad. Global EV sales rose by an astonishing 40% in the first half of 2023 alone, indicating a rising preference for greener sources of mobility. According to analysts, the automotive industry will experience a dramatic transformation when EV sales reach 35–40% of worldwide new car sales by 2030.

Electric cars (EVs) are gaining popularity in India, offering a more ecologically responsible and cost-effective method to travel around congested cities. These cars, with their sleek appearance and smart technology, simplify commuting while also contributing to cleaner, fresher streets and air. It's wonderful to see the government promote EV use, while private entrepreneurs are creating novel models customized to India's specific roads and traffic patterns.

Advancements in battery technology have transformed the electric vehicle (EV) market. Significant advances in lithium batteries since 2008 have resulted in a remarkable 90% drop in the cost of EV batteries. This means that EVs are now more inexpensive, with longer range on a single charge and less worry about running out of gasoline in the midst of a trip. Customers are paying attention since, in addition to improved batteries, advances in EV chargers and smart grids have made recharging easier.

The global growth of the electric vehicle (EV) market is significantly influenced by governments providing incentives and perks to individuals who own electric vehicles. In the present era, the shift toward electric vehicles is driven not only by environmental concerns but also by the increasing emphasis on convenience and accessibility, making the transition to electric vehicles more appealing and attainable for everyone.

As electric vehicles and their charging infrastructure become more integrated with smart grids and renewable energy, there will be a greater emphasis on data exchange between grids and electric vehicles. This will help determine the optimal time to charge the vehicles using smart algorithms that take into account renewable energy generation. Ultimately, this will result in cost optimization and the implementation of charging solutions in offices and homes. With the introduction of GaN and Silicon Carbide-based power electronics, electric vehicles are moving towards higher voltages, reaching as high as 1500 V (Kihara 2024). Higher voltages will bring fast charging closer to reality, while also improving the efficiency of energy in electric vehicles. This improvement will result in longer range and faster charge capabilities.

Further advancements in cell-level technology and their faster charging capabilities will optimize system designs for rapid charge and fast discharge batteries, thus rationalizing prices and ranges for electric cars.

9 Battery Performance

Battery management systems (BMS) use AI algorithms to monitor and regulate crucial metrics including battery performance, health, and charging cycles. Artificial intelligence programs can detect possible problems and improve energy use by studying real-time data from sensors embedded in batteries. This allows battery packs to live longer. This protects the longevity of electric cars while also boosting their overall performance.

In the field of electric car technology, artificial intelligence is crucial for autonomous driving. Artificial intelligence systems analyze data from a variety of sensors, including lidar, radar, and cameras, to better comprehend the vehicle's surroundings and make informed judgments. Self-driving electric cars (EVs) that perform driving tasks can enhance safety, lower accident rates, and increase convenience.

As manufacturers begin on this revolutionary path, coordination among industry stakeholders, governments, and technology suppliers is vital to overcoming hurdles and harnessing the transformative power of AI and automation to drive the future of electric mobility. By embracing innovation while upholding sustainability and inclusion ideals, the automotive sector can lead the path for a brighter, more efficient, and sustainable future driven by electric cars.

10 Smart Cities

Smart urbanism and platform urbanism are complementary approaches to municipal development, with the latter emerging from the former's extensive elaboration over the previous two decades. Smart urbanism is a civic development model that stresses the use of Big Data, digital overflows, and networked technology in municipal functional governance and services. These smart urbanism qualities serve to summarize the essence of platform urbanism as a representation of the effort and process of platformization. According to the platform urbanism, it is a "unique collection of digitally enabled socio-technological assemblages grounded in the civic, enabling the production of new social and material connections, including intermediations and agreements."

Platformization is the penetration of structures, processes, and governmental systems with digital platforms in various sectors and aspects of life. It also involves the reorganization of artistic practices and imaginations around these platforms (Sharifi et al. 2021). This concept has given rise to the idea of a platform society, which involves conflicts between government, industry, and civil society. These conflicts revolve around the responsibility for upholding public values and the common good in a platform society, including issues such as privacy, safety, fairness, accessibility, accountability, and democratic control.

Seoul, South Korea, serves as an example of how the Metaverse can significantly impact urban governance. This megacity has fully embraced the Metaverse concept, offering a range of government services and cultural experiences within a digital environment. Seoul residents will be able to immerse themselves in virtual reality by using VR headsets or AR glasses to explore the city's resources, including virtual social spaces, galleries, and facilities. They can also access various government services such as submitting civil complaints. Intelligent Transportation Systems (ITS) have made substantial progress in enhancing transportation safety, reliability, and efficiency. However, challenges persist in security, data management, and integration.

The arising technology of Metaverse incontrovertibly provides compelling results to these issues. It offers secure communication channels, enables virtual simulations for safe testing and training, and allows centralized data operation through real-time analytics (Singh et al. 2005). While the Metaverse has set up operations across colorful sectors, its eventuality in transportation remains largely unexplored. The miracle of global warming is impacting the Earth's natural balance on a global scale, leading to consequences similar as the melting of glaciers, severe famines, and an increase in timber backfires. also, rising ocean situations are causing an elevated threat of cataracts and dismembering natural rainfall patterns, putting numerous territories and species at threat of extermination. Scientists are warning about the implicit consequences and impact of continuing the current trend of global warming on the terrain, frugality, and health. reactionary energies, a type of natural resource that cannot be renewed, are employed for energy generation through combustion, leading to the product of by-products similar as carbon dioxide (CO_2), carbon monoxide (CO), and other dangerous feasts.

When these feasts are released into the atmosphere, they contribute to the hothouse effect, which in turn leads to ozone sub caste reduction, increased Earth's temperature, and the posterior circumstance of global warming. These emigrations appear from colorful sources, with a significant portion attributed to the transportation sector, counting for nearly 20.9% of total carbon emigrations.

11 Challenges of Electric Vehicles

Electric vehicles (EVs) are being hailed as a sustainable and eco-friendly volition to conventional petrol and diesel vehicles. Hence, EVs hold great pledge in reducing carbon emigrations and achieving sustainable transport in India. Their current fashion ability has also led to the creation of EV-specific insurance programs to cover these vehicles. Still, the wide relinquishment of EVs in the country faces numerous challenges. While switching to electric vehicles may feel economic to numerous, there are several challenges to overcome along the way.

11.1 Largely Cost

One of the major challenges in the transition to electric vehicles in India is the overall cost. Electric vehicles tend to have a advanced original cost than conventional internal combustion machine (ICE) vehicles. The cost of electric vehicle factors similar as batteries and electric motors is presently advanced than that of conventional energy-powered vehicles. This superior charge can suspend numerous prospective consumers, creating electric vehicles smaller accessible. This cost can also mean that other associated costs, similar as insurance for an electric vehicle, will be fairly advanced (Denton and Pells 2023).

11.2 Luxurious Batteries

The price of batteries remnants a major hedge to electric vehicle perpetration. Batteries make up a significant portion of the total cost of an electric vehicle. Although battery prices have come down over the times, they're still a significant expenditure. Precious batteries contribute to the high outspoken cost of electric vehicles, making them less affordable for numerous consumers.

11.3 Battery Charging Transportation

The vacuity of robust Battery Charging Transportation is pivotal for the wide relinquishment of electric vehicles. Still, in India, charging structure remains fairly underdeveloped and limited. The failure of charging stations and the need for long charging times pose challenges for electric vehicle possessors, leading to range anxiety and vexation. Establishing a wide and dependable charging network is essential to encourage further people to switch to electric vehicles.

11.4 Scope Fear

Compass Fear is a term generally used to explain the anxiety that electric vehicle possessors or motorists may feel about the compass their vehicle can travel after charging, whether completely or incompletely. Limited range and lack of charging structure contribute to this concern. India's vast geographic spread adds to the issue of range anxiety, as long distances between charging stations can be a concern for implicit buyers. Adding the range of electric vehicles and expanding charging structure can help palliate this issue.

11.5 Lack of Standardization

The lack of standardization of electric vehicle technology poses a challenge for the assiduity. Different manufacturers use different battery chemistries, charging connectors, and power train configurations (Bhambri and Gupta 2018). This need of chronicity make problems the charging procedure and deters the expansion of suitable charging transportation. Standardization sweats are demanded to insure comity and streamline the charging experience for electric vehicle possessors.

11.6 Coping with Extreme Conditions

India gests extreme rainfall conditions, including hot summers and cold layoffs, which can affect the performance and range of electric vehicle operation. High temperatures can beget battery declination and reduce the overall performance of electric vehicles. Likewise, cold rainfall can affect battery performance and reduce battery life. Developing electric vehicle technologies that can repel and perform optimally in extreme rainfall conditions is critical for successful relinquishment in India.

11.7 Need for Dirt Free Power

The transition to electric vehicles is not entirely sustainable if the electricity used to charge them is produced from fossil energies. In India, where utmost electricity still comes from coal-fired power shops, the environmental benefits of electric vehicles may be lowered. Wide relinquishment of electric vehicles must be accompanied by an increased focus on renewable energy sources to insure cleaner, greener transportation systems.

11.8 The Need for Integrated Approaches

There are sure mineral deposits for electric vehicles like azure, lithium, and nickel. The exploitation of these minerals raises enterprises about unethical mining practices, child labor, and environmental declination. These issues, frequently associated with the term "blood battery", add to the ethical challenges facing the relinquishment of electric vehicles. It's essential that the assiduity prioritizes responsible sourcing and ensures that the minerals used in electric vehicle batteries are sustainably and immorally booby-trapped. Addressing the challenges of electric vehicles in India requires a multi-pronged approach and cooperative sweats from manufacturers, policymakers and stakeholders. By addressing these issues and enforcing probative programs, India can accelerate the relinquishment of electric vehicles, creating a greener, cleaner transport system for a sustainable future. In addition, it's essential to have comprehensive insurance, similar as-bike insurance or electric vehicle insurance, to cover both the vehicle proprietor and the terrain. Also, encouraging the purchase of electric vehicles through duty impulses, subventions, and reduced electric vehicle insurance decorations can play an important part in encouraging relinquishment. Claims are subject to the terms and conditions set out in the electric vehicle insurance policy.

12 Strategies for Adopting Electric Vehicles

12.1 Expanding Charging Infrastructure

Expanding charging infrastructure is an effective way to increase the use of electric vehicles. Charging systems are a key component of using electric vehicles in transportation, but charging electric vehicles on public transportation can be difficult due to the amount of electrical power and time required. Charging at work is another effective way to promote the use of electric vehicles (Sharifi and Khavarian-Garmsir 2023).

12.2 Improved Battery Technology

Improved battery technology allows electric vehicles to store energy for longer periods of time, increasing their efficiency and popularity. Electric vehicle manufacturers are also researching and developing electric vehicle models that can travel further on a single charge.

12.3 Support by Government

Government support can play a key role in transforming the EV market.

12.4 Education of Consumer

Some people may not have paid much attention to EVs, so it is important to educate consumers about the basics and options.

12.5 Exhilarating the Unit

Enabling unit to freely purchase EVs can help states meet their EV adoption goals, reduce vehicle emissions, and increase economic opportunity.

12.6 Collaboration

Successful EV policy depends on collaboration among stakeholders.

13 Impacts of Adopting Electric Vehicles by Metaverses Applications

13.1 Environment

Electric vehicles reduce air pollution and greenhouse gas emissions because they do not emit tailpipe emissions. Using renewable energy sources, such as solar or wind, to power electric vehicles can further reduce the carbon footprint of the transportation sector. Economy: Electric vehicles have lower operating costs than conventional cars.

13.2 *Efficiency*

Electric vehicles are more efficient than conventional cars.

13.3 *Convenience*

Electric vehicles are quieter and smoother than conventional cars because electric motors create less vibration and noise.

14 Enhancement of Electric Vehicles by Metaverse Applications

Buying an electric two-wheeler will never be the same again. No need to take time off from work or household chores. When it comes to buying two-wheelers in India, consumers today have a plethora of options available. The traditional route is to visit authorized dealers, explore the offerings of third-party sellers or avail the convenience of online platforms. Authorized dealers have long been the preferred choice for most people and their popularity across the country is due to the presence of leading two-wheeler manufacturers.

These dealerships offer engaging showrooms where customers can enjoy a hands-on experience of viewing different models, taking test drives and getting expert advice from friendly sales representatives. Once a decision is made, the purchase can be completed quickly at the dealership with financing options available. The MetaVerse concept has the potential to impact various industries, with the electric vehicle (EV) sector being one of the pioneers. The days of going to a showroom or dealership to buy your favorite car are about to be thrown out of the window and consigned to the pages of history in a few years (Song et al. 2023).

MetaVerse has the potential to change the way electric vehicles are sold in India. The latest data from ARAI shows that electric two-wheeler sales are growing at a rapid pace. Registrations of high-speed 2W EV scooters in May 2023 increased by 148% year-on-year and 57% month-on-month to 104,771 units, while the total 2W increased by 11% year-on-year and 21% month-on-month. Tapping into Meta-Verse will only accelerate this process. Immersive Experiences with MetaVerse, automakers and dealers can create immersive virtual showrooms where customers can explore and interact with sales representative avatars, and be taken around the showroom floor and showcased electric vehicle models.

This can improve the EV buying experience by allowing customers to view different electric vehicle models. Customers can customize their cars in MetaVerse without ever visiting a physical showroom. In addition to the immersive experience, MetaVerse can also provide virtual simulations and test drives, allowing potential buyers to experience driving an electric vehicle without being physically present in the vehicle. This can be done through virtual reality (VR) or augmented reality (AR) technology, giving a realistic feel for the performance and functionality of different electric vehicle models.

MetaVerse can provide customers with the ability to personalize and customize their virtual electric vehicle, exploring various options for colors, finishes, accessories, etc. This will provide customers with a more interactive and engaging way to configure the electric vehicle of their choice before making a purchase decision. Democratizing electric vehicle purchasing since there is no need for a physical showroom, the cost of setting up electric two-wheeler dealerships will be much lower as Dealers do not need as much floor space as traditional dealerships. This allows them to break even much faster. Additionally, businesses will not need to maintain as much inventory as a physical store, as most of the initial dealer demonstrations can take place within the MetaVerse.

Similarly, potential buyers can test drive from home as the car will be delivered to their doorstep for them to call. One of the biggest benefits of the MetaVerse is that people in rural areas have as much access to vehicle information as people in urban areas. This is made possible by the technologies that power the MetaVerse. Accessibility to electric vehicles is increasing thanks to the rapid development of the digital economy. Online marketplaces on MetaVerse provide a platform for businesses to introduce and sell their products to a wider audience. This eliminates geographical barriers and allows direct interaction with customers, avoiding intermediaries. The cost–benefit ratio can be passed on to customers. Trained staff MetaVerse can facilitate the training of service and sales staff in rural and semi-urban areas on par with their urban counterparts, providing them with the skills and knowledge needed to provide better service to electric vehicle owners (Shahzad et al. 2022).

This is possible because training programs can be centralized and delivered from one location, ensuring consistent training for sales and service personnel. The virtual communities and social platforms within MetaVerse allow individuals from diverse backgrounds to connect, share experiences and collaborate while discussing various aspects of their respective electric two-wheelers. Within MetaVerse, rural people can engage in discussions, participate in cultural events and learn about different traditions and practices with their urban counterparts, fostering a sense of belonging and inclusion. MetaVerse is no longer a distant dream for anyone in India.

With increasing internet penetration and technological advancements, rural communities can now also harness the power of MetaVerse for development and growth. From education to economic empowerment, social inclusion and access to healthcare, the Metaverse has immense potential to transform the lives of people living in rural areas. As we move forward, it is important to ensure that the infrastructure and policies are in place to support equitable access and participation in the Metaverse for all. By leveraging this digital frontier, we can create a more inclusive and connected India where the benefits of technology extend to every corner of the country. So, let's embrace the Metaverse and empower everyone to tap into its limitless possibilities. Together, we can bridge the digital divide and build a better future for all.

15 Use of Metaverse for Electric Vehicles

The Metaverse is a network of digital twins that mirror their physical counterparts and has the potential to revolutionize many industries, including the automotive industry. Here are some ways the Metaverse could be used for electric vehicles (EVs).

15.1 Design Testing and Optimization

Automakers can test and optimize designs in safe virtual environments.

15.2 Autonomous Vehicle Testing

Autonomous vehicles can be safely tested in real-world environments and traffic situations.

15.3 Immersive Experiences

The Metaverse and augmented reality (AR) could play a major role in in-vehicle information and entertainment systems.

For example, users could overlay content on the outside world or repurpose a car window to display a different world. Car Shopping Users can virtually tour car showrooms, sit inside virtual car models, and view cars from different angles. If they like a car, they can pay with crypto currency and have it delivered to the real world (Bibri 2021).

15.4 Driverless Electric Vehicles

The Metaverse could bring significant benefits to driverless electric vehicles, such as remote monitoring and maintenance, predictive maintenance, improved efficiency, enhanced security, and superior user experiences.

16 Conclusion

EVs are gaining popularity in India, offering a more ecologically responsible and cost-effective method to travel around congested cities. Global EV sales climbed by a staggering 40% in the first half of 2023 alone. By embracing innovation while upholding sustainability and inclusion ideals, the automotive sector can lead the path for a brighter, more efficient, and sustainable future driven by electric cars.

References

Anand A, Bhambri P (2018) Orientation, scale and location invariant character recognition system using neural networks. Int J Theoret Appl Sci 10(1), 106–109. ISSN No. (Print): 0975-1718. ISSN No. (Online): 2249-3247. Published by Research Trend. Retrieved from www.researchtrend.net

Anumba C, Lu Q (2022) Digital twin built environments: future prospects. Digit Twins Built Environ 345–352. https://doi.org/10.1680/dtbe.65802.345

Bhambri P, Gupta OP (2018) Implementing machine learning algorithms for distance based phylogenetic trees. Ph.D. thesis. I. K. Gujral Punjab Technical University, Jalandhar

Bhambri P, Mangat AS (2005) Wireless security. In: Abstract book of UGC sponsored National conference on emerging computing technologies. HRMMV, Jalandhar, p 101

Bhambri P, Singh M (2006) Artificial intelligence. In: National seminar on E-Governance—pathway to progress. SSIET, Dera Bassi, p. 14

Bhambri P, Sinha VK, Jaiswal M (2019) Change in iris dimensions as a potential human consciousness level indicator. Int J Innov Technol Explor Eng 8(9S):517–525. https://doi.org/10.35940/IJITEE.I1082.0789S19

Bhambri P, Sinha VK, Dhanoa IS (2020) Development of cost effective PMS with efficient utilization of resources. J Crit Rev 7(19):781–786. Retrieved from http://www.jcreview.com/index.php?fulltxt=102893&fulltxtj=197&fulltxtp=197-1594892754.pdf

Bibri SE (2021) Data-driven smart eco-cities and sustainable integrated districts: a best-evidence synthesis approach to an extensive literature review. Eur J Futures Res 9(1). https://doi.org/10.1186/s40309-021-00181-4

Denton T, Pells H (2023) Introduction to electric vehicles. In: Electric and hybrid vehicles, pp 1–14. https://doi.org/10.1201/9781003431732-1

Global EV Outlook 2023 (2023) Global EV Outlook. https://doi.org/10.1787/cbe724e8-en

Herrmann F, Rothfuss F (2015) Introduction to hybrid electric vehicles, battery electric vehicles, and off-road electric vehicles. In: Advances in battery technologies for electric vehicles, pp 3–16. https://doi.org/10.1016/b978-1-78242-377-5.00001-7

Jiang D, Huo L, Zhang P, Lv Z (2021) Energy-efficient heterogeneous networking for electric vehicles networks in smart future cities. IEEE Trans Intell Transp Syst 22(3):1868–1880. https://doi.org/10.1109/tits.2020.3029015

Kaur J, Bhambri P, Kaur S (2019) SVM classifier based method for software defect prediction. Int J Anal Exp Model Anal 11(10):2772–2776. Retrieved from http://ijaema.com/Volume-11-Issue-10-October-2019-1/

Kihara D (2024) Editorial: expert opinions in protein bioinformatics: 2022. Front Bioinform 3. https://doi.org/10.3389/fbinf.2023.1338560

Panda SK, Reddy GSM, Goyal SB, Thirunavukkarasu K, Bhambri P, Rao MV, Singh AS, Fakih AH, Shukla PK, Shukla PK, et al (2019) Invention titled "Method for management of scholarship of large number of students based on blockchain" (Application no. 201911034937). Published on

06 Sept 2019 (issue no. 36/2019, page no. 40572). Retrieved from http://ipindia.gov.in/writereaddata/Portal/IPOJournal/1_4785_1/Part-1.pdf

Potluri S, Tiwari PK, Bhambri P, Obulesu O, Naidu PA, Lakshmi L, Kallam S, Gupta S, Gupta B (2019) Invention titled "Method of load distribution balancing for fog cloud computing in IoT environment" (Application no. 201941044511). Published on 29 Nov 2019 (issue no. 48/2019, page no. 56118). Retrieved from http://www.ipindia.nic.in/writereaddata/Portal/IPOJournal/1_4813_1/Part-1.pdf

Shahzad M, Shafiq MT, Douglas D, Kassem M (2022) Digital twins in built environments: an investigation of the characteristics, applications, and challenges. Buildings 12(2):120. https://doi.org/10.3390/buildings12020120

Sharifi A, Khavarian-Garmsir AR (2020) The COVID-19 pandemic: impacts on cities and major lessons for urban planning, design, and management. Sci Total Environ 749:142391. https://doi.org/10.1016/j.scitotenv.2020.142391

Sharifi A, Khavarian-Garmsir AR, Kummitha RK (2021) Contributions of smart city solutions and technologies to resilience against the COVID-19 pandemic: a literature review. Sustainability 13(14):8018. https://doi.org/10.3390/su13148018

Sharifi A, Khavarian-Garmsir AR (2023) Smart city solutions and climate change mitigation: an overview. Urban Clim Adapt Mitig 93–116. https://doi.org/10.1016/b978-0-323-85552-5.00017-8

Singh M, Bhambri P, Kaur K (2005) Network security. In: Proceedings of the national conference on future trends in information technology. SJPMLIET, Radaur, Yamuna Nagar, pp 155–161

Song M, Cheng L, Du M, Sun C, Ma J, Ge H (2023) Charging station location problem for maximizing the space-time-electricity accessibility: a lagrangian relaxation-based decomposition scheme. Expert Syst Appl 222:119801. https://doi.org/10.1016/j.eswa.2023.119801

Zeng T, Zhang C (2023) Editorial: expert opinions in integrative bioinformatics: 2022. Front Bioinform 3. https://doi.org/10.3389/fbinf.2023.1218466

Zheng Y, Shao Z, Shang Y, Jian L (2023) Modeling the temporal and economic feasibility of electric vehicles providing vehicle-to-grid services in the electricity market under different charging scenarios. J Energy Storage 68:107579. https://doi.org/10.1016/j.est.2023.107579

MIX
Papier aus verantwortungsvollen Quellen
Paper from responsible sources
FSC® C105338

If you have any concerns about our products,
you can contact us on
ProductSafety@springernature.com

In case Publisher is established outside the EU,
the EU authorized representative is:
Springer Nature Customer Service Center GmbH
Europaplatz 3, 69115 Heidelberg, Germany

Printed by Libri Plureos GmbH
in Hamburg, Germany